Nondestructive Quality Assessment Techniques for Fresh Fruits and Vegetables

Pankaj B. Pathare • Mohammad Shafiur Rahman
Editors

Nondestructive Quality Assessment Techniques for Fresh Fruits and Vegetables

Springer

Editors
Pankaj B. Pathare
Department of Soils, Water and Agricultural Engineering
Sultan Qaboos University
Al-Khod, Oman

Mohammad Shafiur Rahman
Department of Food Science and Nutrition
Sultan Qaboos University
Al-Khod, Oman

ISBN 978-981-19-5421-4 ISBN 978-981-19-5422-1 (eBook)
https://doi.org/10.1007/978-981-19-5422-1

This Springer imprint is published by the registered company Springer Nature Singapore Pte Ltd.
The registered company address is: 152 Beach Road, #21-01/04 Gateway East, Singapore 189721, Singapore

Preface

Quality and safety of fresh fruits and vegetables are of the utmost importance to consumers, retailers, and processors. Traditional methods of quality evaluation, such as mechanical or chemical approaches, are destructive and time-consuming. In addition, chemical methods need to use many chemicals and many of these may be toxic and difficult to dispose. Nondestructive testing is an emerging technology for the fast and easy testing of external and internal fresh produce quality. In recent years, significant progress has been made on the nondestructive methods to develop and their applications in food quality detection. Due to its rapid and nondestructive benefits, nondestructive testing is becoming a key technical support for promoting the fresh produce business by applying technology in an intelligent way.

This book titled *Nondestructive Quality Assessment Techniques for Fresh Fruits and Vegetables* contains details of the nondestructive approach to feature the present-day trends and existing future opportunities in the fresh food supply chain. First, this book provides an overview of different nondestructive techniques in food quality detection. Then it presents nondestructive methods: monochrome computer vision, imaging techniques, biospeckle laser technique, Fourier transform infrared (FTIR) spectroscopy, hyperspectral imaging, Raman spectroscopy, near infrared (NIR) spectroscopy, X-ray computed tomography, ultrasound, acoustic emission, chemometrics, electronic nose and tongue. Selected applications on each method are also introduced in this book. As a result, the reader can definitely experience a better understanding of how to use nondestructive methods and technologies to detect the quality of fresh fruits and vegetables.

With various interesting topics covered, it is expected that the book will benefit a wide array of audiences including postharvest and food scientists/technologists, industry personnel, and researchers involved in the field of fresh produce quality detection. Besides, the book can serve as a readily accessible reference material for postgraduate students including Ph.D. students, and scientists can extend their knowledge in these research fields. The editors are confident that readers will find this book informative and interesting.

Al-Khod, Muscat, Oman

Pankaj B. Pathare
Mohammad Shafiur Rahman

Acknowledgments

As the editors of this book, we are grateful to all of our research collaborators for their invaluable contributions. We wish to express our sincere gratitude to Sultan Qaboos University, Muscat, Oman, for providing us the opportunity, facilities, and support to execute such an exciting book project. We sincerely acknowledge our parents and teachers for their unconditional contribution in our educational progress and support in life. We acknowledge our spouses and children for their kind support with their encouragement and patience throughout the project. Special thanks to our colleagues and other research team members for their support and encouragement.

Contents

About the Editors

Pankaj B. Pathare is an Assistant Professor of Postharvest Technology at Sultan Qaboos University, Oman. He has graduated with a B.Tech degree (Agricultural Engineering) from Dr. Panjabrao Deshmukh Krishi Vidyapeeth, Akola (India), and PhD (Process & Chemical Engineering) from the University College Cork (Ireland). Before joining at SQU, he worked as researcher at Newcastle University, UK, and Stellenbosch University, South Africa. He has gained expertise on postharvest technology and food engineering which includes quantification of postharvest losses during transportation, mechanical damage of perishables, food agglomeration/granulation, food drying and cooling, and structural design of corrugated packaging. The research results are well documented in 50 original scientific peer-reviewed journal papers. He also worked as supervisor/co-supervisor for ten postgraduate research students. He is a member of the editorial board of several research journals including *PLoS ONE*, *Frontiers in Sustainable Food Systems*, *Open Agriculture*, and *Journal of Food Quality*.

Mohammad Shafiur Rahman is the author/co-author of over 450 technical articles including more than 160 journal papers and 15 books. He is the author of the internationally acclaimed and award-winning *Food Properties Handbook* and editor of the popular *Handbook of Food Preservation* published by CRC Press, Florida. He has initiated the *International Journal of Food Properties* and serves as the founding Editor-in-Chief for more than 20 years. He serves on the editorial boards of 10 international journals and book series. He also serves as Editor-in-Chief for the *Journal of Agricultural and Marine Sciences*, published by Sultan Qaboos University. In 2008, Professor Rahman was ranked among the top five Leading Scientists and Engineers of 57 OIC Member Countries in the Agroscience Discipline. In 2020, he was recognized among the World Top 2% influential scholars (ranked 107 among 48,454 food science researchers), published by Stanford University, USA.

Nondestructive Techniques for Fresh Produce Quality Analysis: An Overview

Pankaj B. Pathare and Mohammad Shafiur Rahman

Abstract Quality and safety of fresh fruits and vegetables have prime importance in various stages of postharvest value chain. Visual examination and destructive procedures have been used to check the quality of fruits and vegetables for decades. Recent technological development has progressed in the many nondestructive techniques for fresh produce quality analysis. Nondestructive techniques are now important because they allow determining simultaneous chemical and physical characteristics of foods without destroying the substance. Nondestructive techniques become feasible to detect surface and interior defects related to fresh product during harvesting and handling operations. Recently, many studies on nondestructive measurements of fresh fruits and vegetables for quality assessment were reported. This chapter provides an overview of the different nondestructive techniques used for fresh produce quality analysis. The use of nondestructive techniques for assessing internal and external quality, such as computer vision, hyperspectral imaging, Raman spectroscopy, X-ray-computed tomography, acoustic techniques, electronic tongue, and electronic nose are discussed. Each technique is highlighted with a brief overview of research and its potential use in fresh produce industry.

Keywords Non-destructive detection · Quality · Computer vision · Hyperspectral imaging · Raman spectroscopy · Acoustic Techniques · Chemometrics · E-nose

P. B. Pathare (✉)
Department of Soils, Water and Agricultural Engineering, College of Agricultural and Marine Sciences, Sultan Qaboos University, Al-Khod, Oman
e-mail: pankaj@squ.edu.om

M. S. Rahman
Department of Food Science and Nutrition, College of Agricultural and Marine Sciences, Sultan Qaboos University, Al-Khod, Oman

P. B. Pathare, M. S. Rahman (eds.), *Nondestructive Quality Assessment Techniques for Fresh Fruits and Vegetables*, https://doi.org/10.1007/978-981-19-5422-1_1

1 Introduction

Fruits and vegetables play a vital part in the human diet due to their nutritional properties, vitamin contents, dietary fiber, phytochemicals, and antioxidants. Food quality refers to the acceptable characteristics of foods to the consumers. It is a crucial aspect of food that determines the nutrition and safety of the meal. External variables like size, shape, color, gloss, consistency, texture, and flavor (taste and odor), as well as internal factors, like chemical composition, physical features, and microbes, are all considered. Among these aspects, certain characteristics, such as appearance, are observed or felt, while others, such as chemical components, require instrumentation to evaluate. External quality parameters include shape, size, skin color, and general appearance, while internal quality parameters include soluble solids content, titratable acidity, pH, starch, and sugar content, carotenoids, sugar, ascorbic acid, total flavonoids, total phenolic, antioxidant activity, and flesh firmness (Abasi et al., 2018; Mesa et al., 2016; Navarro et al., 2015).

Fruits and vegetables with high levels of desired characteristics can provide a high economic value. The reliability and authenticated good quality food are currently one of the most pressing concerns among customers. Fresh produce of good quality should be of the appropriate ripeness, texture, chemical composition, and structure, and be free of defects (e.g., bruises, browning, mold, and insect damage). Internal quality is currently measured on a subsample of the product that has been destructively assessed. To completely analyze foods, many evaluation technologies are being developed or have been widely developed and deployed. Traditional methods of evaluation, such as mechanical or chemical approaches, are both destructive and time-consuming. In addition, chemical methods need to use costly chemicals. Visual quantification of elements affecting fruits and vegetables is difficult, expensive, and easily influenced by subjective physical factors, such as inconsistent judgment and subjective outcomes. The quality assessment was carried out by skilled humans using their senses of touch and sight. This procedure is highly variable and finicky, and investigators may differ in their decisions. Fruit and vegetable analysis for multiple aspect criteria is a continuous task in this type of environment. Furthermore, destructive procedures are known to be more labor-intensive, time-consuming, and inapplicable to inline grading and sorting. They also require specific sample preparation. The evaluation of many fresh produce commodities has resulted in the progress of nondestructive methods, which can be used to replace analytical and destructive analysis. Nondestructive techniques in fresh produce quality assessment have become an urgent need since it offers rapid and accurate measurement. Nondestructive technologies allow determining the quality characteristics of fruits and vegetables at any stage in the value chain without destroying the product.

2 Nondestructive Measurements

Consumer demand for the assurance of quality has attracted the industry in developing rapid, low-cost noninvasive instruments for detecting and monitoring fruit quality. Nondestructive techniques involve noninvasive testing of fruit surfaces without intruding on its quality. Food's physical, structural, mechanical, and chemical qualities are correlated with the nondestructive procedures. Optical, electromagnetic, electrical, and mechanical measurements are used in these technologies (Magwaza et al., 2013). The most common nondestructive methods for detecting damaged fruits are computer vision, biospeckle, spectroscopy techniques, ultrasonic and acoustic techniques, hyperspectral imaging, nuclear magnetic resonance (NMR), magnetic resonance imaging (MRI), X-ray imaging, and thermal imaging (Mahanti et al., 2022).

2.1 Computer Vision Technology

Computer vision technology (i.e., image, interpretation) mimics the human vision in assessing the quality of fruits and vegetables. This information can be used in quality grading sorting machine. Quality inspection of fruits and vegetables using image processing technique involves five steps, namely, image acquisition, preprocessing, image segmentation, feature extraction, and classification. A light, an imaging unit, an image capture board, and the appropriate computer hardware and software are the five essential components of a machine vision system. The computer vision is affordable, quick, economical, hygienic, and consistent (Mahendran et al., 2012; Narendra & Hareesha, 2010). Several studies on computer vision system for the detection of defects in apple, banana, citrus, plum, dates, and mangoes have been reported by the researchers (López-García et al., 2010; Manickavasagan et al., 2017; Patel et al., 2019; Sanaeifar et al., 2016).

Similarly, monochrome camera-based CVS could be more advantageous for fresh produce product quality and assessment because each place on an image during monochrome photography can record and show a varied amount of light, but not a different hue (purity of color). Monochrome camera imaging encompasses all types of black-and-white photography and generates photos with neutral grey tones ranging from black to white (Patel et al., 2021).

2.2 Biospeckle Laser Technique

Biospeckle, a nondestructive optical technique, is used to examine food materials (Pandiselvam et al., 2020). Biospeckle is the phenomenon of exhibiting various types of activity or biological or nonbiological dynamic processes when laser light is incident on a substance (Ansari & Nirala, 2013). A speckle pattern is generated in an observation plane due to the interference of backscattered light. The speckle pattern of biological samples usually has two components: a static pattern caused by the tissue's fixed particles and a variable pattern caused by the tissues' moving sections. The changing speckle pattern, known as biospeckle (Ansari et al., 2012), is a distinguishing property of biological material, whereas inorganic parts are connected with static patterns. Biospeckle activity refers to the degree of mobility of speckle patterns. Biospeckle activity is generated due to the movement of the scattering centers in the tissues, cytoplasmic streaming/protoplasmic streaming, growth, and division of cell during maturation of the fruit, moment of organelle, biochemical reactions, and Brownian motions (Pandiselvam et al., 2020). Furthermore, the activity of biospeckles varies with age, surface characteristics, water, starch, and chlorophyll. Due to light absorption by this pigment, lesser chlorophyll content promotes higher apparent BA and, as a result, shallower light penetration through a tissue (Zdunek & Cybulska, 2011).

2.3 X-Ray CT Imaging

X-ray CT imaging is a novel noninvasive machine vision technique with a lot of promise for detecting inside defects where other machine vision technologies show difficulty. Because X-rays can pass through most items, this technology offers a competitive advantage. The rising interest in the visualization of inner dynamic circumstances among researchers is sparked by the use of computed tomography (CT) technology in the food research field. X-ray CT produces 3D images of the X-ray absorption contrast, clarifying the shape and volume of the fruit while providing greater spatial information on the interior structure and diseases of the fruit than radiographies. (Van De Looverbosch et al., 2020).

The main components of a CT scanner are X-ray tubes, X-ray detectors, and a data acquisition system. These components are placed inside a chamber called a gantry. The gantry can rotate 360° on a circular path, causing the X-ray source and detectors also to rotate. The X-ray source and the detectors are positioned on opposite sides facing each other and can be rotated 360°. The x-ray detectors can measure the amount of radiation passing through the body at any point during this rotation, which allows the body to be scanned from all angles (Saunders & Ohlerth, 2011).

X-ray CT technology has become a useful tool for assessing the quality of agricultural goods in recent years, allowing for a better knowledge of the composition, physicochemical properties, and internal structure of samples (Du et al., 2019). The physical texture and sensory qualities of agricultural goods are determined by their microstructure, which can be studied with a CT scanner. The microstructure of food products, notably porosity, connectivity index, and cell wall thickness distribution, must be carefully studied in order to fully comprehend their mechanical and organoleptic qualities (Aguilera, 2005; van Dalen et al., 2003). Agricultural products benefit from X-ray CT technology because it allows for microstructural examination. Accurate calculations of moisture content, density, texture, and hardness could lead to an improved internal quality evaluation, microstructure observation, and mechanical property measurement (Arendse et al., 2016; Chen et al., 2017; Du et al., 2019).

2.4 *Near Infrared Spectroscopy*

Near infrared spectroscopy is a nondestructive technique that interacts with molecular functional groups containing polar groups such as –OH, C–O and N–H bonds (Blanco & Villarroya, 2002). For fruit and vegetables, NIRS has become one of the most used methods for the assessment of fresh fruit according to their internal quality attributes for authentication and detection of adulteration (Arendse et al., 2017; Fan et al., 2009).

NIR spectroscopy has several advantages when used as an objective analytical tool for quality analysis and control. For instance, it is a rapid technique at a speed compatible with the typical speed of commercial sorting lines, permits repeated scanning of the same sample, reduces waste as measurements can be made directly with minimum sample preparation, and the technique is versatile when coupled with an optical fiber probe and it can operate inline or online for in situ measurements in addition to the available portable instruments (Arendse et al., 2018; Nicolai et al., 2007; Williams et al., 2019).

2.5 *Raman Spectroscopy*

Raman spectroscopy is a technique to detect vibrations in molecules, which is based on the inelastic scattering of light (called Raman scattering). Raman spectroscopy is a vibrational spectroscopy technique that uses high-energy light to interact with molecule vibrations. An excitation source, a wavelength separation device, and a detector are the three basic instrument components in a Raman system. As Raman excitation sources, monochromatic lasers with high power are often used (Qin et al., 2019). Raman techniques can be used to examine intrinsic components of foods, such as proteins, lipids, carbohydrates, fatty acids, and inorganics, for food quality purposes. Extrinsic components, such as adulterants, germs, and pesticides, can also be inspected for food safety purposes. Several researchers have attempted to use Raman spectroscopic techniques in fruit and vegetable for their structural investigation, safety, categorization, and quantification (López-Casado et al., 2007; Pudney et al., 2011; Yang & Ying, 2011).

2.6 *Acoustic Techniques*

The acoustic method is an interesting process for determining the quality of food and agricultural products. It is rapid, cost-effective, and nondestructive. Because of these benefits, instrumental acoustic techniques become more popular as effective tools for evaluating food quality. The acoustic system consists of a sound unit, an absorption system, and procedures for determining phase oscillation, in which the quality of foods is determined by the sounds produced by crushing the food (Demattè et al., 2014). Acoustic approaches can be quickly classed as dispersing or reflecting sound waves, similar to how light waves dissipate and reflect. A transducer sends sonic pulses to the material, which it then reflects. For viewing and assessing final food materials to confirm their safety and assure standard quality, high-quality confirmation methods are urgently needed (Butz et al., 2005; Lewicki et al., 2009).

Acoustic vibration waves can propagate through the product body and provides some internal information about the specimen. The main parameters governing the acoustic vibration properties of the product include natural frequency, propagation velocity, acoustic impedance, and attenuation coefficient among others. These acoustic parameters vary for different products with their varied mechanical structures (Zhang et al., 2018).

2.7 *Chemometrics for Spectral Analysis*

Chemometrics is described as the use of mathematical and statistical approaches to extract and interpret data gathered during a chemical measurement process (Workman et al., 1996). Spectral processing entails reducing dimensionality while retaining useful information and establishing a mathematical relationship between spectra and reference data in order to classify and quantify critical quality criteria (Amigo et al., 2013). Optical properties of food materials, such as absorbance, transmittance, and emission at various wavelengths, produce a spectrum that provides the chemical information of an object. The chemometrics operation involves choosing sensitive wavelengths as well as weighting vectors or beta-coefficients for each wavelength used in the calibration. When working with new data, algorithm developers frequently go through three steps of the algorithm: preprocessing, model development, and validation. To detect the anomaly in the food materials, chemometrics models correlate the picture and/or spectra with the desired attributes. Chemometrics is a subset of machine learning that is mostly used to solve problems in linear analytical chemistry.

2.8 *Electronic Nose*

The electronic nose is based on the human sense of smell and has shown to be a useful instrument for determining the quality of food products. It facilitates the detection and identification of volatile organic compounds (VOC) in food samples. It uses an array of sensors and a sample handling mechanism to differentiate and discriminate complex scents. The sensory array is often made of chemically treated metal oxides that, when exposed to volatile components from the sample, create a unique fingerprint or fragrance signature using a sensor-electronic interface. Using data prediction models and chemometrics, the generated signature of aromatic chemicals is used to classify and quantify the molecules.

Quality control, process monitoring, freshness evaluation, shelf-life study, and authenticity assessment are all possible uses for electronic noses in the food business. Using an e-nose, Ezhilan et al. (2018) assessed pathogen contamination in apples. Chen et al. (2019) used an e-nose to examine the freshness of fresh-cut broccoli throughout storage. E-nose has a lot of potentials, especially for fresh produce, because it uses an array of sensors combined with pattern recognition software for nondestructive food quality inspection.

2.9 Electronic Tongue

The electronic tongue imitates human taste perception and distinguishes itself as presentable. It is made up of a nonspecific array of low-sensitivity chemical sensors that are cross-specific. When combined with appropriate recognition pattern models and/or multivariate data processing techniques, these sensors display partial specificity to respond to distinct substances in the sample solution. The e-tongue is mostly used for liquid sample analysis. Any solid materials must be extracted in an electrolyte that works as a continuous phase and keeps the electrode array in close contact (Lvova, 2016).

Taste recognition by e-tongue works on three levels: receptors (chemo-sensitive metal oxide sensors), transmission (detectors and transducers), and perception (data analytics and chemometrics). Food spoilage requires the precise activity of bacteria and enzymes, so an electronic tongue has been used (Ghasemi-Varnamkhasti et al., 2018). By using online equipment to monitor real-time measurement, the e-tongue would be more advantageous than traditional sensory approaches.

3 Conclusions

Internal and external quality evaluation of fresh fruits and vegetables during their postharvest handling and storage and before their industrial processing is an important step. The advances in nondestructive techniques have made it possible to demonstrate numerous practical applications for fresh fruits and vegetable quality detection. Nondestructive approaches, unlike traditional methods, can handle large amounts of data despite the complexity and diversity of biomaterial characteristics encountered during the development of these applications. Fresh fruit and vegetable color, size, shape, and surface defects can all be detected using computer vision. Nondestructive techniques including computer vision, hyperspectral and multispectral imaging, X-ray imaging, computed tomography, near infrared spectroscopic, and acoustic methods are highlighted. E-sensing techniques such as e-nose and e-tongue have also advanced in real-time quantification of food quality changes.

The commercial value of the equipment to improve the quality standards of food products can determine the success of nondestructive approaches. Although the cost of such equipment is now quite high. Another option is to deliberately seek for ways to develop mini-systems that are similarly effective but less expensive and more easily accessible. The predicted availability of these applications in the coming years could aid in the improvement of current nondestructive techniques, reducing issues and presenting viable answers in the field of food quality detection. The requirement for portable and on-field sensing equipment to check the quality and processing of food products is sought for this reason. However, in order to fulfill the demands of current applications, these nondestructive procedures need to be improved in their cost, precision, and accuracy. Integration of these technologies with other sensing devices could open the way for more accurate fresh produce quality evaluations, which is another area that needs additional research.

References

Abasi, S., Minaei, S., Jamshidi, B., & Fathi, D. (2018). Dedicated non-destructive devices for food quality measurement: A review. *Trends in Food Science & Technology, 78*, 197–205.

Aguilera, J. M. (2005). Why food microstructure? *Journal of Food Engineering, 67*, 3–11.

Amigo, J. M., Martí, I., & Gowen, A. (2013). Hyperspectral imaging and chemometrics: A perfect combination for the analysis of food structure, composition and quality. In F. Marini (Ed.), *Data handling in science and technology* (pp. 343–370). Elsevier.

Ansari, M. Z., Minz, P. D., & Nirala, A. (2012). Fruit quality evaluation using biospeckle techniques. In *2012 1st International Conference on Recent Advances In Information Technology (RAIT)* (pp. 873–876). IEEE.

Ansari, M. Z., & Nirala, A. K. (2013). Biospeckle activity measurement of Indian fruits using the methods of cross-correlation and inertia moments. *Optik, 124*, 2180–2186.

Arendse, E., Fawole, O. A., Magwaza, L. S., Nieuwoudt, H., & Opara, U. L. (2018). Evaluation of biochemical markers associated with the development of husk scald and the use of diffuse reflectance NIR spectroscopy to predict husk scald in pomegranate fruit. *Scientia Horticulturae, 232*, 240–249.

Arendse, E., Fawole, O. A., Magwaza, L. S., Nieuwoudt, H. H., & Opara, U. L. (2017). Development of calibration models for the evaluation of pomegranate aril quality by Fourier-transform near infrared spectroscopy combined with chemometrics. *Biosystems Engineering, 159*, 22–32.

Arendse, E., Fawole, O. A., Magwaza, L. S., & Opara, U. L. (2016). Non-destructive characterization and volume estimation of pomegranate fruit external and internal morphological fractions using X-ray computed tomography. *Journal of Food Engineering, 186*, 42–49.

Blanco, M., & Villarroya, I. (2002). NIR spectroscopy: A rapid-response analytical tool. *TrAC Trends in Analytical Chemistry, 21*, 240–250.

Butz, P., Hofmann, C., & Tauscher, B. (2005). Recent developments in noninvasive techniques for fresh fruit and vegetable internal quality analysis. *Journal of Food Science, 70*, R131–R141.

Chen, H.-Z., Zhang, M., & Guo, Z. (2019). Discrimination of fresh-cut broccoli freshness by volatiles using electronic nose and gas chromatography-mass spectrometry. *Postharvest Biology and Technology, 148*, 168–175.

Chen, S., Xu, L., Yin, J., & Tang, M. (2017). Quantitative characterization of grain internal damage and 3D reconstruction based on micro-CT image processing. *Transactions of the Chinese Society of Agricultural Engineering, 33*, 144–151.

Demattè, M. L., Pojer, N., Endrizzi, I., Corollaro, M. L., Betta, E., Aprea, E., Charles, M., Biasioli, F., Zampini, M., & Gasperi, F. (2014). Effects of the sound of the bite on apple perceived crispness and hardness. *Food Quality and Preference, 38*, 58–64.

Du, Z., Hu, Y., Ali Buttar, N., & Mahmood, A. (2019). X-ray computed tomography for quality inspection of agricultural products: A review. *Food Science & Nutrition, 7*, 3146–3160.

Ezhilan, M., Nesakumar, N., Babu, K. J., Srinandan, C., & Rayappan, J. B. B. (2018). An electronic nose for royal delicious apple quality assessment–a tri-layer approach. *Food Research International, 109*, 44–51.

Fan, G., Zha, J., Du, R., & Gao, L. (2009). Determination of soluble solids and firmness of apples by Vis/NIR transmittance. *Journal of Food Engineering, 93*, 416–420.

Ghasemi-Varnamkhasti, M., Apetrei, C., Lozano, J., & Anyogu, A. (2018). Potential use of electronic noses, electronic tongues and biosensors as multisensor systems for spoilage examination in foods. *Trends in Food Science & Technology, 80*, 71–92.

Lewicki, P. P., Marzec, A., & Ranachowski, Z. (2009). Acoustic properties of foods. In M. S. Rahman (Ed.), *Food properties handbook* (pp. 811–841). CRC Press.

López-Casado, G., Matas, A. J., Domínguez, E., Cuartero, J., & Heredia, A. (2007). Biomechanics of isolated tomato (Solanum lycopersicum L.) fruit cuticles: The role of the cutin matrix and polysaccharides. *Journal of Experimental Botany, 58*, 3875–3883.

López-García, F., Andreu-García, G., Blasco, J., Aleixos, N., & Valiente, J.-M. (2010). Automatic detection of skin defects in citrus fruits using a multivariate image analysis approach. *Computers and Electronics in Agriculture, 71*, 189–197.

Lvova, L. (2016). Electronic tongue principles and applications in the food industry. In M. L. Rodríguez Méndez (Ed.), *Electronic noses and tongues in food science* (pp. 151–160). Academic Press.

Magwaza, L. S., Ford, H. D., Cronje, P. J. R., Opara, U. L., Landahl, S., Tatam, R. P., & Terry, L. A. (2013). Application of optical coherence tomography to non-destructively characterise rind breakdown disorder of 'Nules Clementine' mandarins. *Postharvest Biology and Technology, 84*, 16–21.

Mahanti, N. K., Pandiselvam, R., Kothakota, A., Ishwarya, P., Chakraborty, S. K., Kumar, M., & Cozzolino, D. (2022). Emerging non-destructive imaging techniques for fruit damage detection: Image processing and analysis. *Trends in Food Science & Technology, 120*, 418–438.

Mahendran, R., Jayashree, G., & Alagusundaram, K. (2012). Application of computer vision technique on sorting and grading of fruits and vegetables. *Journal of Food Processing & Technology, 10*, 2157–7110.

Manickavasagan, A., Al-Shekaili, N. H., Al-Mezeini, N. K., Rahman, M. S., & Guizani, N. (2017). Computer vision technique to classify dates based on hardness. *Journal of Agricultural and Marine Sciences, 22*, 36–41.

Mesa, K., Serra, S., Masia, A., Gagliardi, F., Bucci, D., & Musacchi, S. (2016). Seasonal trends of starch and soluble carbohydrates in fruits and leaves of 'Abbé Fétel'pear trees and their relationship to fruit quality parameters. *Scientia Horticulturae, 211*, 60–69.

Narendra, V., & Hareesha, K. (2010). Prospects of computer vision automated grading and sorting systems in agricultural and food products for quality evaluation. *International Journal of Computer Applications, 1*, 1–9.

Navarro, J. M., Botía, P., & Pérez-Pérez, J. G. (2015). Influence of deficit irrigation timing on the fruit quality of grapefruit (Citrus paradisi Mac.). *Food Chemistry, 175*, 329–336.

Nicolai, B. M., Beullens, K., Bobelyn, E., Peirs, A., Saeys, W., Theron, K. I., & Lammertyn, J. (2007). Nondestructive measurement of fruit and vegetable quality by means of NIR spectroscopy: A review. *Postharvest Biology and Technology, 46*, 99–118.

Pandiselvam, R., Mayookha, V. P., Kothakota, A., Ramesh, S. V., Thirumdas, R., & Juvvi, P. (2020). Biospeckle laser technique—A novel non-destructive approach for food quality and safety detection. *Trends in Food Science & Technology, 97*, 1–13.

Patel, K. K., Kar, A., & Khan, M. (2019). Common external defect detection of mangoes using color computer vision. *Journal of the Institution of Engineers (India): Series A, 100*, 559–568.

Patel, K. K., Kar, A., & Khan, M. (2021). Rapid assessment of some physical parameters of mangoes using monochrome computer vision. *Agricultural Research, 10*, 468–482.
Pudney, P. D., Gambelli, L., & Gidley, M. J. (2011). Confocal Raman microspectroscopic study of the molecular status of carotenoids in tomato fruits and foods. *Applied Spectroscopy, 65*, 127–134.
Qin, J., Kim, M. S., Chao, K., Dhakal, S., Cho, B.-K., Lohumi, S., Mo, C., Peng, Y., & Huang, M. (2019). Advances in Raman spectroscopy and imaging techniques for quality and safety inspection of horticultural products. *Postharvest Biology and Technology, 149*, 101–117.
Sanaeifar, A., Bakhshipour, A., & De La Guardia, M. (2016). Prediction of banana quality indices from color features using support vector regression. *Talanta, 148*, 54–61.
Saunders, J., & Ohlerth, S. (2011). CT physics and instrumentation–mechanical design. In T. Schwarz & J. Saunders (Eds.), *Veterinary computed tomography* (pp. 1–8). Wiley.
van Dalen, G., Blonk, H., van Aalst, H., & Hendriks, C. L. (2003). 3-D imaging of foods using X-ray microtomography. *GIT Imaging and Microscopy, 3*, 18–21.
Van De Looverbosch, T., Rahman Bhuiyan, M. H., Verboven, P., Dierick, M., Van Loo, D., De Beenbouwer, J., Sijbers, J., & Nicolaï, B. (2020). Nondestructive internal quality inspection of pear fruit by X-ray CT using machine learning. *Food Control, 113*, 107170.
Williams, P., Manley, M., & Antoniszyn, J. (2019). *Near infrared technology: Getting the best out of light*. African Sun Media.
Workman, J. J., Mobley, P. R., Kowalski, B. R., & Bro, R. (1996). Review of chemometrics applied to spectroscopy: 1985–95, part I. *Applied Spectroscopy Reviews, 31*, 73–124.
Yang, D., & Ying, Y. (2011). Applications of Raman spectroscopy in agricultural products and food analysis: A review. *Applied Spectroscopy Reviews, 46*, 539–560.
Zdunek, A., & Cybulska, J. (2011). Relation of biospeckle activity with quality attributes of apples. *Sensors, 11*, 6317–6327.
Zhang, W., Lv, Z., & Xiong, S. (2018). Nondestructive quality evaluation of agro-products using acoustic vibration methods—A review. *Critical Reviews in Food Science and Nutrition, 58*, 2386–2397.

Monochrome Computer Vision for Detecting Quality Defects of Fruits and Vegetables

Krishna Kumar Patel, Abhimanyu Kalne, and Pankaj B. Pathare

Abstract Defects are a significant barrier in the marketing channel for fresh produce. Defect degrades the quality of fresh produce and causes huge loss in terms of quality, economical value, and reputation of growers or exporters. Timely, accurate, and rapid recognition of defects in the fresh produce can change the whole scenario and fulfills the consumers' needs as today's consumers are more aware of their health and prefer good quality fresh produce. Several techniques have now been developed for the detection of quality defects of fresh produce. Computer vision (CV), a novel technology, is applied for the nondestructive quality assessment of fresh produce and getting momentum. This technology comprises of three main steps, such as image acquisition, image processing, and image analysis/classification. Different lighting and camera setup can further be used for the image acquisition of fresh produce. If a monochrome camera is used for the capturing of images of produce illuminated with visible (RGB) light, the computer vision system is known as monochrome computer vision system. Monochrome camera has been reported better than the color camera CV in many cases. This chapter discusses the fundamental concepts of the monochrome CVS, the components of MCVS, as well as applications and potential of MCVS for detecting quality defects of fresh produce.

Keywords Nondestructive · Monochrome · Computer vision · Quality · Defects · Agricultural produce

K. K. Patel
Department of Agricultural Engineering, Post Graduate College, Ghazipur, Uttar Pradesh, India

A. Kalne
Department of Agricultural Process and Food Engineering, Indira Gandhi Agricultural University, Raipur, Chhattisgarh, India

P. B. Pathare (✉)
Department of Soils, Water & Agricultural Engineering, College of Agricultural & Marine Sciences, Sultan Qaboos University, Muscat, Oman
e-mail: pankaj@squ.edu.om

P. B. Pathare, M. S. Rahman (eds.), *Nondestructive Quality Assessment Techniques for Fresh Fruits and Vegetables*, https://doi.org/10.1007/978-981-19-5422-1_2

1 Introduction

In spite of more awareness among people about healthy eating, today, human health and lives are at more risk around the globe. People are more anxious and they are keen to boost their immunity. The word "immunity boosting" has become more common and prevalent in this current scenario of covid-19. Fruits and vegetables have been an important and integral part of our diets for centuries. Consumption of fruits and vegetables is very important to strengthen the immune system and to protect against disease. Thus, adequate consumption of fruits and vegetables may reduce the severity of the infectious disease and increase the recovery of consumers (Chowdhury et al., 2020). Consumption of at least 400 grams per day of fruits and vegetables is recommended by the WHO (World Health Organization) or five servings of 80 grams each (WHO & FAO, 2003). According to FAOSTAT, about 1837 million tons of fruits and vegetables are produced globally in 2019 (Anonmous, 2021). China and India produce about 30% of the world's vegetables.

The huge postharvest loss and high cost of quality products is a major cause of concern, which makes fruits and vegetables away from the lower and middle-class families, in spite of such global high production. The estimated level of losses of fruit and vegetables along with roots and tubers is about 45%. Pests, microbial infection, natural ripening process, environmental conditions such as heat, drought, and improper postharvest handling are the major cause of postharvest losses (Olayemi et al., 2010). According to Mrema and Rolle (2002), postharvest loss occurs during harvesting, handling, storing, processing, packaging, transportation, and marketing. Different types of quality defects in fruits and vegetables can cause permanent deleterious impacts. Fresh fruits and vegetables, especially, are very sensitive to impacts during transportation and handling which can cause loss in quality and commercial value, regardless of harvest methods (Al-Dairi et al., 2021; Spricigo et al., 2013).

The quality of fresh fruits and vegetables is one of the significant driving forces in the local as well as global market. Quality itself can be defined as the degree of excellence or its suitability for particular use. However, it is a human construct comprising many properties or characteristics encompass sensory properties (appearance, texture, taste, and aroma), nutritive values, chemical constituents, mechanical properties, functional, defects, etc., in terms of fruits and vegetables (Abbott, 1999). As fruits and vegetables are high moisture and perishable agricultural produce, quick quality assessment is required for timely reaching in the market with optimum quality. But, the traditional method (manual) of quality inspection of fruits and vegetables is not only very tedious, difficult, and labor intensive but also very time-consuming task (Patel et al., 2012, 2012). Thus, the traditional methods are not able to fulfill consumers' increased expectations for fruits and vegetables of high quality and safety standards. To achieve today's consumer and market requirements, there is a need for accurate, fast, and objective determination of quality defects of fruits and vegetables. Various real time, nondestructive, or noninvasive image processing technologies have thus far been applied for rapid quality assessment. Computer vision among these nondestructive methods is rapid, accurate, economical, consistent, objective, real time, and more efficient for the inspection of quality defects (external) of fruits and vegetables. The development of computer vision system to detect defects in a several fruits and vegetables, such as apple and banana (Kandi, 2010), citrus fruits (Lopez-Garcıaa et al., 2010), plum (Kandi, 2010), mangoes (Patel et al., 2019, 2019), potato (Hassankhani et al., 2012), tomato (Arjenaki et al., 2013), cucumber (Vakilian & Massah, 2013), and beans (Velioglu & Saglam, 2012).

Computer vision system uses a camera for capturing the images of targeted fruits and vegetables, process the profile image to develop an algorithm for image analysis and measurement or classification, or for quality defect detection using image processing/analysis software. Different types of the cameras such as color (visible RGB) camera, monochrome (visible) camera, ultraviolet camera, and near-infrared can be used for image acquisition as per requirement. Monochrome camera among these has its own importance due to its higher spatial resolution, good sensitivity, acquisition speed, and lower cost. For example, monochrome CVS has been reported very successful in detecting various common external defects (such as black lesion and mechanical damage) of mangoes (Patel et al., 2021a). The accuracy and efficiency of an algorithm developed for MCVS for detecting mango's defects were also recorded very high. In addition, the MCVS has great potential for the evaluation of various physical parameters of mangoes (Patel et al., 2021b). The main objective of this chapter is to explore the fundamental concepts of monochrome CVS, the components of CVS, function as well as application and potential of MCVS for detecting quality defects of fresh produce.

2 Computer Vision System

Computer vision is an integrated mechanical-optical-electronic-software system widely used for examining, monitoring, and controlling various types of applications. In other words, computer vision combines mechanics, optical instrumentation, electromagnetic sensing, and digital video and image processing technology for the construction of explicit and meaningful descriptions of physical objects from images. Computer vision system thus uses optical and noncontact sensing devices for automatically receiving and obtaining targeted information or interpreting images of real things (Ballard & Brown, 1982; Patel et al., 2012). An illumination system, a camera, an image capture board (i.e., frame grabber or digitizer), computer hardware, and software are the basic components of the computer vision system. The mechanical design on a specified computer vision system however can uniquely be structured based on the inspection required for any particular product.

In the 1860s James Clerk Maxwell first used a series of black and white photographs to produce color images. These black and white photographs of an object illuminated by white light were obtained using red (R), green (G), and blue

(B) filters. This study gave the idea of the development of trichromatic (RGB) color photography and imaging (Weber & Menko, 2005). The monochrome CVS is therefore like basic CVS equipped with monochrome camera, illumination system (illumination chamber and visible lighting arrangement for illumination of the object), image capture board, and computer software and hardware.

2.1 Principle of Image Acquisition

The camera is a device having a light-tight box with an aperture to admit light energy focused onto a sensitized film or plate for recording an image of an object on a light-sensitive surface. Digital camera is a combination of lens, sensors (made out of silicon), viewfinder, and a grid of tiny photosites (each called pixels: a contraction of picture element) sensitive to the light. The combination of sensors (also a hardware device), sensitive to the particular type of energy, in the camera system converts the incoming light energy from an object into an electrical signal during the image capturing process. Image sensors are employed either by the CCD (charge-coupled device) cameras or by the CMOS (complementary metal-oxide semiconductor) camera technology. Most digital cameras are based on CCD technology where the device transports the charge across the chip and reads it at one corner of the array. An AD (analog-to-digital) converter is then used to convert each pixel value into a digital value by measuring the amount of charge at each photosites and turns it into binary form. The charge-coupled device produces high-quality image with low level of noise in comparison to the CMOS, which is more susceptible to noise. In CMOS devices, several transistors are used at each pixel to amplify and to move the charge using more traditional wires (Mishra et al., 2017).

Further, the image sensors are of two types: (1) photoelectric image sensors and (2) photographic image sensors. The photoelectric image sensors are further classified as photoconductive (such as CCD and CID) and photoemissive type (such as MCP, EBCCD, and Videocon). However, the cameras can be classified as 2D cameras, 3D cameras, and hyperspectral cameras based on the image format while based on the acquisition type, cameras could be classified as line scan and area scan cameras. Line scan cameras are used when constant object motion and high speeds are needed, for instance, film scanning, food inspection/sorting, medical imaging/microscopy, etc. But, the requirement of more attention to alignment and timing limits the application of line scan cameras in computer vision systems. In contrast, the area scan cameras are most commonly used in computer vision system might be due to easy design, easy image processing, possibility of longer time integration, fixed aspect ratio, etc. About 80% of computer vision applications are monochrome area scan.

From the view of photography, the digital images are of two types: (i) black and white images and (ii) color images. Black and white images are made of different pixels of grayscale with pixel intensity lying between 0 and 255. The zero (0) refers to black and 255 refers to white while the intermediate pixel intensity values refer to the different shades of black and white. Grayscale refers to the range of neutral tonal values from black to white. In contrast, color images are made of colored pixels which capture a much broader range of intensity values than grayscale.

2.2 Visible Imaging System

When sunlight (electromagnetic wave, Fig. 1) passes through the prism spectrum (the band of colors) is produced this includes billions of colors. Human eye can perceive about 7–10 million colors. RGB is the most commonly adopted color system in the CV-based image processing technique. Three monochrome images can be obtained through wide wide-band red (R), green (G), and blue (B) filters. Various types of imaging systems such as (i) UV imaging, (ii) visible imaging, (iii) NIR imaging, and (iv) infrared and far infrared imaging system in computer vision are used for the assessment of external quality of fruits and vegetables. Among these, the application of visible imaging systems in food industries is more common in computer vision system.

Visible imaging systems among these provide information similar to what the human eye would process. The most important application of color camera-based visible imaging systems is the assessment of quality of fruits and vegetables to reduce postharvest loss by judging the physiological disorders and surface blemishes. Various studies have been conducted using visible imaging systems to replace human vision. The visibility of human vision lies between 400 and 780 nm in the portion of electromagnetic spectrum. In this context, the study conducted by Richards (2010) was traditionally centered on visible imaging system and visible light illumination. Based on the visible imaging system and visible lighting regime,

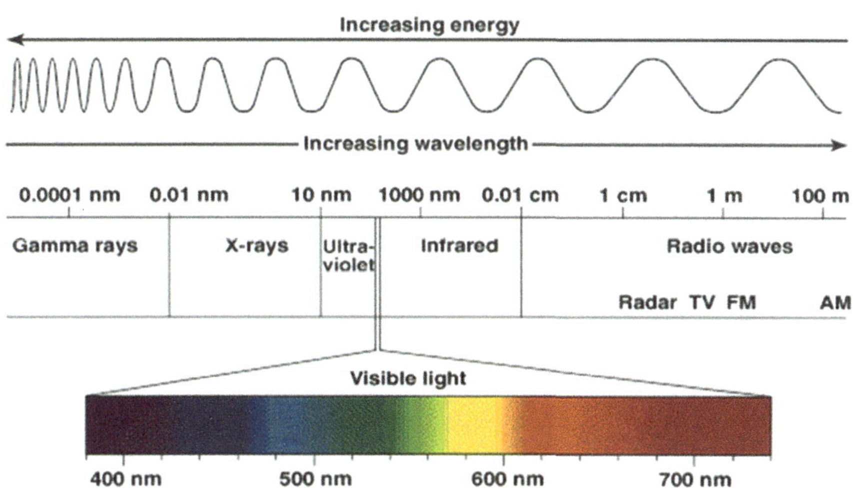

Fig. 1 Electromagnetic spectrum

the physiological peel disorder in citrus fruits caused by mechanical damage was explained by Knight et al. (2002). According to them, the freeze damage disorder oleocellosis (rind-oil spot) is due to depression of tissue surrounding the individual glands when illuminated with visible light. Patel et al. (2019) have also developed a methodology for external defect detection of Indian mangoes using a color (RGB) camera and visible lighting regime. Monochrome imaging is another approach to visible imaging which includes image analysis of black and white images. Monochrome camera can provide a greater range of digital information than color cameras (Marceau, 2009). Considering the importance of monochrome imaging in the quality assessment of food and agricultural produce, several studies on monochrome computer vision system were carried out.

3 Monochrome Computer Vision System

The camera, as discussed above, has significant importance in the computer vision system. The word "mono" refers image with one foreground and one background color. For instance, black (0, 0, 0) on white (255, 255, 255), black (0, 0, 0) on green (0, 255, 0), white (255, 255, 255) on black (0, 0, 0), etc. Figure 2 shows three monochrome images of mango. White, the color of the object is referred to as the foreground color while black the rest color is referred to as the background color. However, this polarity may be inverted with respect to thresholding needs. In such inverted cases the object is displayed with "0" while the background is with a nonzero value of pixel intensity. Monochrome imaging thus can be defined as capturing of image in different shades of a single color or can be created with variations of RGB (red, green, and blue) or any other color in the spectrum. Monochrome image can either be captured by monochrome camera or by attaching physical filter to the lens or monochromatic effect can be applied later post-processing of the acquired color image. The white, black, green, yellow, and purple color can be displayed as (255, 255, 255), (0, 0, 0), (0, 255, 0), (255, 255, 0), and

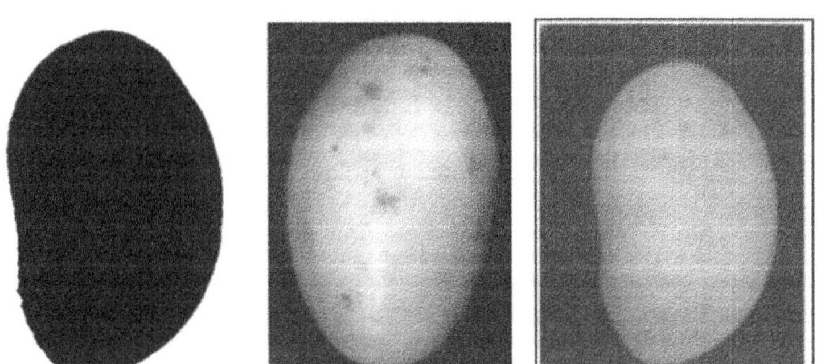

Fig. 2 Monochrome images of mangoes

(255, 0, 255), respectively. At least three coordinates are required to display color image while these three channels with the values 255 (for 8 bit resolution), 1024 (for 10 bit resolution), or 4096 (for 12 bit resolution) can be expressed as a monochrome image.

The basic computer vision system attached with monochrome camera for the capturing of the images is called monochrome CV system. The monochrome camera-based computer vision system can be used for detecting quality defects of fruits and vegetables more efficiently because of its capability of recording and showing each position on an image with a different amount of light and maintaining the purity of color (i.e., hue) during monochrome imaging (Patel et al., 2021a). Monochrome cameras can also provide higher spatial resolution than color cameras resulting from simultaneous usage of all charge-coupled device photodiodes for the acquisition of one image without color mosaic and extrapolation of color information. Monochrome camera also increases sensitivity and acquisition speed, because transmitted light is not diminished by filters necessary for color acquisition. In spite of the above, cost of monochrome cameras are less and affordable than color cameras (Patel et al., 2021a, 2021b; Weber & Menko, 2005). In addition, according to Howison and Menard (2012) the use of color camera is not necessary for the inspection of cracks, scratches, and other defects, where the goal is to discern a difference in lightness on the surface of the object.

Any CVS camera should be more advanced and must be better for easy image processing and analysis, the camera among the five basic components therefore have a significant role in the quality assessment of fruits and vegetables accurately, efficiently, and economically. Camera among the principle components of the computer vision system, as discussed above, has very important and crucial role during the image capturing, processing, and analysis of images. Image acquisition is the first and foremost significant step in this type of image processing technology. Without capturing of an image or good image, no actual or correct image processing is possible by the CV system.

3.1 Monochrome Image

A two-dimensional digital image is represented as a set of finite number "x" and "y." The digital values of "x" and "y" are called pixels. Digital image is made of number of rows and columns of pixels. The function $f(x, y)$ can bc given as: $0 < f(x, y) < \infty$. All binary and grayscale images are monochrome image but all monochrome images are not grayscale. Monochrome image can be made of varying shades of only one color. However, monochrome images can be made of any color. The 8-bit monochrome image can be represented as red, green, and blue (RGB) of 8-bit for each. This type of image has 24 bits/pixels (8 bit for each color band). An image displaying a single color or different shades of single color is therefore known as monochrome image.

3.2 Monochrome Camera

Like the color cameras, the monochrome cameras can also be CCD and CMOS type sensors. The CCD (charge-coupled device) cameras are current devices where charge is collected in the pixels. The charge is then physically shifted on the sensor's surface to the output for sampling. The CCD output is an analog pulse where the charge is proportional to the intensity. Almost all modern monochrome cameras are designed with micro-lenses. Micro-lens effectively increases the quantum efficiency of the pixels. However, it creates angular sensitivity to the incident light ray. The CMOS (complimentary metal-oxide semiconductor) cameras, on the other hand, have completely digital output and are voltage-driven devices where light striking the pixel creates a voltage proportional to intensity. The voltage is sampled directly at the pixel, digitized on the sensors, and cleared for the next image (picture/frame). The CMOS camera is therefore more resistant to smearing or blooming. Blooming is spreading of the charges to the adjacent pixels due to oversaturation of pixels which makes some very bright spots in the image while smearing is a bleeding effect, a common problem in high light intensity setting, caused by saturated pixels and light spilling over into the vertical shift register while clocking out.

Further, signal from monochrome camera is first converted into grayscale and then saved while in a color camera system, the camera control unit converted the composite color signal into parallel analog video (RGB) signals and a synchronous signal. The gray values from the monochrome images can be used for the recognition of objects. A CCD (charged-coupled device) monochrome camera can be used to capture the monochrome image and at a spatial resolution unique to the system, these images can be digitized into 8-bit grayscale images. A monochrome digital camera (model C4742-95, Hamamatsu, Bridgewater, NJ, USA) was used by Weber and Menko (2005) for color image acquisition using captured monochrome images.

3.3 Illumination Chamber and Lighting Regime

Illumination chamber is a confined chamber which restricts the interference of the natural and/or external light during the image acquisition. Illumination chamber made of black painted iron sheet was used by Patel et al. (2021b) to protect samples from external light during the rapid assessment of some physical parameters of mangoes using monochrome CV system (Fig. 3). As, the perfect and uniform illumination of object is very important in order to correct processing and analysis of the images or to extract the required information of the object, carefulness in the selection of lighting regime is of utmost importance. The use of lighting sources such as fluorescent light, metal halide (mercury), quartz halogen-fiber optics, xenon, LED (light emitting diode), and high-pressure sodium, considering the above facts, are more common in computer vision systems. For small to medium-scale inspection, fluorescent, quartz-halogen, and LED are widely used while the metal halide (also

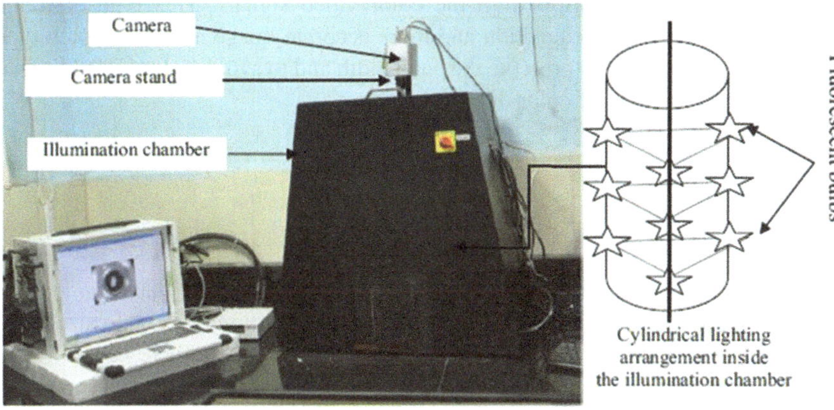

Fig. 3 A glimpse of computer vision/imaging system and lighting arrangement (Patel et al., 2019)

known as mercury), xenon, and high-pressure sodium are typically used in large-scale applications or where there very bright light source is required. Although, LED lighting is not cost-effective for large area lighting. Presently, the use of LED light is predominant in computer vision systems due to high life expectancy with minimum degradation of intensity, flexibility and improved stability, intensity, and cost-effectiveness. In addition, the LED lighting can be structured into a variety of shapes: may be directional or diffuse. Furthermore, the LED lighting is individually addressable with each of the illumination techniques (i.e., back lighting, diffuse lighting: also known as full bright field, bright field: actually partial bright field or directional, and dark field). However, there is a need for a specific light and geometry, or relative placement of camera, sample, and light in some illumination techniques. The main aim of illumination techniques is therefore to maximize the grayscale difference (i.e., contrast) features of interest and surrounding background.

3.4 *Frame Grabber*

The single most important function of frame grabber is converting video signals. Frame grabber is therefore a device that captures and converts analog video signals into digital data. Although digital frame grabbers, which interface to digital cameras and therefore do not convert data, depends on the manufacturer's design goals. This means, no frame grabber is required with digital camera. But, analog frame grabbers come in many sizes, shapes, and types. Although the design and features of frame grabbers can differ, most analog FG shares the basic functional blocks. A wide variety of circuitry and techniques can be used to develop the frame grabber. There are two types of frame grabbers; first frame grabber is generally designed for industrial and scientific applications while the second one is designed for applications in multimedia. Monochrome frame grabbers use chrominance filters in order to remove the color information from the color video signals. This type of frame grabber allows for clean acquisition and more accurate image analysis. Notch type filters are often preferred to preserve the bandwidth and to avoid removing additional image data and removal of color data, only.

4 Detection of Quality Defects of Fruits

The major hindrance to good economical returns of fruits and vegetables in local as well as in global market is quality defects such as discoloration, undersize, misshapes, bruising, molds, and over/under ripeness. However, there are different reasons for external defects including mechanical damage during transportation from plant to supermarket shelves, insect or bird bite, fungal disease, damage due to frost and sunburn (physiological disorder), etc. These quality defects cause various changes in color, texture, size, shape, etc., of fruits and vegetables. Three major surface features (elongated blemish, nondefective area, and patch-like blemishes) were used for the classification of apples by Yang (1993). Yang (1994) presented a new technique for the detection of surface features in apple image by introducing the concept of topographic representation. In his study, Yang (1994) captured images of Golden Delicious and Granny Smith apples having eight types of blemishes (bruise, russet, insect hole, sunscald, wound, bitter pits, scab, and disease) using monochrome CCD camera. A flooding algorithm was adopted and modified to detect the catchment basins, i.e., the features. The patch-like features is treated as one of the catchment basin in apple gray-level landscapes. According to them, the proposed algorithm was clearly capable of detecting the patch-like surface features of apple. Although this technique is developed for apples (Fig. 4a, b), it is also suitable for other fruits with dark surface features (Yang, 1994).

Similarly, Yang and Marchant (1995) also detected accurate blemishes in apple fruit using CCD monochromatic video camera-based computer vision system. In their study, they first applied flooding algorithm technique for the segmentation of the blemishes of the fruit and then a snake algorithm (i.e., an active contour model) was used to refine the segmentation for the improvement of localization and accuracy of the size of blemish detection. The combination of region-based segmentation (i.e., the flooding algorithm) and a contour-based segmentation (i.e., the snake algorithm) technique accorded well to the visual perceptive judgment, even when the boundary was poorly defined. Using median and Gaussian filters, the noise had to be reduced in both techniques. Monochrome image of bicolor apples acquired in the visible part of the spectrum is the basis of working of these techniques. Reflectance variation between the ground color and the blush was very important.

Furthermore, Yang and Tillett (1994) and Yang (1994) linked the gray level to the locally variable apple fruit color using monochrome approach. In their study, Rehkugler and Throop (1986) and Davenel et al. (1988) explained that the defects usually appear darker than the rest part of the apple. However, luminance between healthy tissues and defected tissues varies from one defect to other. As the fruit's shape is curvature, the intensity of gray level decreases from the center to the boundaries of the apple fruit. In addition, some noises exist due, among other things, to lenticels. For defect detection, simple techniques (threshold or background subtraction) give poor results while pattern recognition techniques were unusable due to strong variation in the size and shape of fruits. Global approach for large diffuse defects however was reported best and proposed flooding algorithm to segment patch-like defects such as russet, bruise, and stalk or calyx area by Yang (1994). However, the machine vision system with monochrome camera can also be used for bruise detection and classification of the apple fruit into USDA grades. Most of the surface of the fruit represented by the computer as after digital image is captured by the camera and the gray-level response to the bruised tissue is represented by

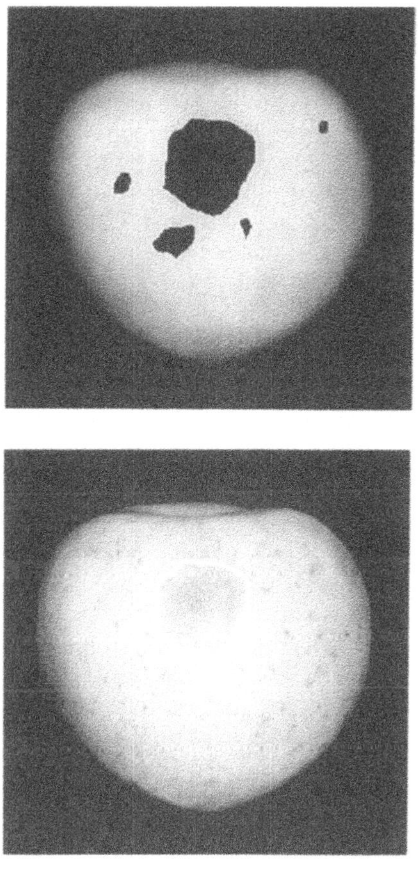

a

b

Fig. 4 (**a**) Results of feature detection for the image in Fig. 1. (*a*) Detected features superimposed on the Gaussian smoothed image. (*b*) The bruise is detected after the average depth thresholding. (**b**) (*a*) The original image of an apple with (1) an insect hole and a stalk area and (2) a russet patch. (*b*) (1, 2): Detected features superimposed on Gaussian smoothed image (1, 2). (*c*) (1, 2): The result of feature detection with the average depth thresholding (Yang, 1994)

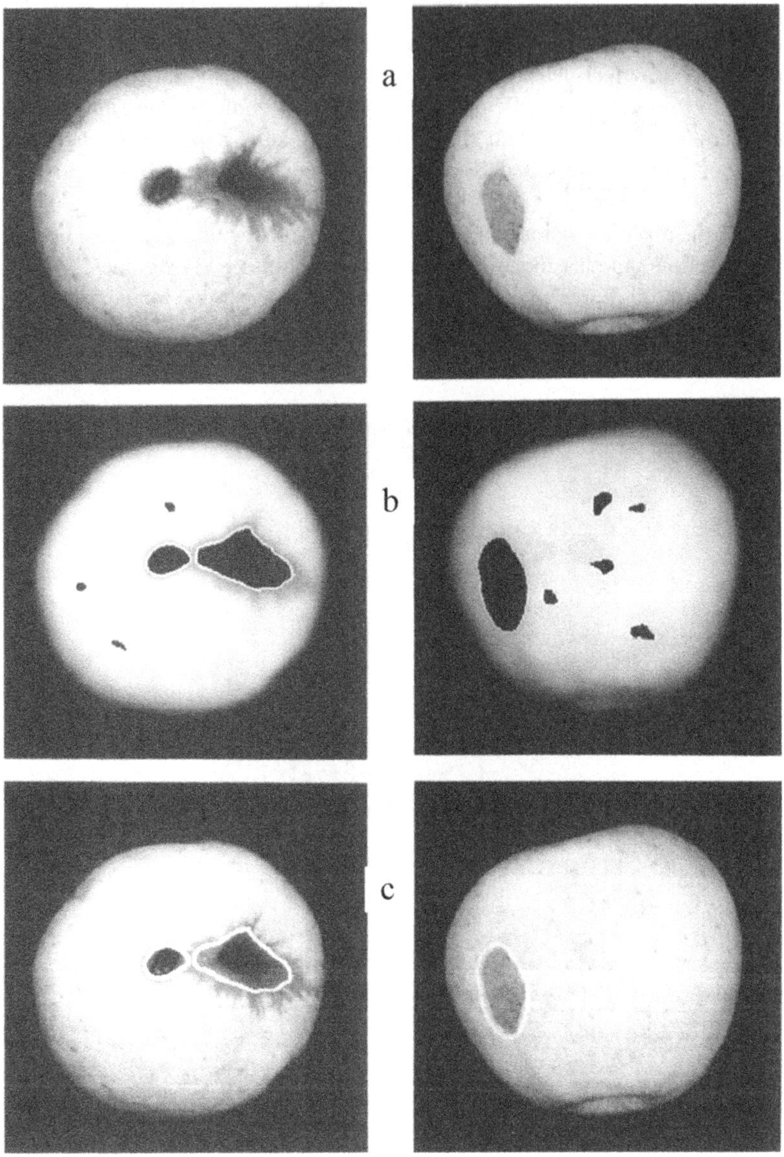

Fig. 4 (continued)

reduced image intensity. Thus, bruise patterns can be determined by image filtering, differencing, binary image thresholding, and measurement of the shape of the areas representative of bruises by using thinness ratios (Rehkugler & Throop, 1986).

A thinning algorithm can also be used for the discrimination between the stem and body of the apples on monochromatic images but the task of calyx and defected

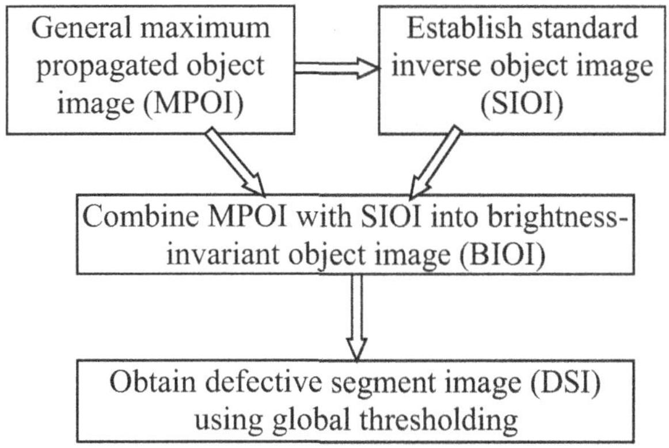

Fig. 5 Four main procedures used in brightness-invariant image segmentation to divide image into defective and good areas (Wen & Tao, 1998)

parts classification might be missed (Pla & Juste, 1995). As the contrast difference among the surrounding sound and defected tissues can be observed using the aspects of a mono-color apple, the quality defect of apple can be detected using monochrome image. For mono color fruits, unsupervised segmentation techniques like flooding algorithms and thresholding on monochrome image provide good results. After analyzing the different techniques applied for the detection of defects, it concludes that in the monochrome approach the conditions impose sophisticated algorithms. This issue generally appears because of fruit-to-fruit and place-to-place (within a fruit) intensity variations. Within a fruit the intensity variations from center to boundary is more common. In color, there exist two problems: (i) difference between fruits which suggests either an algorithm using little a priori information or information independent from the fruit's background color or self-fitting algorithm; (ii) differences in an image which inspire the region-based algorithm. This is the reason why whole color information was not used in most of the research which are in the public domain. Only one or two channels were selected in these studies. For instance, the hue (H) channel does not depend on the light intensity and does not present any specular reflection. However, monochrome images treat less information but are faster to compute (Leemans et al., 1998).

Furthermore, Wen and Tao (1998) developed a novel and practical technique (i.e., brightness-invariant image segmentation) for the online extraction of defective segments from fruit images. This technique accommodates the gradient, reflects characteristics of the curved surfaces of fruits (such as apple), and avoids the defect inspection error due to variability in the natural brightness of fruits. This means the original gradient fruit image is corrected into a nongradient image without losing defect information on apples that may have various brightness and extracted defect segments on fruit using global thresholding. Four main procedures used in brightness-invariant image segmentation that divides an apple image into defective and good areas are given in Fig. 5.

As red delicious presents more difficulties during the defect detection procedure due to serious variations in brightness, red delicious apples were considered in this study. A monochrome camera was equipped with 700 nm long pass filter in machine vision to acquire images of fruits. Image segmentation based on MPOI (maximum propagated object image), SIOI (standard inverse object image), BIOI (brightness-invariant object image), and DSI (defective segment image) was used for the separation of defective and good areas (Table 1). Definition of various object images obtained during brightness-invariant image segmentation has been presented in Table 1 (Wen & Tao, 1998).

The proposed method was reported very effective for large-scale multiple line processing of massive numbers of apples. In another study by Guedalia (1997), visual-based quality sorting method of fruit was developed using three monochrome cameras. He used all three cameras to observe fruit rolling freely on rope. Monochrome cameras were used with an optical device projecting four times the same views with different bandwidths on one CCD. After defect detection, features of apples extracted and fruits were classified based on features extracted. Two methods were used by Guedalia (1997) in his study to summarize segmentation data and then apple fruits were graded into four classes.

Some researchers preferred monochrome camera to study this problem while others preferred to use color cameras. For instance, a 3-CCD camera (XC-003, Sony, matrix type) and a coupled frame grabber (Imascan Chroma, Imagraph) were used by Leemans et al. (1998) for the segmentation of the defects (bitter pit, fungi attack, growth defetcts, bruising, punches, insect holes, russeting, and scab) on golden delicious apples using RGB (red, green, blue) color information. They reported that the resolution has a major influence on the speed of the process, while the pre-filtering is less important. The post-filtration is concerned with monochrome images, therefore it processes less data and computes more quickly. Similarly, Arunachalam et al. (2018) identified defects in fruits using RGB image. In their study, first they converted RGB image into grayscale image (i.e., monochrome image) and the monochrome image was converted into binary image by thresholding technique. Figure 6 shows monochrome and binary image along with the original image. Features required for the identification of defects on fruits were extracted from the binary image by creating a boundary of contours of defective parts on the filtered image.

In the case of mangoes, like other fruits, early defect detection or quality assessment is expected to avoid nutritional as well as economical loss. Since several types of postharvest defects such as mechanical damage, black lesion, latex stain, shriveling, cracks, and scratches limits the market value. The importance of quick online sorting and quality assessment technique is therefore increased more. Considering these facts, Patel et al. (2021a) developed a computer vision system for the detection of defects in mango fruit using monochrome camera. They developed an algorithm and studied the potential of the system on the basis of its performance (Fig. 7).

The performance was evaluated on the basis of accuracy (the ratio of number of defected fruits detected by image processing technique accurately to the number of fruits detected manually, by expert), efficiency (the ratio of correctly processed images concerning the total number of tested images by using the proposed algorithm), and average inspection time (calculated using performance evaluation meter)

Table 1 Definition of various object images obtained during brightness-invariant image segmentation

S.No.	Image Type	Description and operation for calculation
1.	**OOI:** Original object image	An image acquired by camera/filter system.
2.	**SOI:** Separated object image	The objects in the OOI that separated from the background using threshold (T_0). $\mathrm{SOI}(x, y) = \begin{cases} \mathrm{OOI}(x, y) \text{ if } \mathrm{OOI}(x, y) > T_0 \\ 0 \text{ otherwise} \end{cases}$ Where "T_0" is a threshold.
3.	**MIOI:** Maximum intensity object image	An object image obtained by propagating the maximum pixel values within each object in SOI, using recursive morphological operation. $\begin{cases} \mathrm{MIOI}_k(x, y) = \max \begin{Bmatrix} \mathrm{MIOI}_{k-1}(x - x', y - y') + B(x', y') \\ \vert(x - x', y - y') \in D_M; (x' - y') \in D_{B,} \end{Bmatrix} \\ \mathrm{MIOI}_0(x, y) = \mathrm{SOI}(x, y), \\ k = 1, 2, 3, \end{cases}$ Where B is the 3 × 3 structuring element of '0's; D_M and D_B are the domains of MIOI and B, respectively; and $k = 1, 2, 3$ means three processing passes are required before all objects have been completely replaced by their maxima.
4.	**MPOI:** Maximum propagated object image	An object image modified according to various brightness of different fruit objects in the original object image (OOI) acquired by camera system. $\mathrm{MPOI}(x, y) = \frac{C_0}{\mathrm{MIOI}(x, y)} \mathrm{SOI}(x, y)$ Where $C_0 \geq \max[\mathrm{MOIO}(x, y)]$ is constant, say, '255'
5.	**SIOI:** Standard inverse object image	An object image obtained through sample test and image analysis. $\mathrm{SIOI}(x, y) = C_0 - g(r)$, $r \in (x, y) \subset \mathrm{COI}(x, y), \; r = 1, 2, \ldots\ldots, N$, where C_0 is constant, r is particular contour outline 'r' of the 'N' contour outlines and $g(r)$ is the standard gradient values related to the typical gradient of MPOI.
6.	**BOI:** Binary object image	A binary object image is obtained from OOI based on threshold (T_0). $\mathrm{BOI}(x, y) = \begin{cases} 1 \text{ if } \mathrm{OOI}(x, y) > T_0 \\ 2 \text{ Otherwise.} \end{cases}$

(continued)

Table 1 (continued)

S.No.	Image Type	Description and operation for calculation
7.	**COI:** Contour object image	A contour object image is generated from BOI using the recursive morphological operations. $\begin{cases} \mathrm{COI}_k(x, y) = \mathrm{COI}_{k-1}(x, y) + \min \begin{Bmatrix} \mathrm{COI}_{k-1}(x, y) \subset S(x, y); \\ \mathrm{COI}_{k-1}(x, y) \leq 1 \end{Bmatrix}, \\ \mathrm{COI}_0(x, y) = \mathrm{BOI}(x, y) \\ k = 1, 2, \ldots\ldots \max(D)/2, \end{cases}$ where S represents the eight neighboring structure elements and D is object diameter in pixels.
8.	**BIOI:** Brightness-invariant object image	A nongradient plane image, but defective areas have lower gray levels. $\mathrm{BIOI}(x, y) = \mathrm{MPOI}(x, y) + \mathrm{SIOI}(x, y)$
9.	**DSI:** Defective segment object image	An object image that contains all the defective segments on each apple in an image. $\mathrm{DSI} = \begin{cases} \mathrm{BIOI}(x, y) \text{ if } \mathrm{BIOI}(x, y) \leq T \\ 0 \text{ otherwise}, \end{cases}$ where the threshold T is near C_0, which can be calibrated to achieve the desired sensitivity of the system.

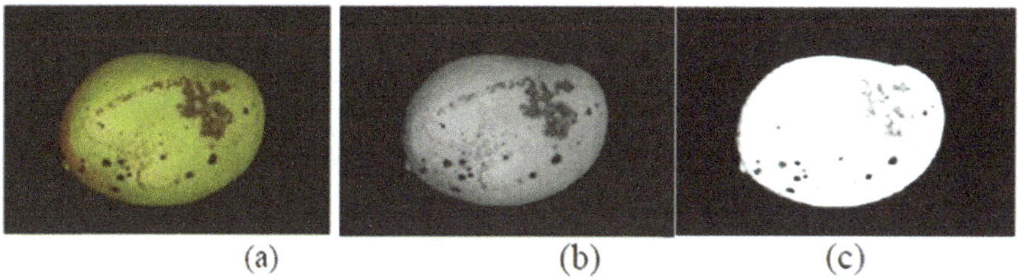

Fig. 6 (**a**) Original image, (**b**) grayscale (i.e., monochrome) image, and (**c**) binary image Arunachalam et al. (2018)

taken during the image processing and analysis. Local threshold method was applied for the correction of the background and defects on the surface of the mangoes were segmented using Otsu (1979) segmentation technique. In their study, they segmented very severely defected, severely defected, minutely defected, and fresh or slightly defected fruits using developed algorithm and reported that the accuracy and efficiency of monochrome CVS were 88.5% and 97.88%, respectively.

As appearance is the first attracting factor and plays a major role in acceptance and purchase for the consumer. The rapid quality assessment of fruits on the basis of parameters such as size, shape, and texture is also of utmost importance for good market return to the mango growers. A monochrome computer vision system can be used for rapid measurement of physical parameters of the mango fruit along with 95% accuracy (Patel et al., 2021b).

According to Patel et al. (2020, 2021b) various morphological attributes (Table 2) such as area, perimeter, Max Feret diameter, Waddel disc diameter, elongation factor, compactness factor, Heywood circulatory factor, and type factor can also be evaluated using particle analysis technique to describe the morphology of the outline and shape of the biological object. For instance, shape factors especially elongation, compactness, and type factors can be used in differentiation and classification of fruits of different cultivars. Fruits with a higher value of elongation factor and compactness factor can be interpreted as more elongated shape and either the shape has protrusion outside or indentations inside of circumference, respectively. In addition, various models based on geometrical properties such as length, width, and thickness, arithmetic and geometrical mean diameters, sphericity, and aspect ratio of mangoes obtained from MCVS (monochrome computer vision system) were used and developed for nondestructive analysis of mass and volume of fruits. Some applications of monochrome computer vision systems for the assessment of fruits and vegetables are summarized in Table 3.

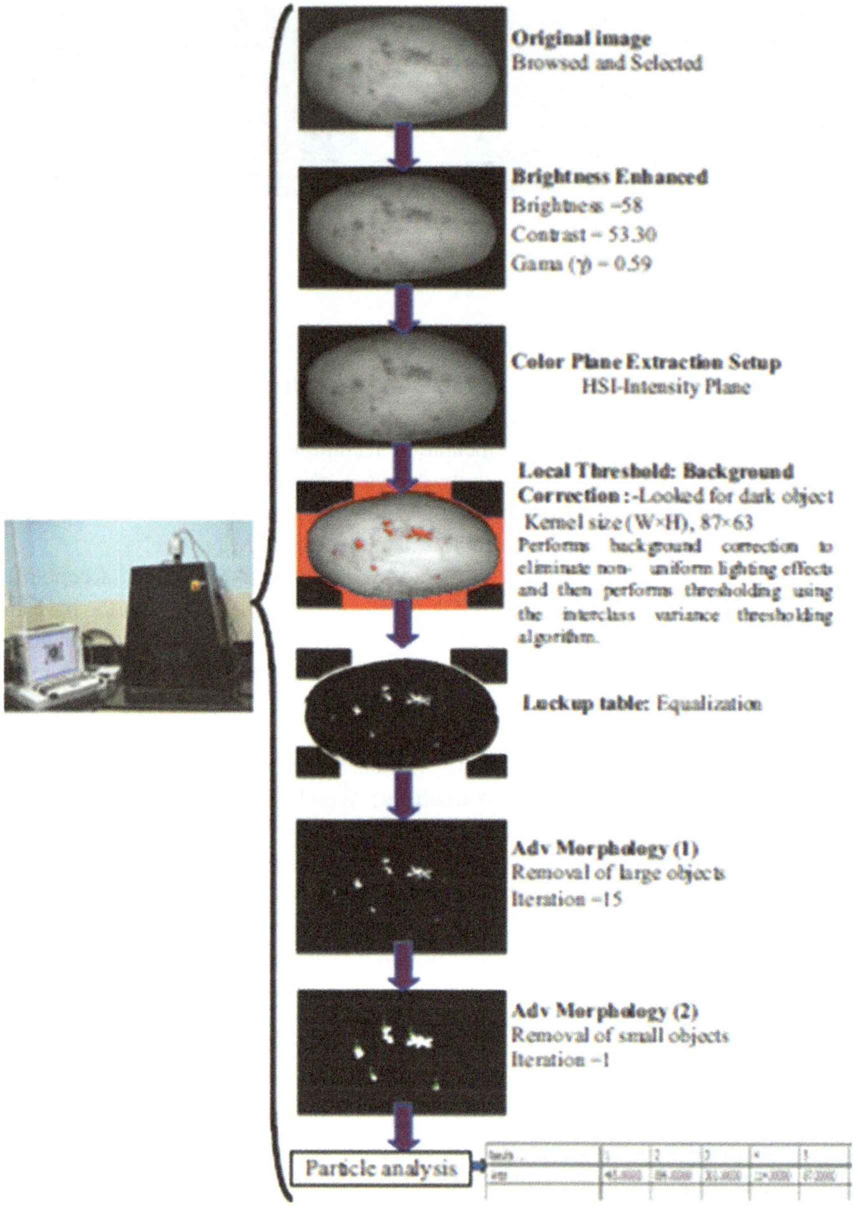

Fig. 7 Algorithm steps for detecting common external defects of mangoes (Patel et al., 2021a)

Table 2 Description of some shape attributes of mangoes evaluated nondestructively

Parameters	Description
Area	Number of pixels within the boundary of object.
Perimeter	The sum of the pixels that form the boundary of the sample.
Max Feret diameter (MFD)	The greatest distance possible between any two points along the boundary of the sample.
Waddel disc diameter (WDD)	The diameter of the disk with the same area as the particle. $\text{WDD} = \frac{\sqrt[2]{\text{Particle area}}}{\sqrt{\pi}}$
Elongation factor	The ratio of the longest segment within an object to the mean length of the perpendicular segments. $\text{Elongation factor} = \frac{\text{Max intercept}}{\text{Mean perpendicular intercept}}$
Compactness factor	The ratio of an object area to the area of the smallest rectangle containing the object. $\text{Compactness factor} = \frac{\text{Particle area}}{\text{Breadth} \times \text{width}}$
Heywood circulatory factor	The ratio of an object perimeter to the perimeter of the circle with the same area. $\frac{\text{Particleperimeter}}{\text{Perimeter of circle with same area as Particle}} = \frac{\text{Particleperimeter}}{\sqrt[2]{\pi \times \text{Particle area}}}$
Type factor	It is a complex factor relating the surface area to the moment of inertia. $\text{Type factor} = \frac{(\text{Particle area})^2}{\sqrt[4\pi]{I_{xx} \times I_{yy}}}$ where, I_{xx} and I_{yy} are moment of inertia with respect to gravity.

4.1 Detection of Quality Defects of Vegetables

Four monochrome CCD cameras and six color CCD cameras were used to develop two sets of machine vision systems for grading eggplants. One set of machine vision contained three color cameras and the other set of machine had two monochrome cameras along with grading conveyor. One set inspect one side of the fruit and other set inspect the opposite side. The eggplant was graded on 180° rotary tray. Monochrome cameras were used to measure the extent of gloss on a fruit surface from its calyx to end side. However, the defect detection and extraction of fruits feature was not easy due to low color contrast of the background. In this context, a NIR-based CCD camera was used to extract fruit features from dark background and to detect low contrast defects (Kondo et al., 2005). In many color camera-based image analysis studied, color images were first converted into grayscale image before image processing and analysis.

Tian et al. (2016) developed a nondestructive testing system based on machine vision technology using a monochrome CCD camera with a pixel of 1292 × 964 and LED light sources. They distinguished normal and black heart of potatoes according to their different transmittance. Initially, adoptive thresholding technique was used during image processing of the collected original transmission image to achieve binary image (Fig. 8). Potato images without background, finally, acquired after extracting contour (as mask image) of potatoes.

Table 3 Summary of some applications of monochrome computer vision for fruits and vegetables

Product	Application	Description	References
Apple	Recognition and classification	• Surface features (elongated blemish, nondefective area, and patch-like blemishes) used.	Yang (1993)
	Blemish inspection	• Flooding algorithm was used for image segmentation.	Yang and Tillett (1994)
		• Neural network was used as a classifier to distinguish the stalk and calyx areas of apple.	
	Detection of patch-like defects	• Flooding algorithm was used.	Yang (1994)
		• Golden delicious and Granny Smith apple foundation effective to the proposed algorithm.	
		• Applicable to a range of different types of blemishes with variable size, shape, and contrast.	
	Accurate blemish detection	• Flooding algorithm for blemish segmentation.	Yang and Marchant (1995)
		• A snake algorithm for refinement.	
Apple	Bruise detection and grading	• By image filtering, differencing, binary image thresholding, and measurement.	
		• Thinness ratios	
Apple	Online extraction of defective segments.	• Brightness-invariant image segmentation technique was used	Wen and Tao (1998)
Apple	Classification	• Used three monochrome cameras.	Guedalia (1997)
		• Defect detected, feature extracted, and graded in four classes.	
		• Neural-based classification.	
Mango	External defect detection	• Local thresholding method was applied for background correction.	Patel et al. (2021b)
		• Interclass variance thresholding algorithm.	
Mango	Physical characterization	• Histogram-based manual thresholding.	Patel et al. (2021a)
		• Used for evaluation of fruit's external diameters (major, intermediate, and minor), morphological attributes, and modeling for nondestructive evaluation of weight and volume.	
Potato	Sorting based on dimension and shape	• Based on the morphology.	McClure and Morrow (1987)
		• Large number of regression relations between major, intermediate, and minor axis.	
Eggplant	Grading	• Four monochrome and six color CCD cameras were used.	Kondo et al. (2005)
		• Two set of machine vision system was used and inspected entire surface of eggplant.	
		• Monochrome camera was used to measure the extent of gloss on fruit surface.	

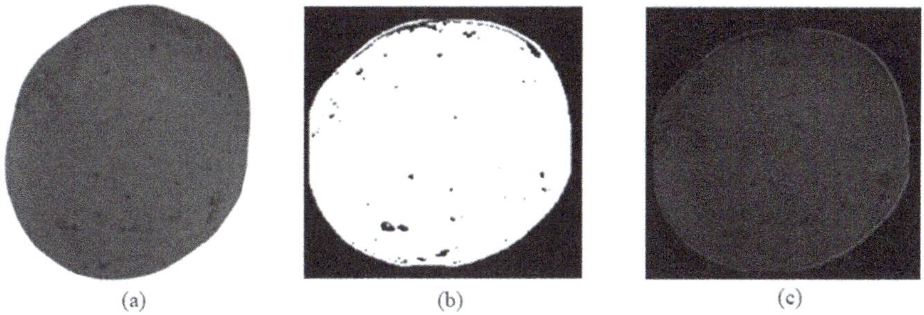

Fig. 8 The potato samples image processing results (**a**) the potato source image after background elimination; (**b**) the potato binary; (**c**) the potato image after black spots eliminated (Tian et al., 2016)

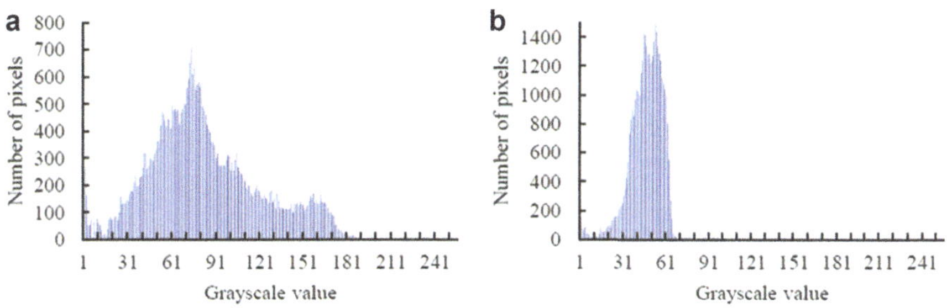

Fig. 9 The grayscale histograms of the normal potato (**a**) and the potato with black heart (**b**) (Tian et al., 2016)

Based on the image, features of the changing trend of the grayscale histogram (Fig. 9) and the average grayscale value were analyzed and reported that the potato with black heart was mainly in low-grayscale region. They achieved higher accuracy during the nondestructive testing of black heart of potatoes.

Similarly, sorting of potatoes based on computer vision system was developed by McClure and Morrow (1987). They detected the size and shape information of white potatoes, and greens and scars of potatoes creases (wrinkles) using monochrome camera. A visual-based quality sorting system for fruit using three monochrome cameras with an optical device was also developed by Guedalia (1997). The optical device they used was projecting four times the same views with different bandwidths on one CCD.

5 Benefits of Monochrome Imaging

There are various benefits of the application of monochrome imaging in nondestructive quality assessment. Some of them are given below:

- Because of responding to all colors of light, monochrome cameras have much better light efficiency
- A monochrome image sensor does not use a color filter array (CFA).
- Because of the absence of color filter array (CFA) and infrared cut filter (IR CF), monochrome camera detects a broader light spectrum.
- Monochrome cameras have high light sensitivity, better contrast and thus generate sharper images with superior resolution.
- Monochrome imaging provides a better signal-to-noise ratio.
- The output of smaller file sizes makes monochrome imaging suitable for a number of applications.
- The overall performance of the monochrome image is even in low lighting conditions.
- Monochrome camera can achieve higher spatial resolution resulting from simultaneous usage of all charged-coupled devices (CCD) photodiodes for the acquisition of one image without color mosaics and the extrapolation of color information.
- Monochrome cameras are generally less expensive than color cameras.
- Monochrome camera provides flawless results.

6 Conclusion

This chapter presents various information related to the fundamentals, developments, and applications of monochrome imaging for detecting defects in fruits and vegetables. Many times monochrome computer vision system has been reported more beneficial in detecting cracks and crevices of fruits and vegetables. Monochrome photography, in spite of, being less expensive, has a high resolution, good sensitivity, and high acquisition speed. But, this study explores that in spite of having good potential in detecting the various types of surface defects of fruits and vegetables very efficiently, the monochrome camera-based CVS is still in the development stage. This chapter thus could be helpful in the new research and provide the basic information on MCVS to the researchers.

References

Abbott, J. A. (1999). Quality measurement of fruits and vegetables. *Postharvest Biology and Technology, 15*, 207–225.

Al-Dairi, M., Pathare, P. B., & Al-Yahyai, R. (2021). Quality chemical and nutritional changes of tomato during postharvest transportation and storage. *Journal of the Saudi Society of Agricultural Sciences, 20*, 401–408.

Anonmous. (2021). *Global fresh fruit and vegetable production stabilizes*. Retrieved from https://www.freshplaza.com/article/9292001/global-fresh-fruit-and-vegetable-production-stabilizes/

Arjenaki, O. O., Moghaddam, P. A., & Motlagh, A. M. (2013). Online tomato sorting based on shape, maturity, size, and surface defects using machine vision. *Turkish Journal of Agriculture and Forestry, 37*, 62–68.

Arunachalam, S., Kshatriya, H. H., & Meena, M. (2018). Identification of defects in fruits using digital image processing. *International Journal of Computer Sciences and Engineering, 6*(10), 637–640.

Ballard, D. A., & Brown, C. M. (1982). *Computer vision*. Prentice-Hall.

Chowdhury, M. A., Hossain, N., Kashem, M. A., Shahid, M. A., & Alam, A. (2020). Immune response in COVID-19: A review. *Journal of Infection and Public Health, 13*(11), 1619–1629.

Davenel, A., Guizard, C. H., Labarre, T., & Sevila, F. (1988). Automatic detection of surface defects on fruit by using a vision system. *Journal of Agricultural Engineering Research, 41*(1), 1–9.

Guedalia, I. D. (1997). Visual based quality sorting of fruit. Sensors for non-destructive testing—Measuring the quality of fresh fruits and vegetables. In *Proceedings from the Sensors for Non-destructive Testing International Conference and Tour, February 18–21, 1997* (pp. 232–238). North Regional Agricultural Engineering Service.

Hassankhani, R., Navid, H., & Seyedarabi, H. (2012). Potato surface defect detection in machine vision system. *African Journal of Agricultural Research, 7*(5), 844–850.

Howison, R., & Menard, B. (2012). Color cameras improve results in vision applications. *Photonics Spectra, 46*(10), 48–51.

Kandi, S. G. (2010). Automatic defect detection and grading of single color fruits using HSV (hue, saturation, value) color space. *Journal of Life Sciences, 4*(7), 39–45.

Knight, T. G., Klieber, A., & Sedgley, M. (2002). Structural basis of the rind disorder oleocellosis in Washington navel orange (Citrus sinensis L. Osbeck). *Annals of Botany, 90*, 765–773.

Kondo, N., Chong, V. K., Ninomiya, K., Ninomiya, T., & Monta, M. (2005). *Application of NIR-color CCD camera to eggplant grading machine*. ASAE Annual International Meeting.

Leemans, V., Magein, H., & Destain, M.-F. (1998). Defects segmentation on "Golden Delicious" apple by using colour machine vision. *Computers and Electronics in Agriculture, 20*, 117–130.

Lopez-Garcıaa, F., Andreu-Garciaa, G., Blascob, J., Aleixosc, N., & Valientea, J. M. (2010). Automatic detection of skin defects in citrus fruits using a multivariate image analysis approach. *Computers and Electronics in Agriculture, 71*, 189–197.

Marceau, F. (2009). *Advance imaging*. Academic Journals and Books at Questa Online Library.

McClure, J. E., & Morrow, C. T. (1987). *Computer vision sorting of potatoes* (ASAE Meeting Paper 87-6501). ASAE.

Mishra, V. K., Kumar, S., & Shukla, N. (2017). Image acquisition and techniques to perform image acquisition. *SAMRIDDHI A Journal of Physical Sciences Engineering and Technology, 9*(1), 21–24.

Mrema, C. G., & Rolle, S. R. (2002). Status of the postharvest sector and its contribution to agricultural development and economic growth. In *9th JIRCAS International Symposium—Value Addition to Agricultural Product* (pp. 13–20). Japan International Research Center for Agricultural Sciences.

Olayemi, F. F., Adegbola, J. A., Bamishaiye, E. I., & Daura, A. M. (2010). Assessment of post-harvest challenges of small scale farm holders of tomatoes, bell and hot pepper in some local government areas of Kano State, Nigeria. *Bayero Journal of Pure and Applied Sciences, 3*, 39–42.

Otsu, N. A. (1979). Threshold selection method from gray-level histogram. *IEEE Transactions on Systems, Man, and Cybernetics, 9*, 62–66.

Patel, K. K., Kar, A., Jha, S. N., & Khan, M. A. (2012). Machine vision system: A tool for on-line sorting and grading of agricultural foods and produces. *Journal of Food Science and Technology, 49*(2), 123–141.

Patel, K. K., Kar, A., & Khan, M. A. (2012). Nondestructive food quality evaluation techniques: Principle and potential. *Agricultural Engineering Today, 36*(4), 29–34.
Patel, K. K., Kar, A., & Khan, M. A. (2019). Common external defect detection of mangoes using color computer vision. *Journal of The Institution of Engineers (India) Series A, 100*, 559–568. https://doi.org/10.1007/s40030-019-00396-6(India)
Patel, K. K., Kar, A., & Khan, M. A. (2020). Development and an application of computer vision system for nondestructive physical characterization of mangoes. *Agricultural Research, 9*, 109–124.
Patel, K. K., Kar, A., & Khan, M. A. (2021a). Monochrome computer vision for detecting common external defects of mango. *Journal of Food Science and Technology, 58*, 4550–4557.
Patel, K. K., Kar, A., & Khan, M. A. (2021b). Rapid assessment of some physical parameters of mangoes using monochrome computer vision. *Agricultural Research, 10*, 468–482.
Patel, K. K., Khan, M. A., & Kar, A. (2019). Potential of reflected UV imaging technique for detection of defects on the surface area of mango. *Journal of Food Science and Technology, 56*(3), 1295–1301.
Pla, F., & Juste, F. (1995). A thinning-based algorithm to characterize fruit stems from profile images. *Computers and Electronics in Agriculture, 13*(4), 301–314.
Rehkugler, G. E., & Throop, J. A. (1986). Apple sorting with machine vision. *Transaction of ASAE, 29*(5), 1388–1397.
Richards, A. (2010). *Reflected ultraviolet imaging for forensics applications*. http://www.company7.com/library/nikon/Reflected_UV_Imaging_for_Forensics_V2.pdf
Spricigo, P. C., Miranda, M., & Ferreira, M. D. (2013). Impacts on fruit from hand- and aided harvests affect postharvest quality of citrus. *Proceedings of the Florida State Horticultural Society, 126*, 225–228.
Tian, F., Peng, Y., & Wei, W. (2016). Nondestructive and rapid detection of potato black heart based on machine vision technology. In M. S. Kim, K. Chao, & B. A. Chin (Eds.), *Sensing for Agriculture and Food Quality and Safety VIII* (Vol. 98640T, pp. 1–7). SPIE (International Society for Optics and Photonics) Proceedings.
Vakilian, K. A., & Massah, J. (2013). An artificial neural network approach to identify fungal diseases of cucumber (Cucumis sativus L.) plants using digital image processing. *Archives of Phytopathology and Plant Protection, 46*(13), 1580–1588.
Velioglu, H. M., & Saglam, O. (2012). Evaluation of insect damage on beans using image processing technology. In *International Conference of Agricultural Engineering CIGR-AgEng, July 8–12, 2012, Valencia, Spain.*
Weber, G. F., & Menko, A. S. (2005). Color image acquisition using a monochrome camera and standard fluorescence filter cubes. *BioTechniques, 38*, 52–56.
Wen, Z., & Tao, Y. (1998). Brightness-invariant image segmentation for on-line fruit defect detection. *Optical Engineering, 37*(11), 2948–2952.
WHO and FAO. (2003). *Diet, nutrition, and the prevention of chronic diseases. Report of a joint WHO/FAO expert consultation*. WHO and FAO.
Yang, Q. (1993). Classification of apple surface features using machine vision and neural networks. *Computers and Electronics in Agriculture, 9*, 1–12.
Yang, Q. (1994). An approach to apple surface feature detection by machine vision. *Computers and Electronics in Agriculture, 11*, 249–264.
Yang, Q., & Marchant, J. A. (1995). Accurate blemish detection with active contour models. *Computers and Electronics in Agriculture, 14*, 77–89.
Yang, Q., & Tillett, R. D. (1994). Apple surface feature for blemish detection. In *Proceedings of Workshop COST 94–System and Operation for Post-harvest Quality, September 1993, Leuven, Belgium* (pp. 93–107).

Imaging Techniques for Evaluation of Ripening and Maturity of Fruits and Vegetables

Hulya Cakmak and Ece Sogut

Abstract Optimal harvesting time of fruits and vegetables is an important factor, which is directly associated with the postharvest quality of the produce and shelf life. Depending on the variety of horticultural products, maturity can be assessed using internal properties like moisture, sugar, starch, oil content, soluble solid content (SSC), titratable acidity (TA), SSC/TA, pH, and firmness, or using external properties like surface or peel color (chlorophyll, carotenoids, lycopene, etc.), size, volume, shape, and peel/flesh ratio that are taken into consideration. The level of maturity for these products is determined by the limits based on the internal and external properties of that specific product. Conventional maturity evaluation methods generally employ destructive analysis; however, an increasing number of studies in the last decade have shown that nondestructive methods have been successfully applied to determine the maturity of produce. Nondestructive methods allow analyzing the raw data extracted from the original image and reconstructing a 3D model of dissected sample for visualization of internal structure. Surface color or the structure of samples is also analyzed with several imaging and image processing techniques in order to determine the maturity levels. Whether the internal or external structure is scrutinized, the compliance of extracted data with destructive maturity or ripening parameters must be clearly verified. Statistical models like artificial neural network, principal component analysis, or machine learning approaches are applied because of reducing the amount of extracted data from imaging analysis and its complexity. In this chapter, the imaging techniques used for determining the maturity or ripening levels of fruits and vegetables are discussed.

Keywords Maturity · Ripening · Firmness · Color · Hyperspectral · 3D

H. Cakmak (✉)
Faculty of Engineering, Department of Food Engineering, Hitit University, Corum, Turkey

E. Sogut
Faculty of Engineering, Department of Food Engineering, Suleyman Demirel University, Isparta, Turkey

P. B. Pathare, M. S. Rahman (eds.), *Nondestructive Quality Assessment Techniques for Fresh Fruits and Vegetables*, https://doi.org/10.1007/978-981-19-5422-1_3

1 Introduction

Maturity and ripening terms are often confused or used interchangeably; maturity refers to development stage of a plant reaching its physiological or horticultural maturation following the cell expansion, whereas ripening follows maturation that includes simultaneous biochemical and morphological changes like softening of the tissue, sugar and pigment accumulation, flavor development, and starch hydrolysis (Brummell et al., 2016; Lamikanra, 2002). Ripening is also a stage placed between maturity and senescence, and irreversible transformations occur during ripening by respiration and ethylene production (Brummell et al., 2016; Hui et al., 2004; Jongen, 2002; Saltveit, 2002). Fruits show different behavior with respect to respiration and ethylene formation and are classified into two main groups depending on their ripening process (Hui et al., 2004). The first group entitled "climacteric fruits & vegetables" continue their ripening after harvest, and detached from vine or tree-like; apple, avocado, banana, cantaloupe, date, kiwi, mango, papaya, peach, pear, persimmon, potato, and tomatoes (Barrett et al., 2005; Brummell et al., 2016; Hui, 2006; Hui et al., 2004; Jongen, 2002; Thompson, 2003). The second group "non-climacteric fruits & vegetables" are harvested mature and do not ripen after harvesting such as artichoke, asparagus, blueberry, broccoli, carrot, celery, cherry, citrus fruits, cucumber, eggplant, grape, green bell pepper, lettuce, pea, pineapple, pomegranate, strawberry, and watermelon (Barrett et al., 2005; Brummell et al., 2016; Hui, 2006; Saltveit, 2002). Climacteric fruits and vegetables produce high amounts of ethylene during ripening and have respiration burst, while non-climacteric plants produce very small amounts of ethylene and similar respiration rate or little decline is observed during ripening (Barrett et al., 2005; Brummell et al., 2016; Chen et al., 2018; Jongen, 2002; Thompson, 2003). Despite having different ethylene production behaviors, both these fruits and vegetables can be classified according to their maturity and ripening stages. Size (volume, weight, etc.), shape, internal or external color, firmness, composition (soluble solids, sugar, starch, oil, acid, tannin, and juice content), specific gravity and surface characteristics are important indices as maturity or ripening evaluation (Barrett et al., 2005; Brummell et al., 2016; Hui, 2006; Reid, 2002; Saltveit, 2002). One or a couple of these indices are required to be examined for determining optimum harvesting time and ripening level thus the postharvest quality of fruits and vegetables.

Total soluble solids (°Brix) or dry matter in avocado, banana, grape, lime, green tomatoes, peach, strawberries, and tomatoes (Benelli et al., 2020; Fatchurrahman et al., 2020; Huang et al., 2021; Minas et al., 2021; Sánchez et al., 2012; Sripaurya et al., 2021; Teerachaichayut & Ho, 2017; Urraca et al., 2016; Vega Díaz et al., 2021), titratable acidity or pH in cherry, olive, and pomegranate (Fashi et al., 2020; Fernández-Espinosa, 2016; Li et al., 2018b), maturity index (SSC/TA) in oranges (Aredo et al., 2019), and BrimA in grapefruit, orange, and pomelo (Gupta et al., 2021, 2022; Ncama et al., 2017) are used as maturity and ripening indices. Besides, firmness/texture in fig, peach, persimmon, and tomatoes (Herrero-Langreo et al., 2012; Mohammadi et al., 2015; Sirisomboon et al., 2012; Sun et al., 2021), internal or external color in apple, apricot, banana, cocoa pods, green bitter gourd, lemon, persimmon, pomelo, raspberries (Gupta et al., 2021; Hsiao et al., 2021; Khojastehnazhand et al., 2019; Kusumiyati et al., 2018; Lockman et al., 2019; Mohammadi et al., 2015; Pourdarbani et al., 2020; Rodríguez-Pulido et al., 2017; Zulkifli et al., 2019), ripening index in mango (Rungpichayapichet et al., 2016) can be other important parameters. There are a few studies in the literature that also employs miscellaneous properties like anthocyanin in blueberries (Cruz, 2020), chlorophyll and carotenoid concentration in mango, papaya, and pomelo (Gupta et al., 2021; Kotwaliwale et al., 2012; Tripathi et al., 2020), pelargonidin-3-glucoside in strawberries (Cho et al., 2021), and oil content in oil palm fruit and olive (Ali et al., 2020; Fernández-Espinosa, 2016; Makky & Soni, 2014) for evaluation of maturity and ripening in fruits and vegetables. Although these indices are selected and evaluated, some of them are not significantly effective in identifying the exact maturity or ripening stage because of almost overlapping values of tested parameters. Table 1 summarizes the upper and lower limits of destructive and nondestructive (surface color) maturity indices of some fruit samples.

Vegetable harvesting maturity is more complex than fruits and depends on the skills of the farmer, and therefore it is rather subjective. For example, asparagus, broccoli, carrot, cauliflower, celery, eggplant, leek, okra, pumpkin, radish, and zucchini maturity is assessed by their size, shape, or thickness, whereas tenderness or firmness is important for cabbage, lettuce, mushroom, peas, pepper, and pumpkin (Kader, 2011; Reid, 2002;Singh et al., 2018 ; Thompson, 2003). The harvesting maturity of onion bulbs and potatoes are determined by the drying of foliage, as haulm color turns yellow and wither (Singh et al., 2018; Thompson, 2003). Leafy or green vegetables like broccoli, chicory, cucumber, endive, kale, lettuce, mustard, pak choy, and spinach maturity is mostly related to their color by the presence of chlorophylls and carotenoids (de Azevedo & Rodriguez-Amaya, 2005; de Azevedo-Meleiro & Rodriguez-Amaya, 2005; do Nascimento Nunes, 2008; Kader, 2011; Singh et al., 2018; Thompson, 2003).

Internal factors that are employed in maturity and ripening evaluation are acquired via destructive techniques, visual inspection of surface color, and other external factors by subjective assessment methods. These are often expensive, laborious, and they have low repeatability and precision (Cakmak, 2019; Sun, 2010). In order to overcome these challenges and prevent losses associated with performing destructive analysis, nondestructive imaging techniques, such as spectral imaging (UV, visible, near-infrared, or infrared range), fluorescence spectroscopy, 3D-imaging (X-ray, computed tomography, nuclear magnetic resonance imaging), biospeckle methods offers cost-effective, objective, repeatable, waste free, real-time, automatic, and portable/stationary measurements that are the alternatives with minimal or no preparation requirement of sample before analysis (Abasi et al., 2018; Hitzmann & Ahmad, 2017; Irudayaraj & Reh, 2008; Kotwaliwale et al., 2014; Lakshmi et al., 2017; Su & Sun, 2018; Sun, 2010).

Table 1 Significant parameters in destructive maturity/ripening evaluation of fruits

Sample	Maturity indices[a]	Maturity/ ripening stage	Limits of maturity/ripening stage	Reference
Apricot	L*, b*, G, gray scale	Unripe-ripe-overripe	L*: −4.2-23.9, b*: 219.9–191.8, G:141.4–220.3, gray scale: 77.1–88.6	Khojastehnazhand et al. (2019)
Banana	F, SSC, M	Stage 1-2–3-4–5-6	F: 10.0–1.1, SSC: 13.9–15.8, M: 74.9–85.3	Rajkumar et al. (2012)
Banana (Berangan)	a*, F, SSC	Stage 2-3–4-5–6-7	a*: −13.9-13.3, F: 8.6–5.8, SSC: 2.3–8.6	Zulkifli et al. (2019)
Bananito	F, SSC, a*, h°	Stage 2-4-6	F: 15.2–5.1, SSC: 7.5–17.9, a*: −13.68-3.1, h°: 111.3–85.9	Pu et al. (2019)
Blueberry	L*, b*, A	Class 1-2-3-4	L*: 19.9–45.9, b*: −2.5-26.5, A: 105–60	Cruz (2020)
Cherry	SSC, pH	Stage 1-2-3	SSC: 10.5–17.6, pH: 3.5–3.8	Li et al. (2018a)
Cocoa pods	F, a*, C, h°	Unripe-ripe-overripe	F: 10.4–6.7, a*: 4.9–12.2, C: 28.5–46.6, h°: 88.0–70.7	Lockman et al. (2019)
Mango	SSC, RPI, F	Unripe-half ripe-ripe-overripe	SSC: ~11–22.5, RPI: 29.0–27.7, F: ~32–8	Rungpichayapichet et al. (2016)
Oil palm fruit	O, L*, a*, b*	Unripe-ripe-overripe	O: 15.2–41.2, L*: 34.6–40.2, a*: 16.8–32.1, b*: 15.4–31.1	Ali et al. (2020)
Pear	F, SSC	Unripe-ripe-overripe	F: 45.7–12.2, SSC: 10.4–17.8	Khodabakhshian and Emadi (2017)
Pear (conference)	F, SSC, M	Unripe-overripe	F: 181.4–112.2, SSC: 11.9–14.2, M: 81.7–83.6	Adebayo et al. (2017)
Persimmon	E, SSC, b*	Unripe-ripe-overripe	E: 0.054–0.029, SSC: 17.2–21.4, b*: 51.4–40.3	Mohammadi et al. (2015)
Pomegranate	SSC, pH, TA	Stage 1-2-3-4	SSC: 15.1–18.5, pH: 3.1–3.4, TA: 0.9–0.7	Khodabakhshian et al. (2017)
Tomatoes (Momotaro)	SSC, F	Mature green-pink-red	SSC: 4.1–5.6, F: 11.2–2.4	Sirisomboon et al. (2012)

A anthocyanin content, *a** redness/greenness, *AIS* alcohol insoluble solid, *b** yellowness/blueness, *C* chroma, *E* elasticity, *F* firmness, *G* green image channel, *h°* hue angle, *L** lightness, *M* moisture, *O* Oil %, *RPI* ripening index, *TA* titratable acidity, *SSC* total soluble solid
[a] Only significant parameters in maturity/ripening stages are tabulated

2 Image Acquisition and Processing

Nondestructive imaging techniques for the classification of fruits and vegetables according to their ripening and maturity stages include several consecutive steps, and at the beginning, image acquisition is performed by digital cameras, Raspberry Pi camera modules, near infrared (NIR) cameras, charged-coupled devices (CCD), complementary metal-oxide semiconductor sensor (CMOS), indium-gallium-arsenide (InGaAs) cameras and sensors, magnetic resonance imaging or computed tomography (Cho et al., 2021; Du & Sun, 2004; Fatchurrahman et al., 2020; Harel et al., 2020; Lakshmi et al., 2017; Park & Lu, 2015; Pathmanaban et al., 2019; Pourdarbani et al., 2020; Russ, 2005; Sanchez et al., 2020; Sun, 2010; Villaseñor-Aguilar et al., 2020; Zhuang et al., 2019). For improving the image quality and sensitivity, illumination of the sample is generally required and provided by UV lamp, halogen tungsten lamps, fluorescent lamps, laser diodes or light emitting diodes (LED) and samples are placed in a dark chamber or room (Ali et al., 2020; Fatchurrahman et al., 2020; Garillos-Manliguez & Chiang, 2021; Gupta et al., 2022; Harel et al., 2020; Lockman et al., 2019; Park & Lu, 2015; Pourdarbani et al., 2020; Pu et al., 2016; Sanchez et al., 2020; Sripaurya et al., 2021; Sun, 2010; Sun et al., 2021; Vega Díaz et al., 2021). Following the image acquisition, preprocessing steps like filtering (average, Gaussian, median, Savitzky–Golay) and segmentation (thresholding, edge-based segmentation, region-based segmentation, classification-based segmentation) are applied for eliminating the background of the image and removing noise, improving contrast and sharpening the edges, and smoothing the image (Du & Sun, 2004; Harel et al., 2020; Hsiao et al., 2021; Huang et al., 2021; Muthulakshmi & Renjith, 2020; Park & Lu, 2015; Ropodi et al., 2016; Russ, 2005; Sanchez et al., 2020; Sripaurya et al., 2021; Sun, 2010; Zhuang et al., 2019). From segmented images, several features can be extracted after choosing the optimal region of interest (ROI) rather than analyzing whole image for reaching the proposed classification (Khodabakhshian & Emadi, 2017; Sun, 2010; Zhu et al., 2017). These could be size, shape, color, texture features, or spectral and spatial features belonging to hyperspectral images for reducing the quantity of data (Du & Sun, 2004; Harel et al., 2020; Park & Lu, 2015; Sun, 2010). Color-based features are expressed differently by their color channels, and the images may be given with red-green-blue (RGB) color space, hue-saturation-intensity (HSI) space, hue-saturation-value (HSV), luminance-bandwidth-chrominance (YUV), or L*a*b* space (Harel et al., 2020; Hitzmann & Ahmad, 2017; Mohammadi et al., 2015; Park & Lu, 2015; Russ, 2005). Moreover, statistical approaches such as principal component analysis (PCA), linear discriminant analysis (LDA), quadratic discriminant analysis (QDA), soft independent modeling of class analogy (SIMCA), partial least square discriminant analysis (PLS-DA), naïve Bayes, support vector machine (SVM), adaptive neuro-fuzzy inference (ANFIS), artificial neural network (ANN), ANN with genetic algorithm (ANN-GA), with particle swarm optimization (ANN-PSO) or with firefly algorithm (ANN-FA), k-nearest neighbors (KNN), random forest, decision tree, fuzzy clustering, and extreme learning machine (ELM) classifier are employed for selecting the important features and classification of the samples (Ali et al., 2020; Fashi et al., 2020; Harel et al., 2020; Huang et al., 2021; Khodabakhshian et al., 2017; Mohapatra et al., 2017; Muthulakshmi & Renjith, 2020; Park & Lu, 2015; Pourdarbani et al., 2020; Pu et al., 2019; Sanchez et al., 2020; Zhuang et al., 2019; Zulkifli et al., 2019).

3 Classification According to SSC or Dry Matter Content

3.1 SSC or Dry Matter

Fruit and vegetable maturity and optimum eating or processing quality are mostly based on the soluble solid content of crop which is correlated with flavor/aroma development by degradation of organic acids into sugars, firmness, and color development (Barrett et al., 2005; Benelli et al., 2020; Huang et al., 2021; Khodabakhshian et al., 2017; Minas et al., 2021; Mohammadi et al., 2015; Pathmanaban et al., 2019; Pereira et al., 2018; Srivastava & Sadistap, 2018; Walsh et al., 2020). Soluble solid is particularly associated with the degree of sweetness or sugar content in many fruits and vegetables like banana, carrot, citrus fruits, grape, mango, melon, and tomatoes (Beghi et al., 2018; Gupta et al., 2022; Lu et al., 2015; Nordey et al., 2017; Sripaurya et al., 2021; Urraca et al., 2016; Xiao et al., 2018). For example, conversion of starch into sugars occurs in the later stages of ripening of banana and mango, while accumulation of sugars in grape or pomelo increases by ripening of the fruit (Gupta et al., 2021; Marques et al., 2016; Sripaurya et al., 2021; Xiao et al., 2018). Thus, the level of maturity of crops can be monitored easily with respect to analyzing the total soluble solid or dry matter content and correlating them with extracted nondestructive image analysis data.

The maturity of tomato is visually assessed by its skin color due to the degradation of chlorophylls into carotenoids mostly lycopene and beta-carotene, which results in bright red color development at fully mature stage (Fatchurrahman et al., 2020; Huang et al., 2021; Kasampalis et al., 2020). It is also possible to detect the maturity level of tomatoes objectively by employing nondestructive techniques. Although tomatoes have quite complex and heterogeneous internal structure by the presence of pericarp, locule and seeds, emittance and transmittance of incident light by these different parts reflect the internal and external characteristics of fruit (Fatchurrahman et al., 2020; Huang et al., 2021). Recent study by Huang et al. (2021) provided an insightful observation about tomato maturity classification by the transmittance behavior of fruit at vertical and horizontal stem-end orientation. Although there was no significant difference in total soluble solids of tomatoes at different maturity stages namely green, turning, pink, light red and red, the authors have predicted the SSC of tomatoes by proposed partial least square regression model using full spectral data ($R^2_{\text{prediction}} = 0.75$). However, the classification accuracy of the model was improved by the reduction of maturity stages from 5 to 3 classes as; immature, intermediate (turning and pink), and mature (light and red) and prediction performance was increased (R^2-immature: 0.91, R^2-intermediate: 0.95 and R^2-mature: 0.89). Since chlorophyll is a distinctive compound for tomato maturity, fluorescence imaging was used for evaluating the level of maturity in green tomatoes (Fatchurrahman et al., 2020). Gray values extracted from hyperspectral images at 690 nm of immature and mature green tomatoes were analyzed in order to classify the samples by the emissions of chlorophyll presence. The proposed model based on gray values of hyperspectral images had very high accuracy for maturity classification, especially for immature green tomatoes, SSC along with the TA and pH was defined as potential maturity indicators by the authors.

3.2 Hyperspectral Imaging

Hyperspectral imaging method is advantageous over only vision-based or spectroscopic techniques by providing spatial and spectral information together (Liu et al., 2017; Sun, 2010). Therefore, recent studies showed that hyperspectral imaging had a great potential not only to be used in maturity classification but also predicting the soluble solid distribution in apple, avocado, grape, kiwi, lime, pear, and persimmon (Benelli et al., 2020; Khodabakhshian & Emadi, 2017; Mo et al., 2017; Munera et al., 2017; Teerachaichayut & Ho, 2017; Vega Díaz et al., 2021; Zhu et al., 2017). The accuracy of maturity classification is dependent on selecting the optimum ROI, wavelength or spectral band range, extracted features, and statistical models representing the classification (Cakmak, 2019; Mahesh et al., 2015; Sun, 2010; Teerachaichayut & Ho, 2017; Vega Díaz et al., 2021). For instance, avocado dry matter content prediction and calibration were tested with multiple linear, principal component analysis, partial least square, and support vector machine by Vega Díaz et al. (2021). The model accuracy for predicting dry matter of avocados was improved by selecting the average spectrum data extracted from both sides of the fruit instead of using only one side. Besides, SVM regression analysis had the highest prediction efficiency by selecting the relevant wavelength for classification.

Portable hyperspectral camera devices allow monitoring of maturity in-field or on-tree fruits. Cabernet Sangiovese red grapes ripening in the vineyard was evaluated by Benelli et al. (2020) using portable push-broom hyperspectral camera device, and raw spectral was processed by PLS that involved Savitzky–Golay smoothing and SNV method. The prediction performance of the SSC was acceptable, since the R^2-calibration and R^2-validation were 0.78 and 0.75, respectively. Similar to this study, SNV pretreatment of extracted hyperspectral data resulted in better classification for each maturity stage (Stage 1-2–3-4) of pomegranates (Khodabakhshian et al., 2017). Also, R^2-prediction of SSC by PLS models reached the highest value (0.92) by employing SNV and median filter together. Nevertheless, hyperspectral imaging is not always successful in the prediction of SSC and corresponding maturity level. As given in the study of Aredo et al. (2019), SSC of intact and half oranges was analyzed by PLS regression and the proposed full model was unable to predict the SSC of intact oranges (R^2-pred.: 0.27). Although prediction performance of the model was higher for half oranges (R^2-pred.: 0.76), it was more successful in estimating the skin color parameters.

3.3 X-Ray Computed Tomography (CT)

X-ray computed tomography and micro-computed tomography (μCT) reconstruct 3D geometric models from 2D absorption images while rotating the scanned object based on penetration of X-rays inside the material (Kotwaliwale et al., 2012; Muziri et al., 2016; Schoeman et al., 2016). Gray scale values or intensity of the X-ray image provide information about the object's porosity, such as lower intensity, represents the higher number of pores and air spaces, whereas the higher intensity shows higher absorption and presence of the compact cells and high tissue density (Cantre et al., 2014; Diels et al., 2017; Kotwaliwale et al., 2012; Muziri et al., 2016; Schoeman et al., 2016). Although the studies with X-ray CT are mostly focused on the detection of internal injuries, bruises, or textural imperfections, it is also used for monitoring the ripening of fruits or imaging of mature fruits (Kotwaliwale et al., 2012; Muziri et al., 2016; Schoeman et al., 2016; Xiao et al., 2021). In the study of Kotwaliwale et al. (2012), postharvest ripening behavior of Chousa cultivar mango was observed with X-ray CT and the relation of CT numbers (Hounsfield unit) with SSC, pH, TA, and carotenoid content was evaluated. Depending on the ripening of mango, uniformity in grayness value decreased which resulted in lower CT values. The regression model that was based on the CT values demonstrated that SSC of mango can be predicted according to CT numbers with a good correlation ($R^2 = 0.82$). However, health and safety-related issues related to X-rays leakage and time-consuming scanning and image processing steps make this method less favorable among other nondestructive methods (Arendse et al., 2018; Diels et al., 2017).

3.4 Nuclear Magnetic Resonance (NMR)

Nuclear magnetic resonance (NMR) and magnetic resonance imaging methods (MRI) analogous to X-ray CT are able to reconstruct 3D images and reveal detailed information about proton density, and relaxation times (Cakmak, 2019; Musse et al., 2009; Zhang & McCarthy, 2012, 2013). The possible changes at cellular level are clearly identified by relaxation times, and water mobility within the sample reflects how the texture changes by ripening or postharvest quality (Patel et al., 2015; Suchanek et al., 2017; Zhang & McCarthy, 2012). Spin-lattice relation time (T_1) and spin-spin relaxation time (T_2) are sensitive to the variations of internal tissues and SSC increase due to ripening is correlated with these parameters (Cakmak, 2019; Musse et al., 2009; Zhang & McCarthy, 2013). Zhang and McCarthy (2013) have studied the quality parameters like SSC, TA, SSC/TA, and pH of fresh and stored pomegranate (3 months) and monitored the variations with low filed NMR (0.04 Tesla) and MRI (1 Tesla). SSC of pomegranates decreased with the storage, which resulted in lower T_2 values; however, the correlation between SSC and T_2 was rather poor and unable to predict the SSC variation of pomegranates by MRI.

Interestingly, these methods are also applicable for observing the ripening of the fruits having complex internal structures like presence of seeds, and tissues having different textures such as gel-like locular tissue, vascular tissue, and pericarp of tomato and persimmon (Clark & MacFall, 2003; Musse et al., 2009; Zhang & McCarthy, 2012). These differences were clearly seen on T_1 and T_2 measurements of the mesocarp and vascular tissue of persimmon, although their tendency during sampling period was almost similar during ripening (Clark & MacFall, 2003). Besides, T_2-weighted imaging of tomato pericarp was referred as a good indicator of tomato maturity assessment (Zhang & McCarthy, 2012).

4 Classification According to TA and PH

The acidity being one of the most important internal parameter affects ripening/ maturity stage and eating quality by flavor development together with soluble solids/ sugars (Gupta et al., 2022; Lamikanra, 2002; Saltveit, 2002). Organic acids like ascorbic, citric, malic, and tartaric acid is inherent in variety of crop, and their variation with respect to ripening might be rather different (Lamikanra, 2002; Nogales-Bueno et al., 2014; Saltveit, 2002; Singh et al., 2018). For most of the fruits, titratable acidity gradually decreases with the ripening process; in contrast, acidity increases with time during maturity when the crop is still on the plant (Gupta et al., 2022; Minas et al., 2021; Saltveit, 2002). Especially for cherry, citrus fruits, kiwi, persimmon, pomegranate, and tomatoes, TA, SSC/TA, BrimA (SSC-k(TA)), and pH are specified as important maturity indices (Gupta et al., 2022; Li et al., 2018b; Minas et al., 2021; Ncama et al., 2017; Reid, 2002). Sometimes optimum eating quality of some crops is well-defined; for instance, maximum 0.8% TA is recommended for strawberries (Sánchez et al., 2012). Nevertheless, standard deviation of titratable acidity depending on the ripening stage of fruit might be high with no distinct variation is observed, so it is not practical to give a specific limit for ripening stage of some varieties (Minas et al., 2021; Mohammadi et al., 2015). Depending on this behavior, researchers largely employ and correlate SSC, dry matter content, firmness, or color-based features (R, G, B, L*a*b*, gray scale) with image processing data for predicting the maturity stage of the fruits and vegetables (Minas et al., 2021; Mohammadi et al., 2015; Pieczywek et al., 2018).

Hyperspectral imaging has quite remarkable results by inspection of maturity stages involving pH and titratable acidity values as predicting parameters (Nogales-Bueno et al., 2014; Pu et al., 2016). It is worth mentioning that the underlying reason of this achievement is selecting the proper spectral bands by successful image processing methods. In the study of Pu et al. (2016), maturity of lychees was evaluated by visible/short-wave NIR (600–1000 nm) and long-wave NIR (1000–2500 nm) hyperspectral imaging system with partial least square regression (PLSR) model for selection of optimal spectral band. The maturity prediction performance of visible/short-wave NIR was rather acceptable for prediction of pH with respect to maturity (r_p = 0.701); however, it was further improved by employing the long-wave NIR region ($r_p = 0.745$). Selecting the optimal wavelength for aforementioned spectral bands was the most important step of prediction capacity and overall the classification accuracy of long-wave NIR (96.9%) was found higher than visible/short-wave NIR (90.6%).

Zalema (white) and Tempranillo (red) grape cultivar maturity evaluation also were performed with a hyperspectral imaging system by recording reflectance values between 950 and1650 nm regions (Nogales-Bueno et al., 2014). Although the average spectra of the grape cultivars are almost overlapping, SNV pretreatment helped to improve the results of calibration set. The authors have stated that the global model representing the spectral data red and white grape cultivars together performed better than the model for white grape only, although the model including only the data of red grape was better among all. The global or individual models were substantially successful in predicting the pH of both grape cultivars having the lowest standard error prediction even better than prediction capacity of total phenols, SSC, and titratable acidity (Nogales-Bueno et al., 2014).

pH is also used as a grading parameter of pomegranate for using in different product types instead of ripening evaluation. For this purpose, Fashi et al. (2020) recorded the pomegranate surface images by digital camera and the extracted color features from processed images were used for estimating the pH of the pomegranates. They proposed algorithms and three models (ANFIS, ANN, and response surface methodology) for predicting the pH upon the selected color (V value from HSV image, Y value in YUV image) and geometric features (roughness and crown diameter of pomegranate). Even though the ANFIS model (R^2-test: 0.987) was designated as the best model for grading pomegranates with respect to pH, the range of pH, or the level of maturity of the pomegranates was not given in the study (Fashi et al., 2020).

5 Classification According to Textural Properties

Internal quality parameters such as textural properties cannot be directly determined by visual inspection thus require destructive analysis (Lu et al., 2020). However, in recent years, the detection of textural parameters through nondestructive techniques such as image analysis has become an important subject for postharvest quality and preharvest maturity assessment of fruits and vegetables.

Cellular structure of horticultural products determines their physicochemical properties such as texture, color, and sugar content. Cell walls include a complex structure consisting of hemicellulose, protein, pectin, cellulose fibrils, and are enclosed in the individual living cells by the middle lamella formed mainly by pectin (Cáez-Ramírez et al., 2018). The pressure, structure, and composition of these walls and layers detained by middle lamella affect the texture of the fruits and vegetables (Pathmanaban et al., 2019). For instance, tomatoes include 93–95% water and 5–7% total solids, of which 80–90% are soluble and 10–20% are insoluble solids, and the texture of tomato greatly depends on its insoluble solids, which are derived from cell walls. The knowledge of three-dimensional network of plant cell walls is considered as an important point as it determines the perception of firmness, consistency, juiciness, etc., in fruit and vegetable tissues (Waldron et al., 2003).

Fresh fruits and vegetables show firmer structure than stored or damaged products (Fathizadeh et al., 2020). In the ripening stages, cell walls start to disrupt leading to softening of tissue due to the changes in the integrity of middle lamella and thus tissue morphology. Image analysis describes the relationship between those mentioned changes and obtained data such as the loss of firmness, development of softening, and SSC with skin texture data (Cáez-Ramírez et al., 2018). The firmness of horticultural products is both related to juiciness and freshness/ripeness, which are important factors determining the harvest time. There is a direct relationship between firmness with juiciness and maturity, while there is an inverse relationship between softness and mealiness. Besides, the loss of firmness is relevant to low water content, cell wall modification, growth regulators, or product anatomy (Fathizadeh et al., 2020). In conclusion, the softening results in textural changes through the cell wall depolymerization and middle lamella dissolution thus leading to changes in the rigidity of fruits and vegetables. Studies showed that scattering and optical absorption properties are associated with microstructure/composition and textural properties of the product (Cen et al., 2013).

Imaging techniques, such as hyperspectral imaging, spectroscopy, and imaging methods, are combined to produce both spatial and spectral information about the product (Rahman et al., 2018). In this technique, the appearance of the fruits and vegetables are measured through image processing, and physicochemical properties including firmness are measured through spectral information. The collected hyperspectral images for internal quality analysis (firmness) include high amounts of data composed of one spectral dimension and two spatial dimensions, which then firmness or other quality attributes are predicted by extracting the spectral information and reducing large data matrices with multivariate methods (Fan et al., 2015).

The measurement of changes in the relevant spectrum provides correlation with firmness as those measured spectrum are relevant to the composition and structure of tested vegetables or fruits. Thus, the scatter mode of HSI technique has been generally used for firmness. Innovative hyperspectral scattering imaging technique based on light scattering principles was developed for determining the internal quality of fruits and vegetables (Vanoli et al., 2020; Wang et al., 2020). In the electromagnetic spectrum, the spectra obtained between 750–1900 nm (short wave infrared region) are related to the vibration combination of C-H, O-H, and N-H bonds that are associated with structural components of organic molecules (Williams & Norris, 2001). Thus, Rahman et al. (2018) studied HSI technique at 1000–1600 nm to obtain firmness and sweetness index values with a variable selection mode and found high correlation coefficients (0.74 and 0.82).

Imaging techniques used in texture measurements of horticultural crops include MRI, fluorescence, hyperspectral imaging techniques, reflectance/domain reflectance/Raman/waveguide/infrared spectroscopy, and light backscattering images. These techniques have been used to evaluate firmness, mealiness, softening, tenderness, and hardness indexes. Different imaging techniques have been applied for the maturity assessment of papaya (Cáez-Ramírez et al., 2018), pear (Fan et al., 2015), banana (Hashim et al., 2018), apple (Mendoza et al., 2012), and sweet cherry (Pullanagari & Li, 2021) by correlating firmness or other textural parameters.

Hyperspectral imaging technique generates scattering images using highly focused light beams and provides a relation between the texture of tested tissue and light scattering. Lu and Peng (2006) used hyperspectral scattering to measure the firmness of peaches and built firmness prediction models by applying step-wise MLR to the parameters of scattering profiles for "Red Haven" and "Coral Star's" cultivar peaches. Similarly, Lu (2007) studied hyperspectral imaging coupled with ANN models to evaluate the firmness and SSC of "Golden Delicious" and "Delicious" cultivar apples and reported relatively poor predictions for "Delicious" which might be attributed to the shape of apple, leading to inconsistency of scattering measurements. The authors also stated that the light scattering technique would be more suitable for firmness prediction than SSC. Rungpichayapichet et al. (2017) used HSI to determine the firmness of mangoes with other quality attributes in a spectral range of 450–998 nm and PLSR approach. The prediction performance of the model analyzing the firmness with R^2, bias, and RMSE was recorded as the best model when compared to other test parameters (TA and SSC). Image textural features have also been used to detect the maturity of different fruits and vegetables by collecting their morphological and structural properties integrated with other imaging techniques. Mendoza et al. (2011) used image textural features and hyperspectral techniques to evaluate the firmness of apples, which improved the prediction accuracies. Mendoza et al. (2012) also stated that integrating the hyperspectral image extracted data with those obtained by other spectral techniques such as NIR improved the prediction capacity. Mendoza et al. (2014) further applied hyperspectral imaging to grade apples with respect to their firmness and SSC. Two quality grades (i.e., "Premium" and "Regular") grading accuracies of 78–98% for firmness, and 62–92% for SSC were achieved with an image acquisition speed of 0.5 fruit per second. Liquid crystal tunable filter-based hyperspectral techniques, which operate under wide-field illumination instead of focused light, have been studied to measure the firmness of strawberries (Nagata et al., 2004). MLR and NIR hyperspectral imaging systems were also reported as useful firmness prediction tools for strawberries (Tallada et al., 2006).

In the laser backscattering (LBI) technique, laser light interacts with the product depending on the chemical composition including pigments, starches, water content, soluble solids, and different scattering response with microstructural composition and cellular matrices (Hussain et al., 2018). On the other hand, the density changes of the sample can be determined by soft X-ray imaging technique using X-ray beams that are partially absorbed by the product while the remaining analyzed by a camera (Hussain et al., 2018). Ali et al. (2017) used LBI with a PLSR model in a spectral range between 400 and 1800 nm for monitoring the firmness, pH, SSC, and moisture of watermelons and found higher correlation values for maturation stages. Besides, Mollazade et al. (2015) combined multispectral imaging and LBI techniques with different models including ANN to test the elasticity of tomatoes with other quality attributes and reported high coefficient of determination values.

High-quality 2D or 3D images can be obtained by MRI, which selectively observes different protons in the product such as water, lipids, and sugars by quantifying/identifying proton signals at spatial area (Hussain et al., 2018). Paniagua et al. (2013) studied the softening behavior of blueberries by using MRI for water distribution and correlated with loss of firmness. The authors observed that water loss patterns detected by MRI provided useful information on the firmness of blueberries during postharvest.

Infrared thermography (IRT) technique produces 2D thermal images giving temperature distribution by converting emitted thermal energy to current or voltage (Hussain et al., 2018). Gonçalves et al. (2016) studied IRT technique for evaluating the maturity of guavas and found that thermography was effective in the detection of maturity leading to tissue softening, which was confirmed by mass loss and firmness utilized by IRT.

6 Classification According to Color Parameters

The color of food products is one of the external quality parameters with other attributes such as shape, size, volume, peel/flesh ratio, and defects. Conventional visual inspection is readily used to observe the ripening stages of different horticultural crops by evaluating their size and shape; however, color and presence of defects should be evaluated through more complicated systems such as hyperspectral imaging. Surface or sub-surface properties and defects inside the fruit or vegetable tissues are detected by reflectance, transmittance, and interactance modes of hyperspectral imaging (Cakmak, 2019; Sun, 2010). Imaging techniques generally process the reflected light from the target through very fine wavelengths, which expand the limits of conventional colorimetry for determination of abovementioned pigments' color. The major applications using imaging techniques to screen the maturity or ripening of horticultural products with external properties are listed in Table 1.

Fruits and vegetables are perishable products that the color retention during postharvest handling is a critical parameter and an important quality indicator (Lu et al., 2020). The color of a fruit or vegetable is formed by the pigments naturally found in them. These pigments are the fat-soluble pigments such as carotenoids (yellow, orange, and red) and chlorophylls (green) and the water-soluble pigments such as anthocyanins (red, blue), betalains (red), and flavonoids (yellow). The water-soluble black, brown, and gray colored pigments are also generated due to the enzymatic and nonenzymatic browning reactions (Pathmanaban et al., 2019). The changes in the color of fruits and vegetables are related to the chemical and physical changes occurring during the ripening stages. Thus, maturity classification data generally utilize color as the main feature due to its high correlation with acidity, sugar content, and flavor. During the ripening, the color of a product changes because of chlorophyll degradation and the production of secondary metabolites such as flavonoids and carotenoids (Li et al., 2018a). Depending on the variety of fruit and vegetables different pigments start to accumulate after chlorophyll degradation (transforming chloroplasts into chromoplasts) and thus determine the color of the mature product. For instance, accumulation of lycopene and β-carotene for tomatoes, pelargonidin-3-glucoside (anthocyanin) for strawberries, carotenoids for mangoes, green vegetables such as broccoli, citrus, papaya, cyanidin-3-O-β-D-glucopyranoside for gulupa (purple passion) fruit, carotenoids and flavonoid glucosides for plums, anthocyanins for cherry, raspberry, and red grape have been used for maturity assessment by evaluating color changes such as from green to red, purple, yellow, brilliant red, blue and loss of green (Cho et al., 2021; Li et al., 2018a; Makino & Kousaka, 2020).

Fruits and vegetables have complex structures/molecules with different sizes depending on their maturation levels. For instance, green fruits are rich in large molecules such as starch chains, pectin, organic acids, and amino acids which are hydrolyzed to small molecules such as glucose during the ripening (Nassif et al., 2012). Thus, the changes in constituents affect their optical properties, as well as absorption and scattering behavior. In this manner, different maturation stages can be monitored using imaging techniques collecting spectral and spatial information about the product (Nassif et al., 2012). Therefore, different imaging techniques have been applied for the maturity assessment of tomatoes (Baek et al., 2020; Hoffmann et al., 2015), apple (Betemps et al., 2012; Cen et al., 2013), cherry (Taghadomi-Saberi et al., 2015), papaya (Garillos-Manliguez & Chiang, 2021), broccoli (Makino & Kousaka, 2020), banana (Xie et al., 2018), grapes (Agati et al., 2013), and orange (Taghadomi-Saberi et al., 2019) by the relation between color change and ripening. In these techniques, hyperspectral imaging technique for color evaluation related to ripening stages, sample is scanned to generate a spatial picture (hyperspectral cube) with a series of images at different wavelengths including spectroscopic and spatial information for each pixel (Cakmak, 2019). The obtained 3D-hyper cube is further analyzed to determine the minor/major changes in fruits and vegetables during maturation. Therefore, a hyperspectral image can be used to determine color transformation to detect the maturation level of horticultural products. Besides, chlorophyll fluorescence is one of the promising techniques used for the quality assessment of horticultural products (DeEll & Toivonen, 2003). Chlorophylls inside a plant tissue absorb short wavelength light and then the physicochemical activities of chlorophylls start (fluorescing). In this technique, maxima and minima fluorescence are measured as a function of photochemical changes, which provide information about physiological changes (color, etc.) of plant tissue for discrimination (Noh & Lu, 2007).

The assessment of ripeness in various fruits and vegetables has been performed using different color indices such as a*, b*, the ratio of a* and b*, hue angle, lightness, and chroma (Lockman et al., 2019; Pourdarbani et al., 2020; Wu & Sun, 2013). For instance, it was shown that a*/b* color values of tomatoes had a positive correlation with lycopene concentration (Brandt et al., 2006), whereas it has been reported that the ripeness in peaches was related to the b* values (Ferrer et al., 2005). On the other hand, lightness, hue, and chroma values were successfully used to evaluate the ripeness of citrus and guavas (Gupta et al., 2021; Mercado-Silva et al.,

1998). However, the color measurement of a single fruit or vegetable product with a conventional colorimeter is not adequate for mapping the whole area and predicting the maturity, and thus requires data correlation by statistical tests using multiple variables (Li et al., 2018a). Besides, color imaging techniques are converting photons to electrical signals, which are received by a camera with suitable sensors. These techniques filter the light to R, G, and B values providing a prediction on the ripening stage of the product (Khojastehnazhand et al., 2019; Leon et al., 2006). Different statistical methods such as fuzzy logic, K-means, K-nearest neighbor, Gustafson–Kessel algorithms, discriminant analysis, principal component, and neural networks have been applied to transform the RGB values from images to L*a*b* values of tomatoes (El-Bendary et al., 2015; Takahashi et al., 2013; Ukirade, 2014), apples (Dadwal & Banga, 2012), blueberry (Li et al., 2014), cherry (Taghadomi-Saberi et al., 2015), and mushrooms (Taghizadeh et al., 2011). On the other hand, hyperspectral imaging provides abundant, well-resolved spectral information and is suitable for more precise color measurements than RGB imaging. The flesh and skin color of pickling cucumbers were determined by using different imaging modes of hyperspectral imaging technique by Ariana and Lu (2008). The authors reported that reflectance mode was better in measuring the skin color indexes while the flesh color was not successfully measured with all modes. In a later study, the authors reported on measuring the surface color of pickles by integrating hyperspectral reflectance imaging data over the 500–675 nm range, instead of proposing predictive models (Ariana & Lu, 2010). Similarly, van Roy et al. (2017) used hyperspectral imaging method for vine tomatoes and reported poor color correlations, which might be due to the sensitivity to intensity variation, fruit shape, and surface glossiness. However, the authors found that multivariate modeling performed better than direct method.

Raman imaging (RI) provides pixel-based information such as mapping morphology or composition and chemical images by integrating digital imaging and Raman spectroscopy (Hussain et al., 2018). Qin et al. (2012) used RI method for tomatoes to monitor the internal maturity with a spectral range of 200–2500 cm^{-1}. The maturity phases for those examined samples were as green (immature), green (mature), breaker, turning, pink, light red, and red. Peaks related to carotenoid content were assessed and the peak shifted at 1525–1513 cm^{-1} range was correlated with the decrease in lutein and carotene by the deposition of lycopene during ripening.

The basis of fluorescence imaging (FI) technique lies on that organic substituents emit unique fluorescence when excited by visible light and electromagnetic radiations (Hussain et al., 2018). For instance, during the ripening of tomatoes, main fluorescence bands such as flavonoids at 300, 450, and 500 nm, anthocyanins at 330, 380, and 450 nm, chlorophyll at 680 nm, and carotene between 420 and 580 nm were successfully determined by FI (Lai et al., 2007). Infrared pattern emitted by a product is measured through a thermal imaging (TI) technique including an active system with an external thermal source for imaging (Hussain et al., 2018). The heat capacity of mangoes was measured for levels of maturity as unripe-ripe-overripe covering 7500–13,000 nm, and blue, black, brighter colors were obtained for immature, mature, and overripe samples by TI (Sumriddetchkajorn & Intaravanne,

2013). Surface measurements or evaluations require high penetration where microwave imaging (MI) technique can achieve 1–30 cm depth. It was argued that some fruits or vegetables can develop mature color even if they are not soft or full of juice. Thus, evaluating water content by MI technique which produces microwave at low doses passing through the tissue while blocking by water, instead of color evaluation for maturity as an alternative.

The quality attributes distribution maps are also generated at pixel level to monitor postharvest or preharvest quality parameters. The distribution of carotenoids, lycopene, chlorophyll, and anthocyanins found in core or pericarp of fruits and vegetables were mapped by hyperspectral imaging techniques to follow the ripening stages. Lleó et al. (2011) studied the chlorophyll absorption peak for rapid assessment of maturity in peaches and found that chlorophyll absorption was the most important parameter to discriminate between non-ripened and ripened samples. Similarly, Qin et al. (2011) used hyperspectral Raman imaging technique and mapping lycopene distribution for ripening stages. Polder et al. (2004) studied hyperspectral imaging for determination of carotene (including lycopene, lutein, and β-carotene) and chlorophyll distribution while ripening of tomatoes. The authors have built pixel-level predictive models by randomly selecting 200 pixels per sample and reported that hyperspectral imaging in the visible range of 396–736 nm was found to be better results over conventional imaging for classifying tomatoes into five ripeness stages based on LDA models. In the pixel-level modeling studies, dense sampling points are important due to obtaining ground-truth reference values and also lengthy model calibration processes.

7 Conclusion

Different imaging and image processing techniques with relevant data used for the maturity or ripening evaluation of horticultural products have been reviewed in this chapter. Imaging techniques have a great potential to be used in the monitoring maturity/ripening stages of fruits and vegetables accurately and rapidly. However, the performance of imaging techniques should be improved by minimizing their disadvantages and hurdles. Limitations related to the image acquisition are poor imaging of the sample by lack of proper illumination, positioning of the camera/imaging system and light source, or scattering due to the uneven geometry/shape of the sample. Following the image acquisition, selection of ROI, image processing steps, and selection of important features with compatible statistical methods are other critical points limiting the use of imaging techniques for maturity and ripening assessment. Complex nature of chemical and textural changes that occur during ripening may also decrease the correlation with provided models extracted from image features.

Researchers have been focused on the improvement of the image processing period of data evaluation and algorithms and making them more suitable for real-time, on-tree, or in-field applications with relatively lower costs for commercial use.

For instance, the HSI technique can both process spectroscopy and 2D imaging while being integrated with a moving platform. However, large data sets require more time to obtain main features from the spectral images and high expenses make this technique less suitable for commercial applications. Similarly, the RI technique has some hurdles such as weak scattering, high background fluorescence interference, lack of efficient substrates with uniform and high-density hot spots, and high cost. Besides, the dependence of obtaining accurate results for horticultural products to their thermal behavior stability is one of the obstacles in the TI technique.

Every imaging method has its own advantages and bottlenecks for visualization of internal or external quality parameters of fruits and vegetables; however, integration of imaging technologies for pre- and postharvest quality evaluation stages will provide more objective, accurate, and repeatable results that reduce labor, and potential costs related with handling, destructive analysis, or standardization.

References

Abasi, S., Minaei, S., Jamshidi, B., & Fathi, D. (2018). Dedicated non-destructive devices for food quality measurement: A review. *Trends in Food Science & Technology, 78*, 197–205.

Adebayo, S. E., Hashim, N., Hass, R., Reich, O., Regen, C., Münzberg, M., et al. (2017). Using absorption and reduced scattering coefficients for non-destructive analyses of fruit flesh firmness and soluble solids content in pear (Pyrus communis 'conference')—An update when using diffusion theory. *Postharvest Biology and Technology, 130*, 56–63.

Agati, G., D'Onofrio, C., Ducci, E., Cuzzola, A., Remorini, D., Tuccio, L., et al. (2013). Potential of a multiparametric optical sensor for determining in situ the maturity components of red and white Vitis vinifera wine grapes. *Journal of Agricultural and Food Chemistry, 61*(50), 12211–12218.

Ali, M. M., Hashim, N., Bejo, S. K., & Shamsudin, R. (2017). Quality evaluation of watermelon using laser-induced backscattering imaging during storage. *Postharvest Biology and Technology, 123*, 51–59.

Ali, M. M., Hashim, N., & Hamid, A. S. A. (2020). Combination of laser-light backscattering imaging and computer vision for rapid determination of oil palm fresh fruit bunches maturity. *Computers and Electronics in Agriculture, 169*, 105235.

Aredo, V., Velásquez, L., Carranza-Cabrera, J., & Siche, R. (2019). Predicting of the quality attributes of orange fruit using hyperspectral images. *Journal of Food Quality and Hazards Control, 6*, 82–92.

Arendse, E., Fawole, O. A., Magwaza, L. S., & Opara, U. L. (2018). Non-destructive prediction of internal and external quality attributes of fruit with thick rind: A review. *Journal of Food Engineering, 217*, 11–23.

Ariana, D. P., & Lu, R. (2008). Quality evaluation of pickling cucumbers using hyperspectral reflectance and transmittance imaging—Part II. Performance of a prototype. *Sensing and Instrumentation for Food Quality and Safety, 2*(3), 152–160.

Ariana, D. P., & Lu, R. (2010). Evaluation of internal defect and surface color of whole pickles using hyperspectral imaging. *Journal of Food Engineering, 96*(4), 583–590.

Baek, S., Lim, J., Lee, J. G., McCarthy, M. J., & Kim, S. M. (2020). Investigation of the maturity changes of cherry tomato using magnetic resonance imaging. *Applied Sciences, 10*(15), 5188.

Barrett, D. M., Somogyi, L., & Ramaswamy, H. S. (Eds.). (2005). *Processing fruits: Science and technology* (2nd ed.). CRC Press.

Beghi, R., Giovenzana, V., Tugnolo, A., & Guidetti, R. (2018). Application of visible/near infrared spectroscopy to quality control of fresh fruits and vegetables in large-scale mass distribution channels: A preliminary test on carrots and tomatoes. *Journal of the Science of Food and Agriculture, 98*(7), 2729–2734.

Benelli, A., Cevoli, C., & Fabbri, A. (2020). In-field Vis/NIR hyperspectral imaging to measure soluble solids content of wine grape berries during ripening. In *2020 IEEE International Workshop on Metrology for Agriculture and Forestry (MetroAgriFor)* (pp. 99–103). IEEE.

Betemps, D. L., Fachinello, J. C., Galarça, S. P., Portela, N. M., Remorini, D., Massai, R., & Agati, G. (2012). Non-destructive evaluation of ripening and quality traits in apples using a multiparametric fluorescence sensor. *Journal of the Science of Food and Agriculture, 92*(9), 1855–1864.

Brandt, S., Pék, Z., Barna, É., Lugasi, A., & Helyes, L. (2006). Lycopene content and colour of ripening tomatoes as affected by environmental conditions. *Journal of the Science of Food and Agriculture, 86*(4), 568–572.

Brummell, D., Atkinson, R., Burdon, J. N., Patterson, K. J., & Schaffer, R. (2016). Fruit growth, ripening and post-harvest physiology. In B. J. Atwell, P. E. Kriedemann, & C. G. N. Turnbull (Eds.), *Plants in action* (pp. 350–380). Macmillan.

Cáez-Ramírez, G., Alamilla-Beltrán, L., & Gutiérrez-López, G. F. (2018). Morphometric analysis and tissue structural continuity evaluation of senescence progression in fresh cut papaya (Carica papaya L.). *Journal of Food Engineering, 216*, 107–119.

Cakmak, H. (2019). Assessment of fresh fruit and vegetable quality with non-destructive methods. In C. M. Galanakis (Ed.), *Food quality and shelf life* (pp. 303–331). Academic Press.

Cantre, D., Herremans, E., Verboven, P., Ampofo-Asiama, J., & Nicolaï, B. (2014). Characterization of the 3-D microstructure of mango (Mangifera indica L. cv. Carabao) during ripening using X-ray computed microtomography. *Innovative Food Science & Emerging Technologies, 24*, 28–39.

Cen, H., Lu, R., Mendoza, F., & Beaudry, R. M. (2013). Relationship of the optical absorption and scattering properties with mechanical and structural properties of apple tissue. *Postharvest Biology and Technology, 85*, 30–38.

Chen, Y., Grimplet, J., David, K., Castellarin, S. D., Terol, J., Wong, D. C., et al. (2018). Ethylene receptors and related proteins in climacteric and non-climacteric fruits. *Plant Science, 276*, 63–72.

Cho, J. S., Lim, J. H., Park, K. J., Choi, J. H., & Ok, G. S. (2021). Prediction of pelargonidin-3-glucoside in strawberries according to the postharvest distribution period of two ripening stages using VIS-NIR and SWIR hyperspectral imaging technology. *LWT, 141*, 110875.

Clark, C. J., & MacFall, J. S. (2003). Quantitative magnetic resonance imaging of 'Fuyu'persimmon fruit during development and ripening. *Magnetic Resonance Imaging, 21*(6), 679–685.

Cruz, J. O. (2020). Terahertz time-domain spectroscopy (THz-TDS) for classification of blueberries according to their maturity. In *2020 IEEE Engineering International Research Conference (EIRCON)* (pp. 1–4). IEEE.

Dadwal, M., & Banga, V. K. (2012). Estimate ripeness level of fruits using RGB color space and fuzzy logic technique. *International Journal of Engineering and Advanced Technology, 2*(1), 225–229.

de Azevedo, C. H., & Rodriguez-Amaya, D. B. (2005). Carotenoid composition of kale as influenced by maturity, season and minimal processing. *Journal of the Science of Food and Agriculture, 85*(4), 591–597.

de Azevedo-Meleiro, C. H., & Rodriguez-Amaya, D. B. (2005). Carotenoids of endive and New Zealand spinach as affected by maturity, season and minimal processing. *Journal of Food Composition and Analysis, 18*(8), 845–855.

DeEll, J. R., & Toivonen, P. M. (Eds.). (2003). *Practical applications of chlorophyll fluorescence in plant biology*. Springer Science & Business Media. https://doi.org/10.1007/978-1-4615-0415-3

Diels, E., van Dael, M., Keresztes, J., Vanmaercke, S., Verboven, P., Nicolai, B., et al. (2017). Assessment of bruise volumes in apples using X-ray computed tomography. *Postharvest Biology and Technology, 128*, 24–32.

do Nascimento Nunes, M. C. (2008). *Color atlas of postharvest quality of fruits and vegetables*. John Blackwell Publishing.

Du, C. J., & Sun, D. W. (2004). Recent developments in the applications of image processing techniques for food quality evaluation. *Trends in Food Science & Technology, 15*(5), 230–249.

El-Bendary, N., El Hariri, E., Hassanien, A. E., & Badr, A. (2015). Using machine learning techniques for evaluating tomato ripeness. *Expert Systems with Applications, 42*(4), 1892–1905.

Fan, S., Huang, W., Guo, Z., Zhang, B., & Zhao, C. (2015). Prediction of soluble solids content and firmness of pears using hyperspectral reflectance imaging. *Food Analytical Methods, 8*(8), 1936–1946.

Fashi, M., Naderloo, L., & Javadikia, H. (2020). Pomegranate grading based on pH using image processing and artificial intelligence. *Journal of Food Measurement and Characterization, 14*(6), 3112–3121.

Fatchurrahman, D., Amodio, M. L., de Chiara, M. L. V., Chaudhry, M. M. A., & Colelli, G. (2020). Early discrimination of mature-and immature-green tomatoes (Solanum lycopersicum L.) using fluorescence imaging method. *Postharvest Biology and Technology, 169*, 111287.

Fathizadeh, Z., Aboonajmi, M., & Beygi, S. R. H. (2020). Nondestructive firmness prediction of apple fruit using acoustic vibration response. *Scientia Horticulturae, 262*, 109073.

Fernández-Espinosa, A. J. (2016). Combining PLS regression with portable NIR spectroscopy to on-line monitor quality parameters in intact olives for determining optimal harvesting time. *Talanta, 148*, 216–228.

Ferrer, A., Remón, S., Negueruela, A. I., & Oria, R. (2005). Changes during the ripening of the very late season Spanish peach cultivar Calanda: Feasibility of using CIELAB coordinates as maturity indices. *Scientia Horticulturae, 105*(4), 435–446.

Garillos-Manliguez, C. A., & Chiang, J. Y. (2021). Multimodal deep learning and visible-light and hyperspectral imaging for fruit maturity estimation. *Sensors, 21*(4), 1288.

Gonçalves, B. J., de Oliveira Giarola, T. M., Pereira, D. F., Boas, E. V. D. B. V., & de Resende, J. V. (2016). Using infrared thermography to evaluate the injuries of cold-stored guava. *Journal of Food Science and Technology, 53*(2), 1063–1070.

Gupta, A. K., Medhi, M., Chakraborty, S., Yumnam, M., & Mishra, P. (2021). Development of rapid and non-destructive technique for the determination of maturity indices of pomelo fruit (Citrus grandis). *Journal of Food Measurement and Characterization, 15*(2), 1463–1474.

Gupta, A. K., Pathak, U., Tongbram, T., Medhi, M., Terdwongworakul, A., Magwaza, L. S., et al. (2022). Emerging approaches to determine maturity of citrus fruit. *Critical Reviews in Food Science and Nutrition, 62*, 5245–5266.

Harel, B., Parmet, Y., & Edan, Y. (2020). Maturity classification of sweet peppers using image datasets acquired in different times. *Computers in Industry, 121*, 103274.

Hashim, N., Adebayo, S. E., Abdan, K., & Hanafi, M. (2018). Comparative study of transform-based image texture analysis for the evaluation of banana quality using an optical backscattering system. *Postharvest Biology and Technology, 135*, 38–50.

Herrero-Langreo, A., Fernández-Ahumada, E., Roger, J. M., Palagós, B., & Lleó, L. (2012). Combination of optical and non-destructive mechanical techniques for the measurement of maturity in peach. *Journal of Food Engineering, 108*(1), 150–157.

Hitzmann, B., & Ahmad, M. H. (Eds.). (2017). *Measurement, modeling and automation in advanced food processing* (Vol. 161). Springer.

Hoffmann, A. M., Noga, G., & Hunsche, M. (2015). Fluorescence indices for monitoring the ripening of tomatoes in pre-and postharvest phases. *Scientia Horticulturae, 191*, 74–81.

Hsiao, W. T., Kuo, W. C., Lin, H. H., & Lai, L. H. (2021). Assessment and feasibility study of lemon ripening using x-ray image of information visualization. *Applied Sciences, 11*(7), 3261.

Huang, Y., Dong, W., Chen, Y., Wang, X., Luo, W., Zhan, B., et al. (2021). Online detection of soluble solids content and maturity of tomatoes using Vis/NIR full transmittance spectra. *Chemometrics and Intelligent Laboratory Systems, 210*, 104243.

Hui, Y. H. (Ed.). (2006). *Handbook of fruits and fruit processing*. Blackwell Publishing Ltd.

Hui, Y. H., Ghazala, S., Graham, D. M., Murrell, K. D., & Nip, W.-K. (Eds.). (2004). *Handbook of vegetable preservation and processing* (1st ed.). Marcel Dekker.

Hussain, A., Pu, H., & Sun, D. W. (2018). Innovative nondestructive imaging techniques for ripening and maturity of fruits–a review of recent applications. *Trends in Food Science & Technology, 72*, 144–152.

Irudayaraj, J., & Reh, C. (Eds.). (2008). *Nondestructive testing of food quality* (Vol. 18). Wiley.

Jongen, W. (Ed.). (2002). *Fruit and vegetable processing: Improving quality*. Elsevier.

Kader, A. A. (2011). *Postharvest technology of horticultural crops* (Vol. 3529, 3rd ed.). University of California Agriculture and Natural Resources.

Kasampalis, D. S., Tsouvaltzis, P., & Siomos, A. S. (2020). Chlorophyll fluorescence, non-photochemical quenching and light harvesting complex as alternatives to color measurement, in classifying tomato fruit according to their maturity stage at harvest and in monitoring postharvest ripening during storage. *Postharvest Biology and Technology, 161*, 111036.

Khodabakhshian, R., & Emadi, B. (2017). Application of Vis/SNIR hyperspectral imaging in ripeness classification of pear. *International Journal of Food Properties, 20*(Suppl. 3), S3149–S3163.

Khodabakhshian, R., Emadi, B., Khojastehpour, M., Golzarian, M. R., & Sazgarnia, A. (2017). Non-destructive evaluation of maturity and quality parameters of pomegranate fruit by visible/near infrared spectroscopy. *International Journal of Food Properties, 20*(1), 41–52.

Khojastehnazhand, M., Mohammadi, V., & Minaei, S. (2019). Maturity detection and volume estimation of apricot using image processing technique. *Scientia Horticulturae, 251*, 247–251.

Kotwaliwale, N., Kalne, A., & Singh, K. (2012). Monitoring of mango (Mangifera indica L.) (cv.: Chousa) ripening using X-ray computed tomography. In *2012 Sixth International Conference on Sensing Technology (ICST)* (pp. 326–330). IEEE.

Kotwaliwale, N., Singh, K., Kalne, A., Jha, S. N., Seth, N., & Kar, A. (2014). X-ray imaging methods for internal quality evaluation of agricultural produce. *Journal of Food Science and Technology, 51*(1), 1–15.

Kusumiyati, K., Sutari, W., Hamdani, J. S., Mubarok, S., Sitepu, R. B., & Oktavia, A. R. (2018). Non-destructive measurement of green bitter gourd quality component using near infrared spectroscopy (NIRS). *Science and Technology Indonesia, 3*(2), 59–65.

Lai, A., Santangelo, E., Soressi, G. P., & Fantoni, R. (2007). Analysis of the main secondary metabolites produced in tomato (Lycopersicon esculentum, mill.) epicarp tissue during fruit ripening using fluorescence techniques. *Postharvest Biology and Technology, 43*(3), 335–342.

Lakshmi, S., Pandey, A. K., Ravi, N., Chauhan, O. P., Gopalan, N., & Sharma, R. K. (2017). Non-destructive quality monitoring of fresh fruits and vegetables. *Defence Life Science Journal, 2*(2), 103–110.

Lamikanra, O. (Ed.). (2002). *Fresh-cut fruits and vegetables: Science, technology, and market*. CRC press.

Leon, K., Mery, D., Pedreschi, F., & Leon, J. (2006). Color measurement in L∗ a∗ b∗ units from RGB digital images. *Food Research International, 39*(10), 1084–1091.

Li, H., Lee, W. S., & Wang, K. (2014). Identifying blueberry fruit of different growth stages using natural outdoor color images. *Computers and Electronics in Agriculture, 106*, 91–101.

Liu, Y., Pu, H., & Sun, D. W. (2017). Hyperspectral imaging technique for evaluating food quality and safety during various processes: A review of recent applications. *Trends in Food Science & Technology, 69*, 25–35.

Li, B., Lecourt, J., & Bishop, G. (2018a). Advances in non-destructive early assessment of fruit ripeness towards defining optimal time of harvest and yield prediction—A review. *Plants, 7*(1), 3.

Li, X., Wei, Y., Xu, J., Feng, X., Wu, F., Zhou, R., et al. (2018b). SSC and pH for sweet assessment and maturity classification of harvested cherry fruit based on NIR hyperspectral imaging technology. *Postharvest Biology and Technology, 143*, 112–118.

Lleó, L., Roger, J. M., Herrero-Langreo, A., Diezma-Iglesias, B., & Barreiro, P. (2011). Comparison of multispectral indexes extracted from hyperspectral images for the assessment of fruit ripening. *Journal of Food Engineering, 104*(4), 612–620.

Lockman, N. A., Hashim, N., & Onwude, D. I. (2019). Laser-based imaging for cocoa pods maturity detection. *Food and Bioprocess Technology, 12*(11), 1928–1937.

Lu, J., Qi, S., Liu, R., Zhou, E., Li, W., Song, S., & Han, D. (2015). Nondestructive determination of soluble solids and firmness in mix-cultivar melon using near-infrared CCD spectroscopy. *Journal of Innovative Optical Health Sciences, 8*(06), 1550032.

Lu, R. (2007). Nondestructive measurement of firmness and soluble solids content for apple fruit using hyperspectral scattering images. *Sensing and Instrumentation for Food Quality and Safety, 1*(1), 19–27.

Lu, R., & Peng, Y. (2006). Hyperspectral scattering for assessing peach fruit firmness. *Biosystems Engineering, 93*(2), 161–171.

Lu, Y., Saeys, W., Kim, M., Peng, Y., & Lu, R. (2020). Hyperspectral imaging technology for quality and safety evaluation of horticultural products: A review and celebration of the past 20-year progress. *Postharvest Biology and Technology, 170*, 111318.

Mahesh, S., Jayas, D. S., Paliwal, J., & White, N. D. G. (2015). Hyperspectral imaging to classify and monitor quality of agricultural materials. *Journal of Stored Products Research, 61*, 17–26.

Makino, Y., & Kousaka, Y. (2020). Prediction of degreening velocity of broccoli buds using hyperspectral camera combined with artificial neural networks. *Food, 9*(5), 558.

Makky, M., & Soni, P. (2014). In situ quality assessment of intact oil palm fresh fruit bunches using rapid portable non-contact and non-destructive approach. *Journal of Food Engineering, 120*, 248–259.

Marques, E. J. N., de Freitas, S. T., Pimentel, M. F., & Pasquini, C. (2016). Rapid and non-destructive determination of quality parameters in the 'Tommy Atkins' mango using a novel handheld near infrared spectrometer. *Food Chemistry, 197*, 1207–1214.

Mendoza, F., Lu, R., Ariana, D., Cen, H., & Bailey, B. (2011). Integrated spectral and image analysis of hyperspectral scattering data for prediction of apple fruit firmness and soluble solids content. *Postharvest Biology and Technology, 62*(2), 149–160.

Mendoza, F., Lu, R., & Cen, H. (2012). Comparison and fusion of four nondestructive sensors for predicting apple fruit firmness and soluble solids content. *Postharvest Biology and Technology, 73*, 89–98.

Mendoza, F., Lu, R., & Cen, H. (2014). Grading of apples based on firmness and soluble solids content using Vis/SWNIR spectroscopy and spectral scattering techniques. *Journal of Food Engineering, 125*, 59–68.

Mercado-Silva, E., Benito-Bautista, P., & de los Angeles García-Velasco, M. (1998). Fruit development, harvest index and ripening changes of guavas produced in Central Mexico. *Postharvest Biology and Technology, 13*(2), 143–150.

Minas, I. S., Blanco-Cipollone, F., & Sterle, D. (2021). Accurate non-destructive prediction of peach fruit internal quality and physiological maturity with a single scan using near infrared spectroscopy. *Food Chemistry, 335*, 127626.

Mo, C., Kim, M. S., Kim, G., Lim, J., Delwiche, S. R., Chao, K., et al. (2017). Spatial assessment of soluble solid contents on apple slices using hyperspectral imaging. *Biosystems Engineering, 159*, 10–21.

Mohammadi, V., Kheiralipour, K., & Ghasemi-Varnamkhasti, M. (2015). Detecting maturity of persimmon fruit based on image processing technique. *Scientia Horticulturae, 184*, 123–128.

Mohapatra, A., Shanmugasundaram, S., & Malmathanraj, R. (2017). Grading of ripening stages of red banana using dielectric properties changes and image processing approach. *Computers and Electronics in Agriculture, 143*, 100–110.

Mollazade, K., Omid, M., Akhlaghian Tab, F., Rezaei Kalaj, Y., & Mohtasebi, S. S. (2015). Data mining-based wavelength selection for monitoring quality of tomato fruit by backscattering and multispectral imaging. *International Journal of Food Properties, 18*(4), 880–896.

Munera, S., Besada, C., Aleixos, N., Talens, P., Salvador, A., Sun, D. W., Cubero, S., & Blasco, J. (2017). Non-destructive assessment of the internal quality of intact persimmon using colour and VIS/NIR hyperspectral imaging. *LWT-Food Science and Technology, 77*, 241–248.

Musse, M., Quellec, S., Cambert, M., Devaux, M. F., Lahaye, M., & Mariette, F. (2009). Monitoring the postharvest ripening of tomato fruit using quantitative MRI and NMR relaxometry. *Postharvest Biology and Technology, 53*(1–2), 22–35.

Muthulakshmi, A., Renjith, P. N. (2020). Classification of durian fruits based on ripening with machine learning techniques. In *2020 3rd International Conference on Intelligent Sustainable Systems (ICISS)* (pp. 542–547). IEEE.

Muziri, T., Theron, K. I., Cantre, D., Wang, Z., Verboven, P., Nicolai, B. M., & Crouch, E. M. (2016). Microstructure analysis and detection of mealiness in 'Forelle'pear (Pyrus communis L.) by means of X-ray computed tomography. *Postharvest Biology and Technology, 120*, 145–156.

Nagata, M., Tallada, J. G., Kobayashi, T., Cui, Y., & Gejima, Y. (2004). Predicting maturity quality parameters of strawberries using hyperspectral imaging. In *2004 ASAE Annual Meeting* (p. 1). American Society of Agricultural and Biological Engineers.

Nassif, R., Pellen, F., Magné, C., Le Jeune, B., Le Brun, G., & Abboud, M. (2012). Scattering through fruits during ripening: Laser speckle technique correlated to biochemical and fluorescence measurements. *Optics Express, 20*(21), 23887–23897.

Ncama, K., Opara, U. L., Tesfay, S. Z., Fawole, O. A., & Magwaza, L. S. (2017). Application of Vis/NIR spectroscopy for predicting sweetness and flavour parameters of 'Valencia' orange (Citrus sinensis) and 'star Ruby' grapefruit (citrus x paradisi Macfad). *Journal of Food Engineering, 193*, 86–94.

Nogales-Bueno, J., Hernández-Hierro, J. M., Rodríguez-Pulido, F. J., & Heredia, F. J. (2014). Determination of technological maturity of grapes and total phenolic compounds of grape skins in red and white cultivars during ripening by near infrared hyperspectral image: A preliminary approach. *Food Chemistry, 152*, 586–591.

Noh, H. K., & Lu, R. (2007). Hyperspectral laser-induced fluorescence imaging for assessing apple fruit quality. *Postharvest Biology and Technology, 43*(2), 193–201.

Nordey, T., Joas, J., Davrieux, F., Chillet, M., & Léchaudel, M. (2017). Robust NIRS models for non-destructive prediction of mango internal quality. *Scientia Horticulturae, 216*, 51–57.

Paniagua, A. C., East, A. R., Hindmarsh, J. P., & Heyes, J. (2013). Moisture loss is the major cause of firmness change during postharvest storage of blueberry. *Postharvest Biology and Technology, 79*, 13–19.

Park, B., & Lu, R. (Eds.). (2015). *Hyperspectral imaging technology in food and agriculture*. Springer.

Patel, K. K., Khan, M. A., & Kar, A. (2015). Recent developments in applications of MRI techniques for foods and agricultural produce—An overview. *Journal of Food Science and Technology, 52*(1), 1–26.

Pathmanaban, P., Gnanavel, B. K., & Anandan, S. S. (2019). Recent application of imaging techniques for fruit quality assessment. *Trends in Food Science & Technology, 94*, 32–42.

Pereira, L. F. S., Barbon, S., Jr., Valous, N. A., & Barbin, D. F. (2018). Predicting the ripening of papaya fruit with digital imaging and random forests. *Computers and Electronics in Agriculture, 145*, 76–82.

Pieczywek, P. M., Nowacka, M., Dadan, M., Wiktor, A., Rybak, K., Witrowa-Rajchert, D., & Zdunek, A. (2018). Postharvest monitoring of tomato ripening using the dynamic laser speckle. *Sensors, 18*(4), 1093.

Polder, G., Van Der Heijden, G. W. A. M., Van der Voet, H., & Young, I. T. (2004). Measuring surface distribution of carotenes and chlorophyll in ripening tomatoes using imaging spectrometry. *Postharvest Biology and Technology, 34*(2), 117–129.

Pourdarbani, R., Sabzi, S., Kalantari, D., Paliwal, J., Benmouna, B., García-Mateos, G., & Molina-Martínez, J. M. (2020). Estimation of different ripening stages of Fuji apples using image processing and spectroscopy based on the majority voting method. *Computers and Electronics in Agriculture, 176*, 105643.

Pu, H., Liu, D., Wang, L., & Sun, D. W. (2016). Soluble solids content and pH prediction and maturity discrimination of lychee fruits using visible and near infrared hyperspectral imaging. *Food Analytical Methods, 9*(1), 235–244.

Pu, Y. Y., Sun, D. W., Buccheri, M., Grassi, M., Cattaneo, T. M., & Gowen, A. (2019). Ripeness classification of bananito fruit (Musa acuminata, AA): A comparison study of visible spectroscopy and hyperspectral imaging. *Food Analytical Methods, 12*(8), 1693–1704.

Pullanagari, R. R., & Li, M. (2021). Uncertainty assessment for firmness and total soluble solids of sweet cherries using hyperspectral imaging and multivariate statistics. *Journal of Food Engineering, 289*, 110177.

Qin, J., Chao, K., & Kim, M. S. (2011). Investigation of Raman chemical imaging for detection of lycopene changes in tomatoes during postharvest ripening. *Journal of Food Engineering, 107*(3–4), 277–288.

Qin, J., Chao, K., & Kim, M. S. (2012). Nondestructive evaluation of internal maturity of tomatoes using spatially offset Raman spectroscopy. *Postharvest Biology and Technology, 71*, 21–31.

Rahman, A., Park, E., Bae, H., & Cho, B. K. (2018). Hyperspectral imaging technique to evaluate the firmness and the sweetness index of tomatoes. *Korean Journal of Agricultural Science, 45*(4), 823–837.

Rajkumar, P., Wang, N., Elmasry, G., Raghavan, G. S. V., & Gariepy, Y. (2012). Studies on banana fruit quality and maturity stages using hyperspectral imaging. *Journal of Food Engineering, 108*(1), 194–200.

Reid, M. S. (2002). Maturation and maturity indices. In A. A. Kader (Ed.), *Postharvest technology of horticultural crops* (3rd ed., pp. 55–62). University of California Agriculture and Natural Resources.

Rodríguez-Pulido, F. J., Gil-Vicente, M., Gordillo, B., Heredia, F. J., & González-Miret, M. L. (2017). Measurement of ripening of raspberries (Rubus idaeus L) by near infrared and colorimetric imaging techniques. *Journal of Food Science and Technology, 54*(9), 2797–2803.

Ropodi, A. I., Panagou, E. Z., & Nychas, G. J. (2016). Data mining derived from food analyses using non-invasive/non-destructive analytical techniques; determination of food authenticity, quality & safety in tandem with computer science disciplines. *Trends in Food Science & Technology, 50*, 11–25.

Rungpichayapichet, P., Mahayothee, B., Nagle, M., Khuwijitjaru, P., & Müller, J. (2016). Robust NIRS models for non-destructive prediction of postharvest fruit ripeness and quality in mango. *Postharvest Biology and Technology, 111*, 31–40.

Rungpichayapichet, P., Nagle, M., Yuwanbun, P., Khuwijitjaru, P., Mahayothee, B., & Müller, J. (2017). Prediction mapping of physicochemical properties in mango by hyperspectral imaging. *Biosystems Engineering, 159*, 109–120.

Russ, J. C. (2005). *Image analysis of food microstructure*. CRC press.

Saltveit, M. E. (2002). Mature fruit vegetables. In J. K. Brecht & J. A. Bartz (Eds.), *Postharvest physiology and pathology of vegetables* (pp. 731–756). CRC Press.

Sánchez, M. T., De la Haba, M. J., Benítez-López, M., Fernández-Novales, J., Garrido-Varo, A., & Pérez-Marín, D. (2012). Non-destructive characterization and quality control of intact strawberries based on NIR spectral data. *Journal of Food Engineering, 110*(1), 102–108.

Sanchez, P. D. C., Hashim, N., Shamsudin, R., & Nor, M. Z. M. (2020). Applications of imaging and spectroscopy techniques for non-destructive quality evaluation of potatoes and sweet potatoes: A review. *Trends in Food Science & Technology, 96*, 208–221.

Schoeman, L., Williams, P., du Plessis, A., & Manley, M. (2016). X-ray micro-computed tomography (μCT) for non-destructive characterisation of food microstructure. *Trends in Food Science & Technology, 47*, 10–24.

Singh, B., Singh, S., & Koley, T. K. (Eds.). (2018). *Advances in postharvest technologies of vegetable crops*. Apple Academic Press.

Sirisomboon, P., Tanaka, M., Kojima, T., & Williams, P. (2012). Nondestructive estimation of maturity and textural properties on tomato 'Momotaro' by near infrared spectroscopy. *Journal of Food Engineering, 112*(3), 218–226.

Sripaurya, T., Sengchuai, K., Booranawong, A., & Chetpattananondh, K. (2021). Gros Michel banana soluble solids content evaluation and maturity classification using a developed portable 6 channel NIR device measurement. *Measurement, 173*, 108615.

Srivastava, S., & Sadistap, S. (2018). Non-destructive sensing methods for quality assessment of on-tree fruits: A review. *Journal of Food Measurement and Characterization, 12*(1), 497–526.

Su, W. H., & Sun, D. W. (2018). Multispectral imaging for plant food quality analysis and visualization. *Comprehensive Reviews in Food Science and Food Safety, 17*(1), 220–239.

Suchanek, M., Kordulska, M., Olejniczak, Z., Figiel, H., & Turek, K. (2017). Application of low-field MRI for quality assessment of 'Conference'pears stored under controlled atmosphere conditions. *Postharvest Biology and Technology, 124*, 100–106.

Sumriddetchkajorn, S., & Intaravanne, Y. (2013, June). Two-dimensional fruit ripeness estimation using thermal imaging. In *ICPS 2013: International Conference on Photonics Solutions* (Vol. 8883, p. 88831C). International Society for Optics and Photonics.

Sun, D. W. (Ed.). (2010). *Hyperspectral imaging for food quality analysis and control*. Elsevier.

Sun, R., Zhou, J. Y., & Yu, D. (2021). Nondestructive prediction model of internal hardness attribute of fig fruit using NIR spectroscopy and RF. *Multimedia Tools and Applications, 80*, 21579–21594.

Taghadomi-Saberi, S., Masoumi, A. A., Sadeghi, M., & Zekri, M. (2019). Integration of wavelet network and image processing for determination of total pigments in bitter orange (Citrus aurantium L.) peel during ripening. *Journal of Food Process Engineering, 42*(5), e13120.

Taghadomi-Saberi, S., Omid, M., Emam-Djomeh, Z., & Faraji-Mahyari, K. H. (2015). Determination of cherry color parameters during ripening by artificial neural network assisted image processing technique. *Journal of Agricultural Science and Technology, 17*, 589–600.

Taghizadeh, M., Gowen, A. A., & O'Donnell, C. P. (2011). Comparison of hyperspectral imaging with conventional RGB imaging for quality evaluation of Agaricus bisporus mushrooms. *Biosystems Engineering, 108*(2), 191–194.

Takahashi, N., Maki, H., Nishina, H., & Takayama, K. (2013). Evaluation of tomato fruit color change with different maturity stages and storage temperatures using image analysis. *IFAC Proceedings Volumes, 46*(4), 147–149.

Tallada, J. G., Nagata, M., & Kobayashi, T. (2006). Non-destructive estimation of firmness of strawberries (Fragaria× ananassa Duch.) using NIR hyperspectral imaging. *Environmental Control in Biology, 44*(4), 245–255.

Teerachaichayut, S., & Ho, H. T. (2017). Non-destructive prediction of total soluble solids, titratable acidity and maturity index of limes by near infrared hyperspectral imaging. *Postharvest Biology and Technology, 133*, 20–25.

Thompson, A. K. (2003). *Fruit and vegetables: Harvesting, handling and storage*. Blackwell Publishing Ltd.

Tripathi, A., Baran, C., Jaiswal, A., Awasthi, A., Uttam, R., Sharma, S., et al. (2020). Investigating the carotenogenesis process in papaya fruits during maturity and ripening by non-destructive spectroscopic probes. *Analytical Letters, 53*(18), 2903–2920.

Ukirade, N. S. (2014). Color grading system for evaluating tomato maturity. *International Journal of Research in Management, Science & Technology, 2*(1), 41–45.

Urraca, R., Sanz-Garcia, A., Tardaguila, J., & Diago, M. P. (2016). Estimation of total soluble solids in grape berries using a hand-held NIR spectrometer under field conditions. *Journal of the Science of Food and Agriculture, 96*(9), 3007–3016.

van Roy, J., Keresztes, J. C., Wouters, N., De Ketelaere, B., & Saeys, W. (2017). Measuring colour of vine tomatoes using hyperspectral imaging. *Postharvest Biology and Technology, 129*, 79–89.

Vanoli, M., Van Beers, R., Sadar, N., Rizzolo, A., Buccheri, M., Grassi, M., et al. (2020). Time-and spatially-resolved spectroscopy to determine the bulk optical properties of 'Braeburn'apples after ripening in shelf life. *Postharvest Biology and Technology, 168*, 111233.

Vega Díaz, J. J., Sandoval Aldana, A. P., & Reina Zuluaga, D. V. (2021). Prediction of dry matter content of recently harvested 'Hass' avocado fruits using hyperspectral imaging. *Journal of the Science of Food and Agriculture, 101*(3), 897–906.

Villaseñor-Aguilar, M. J., Sánchez-Bravo, M. G., Padilla-Medina, J. A., Vázquez-Vera, J. L., Guevara-González, R. G., García-Rodríguez, F. J., & Barranco-Gutiérrez, A. I. (2020). A maturity estimation of bell pepper (Capsicum annuum L.) by artificial vision system for quality control. *Applied Sciences, 10*(15), 5097.

Waldron, K. W., Parker, M. L., & Smith, A. C. (2003). Plant cell walls and food quality. *Comprehensive Reviews in Food Science and Food Safety, 2*(4), 128–146.

Walsh, K. B., Blasco, J., Zude-Sasse, M., & Sun, X. (2020). Visible-NIR 'point'spectroscopy in postharvest fruit and vegetable assessment: The science behind three decades of commercial use. *Postharvest Biology and Technology, 168*, 111246.

Wang, Z., Van Beers, R., Aernouts, B., Watte, R., Verboven, P., Nicolai, B., & Saeys, W. (2020). Microstructure affects light scattering in apples. *Postharvest Biology and Technology, 159*, 110996.

Williams, P., & Norris, K. H. (2001). *Near-infrared technology in the agricultural and food industries* (2nd ed.). American Association of Cereal Chemists.

Wu, D., & Sun, D. W. (2013). Advanced applications of hyperspectral imaging technology for food quality and safety analysis and assessment: A review—Part I: Fundamentals. *Innovative Food Science & Emerging Technologies, 19*, 1–14.

Xiao, H., Li, A., Li, M., Sun, Y., Tu, K., Wang, S., & Pan, L. (2018). Quality assessment and discrimination of intact white and red grapes from Vitis vinifera L. at five ripening stages by visible and near-infrared spectroscopy. *Scientia Horticulturae, 233*, 99–107.

Xiao, Z., Stait-Gardner, T., Willis, S. A., Price, W. S., Moroni, F. J., Pagay, V., et al. (2021). 3D visualisation of voids in grapevine flowers and berries using X-ray micro computed tomography. *Australian Journal of Grape and Wine Research, 27*(2), 141–148.

Xie, C., Chu, B., & He, Y. (2018). Prediction of banana color and firmness using a novel wavelengths selection method of hyperspectral imaging. *Food Chemistry, 245*, 132–140.

Zhang, L., & McCarthy, M. J. (2012). Measurement and evaluation of tomato maturity using magnetic resonance imaging. *Postharvest Biology and Technology, 67*, 37–43.

Zhang, L., & McCarthy, M. J. (2013). Assessment of pomegranate postharvest quality using nuclear magnetic resonance. *Postharvest Biology and Technology, 77*, 59–66.

Zhu, H., Chu, B., Fan, Y., Tao, X., Yin, W., & He, Y. (2017). Hyperspectral imaging for predicting the internal quality of kiwifruits based on variable selection algorithms and chemometric models. *Scientific Reports, 7*(1), 1–13.

Zhuang, J., Hou, C., Tang, Y., He, Y., Guo, Q., Miao, A., et al. (2019). Assessment of external properties for identifying banana fruit maturity stages using optical imaging techniques. *Sensors, 19*(13), 2910.

Zulkifli, N., Hashim, N., Abdan, K., & Hanafi, M. (2019). Application of laser-induced backscattering imaging for predicting and classifying ripening stages of "Berangan" bananas. *Computers and Electronics in Agriculture, 160*, 100–107.

Application of Biospeckle Laser Technique for Assessment of Fruit Quality

O. J. Sujayasree, R. Pandiselvam, A. K. Chaitanya, and Anjineyulu Kothakota

Abstract Evaluation of fruit quality at various phases of the postharvest is imperative to offer superior quality to consumers. Biospeckle laser technology, an opto-electronic approach has numerous merits, such as low-cost, rapid, and nondestructive. In contrast to other multispectral approaches, laser light can portray biologically active sections of the samples tested by estimating the differences in the light interferences. Based on the plant metabolism, it brings forth information about chemical and physical quality ascribed to tissues. Quality is evaluated by analyzing the "cross-correlation function and the temporal history of speckle pattern" obtained as a result of biological processes in fruits. This technology exhibits potential applications of the measurement of fruit quality indicators such as firmness, moisture content, ripening degree, juiciness, maturity indices, acidity, starch content, soluble solid content, pigment transformations, respiration rate, senescence patterns, shelf life predictions, damages/bruises, diseases, and pesticide residues. Hence, it is well utilized as an effective noninvasive testing tool in fruit crops like apple, orange, mango, guava, tomato, and macaw palm fruit. Despite its wide range of applications, there is a need for robust evaluation of the significance of biological speckle in fruit physiology to standardize and commercialize this technology.

O. J. Sujayasree
Division of Post-Harvest Technology, ICAR-Indian Agricultural Research Institute, New Delhi, Delhi, India

R. Pandiselvam (✉)
Physiology, Biochemistry and Post-Harvest Technology Division, ICAR-Central Plantation Crops Research Institute, Kasaragod, Kerala, India
e-mail: r.pandiselvam@icar.gov.in

A. K. Chaitanya
Department of Genetics and Plant Breeding, Lovely Professional University, Phagwara, Punjab, India

A. Kothakota
Agro-Processing & Technology Division, CSIR-National Institute for Interdisciplinary Science and Technology (NIIST), Trivandrum, Kerala, India

P. B. Pathare, M. S. Rahman (eds.), *Nondestructive Quality Assessment Techniques for Fresh Fruits and Vegetables*, https://doi.org/10.1007/978-981-19-5422-1_4

Keywords Laser biospeckle · Fruit · Quality evaluation · Nondestructive · Cross-correlation · Speckle grams

1 Introduction

The quality of the fresh fruit affects its market value and the consumer's choice. Hence, quality inspection of fruits has profound importance. The internal quality of fruits is becoming more and more significant rather than its external attributes, and consumer trend is to choose fruits of high quality, which have a typical texture and taste, free of contaminants and nutritional and health-promoting constituents (Rizzolo et al., 2010). Thus, assurance of safety and quality of the fresh horticultural harvests from farm to fork has got immense importance due to its perishability at each stage of handling, transportation, distribution, and processing (Bobelyn et al., 2010). The deterioration parameters that affect the fresh produces are causing their spoilage such as physical, biological, physiological, microbiological, and biochemical aspects that attributes to quality loss (Ramos et al., 2013).

Physical causes affecting quality loss of a product are their rheology and moisture percentage change, wilting, wounding, bruising, bursting, crushing, cutting, freezing, drying, etc. (Ramos et al., 2013). Also, cellular decompartmentalization/derealization of enzymes and substrates paves way to various biochemical deteriorations due to injury stresses, such as browning, off-flavors, and texture breakdown (Mesa et al., 2016). Biochemical fluxes such as polyphenol oxidase and phenol peroxidase mediated enzymatic browning, green tissues discoloration by chlorophyll degradation involved and coloring processes in particular crops like secondary browning in apples, ripening, shelf life and senescence pattern related changes, such as declining cell wall strength and reduced intercellular adhesion (Lee et al., 2015; Mesa et al., 2016; Romero et al., 2009). Also, the biological interventions by various postharvest pathogens and pests can also affect the appearance, quality, and sensory distinctions of fresh fruits.

The rising need for the best quality commodities gives an urgent call for additional innovative postharvest technologies. Hence, the fruit quality and its assessment techniques turn out to be extremely crucial. This demands the requirement for speedy and nondestructive quality assessments of fresh produce. With the advancement in information technology, the computer vision-based prototype identification and imaging practice seem to be a fully fledged system for safety and quality evaluation of various fresh produces (Bhargava & Bansal, 2018). Due to easiness in operation, various mechanical, electrochemical, and electromagnetic nondestructive quality evaluation techniques are prevalent for the detection of textural, appearance, and compositional characteristics (Pandiselvam et al., 2020; Zdunek et al., 2014). Maximum prevailing techniques are destructive and the current advancements in scientific and computer knowhow have paved the way for the growth of nondestructive methods such as "visible or near infrared spectroscopy, ultrasonics, acoustics, electronic nose technology, hyperspectral imaging" for analyzing the physical and chemical quality attributes of fresh produce (Samuel et al., 2017).

But it possesses a wide range of technical challenges in quick and precise detection of internal quality as a result of the intricate mechanical, physical, and chemical characteristics of fruits. These techniques are used for mass handling of commodity as a result of its instantaneous nature and can simultaneously determine many parameters (Abasi et al., 2018; Samuel et al., 2017; Zdunek et al., 2014). Seemingly various on-farm nondestructive methods are followed in developed countries. But developing nations, because of technical glitches and scarcity of advancement of equipment accessibility, have limited their larger use in the quality assessment of fruits (Abasi et al., 2018). Hence, the advancement of a precise, consistent, and nondestructive approach for the quality assessment at preharvest and packaging stage is crucial to assure the quality excellence of fruits.

In this context, biospeckle laser imaging method, which is rapid, noninvasive, handy, and economical, has its relevance in assuring the safety and quality of the fresh commodities by assessing the biological tissue (Retheesh et al., 2016). The biospeckle aspect occurs when the target tissue is exposed to coherent laser light. As a result, it produces deviations in the speckle configuration as light and dark regions due to a huge array of relations between the coherent light and complicated biological tissue, that are documented in a certain frame of time (Lu et al., 2020; Retheesh et al., 2016). The light variations are caught by a camera and thereby shifted towards a computer. The speckles of back and forward scattering are employed to assess the biospeckle activity (Omidi-Arjenaki et al., 2019). The laser light goes to the exterior layer of the target and the light reflected is caught in the back-scattering setup, whereas the laser light progresses the target and extends the spin hole of the camera in the forward-scattering set-up. The moving and static fragments of the tissues generate a fluctuating and static pattern respectively in the speckle configuration of the biological samples. The variable speckle pattern is a unique characteristic of the biological substance and is termed as biospeckle although the elements other than the organic are correlated with stationary forms (Ansari et al., 2012; Ansari et al., 2018). The motion of the speckle pattern is defined as the biospeckle activity. The laser light of over 600 nm wavelength should be applied for illumination purpose in biospeckle instruments for most agricultural or horticultural commodities having hard surfaces (Ansari et al., 2018; Ansari & Nirala, 2012; Ansari & Nirala, 2013a). The physical movement of particles inside cells provides variations in the light absorption by tissue pigments which gives the apparent action of biospeckles. Hence, it gives knowledge of several living activities within the cells, biological and physical characteristics of the targets, and has all-inclusive utilization for assurance of food quality and safety (Ansari & Nirala, 2013a; Braga et al., 2011; Braga Jr et al., 2007).

Numerous studies have adopted in the horticultural usage of laser biospeckle technique like correlating the biospeckle action to the pigment (chlorophyll) (Zdunek & Herppich, 2012), bruise assessment (Costa, Pinto, et al., 2017; He et al., 2021; Kumari & Nirala, 2016; Pajuelo et al., 2003; Romo & Yoxall, 2005; Yan et al., 2017), shelf life study on injured and fresh parts (Kumari & Nirala, 2016), categorizing the mealy and non-mealiness of the apples (Arefi et al., 2016; Kurenda et al., 2012), for determining age of lemons (Ansari & Nirala, 2016), for maturation examination of mango (Retheesh et al., 2016), for disease examination of apple bull's eye rot and quality of the fruits (Adamiak et al., 2012), to evaluate the storage life of the commodity (Abou Nader et al., 2019; Costa, de Carvalho Pinto, et al., 2017; Pieczywek et al., 2018; Rabelo et al., 2005; Zdunek & Cybulska, 2011), and efficient approach to analyze the metabolism and biochemical vagaries occurring within the cells and tissues of the produce (Bergkvist, 1997; Rahmanian et al., 2020; Romero et al., 2009; Skic et al., 2016). Here, we summarized the existing phase in advancement of biospeckle technology and its usage, especially for fruit quality assessment.

2 Working Principles and Their Operations

2.1 Device Components

It requires apparatus such as laser light (a diode laser), a detector like Charge-Coupled Devices (CCDs) camera (commonly highly defocused using a lens), and a computer with a frame apparatus, which can capture a series of pictures in a persistent time interval. The biospeckle activity is estimated from the time series images using various mathematical methods thus helping in fast and nondestructive sampling applications. Figure 1 depicts the experimental setup for biospeckle activity measurement. Greater stationary speckles from the exterior are light incidence angle prevalent; however, lesser dynamic speckles consequential from the interior of

Fig. 1 Experimental setup for biospeckle activity measurement

material are angle independent. The biospeckle computation procedure must ponder parameters such as "wavelength of laser, the distance and angle amongst the targets and the detector, polarization, light intensity and the detector objective aperture." When the coherent laser light illuminates a diffuse target, the backscattered light will form a light and dark interference pattern on the detector which is termed as speckle pattern. The speckle phenomenon will vary over time based on the movement in the object and helps in visualizing the tissues.

2.2 *Measurement Methods and Spatial Analysis*

Universal estimates of speckle action are of three types such as "spatial and temporal contrast, time history of the speckle pattern (THSP) and spatial-temporal speckle correlation technique." The biospeckle activity can be examined by numerous techniques, but majority measures in terms of "correlation of irradiance as a function of time." Fuji's technique (Fujii et al., 1985), Weighted Generalized Difference (WGD), and Generalized Difference (GD) (da Silva et al., 2008) approaches reveal the biological action patterns; however, Inertia Moment (IM) (Ansari et al., 2016; Arizaga et al., 1999) and Absolute Value Difference (AVD) (Cardoso et al., 2011) are the numerical means of application.

In context to the abovementioned techniques, diagnostic methodologies are further categorized as "qualitative and quantitative" techniques. The quantitative analytical approaches are "Absolute Value Difference, Cross-correlation and Inertia Moment approaches" whereas "Fujii (Rabal et al., 2018), temporal difference (TD) (Arizaga et al., 2002), laser speckle temporal contrast analysis (LASTCA), laser speckle contrast analysis (LASCA) (Briers, 1975a; Briers, 1978; Briers, 2007), and generalized differences" are generally used for qualitative assessment. Mathematical study of speckle phenomenon in time majorly utilizes Temporal History of Speckle Pattern (THSP) (Oulamara et al., 1989), that are collections of the pixels in due course of time.

2.3 *Spatial Analysis of Biospeckle Activity*

Fujii: This approach utilizes scanning process to examine the image action. Recently this is extensively utilized in biospeckle-based study, even though it was developed for the blood flow measurement. This technique generates a plan of action depending upon the total absolute differences among the intensities of successive image frames; however, mean of each frame estimate is calculated. Nevertheless, the weighing technique recommended by Fujii signifies a nonlinear correlation, which creates gradually fewer stable regions of lower activity where finder noise is strengthened and may be incorrectly interpreted as a signal motion (Retheesh et al., 2018).

Generalized difference (GD): It is described as the aggregated total of absolute values of the variations amongst the pixel strengths between every frame. This technique was created as an option to Fuji's method. The weighting process is eliminated at the basic version of this algorithm (Minz & Nirala, 2016).

Absolute value difference (AVD): This method was the "first statistic moment order" that could be an alternative for the Inertia Moment technique. The prime basis of AVD is that the total difference is the key data examined and square used in the IM technique can strengthen the variations over time in a prejudiced way (Braga et al., 2011). In few cases, AVD method displays improved biospeckle action analysis, especially while THSP matrix does not include information at elevated intensities.

Laser speckle contrast analysis (LASCA): In common, this algorithm is dependent on the computations of "spatial or temporal" difference over a local, square window of MxM pixels. The window is shifted alongside the speckle phenomenon and the estimated contrast value is allotted to the center pixel. A lower statistical validity can be there for the smaller window whereas larger windows lower the actual resolution of the image. This technique has several differences. In a very fundamental arrangement of LASCA, utilizing a movable window the contrast at every site is assessed through specific speckle form image. LASCA is evaluated from group of frames for spatially obtained contrast (Braga Júnior, 2017).

Motion History Image (MHI): It is a "real-time imaging" system which forms a series of progress depending on the latest movement by an image progression. MHI technique is recommended as an alternate to the usual virtual techniques of biospeckle image assessment. Here, the "pixel intensity" is a function of the movement recorded at specific site, whereas more latest motion is parallel to the brighter values. The images deposited within the buffer are primarily treated (Godinho et al., 2012). The processing technique comprises of "threshold of the silhouette" developed by obtaining the differences between two consecutive images. By utilizing this technique of weighing the threshold images deposited within the buffer, ultimate MHI image is produced with regard to "lifetime" of every image. When contrasted to LASCA, MHI revealed superior findings while assessing biological and nonbiological targets (Aboonajmi & Faridi, 2016). Moreover, Motion History Image delivered similar findings to the prevailing offline methods like Fujii's and GD.

Inertia moment (IM): It is the frequently utilized mathematical approach centered upon initiation of the specific matrix identified as the THSP subsequently with the co-occurrence matrix (COM) that could be an intermediate image. The IM of the co-event association was appropriate in assessing distribution of "M" values across through major diagonal. It is considered as the sum of the matrix numbers multiplied by the squared separation from the main diagonal. It is predictable that a higher inertia moment recognizes greater biospeckle action. Assessments with a simulated THSP exhibited that the ascertained IM diminishes with a rise in the speckle length beyond 33% window size, and the IM attains its saturation circumstances (Ansari & Nirala, 2013b).

3 Analyses in Spectral Domains

Because of the intricacy of biological substances, alternate methods of biospeckle action assessments, based on spectra domains have been established. Alteration of the biospeckle signals to the frequency domain could be achieved promptly by the "Fourier transform," or exclusively by the "wavelet transform" that permits formation of novel markers associated with physicochemical phenomenon. The alteration to spectral domain can be done along with conventional biospeckle laser approaches such as "Fujii, GD and THSP" (Rabelo et al., 2011). It leads to the spatial assessment of biospeckle action by characteristic indicators, such as IM, among preferred spectra. The spectral attributes of biospeckle action demonstrated favorable capability to differentiate processes associated with a_w (water activity) of corn and bean seeds, partition of embryo and endosperm in corn seeds and surveillance, in frequency domain, maturing of tomato, apple, and pear (Pra et al., 2014).

As there is an utmost necessity to deliver the best quality and safe products to consumers, evaluation of the fruit quality at various stages of pre- and postharvest conditions is significant. Hence, biospeckle technique which measures the speckle activity has close relationships with the physiological traits of the fresh agricultural/horticultural products like cytoplasmic streaming and organelles size. Studies demonstrated that responsible factors for a particular biospeckle activity is the progressions associated with movement of the scattering centers in the tissue, such as cytoplasmic streaming, organelle movement, cell growth and division during fruit maturation, and biochemical reactions (González-Peña et al., 2016). As well, Brownian motions would be considered as a source of biospeckle activity. Also, the physicochemical indicators could be judged, such as the peel thickness, ripeness, acidity, soluble solid content, starch content, water content, and respiration rate. Rather than the physical composition of the tissue which significantly changes during postharvest ripening of fruits, such as starch and pectin degradation, chlorophyll degradation, postharvest water evaporation, particle mobility, etc., could also be evaluated (Briers, 1975b). Bruising parameters, such as maturity, turgor, damage, aging, and mechanical properties can also be evaluated by biospeckle activity (Du et al., 2020). Studies exhibited that decaying of a tissue caused by age, illness or infection, or mechanical damages like impact, vibration/compression, relates to lower biospeckle activity.

4 Application of Biospeckle Technique in Assessing Fruit Quality

As the laser biospeckle technique is harmless, risk-free, chemical-free, and nondestructive, it is regarded as a precious approach for examining a variety of biological activities in crops. Its extensive applications are used in all fields of research from medicine to agricultural commodities including fruits and vegetables (Bhargava &

Bansal, 2018; Pandiselvam et al., 2020; Zdunek et al., 2014). This review of speckle technology mainly focused on the researches like assessment of fruit quality, monitoring of ripening, shelf life and senescence pattern in postharvest storage period, and prediction of the harvest window of fruit crops. Also focuses on biochemical processes and metabolism-related changes during fruit maturation that comprises starch content, pigment transformations, and firmness. Also, focuses on the detection of mechanical defects for the distinction of integral and damaged areas in fruits, and identification of disease occurrence in fruits at the early stage of infections helping in reducing the postharvest losses occurring in fruits and increasing the quality and shelf life (Table 1). Many qualitative and quantitative parameters can be analyzed using biospeckle technique, among them STD (standard deviation) is the greatest among the various proposed algorithms aimed at knowing the qualitative along with the quantitative difference and assessment of ripening stages of the fruits by laser biospeckle technique (Kumari & Nirala, 2016).

4.1 Monitoring of Ripening, Shelf Life, and Senescence Pattern in Postharvest Storage Period

Various textural, structural, physiological, and biochemical process changes as the crop matures. The ripening losses were a great threat to industries and stakeholders. The crop ripening directly affects consumer acceptance as it mainly affects change in odor. The ripening leads to an increase in the carotenoid content, which alters the chemical composition of the fruits thereby degrading the cell wall and making it soften. Pectin enzyme degradation is also a prime source responsible for softening and deterioration of cell structure. The optical absorption and scattering properties of tomatoes during ripening can be measured to analyze the physiological changes (Zhu et al., 2015). In this context, the biospeckle laser technique is experimented to justify the correlation between biospeckle activity and quality indices in fruit crops in various researches.

The ripening of tomatoes was judged by two different speckle activity metrics at wavelengths 640 nm and 830 nm respectively for two cultivars. In this case study, the authors depicted that the progression of maturity increases with an increase in speckle action mostly at 640 nm than at 830 nm rapidly. However, in both cultivars, advancing maturation development corresponded with the enhancement in speckle activity (Pieczywek et al., 2018; Retheesh et al., 2016).

Generally, technique of biospeckle activity increased with the decrease in firmness and chlorophyll content, while increases in carotenoid content corresponded to an increase in biospeckle activity. Standard linear regression models were evaluated based on the average of speckle action amounts and optical factors with chlorophyll, firmness, and carotenoid content. The biospeckle activity assessed for two cultivars of tomato portrayed a very strong correlation along with the firmness of its flesh (Qing et al., 2007). Another similar study in lemons to determine the age through

Table 1 Application of biospeckle technique in the assessment of various fruit quality

Name of the fruit	Sample size and number of images/ frame	Algorithms used	Result	References
Apple	500 images (300 × 300 pixels) of bruised apple	LASCA, WGD, and Konishi method	Bruise detection	Passoni et al. (2005)
	15 frame per second (fps) rate, which resulted in 210 frames. The image resolution was 340 × 160 pixels	Correlation coefficient was calculated using the Matlab R2010a software	Biospeckle activity linearly decreases with increasing chlorophyll content	Zdunek and Cybulska (2011)
	512 frames (480 × 320 pixels) of 50 undamaged apples	GD, Fujii, and LASTCA methods	Less biospeckle activity in damaged area	Yan et al. (2017)
	The image size was 640 × 480 pixels, A stack of uncompressed images (BMP, 8 bits) was recorded for 14 s at a rate of 15 fps	Correlation coefficient was calculated using the Matlab R2010a software	Quality attributes change significantly during storage	Zdunek and Cybulska, (2011)
	• The imaging source Europe GmbH, Bremen, (Germany) lasting 4 s with 15 frames per second (fps) • The image resolution was 320 × 240 pixels, which corresponded to an observation area of 7.5 mm^2 • The image exposure time of the CCD camera was 1/250	"Corrcoef" function, as a cross-correlation coefficient	Natural biochemical changes post-HHP changes	Kurenda et al. (2012)
	• Light scattered through the sample slab is collected by a CCD camera (Hamamatsu photonics, Hamamatsu, Japan, C8484-	• To the scattering particle's mean square displacement (MSD)	• Speckle activity at short time scales	Abou Nader et al. (2019)

(continued)

Table 1 (continued)

Name of the fruit	Sample size and number of images/ frame	Algorithms used	Result	References
	05G01, 1344 × 1024 pixels			
	• 6.45 m × 6.45 m pixel size placed ~35 cm from the slab (~0.14 magnification, 216 pixels corresponding to 1 cm on the sample)	• Correlation coefficient of biospeckle activity (BA) value	• Correlations between speckle parameters and the ratio of apples' firmness to their soluble solids content reveal significant links despite the unknown fruit's origin, harvest date, and storage history	
	• Speckle grams of size 512 × 512 were obtained by a CMOS camera with the focusing lens arrangement	Cross-correlation function and the temporal history of speckle grams	• Assess the quality of certain seasonal fruits/vegetables such as apples, mangoes, guavas, oranges, and cucumbers	Samuel et al. (2017)
	• These images were stored as an 8-bit gray level		• Mechanical damage of fruits was numerically assessed using the methods of IM and cross-correlation	
			• The damaged regions were effectively identified by activity maps generated using the generalized differences method	
			• The technique was extended to identify pesticides in vegetables	
	10 composite images of 512 × 512 pixels formed by stacking consecutive	Temporal history of speckles patterns (THSP), the moment of inertia of	Study of impact on apples and the analysis of bruises and physical properties with quality factors	Pajuelo et al. (2003)

(continued)

Table 1 (continued)

Name of the fruit	Sample size and number of images/frame	Algorithms used	Result	References
	columns digitized to 256 gray levels	the co-occurrence matrix (COM)		
	Image resolution was 320 × 240 pixels that corresponded to an observation area of about 3 × 2 mm	Correlation coefficient of biospeckle activity (BA) value	• Optimum harvest window (OHW), firmness and standard quality attributes (firmness, acidity, starch, soluble solids content, Streif index), and physiological parameters (respiration and ethylene emission)	Skic et al. (2016)
	256 images of dynamic speckle of 200 × 200 pixels	Co-occurrence matrix, biospeckle activity (BA) value, AVD, IM, GD, parameterized Fujii, parameterized generalized difference, GLCM, GSD, AGD, and parameterized global average Fujii	• Relatively lesser biospeckle activity in the damaged region of the apple compared to fresh region • Biospeckle activity maximum for the inertia moment method. AGD gives the maximum overall difference in biospeckle activity	Kumari and Nirala, (2016)
	• 100 apples per experiment. Biospeckle movies lasting at 4 s recorded in uncompressed AVI film at a 15 fps rate • Image resolution was 320 × 240 pixels	Correlation coefficient of biospeckle activity (BA) value	Bull's eye rot disease detection	Adamiak et al. (2012)
	• Sequences of 256 images of dynamic speckle of 200 × 200 pixels were captured by a digital color CMOS camera (Basler 1300–32 fc) at the	Co-occurrence matrix, inertia moment, absolute value difference, generalized difference, parameterized Fujii, biospeckle activity (BA) value, granulometric size	• Maximum for the inertia moment method (521.99) • AGD gives the maximum overall difference in biospeckle activity (42.35)	Kumari and Nirala (2016)

(continued)

Table 1 (continued)

Name of the fruit	Sample size and number of images/ frame	Algorithms used	Result	References
	rate of 12.5 frames per second	distribution (GSD), and gray-level co-occurrence matrix (GLCM), as well as three new proposed algorithms namely parameterized generalized difference, alternative generalized difference (AGD), and parameterized global average Fujii, have been used for qualitative and quantitative analysis	• BA and GLCM are used to distinguish between bruised and fresh Indian apples	
	20 frames per second until 500 frames	Inertia moment, absolute value of differences (AVD), and autocorrelation method	Classified mealy and non-mealy apples. Biospeckle activity was higher for fresh apples. Good results were achieved at a biospeckle feature of 780 nm	Arefi et al. (2016)
	Twenty apples per experiment. The camera captured uncompressed video sequences at 60 frames per second	Correlation coefficient of BA, Fujii index, and moment of inertia (IM) frequency-based analysis	• Visual inspection detected the Bull's eye rot disease 4–5 days after inoculation • Biospeckle activity enables significant detection of infected areas as early as 2 days after inoculation	Pieczywek et al. (2018)
Minimally processed apples	Pixel size of 3.75 × 3.75 μm recorded at an angle of 25°. The sampling time and shutter speed of camera were at 0.08 s and 9 ms, respectively	Autocovariance method	The degree of polarization and speckle grain size obtained from the speckled image contain the physicochemical information of sample	Minz and Nirala (2016)

(continued)

Table 1 (continued)

Name of the fruit	Sample size and number of images/ frame	Algorithms used	Result	References
Apple, pear, and guava	Recording time was equal to 15 s with the frame rate equal to 20 fps	Cross-correlation coefficient difference	Quality evaluation of different Indian fruits namely apple, pear, and guava	Ansari and Nirala (2012)
Oranges, mangoes, apples, guavas, and cucumbers	Dynamic specklegrams of size 512 × 512 were obtained by a CMOS camera	Cross-relationship and the time history of speckle patterns	Detection of damaged areas and pesticides in vegetables	Samuel et al. (2017)
Macaw palm	128 successive images of 8 bits relative to the biospeckle patterns were collected in intervals of 0.08 s, with sampling frequency limited to 0–12.50 Hz	Using the algorithm of the absolute values of difference	Biological activity showed a significant linear ratio ($R^2 = 0.913$) with the maturation of fruits	Costa, de Carvalho Pinto, et al. (2017)
Lemon fruit	256 images 1294 × 964 pixels with frame rate of 32 fps and pixel size 3.75 × 3.75	Co-occurrence matrix (COM), modified co-occurrence matrix, temporal history speckle pattern (THSP)	Determine the age of the lemon fruit from the observation of its dynamic speckle pattern and analyzing maturation patterns	Ansari and Nirala (2016)
Green orange	3.75 μm × 3.75 μm pixel size was used to record subjective biospeckle images at 32 frames per second	Fujii method, temporal difference, laser speckle contrast analysis, and motion history image (MHI)	MHI algorithm accurately detects the scar area	Retheesh et al. (2018)
Orange	Resolution and frame rate of 5 megapixels (1920 × 1080 pixels) and 30 fps, respectively. A set of 41 images were recorded related to dynamic speckle over a period of 1.4	Fujii algorithm and numerical procedure using features extracted from the COM matrix	Chilling and freezing assessments of oranges	Rahmanian et al. (2020)
	0.08 s and a composite image of 512 × 512 pixel size was generated by storing consecutive	Autocorrelation function and the modified occurrence matrix were used with statistical cumulant and the	Quality and senescence indicators for the specimens and were compared with other parameters, such as total soluble	Rabelo et al. (2005)

(continued)

Table 1 (continued)

Name of the fruit	Sample size and number of images/ frame	Algorithms used	Result	References
	columns digitized to 8-bit gray levels	moment of inertia, calculated from the modified occurrence matrix	solids, total acidity, the penetration force, and the storage period	
Mango	512 sequential images of the biospeckle pattern with a pixel resolution of 512 × 512	Inertia moment from the temporal history of speckle pattern (THSP) and by calculating the 2D (two-dimensional) cross-correlation function	Monitoring the biological activity of fruit samples	Retheesh et al. (2016)
Tomato	25 frames per second. For each state of the phenomenon under study, 512 successive images of the dynamic speckle pattern were taken and a fixed column of each image (i.e., the central column) was grabbed. With these 512 columns, an image of 512 × 512 pixels was built and digitized in 256 gray levels	Autocorrelation functions of the intensity fluctuation	Fruit ripening of each tomato and shelf life prediction of fruit and vegetables	Romero et al. (2009))
	The resolution of monochrome sensor was equal to 640 × 480 (H V) pixels with a 5.6 × 5.6 m pixel size.	Temporal changes of the biospeckle pattern-correlation coefficient	Maturation monitoring and the prediction of the maturity indices of tomatoes, proving the possibility of practical implementation of this method for the determination of the maturity stage of tomatoes	Pieczywek et al. (2018)
Strawberries	≥2 number of frames per observation	Generalized differences (GD) and Fujii	Both the algorithm allowed for early detection of fungi colonies growing	Mulone et al. (2013)

(continued)

Table 1 (continued)

Name of the fruit	Sample size and number of images/ frame	Algorithms used	Result	References
			onto the surface of the fruit and distinguishing the good quality and advanced matured fruits	
	NA	Dynamic speckle pattern analysis	The age from observation of its dynamic speckle pattern	Mulone et al. (2013)
		Partial least squares (PLS)	Native phenolic compounds	Wulf et al. (2008)
Blueberry	NA	The laser air-puff	The firmness index derived from the laser air-puff tester achieved a significant correlation with the firmness values measured by the Firmtech	Li et al. (2011)
Mulberry	NA	Principal component analysis (PCA) and partial least square regression (PLSR)	This research confirmed the feasibility of using LIBS and HSI for the rapid detection of thiophanate-methyl residue on mulberry fruit	Wu et al. (2019)

biospeckle laser technique and the speckle pattern was analyzed through the algorithms by IM and Spatial-temporal correlation coefficient. They depicted that fruit quality gradually declines with change in dynamic speckle activity and thereby analyzing maturation patterns (Ansari & Nirala, 2016).

A study for evaluation of processing algorithms like "histograms, RGB and HSB color space, autocovariance, inertia moment (IM), absolute value difference, granulometric size distribution, Gray level co-occurrence matrix, and three new proposed algorithms, namely parameterized geometric mean of generalized difference (GD), image sequencing mean of parameterized GD, and squared temporal difference (STD)" in Indian climacteric fruits (Mango, Banana, Guava) for knowing the ripening changes and maturity (Ansari & Nirala, 2016). They depicted that out of all the algorithms used IM is the best for assessment of differentiation between maturity and ripe phases with the greatest BA variation. Results also depicted that

STD is the most excellent approach for obtaining both quantitative and qualitative differences as well as assessment of the ripening of fruits (Ansari & Nirala, 2016).

4.2 Prediction of the Harvest Window of Fruit Crops

Biospeckle technology is the quickest method for improving the harvesting and storage of fruits was reported in several studies. Harvesting the crop at the right time and correct maturity stages are very important for better shelf life and market demand. So, the determination of OHW (Optimum Harvest Window) is an important phenomenon in the agro-food chain (Romero et al., 2009). In this, the researchers studied "Ligol" and "Szampion" apple cultivars for analyzing the optimum harvest window by the application of biospeckle technology. Physiological parameters, such as ripening hormone ethylene emission and respiration rate as well as other quality attributes were studied using the "Correlation coefficient algorithms" of biospeckle activity value. The results predicted that OHW in these cultivars of apples was due to the decline of biospeckle activity before the harvesting stages and depends on the selectivity (Skic et al., 2016). Studies elucidated that the greater correlation between speckle grain size and chlorophyll deprivation can result in a direct positive association among the reductions in both. It also entails that speckle grain size assessment during fruit ripening of pear may be utilized as a non-harmful assessment of fruit maturity (Nassif et al., 2012).

4.3 Biochemical Processes and Metabolism-Related Changes during Fruit Maturation

In a study, the assessment of soluble solid content, starch content, and firmness were done. Each parameter was estimated separately on individual apples. Biochemical changes like TSS, starch substance, and firmness have a direct relation to growth, development, and fruit maturity. The result concluded that there is an increase in the soluble solid matter and a slender decrease in acid and starch matter with an increase in the speckle activity. It resulted in a great and high correlation among the biospeckle activity and biochemical parameters in the study revealing that biospeckle laser technique has a terrific possibility to be utilized before the harvesting stages for nondestructive evaluation (Arefi et al., 2016; Skic et al., 2016; Zdunek & Herppich, 2012).

A significant correlation concerning biospeckle activity, firmness, starch index, Streif Index, and TSS were obtained, elucidating that observing biospeckle activity during the growing period might be used through other maturity indices for optimum harvest window assessment (Arefi et al., 2016; Romero et al., 2009). Tomato quality is concerned with several factors, such as firmness, appropriate degree of ripeness, free of defects, and so on. The results on tomatoes revealed a typical decline in firmness, which arose during shelf life. Maturation contrasted to chlorophyll as well as carotenoids content, firmness was forecasted with maximum accuracy using the biospeckle methodology and standard optical measurements (Gautier et al., 2008).

4.4 Detection of Mechanical Defects

Bruising was major mechanical damage caused in many of the horticultural specimens, particularly fruits. These were not easily seen with the naked eye during the very early stages. In order to control, detection is necessary for better shelf life and consumer acceptance. Biospeckle activity is a nondestructive mechanism to evaluate such conditions (Pajuelo et al., 2003). It can distinguish the intact and damaged areas of fruits.

Depicted that with the use of algorithms like the moment of inertia of the co-occurrence matrix (COM) and autocorrelation function internal structural variations caused by bruising were able to be detected and provide reliable outcomes with significant differences by two different methods, Weighted Generalized Differences (WGD) and Laser speckle contrast analysis (LASCA). These application methods can be employed in various fruits, like apples and other fruits for assessing damages caused by bruising during various processing, packaging, handling, and storage phases for developing bruising resistance (Pajuelo et al., 2003; Romo & Yoxall, 2005).

Assessment of fruit quality is the prime focus of various researchers on important crops like apple, guava, oranges, and pears. Research has been done on Indian fruits, specifically apple, pear, and guava by assessing through the method of algorithm cross-correlation coefficient difference using biospeckle recording time equal to 15 s with the frame rate equals to 20 fps. Quality evaluation parameters like storage time and shelf life were a basic part of this study and the values of the algorithm depicted that coefficient value variation reduces along with an increase in shelf life and storage time, which was the same as the results of the puncture test. In the same comparative study, the value of pears has a high cross-correlation coefficient and the average value is lower in guava than in the other two crops (Ansari & Nirala, 2012; Ansari & Nirala, 2013a).

The assessment of quality attributes studied in seasonal fruits, like apples, mangoes, guavas, oranges, and some vegetable crops by using the biospeckle and speckle grams was recorded under various conditions of artificial using heat and mechanical effects to change various quality parameters like bruising, changes in texture at various ripening stages to observe the activity using the algorithms IM and cross-correlation. The dynamic speckle activity decreases gradually by showing a correlation of 50 and 75% before and after the damage which was an application of biospeckle technology for stakeholders and industries for increasing the shelf life and quality protection. They also resulted that live correlation plotting is most precise for more time minimizing local disturbances (Samuel et al., 2017).

4.5 *Identification of Disease Occurrence*

There has been a substantial improvement in the deployment of biospeckle as a means for analyzing various disease developments. Research elucidated that the prevalence of bruising in apple crops may lead to 50% loss, with an average span of 10–25%. Many fruits like apples are subjected to various mechanical damages like bruising, puncturing, fruit cracking, and various other disorders by pathogens or nutritional disorders during a wide range of package operations before reaching the consumer from harvesting, transport, cleaning, sorting, and grading, packaging, and cooling, availability in the market, etc. Out of all these, major studies depicted that bruising is the most important and major form of mechanical damage causing low yields by making the tissue soften and thereby increasing the rotting, which leads to a substantial decline in profits (Studman et al., 1997). So, in this regard, various researchers identified mechanisms for early detection using biospeckle laser technology. A work on apple crop revealed that diseased symptoms can be identified as early stage. They proved it by artificial inoculation of fungus called *Peziculamalicorticus* which causes the "bull's eye rot." The researchers have created an approach using the biospeckle activity, chlorophyll fluorescence, and assessment of the hyperspectral images that had taken 2 days for the significant identification of the symptoms, although the regular visual assessment had taken 4–5 days after inoculation. The spatial images of the biospeckle activity remained excellent in describing the advancement of disease contrasted to the hyperspectral imaging systems. Correspondingly, it has assisted to recognize the development of disease appropriately ahead contrasted to the chlorophyll fluorescence method (Pieczywek et al., 2018).

Biospeckle technology application in early detection of apple bruising has also been reported where an experiment made by them on 50 artificially defected apples was analyzed using the algorithms Fujii, GD, and LSTCA and intensity of the biospeckle phenomenon was calculated. However, three algorithms were depicted as higher the damage caused, lesser the biospeckle activity, and vice versa. Consequently, early bruising can be uncovered by biospeckle phenomenon, which is swift and harmless (Yan et al., 2017).

5 Limitations and Future Directions

Despite the fact that there is an extensive range of adoptability of this biospeckle laser method as a nondestructive means in quality and safety examination, there exist a few challenges and limitations that deter its wider adoption. The lack of efficient calibration or sorting models is an imperative hindrance for agro-industry due to the complex interaction between light and a biological substance, where elastic and nonelastic scattering happens, as well as influenced by "biochemical absorption." Moreover, biospeckles have a physical particle drive that is hard to enumerate for the development of calibration schemes. The sensitivity of the technique for the particle motion accomplishes this technique chiefly amenable for observing processes associated with movement of water within the vessels, that happens comparatively quicker than courses like polysaccharide/degradation of pigments and hence it is not influenced by these biochemical deviations. Another issue is its sensitive responsiveness to peripheral vibrations resulting in a large de-correlation in biospeckle activity. Need of standardized protocols for directing the analysis should be framed. Other attributes that make limited applications are varied surface characteristics of agro-produce and the penetration depth of laser on biological samples. Also, to find out the quality-related measures that are accountable for speckle activity, there is a need to work on aspects like frequency examination of biospeckle action to de-convolute signals for definite biological processes (Ansari & Nirala, 2013a; Pandiselvam et al., 2020; Zdunek et al., 2014).

6 Conclusion

In fact, the biospeckle laser technique is a nondestructive quality evaluation tool that encompasses a simple and flexible experimental system. The chief adoption comprises of observing the maturation and aging progression in fresh harvests and the recognition of disease and blemishes. The basis of the biospeckle action in the biological material are biochemical changes and intracellular progressions like "organelle movement, cytoplasmic streaming, cell growth and division, Brownian motion" and so on. Research exhibited that biospeckle activity associates with physiological vagaries during growth, development, ripening, defects, and diseases. In this context, one can affirm that this process has witnessed widespread claims as a nondestructive process in safety and quality assessment of the food stuffs or as an indicator of the stage of maturity. The various methods utilized for analyzing and interpreting biospeckles quantitatively to obtain meaningful results are the "absolute value difference (AVD), cross-correlation and inertia moment (IM) means." On the other hand, methods like "Fujii, temporal difference (TD), laser speckle contrast analysis (LASCA), laser speckle temporal contrast analysis (LASTCA) and generalized differences (GD)" are normally used as qualitative approaches. Furthermore, its application in other areas of food processing for ensuring quality and safety, such as evaluation of pesticidal residues and microbial colonization on the fruit surface, measuring thickness of edible coating on fruit surfaces, accessing the physicochemical properties of processed products, browning effect on fruits, especially on minimally processed ones, moisture studies on dried products to predict its storability and so on. Therefore, this nondestructive quality evaluation tool is quite well applied on fresh horticultural products. But still, this method is under development in the field of quality evaluation to make it utilized on commercial scale. Yet, industry scale adoption necessitates additional crop-specific scientific studies to standardize the calibration/classifications models particular to that produce.

References

Abasi, S., Minaei, S., Jamshidi, B., & Fathi, D. (2018). Dedicated non-destructive devices for food quality measurement: A review. *Trends in Food Science & Technology, 78*, 197–205.

Aboonajmi, M., & Faridi, H. (2016, February). Nondestructive quality assessment of Agro-food products. In *Proceedings of the 3rd Iranian International NDT Conference*, IRNDT 2016-A13105: 1–9. https://www.ndt.net/search/docs.php3?id=19134

Abou Nader, C., Tualle, J. M., Tinet, E., & Ettori, D. (2019). A new insight into biospeckle activity in apple tissues. *Sensors, 19*(3), 497.

Adamiak, A., Zdunek, A., Kurenda, A., & Rutkowski, K. (2012). Application of the biospeckle method for monitoring Bull's eye rot development and quality changes of apples subjected to various storage methods—Preliminary studies. *Sensors, 12*(3), 3215–3227.

Ansari, M. Z., Minz, P. D., & Nirala, A. K. (2012, March). Fruit quality evaluation using biospeckle techniques. In *2012 1st International Conference on Recent Advances in Information Technology (RAIT)* (pp. 873–876). IEEE.

Ansari, M. Z., Mujeeb, A., & Nirala, A. K. (2018). Assessment of biological leaf tissue using biospeckle laser imaging technique. *Laser Physics, 28*(6), 065–608.

Ansari, M. Z., & Nirala, A. K. (2012). Biospeckle techniques in quality evaluation of Indian fruits. *World Academy of Science, Engineering and Technology, 6*, 11–20.

Ansari, M. Z., & Nirala, A. K. (2013a). Biospeckle activity measurement of Indian fruits using the methods of cross-correlation and inertia moments. *Optik, 124*(15), 2180–2186.

Ansari, M. Z., & Nirala, A. K. (2013b). Assessment of bio-activity using the methods of inertia moment and absolute value of the differences. *Optik, 124*(6), 512–516.

Ansari, M. Z., & Nirala, A. K. (2016). Assessment of biospeckle activity of lemon fruit. *Agricultural Engineering International: CIGR Journal, 18*(2), 190–200.

Ansari, M. Z., Ramírez-Miquet, E. E., Otero, I., Rodríguez, D., & Darias, J. G. (2016). Real time and online dynamic speckle assessment of growing bacteria using the method of motion history image. *Journal of Biomedical Optics, 21*(6), 066006.

Arefi, A., Moghaddam, P. A., Hassanpour, A., Mollazade, K., & Motlagh, A. M. (2016). Non-destructive identification of mealy apples using biospeckle imaging. *Postharvest Biology and Technology, 112*, 266–276.

Arizaga, R. A., Cap, N. L., Rabal, H. J., & Trivi, M. (2002). Display of local activity using dynamical speckle patterns. *Optical Engineering, 41*, 287–294.

Arizaga, R., Trivi, M., & Rabal, H. (1999). Speckle time evolution characterization by the co-occurrence matrix analysis. *Optics & Laser Technology, 31*(2), 163–169.

Bergkvist, A. (1997). Biospeckle-based study of the line profile of light scattered in strawberries. *Lund Reports in Atomic Physics, LRAP-220*, 1–62.

Bhargava, A., & Bansal, A. (2018). Fruits and vegetables quality evaluation using computer vision: A review. *Journal of King Saud University - Computer and Information Sciences, 33*(3), 243–257.

Bobelyn, E., Serban, A. S., Nicu, M., Lammertyn, J., Nicolai, B. M., & Saeys, W. (2010). Postharvest quality of apple predicted by NIR-spectroscopy: Study of the effect of biological variability on spectra and model performance. *Postharvest Biology and Technology, 55*(3), 133–143.

Braga, R. A., Jr., Horgan, G. W., Enes, A. M., Miron, D., Rabelo, G. F., & Barreto Filho, J. B. (2007). Biological feature isolation by wavelets in biospeckle laser images. *Computers and Electronics in Agriculture, 58*(2), 123–132.

Braga Júnior, R. A. (2017). When noise became information: State-of-the-art in biospeckle laser. *Ciência e Agrotecnologia, 41*(4), 359–366.

Braga, R. A., Nobre, C. M. B., Costa, A. G., Sáfadi, T., & Da Costa, F. M. (2011). Evaluation of activity through dynamic laser speckle using the absolute value of the differences. *Optics Communications, 284*(2), 646–650.

Briers, J. D. (1975a). A note on the statistics of laser speckle patterns added to coherent and incoherent uniform background fields, and a possible application for the case of incoherent addition. *Optical and Quantum Electronics, 7*(5), 422–424.

Briers, J. D. (1975b). Wavelength dependence of intensity fluctuations in laser speckle patterns from biological specimens. *Optics Communications, 13*(3), 324–326.

Briers, J. D. (1978). The statistics of fluctuating speckle patterns produced by a mixture of moving and stationary scatterers. *Optical and Quantum Electronics, 10*(4), 364–366.

Briers, J. D. (2007). Laser speckle contrast imaging for measuring blood flow. *Optica Applicata, 37*, 139–152.

Cardoso, R. R., Costa, A. G., Nobre, C. M. B., & Braga, R. A., Jr. (2011). Frequency signature of water activity by biospeckle laser. *Optics Communications, 284*(8), 2131–2136.

Costa, A. G., de Carvalho Pinto, F. D. A., Junior, R. B. A., Motoike, S. E. Y., & Gracia, L. I. M. N. (2017). Determination of macaw fruit harvest period by biospeckle laser technique. *African Journal of Agricultural Research, 12*(9), 674–683.

Costa, A. G., Pinto, F. A., Braga, R. A., Motoike, S. Y., & Gracia, L. (2017). Relationship between biospeckle laser technique and firmness of Acrocomia aculeata fruits. *Revista Brasileira de Engenharia Agrícola e Ambiental, 21*(1), 68–73.

da Silva, E. R., da Silva Junior, E., Júnior, M. F., da Silva Lannes, S. C., & Muramatsu, M. (2008, April). Application of biospeckle phenomenon on monitoring of leavening process in breadmaking. *AIP Conference Proceedings, 992*(1), 309–313.

Du, Z., Zeng, X., Li, X., Ding, X., Cao, J., & Jiang, W. (2020). Recent advances in imaging techniques for bruise detection in fruits and vegetables. *Trends in Food Science & Technology, 99*, 133–141.

Fujii, H., Asakura, T., Nohira, K., Shintomi, Y., & Ohura, T. (1985). Blood flow observed by time-varying laser speckle. *Optics Letters, 10*(3), 104–106.

Gautier, H., Diakou-Verdin, V., Bénard, C., Reich, M., Buret, M., Bourgaud, F., et al. (2008). How does tomato quality (sugar, acid, and nutritional quality) vary with ripening stage, temperature, and irradiance? *Journal of Agricultural and Food Chemistry, 56*(4), 1241–1250.

Godinho, R. P., Silva, M. M., Nozela, J. R., & Braga, R. A. (2012). Online biospeckle assessment without loss of definition and resolution by motion history image. *Optics and Lasers in Engineering, 50*(3), 366–372.

González-Peña, R. J., Pérez-Montoyo, H., Braga, R. A., & Viana, D. C. (2016). Assessment of biological activity in RAW 264.7 cell line stimulated with lipopolysaccharide using dynamic laser speckle. *Applied Physics B, 122*(11), 1–8.

He, Y., Xiao, Q., Bai, X., Zhou, L., Liu, F., & Zhang, C. (2021). Recent progress of nondestructive techniques for fruits damage inspection: A review. *Critical Reviews in Food Science and Nutrition, 62*, 5476–5494.

Kumari, S., & Nirala, A. K. (2016). Biospeckle technique for the non-destructive differentiation of bruised and fresh regions of an Indian apple using intensity-based algorithms. *Laser Physics, 26*(11), 115601.

Kurenda, A., Adamiak, A., & Zdunek, A. (2012). Temperature effect on apple biospeckle activity evaluated with different indices. *Postharvest Biology and Technology, 67*, 118–123.

Lee, S., Lohumi, S., Lim, H. S., Gotoh, T., Cho, B. K., Kim, M. S., & Lee, S. H. (2015). Development of a detection method for adulterated onion powder using Raman spectroscopy. *Journal of the Faculty of Agriculture, Kyushu University, 60*(1), 151–156.

Li, C., Luo, J., & MacLean, D. (2011). A novel instrument to delineate varietal and harvest effects on blueberry fruit texture during storage. *Journal of the Science of Food and Agriculture, 91*(9), 1653–1658.

Lu, Y., Saeys, W., Kim, M., Peng, Y., & Lu, R. (2020). Hyperspectral imaging technology for quality and safety evaluation of horticultural products: A review and celebration of the past 20-year progress. *Postharvest Biology and Technology, 170*, 111–318.

Mesa, K., Serra, S., Masia, A., Gagliardi, F., Bucci, D., & Musacchi, S. (2016). Seasonal trends of starch and soluble carbohydrates in fruits and leaves of 'Abbé Fétel'pear trees and their relationship to fruit quality parameters. *Scientia Horticulturae, 211*, 60–69.

Minz, P. D., & Nirala, A. K. (2016). Laser speckle technique to study the effect of chemical pre-treatment on the quality of minimally processed apples. *Laser Physics, 26*(4), 045602.

Mulone, C., Budini, N., Vincitorio, F. M., Freyre, C., Díaz, A. L., & Rego, A. R. (2013, November). Analysis of strawberry ripening by dynamic speckle measurements. In *8th Iberoamerican Optics Meeting and 11th Latin American Meeting on Optics, Lasers, and Applications* (Vol. 8785, p. 87851X). International Society for Optics and Photonics.

Nassif, R., Pellen, F., Magne, C., Le Jeune, B., Le Brun, G., & Abboud, M. (2012, September). Laser speckle dynamic for monitoring fruits maturation. In *Speckle 2012: V International Conference on Speckle Metrology* (Vol. 8413, p. 84131G). International Society for Optics and Photonics.

Omidi-Arjenaki, O., Ghanbarian, D., Naderi-Boldaji, M., & Mollazadeh, K. (2019). Use of biospeckle imaging to measure mechanical properties of apples in the non-invasive manner. *Journal of Agricultural Machinery, 9*(1), 73–86.

Oulamara, A., Tribillon, G., & Duvernoy, J. (1989). Biological activity measurement on botanical specimen surfaces using a temporal decorrelation effect of laser speckle. *Journal of Modern Optics, 36*(2), 165–179.

Pajuelo, M., Baldwin, G., Rabal, H., Cap, N., Arizaga, R., & Trivi, M. (2003). Bio-speckle assessment of bruising in fruits. *Optics and Lasers in Engineering, 40*(1–2), 13–24.

Pandiselvam, R., Mayookha, V. P., Kothakota, A., Ramesh, S. V., Thirumdas, R., & Juvvi, P. (2020). Biospeckle laser technique–A novel non-destructive approach for food quality and safety detection. *Trends in Food Science & Technology, 97*, 1–13.

Passoni, I., Dai Pra, A., Rabal, H., Trivi, M., & Arizaga, R. (2005). Dynamic speckle processing using wavelets-based entropy. *Optics Communications, 246*(1–3), 219–228.

Pieczywek, P. M., Nowacka, M., Dadan, M., Wiktor, A., Rybak, K., Witrowa-Rajchert, D., & Zdunek, A. (2018). Postharvest monitoring of tomato ripening using the dynamic laser speckle. *Sensors, 18*(4), 1093.

Pra, A. L. D., Passoni, I. L., SFendra, G. H., Trivi, M., & Rabal, H. (2014). Biospeckle signal descriptors: A performance comparison. *arXiv preprint arXiv: 1411.7522.*

Qing, Z., Ji, B., & Zude, M. (2007). Predicting soluble solid content and firmness in apple fruit by means of laser light backscattering image analysis. *Journal of Food Engineering, 82*(1), 58–67.

Rabal, H. J., & Braga, R. A., Jr. (2018). *Dynamic laser speckle and applications*. CRC press.

Rabelo, G. F., Braga Júnior, R. A., & Fabbro, I. (2005). Laser speckle techniques in quality evaluation of orange fruits. *Revista Brasileira de Engenharia Agrícola e Ambiental, 9*(4), 570–575.

Rabelo, G. F., Enes, A. M., Junior, R. A. B., & Dal Fabbro, I. M. (2011). Frequency response of biospeckle laser images of bean seeds contaminated by fungi. *Biosystems Engineering, 110*(3), 297–301.

Rahmanian, A., Mireei, S. A., Sadri, S., Gholami, M., & Nazeri, M. (2020). Application of biospeckle laser imaging for early detection of chilling and freezing disorders in orange. *Postharvest Biology and Technology, 162*, 111118.

Ramos, B., Miller, F. A., Brandão, T. R. S., Teixeira, P., & Silva, C. L. M. (2013). Fresh fruits and vegetables—An overview on applied methodologies to improve its quality and safety. *Innovative Food Science & Emerging Technologies, 20*, 1–15.

Retheesh, R., Ansari, M. Z., Radhakrishnan, P., & Mujeeb, A. (2018). Application of qualitative biospeckle methods for the identification of scar region in a green orange. *Modern Physics Letters B, 32*(09), 1850113.

Retheesh, R., Samuel, B., Radhakrishnan, P., Nampoori, V. P. N., & Mujeeb, A. (2016). Use of laser biospeckle for the evaluation of fruit ripening. *Journal of Pure Applied and Industrial Physics, 6*(5), 65–70.

Rizzolo, A., Vanoli, M., Spinelli, L., & Torricelli, A. (2010). Sensory characteristics, quality and optical properties measured by time-resolved reflectance spectroscopy in stored apples. *Postharvest Biology and Technology, 58*(1), 1–12.

Romero, G. G., Martinez, C. C., Alanis, E. E., Salazar, G. A., Broglia, V. G., & Alvarez, L. (2009). Bio-speckle activity applied to the assessment of tomato fruit ripening. *Biosystems Engineering, 103*(1), 116–119.

Romo, E. R., & Yoxall, A. (2005). *An experimental, analytical and numerical analysis of apple bruising* [PhD thesis].

Samuel, B., Retheesh, R., Ansari, M. Z., Nampoori, V. P. N., Radhakrishnan, P., & Mujeeb, A. (2017). Cross-correlation and time history analysis of laser dynamic specklegram imaging for quality evaluation and assessment of certain seasonal fruits and vegetables. *Laser Physics, 27*(10), 105–601.

Skic, A., Szymańska-Chargot, M., Kruk, B., Chylińska, M., Pieczywek, P. M., Kurenda, A., & Rutkowski, K. P. (2016). Determination of the optimum harvest window for apples using the non-destructive biospeckle method. *Sensors, 16*(5), 661.

Studman, C. J., Brown, G. K., Timm, E. J., Schulte, N. L., & Vreede, M. J. (1997). Bruising on blush and non-blush sides in apple-to-apple impacts. *Transactions of ASAE, 40*(6), 1655–1663.

Wu, D., Meng, L., Yang, L., Wang, J., Fu, X., Du, X., et al. (2019). Feasibility of laser-induced breakdown spectroscopy and hyperspectral imaging for rapid detection of thiophanate-methyl residue on mulberry fruit. *International Journal of Molecular Sciences, 20*(8), 2017.

Wulf, J. S., Rühmann, S., Rego, I., Puhl, I., Treutter, D., & Zude, M. (2008). Nondestructive application of laser-induced fluorescence spectroscopy for quantitative analyses of phenolic compounds in strawberry fruits (Fragaria x ananassa). *Journal of Agricultural and Food Chemistry, 56*(9), 2875–2882.

Yan, L., Liu, J., & Men, S. (2017). The biospeckle method for early damage detection of fruits. *Modern Physics Letters B, 31*(19–21), 1740034.

Zdunek, A., Adamiak, A., Pieczywek, P. M., & Kurenda, A. (2014). The biospeckle method for the investigation of agricultural crops: A review. *Optics and Lasers in Engineering, 52*, 276–285.

Zdunek, A., & Cybulska, J. (2011). Relation of biospeckle activity with quality attributes of apples. *Sensors, 11*(6), 6317–6327.

Zdunek, A., & Herppich, W. B. (2012). Relation of biospeckle activity with chlorophyll content in apples. *Postharvest Biology and Technology, 64*(1), 58–63.

Zhu, Q., He, C., Lu, R., Mendoza, F., & Cen, H. (2015). Ripeness evaluation of 'sun bright' tomato using optical absorption and scattering properties. *Postharvest Biology and Technology, 103*, 27–34.

Application of Spectroscopy for Assessing Quality and Safety of Fresh Horticultural Produce

Khayelihle Ncama and Lembe Samukelo Magwaza

Abstract Recently, quality and safety of fruits and vegetables are a great concern to consumers. It is of great importance for the horticultural industry to produce high-quality and safe products for consumption. The main quality attributes required by the consumers are visual appearance, such as color, glossiness, surface texture, size, and absence of blemishes. There is also an increasing awareness on the chemical residues that threaten the health of consumers. Fresh produce is susceptible to a wide range of defects before harvest, at harvest, and during postharvest operations. Traditional detection of defects relies on manual and visual inspections. The majority of traditional quality assessment methods are destructive and offline in nature. Over the past few decades, research has focused on the development of diverse noncontact, rapid, eco-friendly, and accurate methods for examination of fruits and vegetables. The application of nondestructive spectroscopic techniques has gained popularity. Application of spectroscopy has been extended to the safety assessment and monitoring quality. This chapter entails the success of spectroscopic applications in fresh fruits and vegetables. It further discusses the measurement or operation principles, major components, steps of measurement, data analysis, important factors that need to be considered for good results, and applications of mid- and near infrared spectroscopy in fresh horticultural produce.

Keywords Nondestructive technology · Fruit quality · Vegetable quality · Vis-NIRS

K. Ncama
Department of Crop Science, North-West University, Mmabatho, South Africa

Food Security and Safety, Faculty of Natural and Agricultural Sciences, North-West University, Mmabatho, South Africa

L. S. Magwaza (✉)
Discipline of Crop and Horticultural Science, University of KwaZulu-Natal, Scottsville, South Africa
e-mail: magwazal@ukzn.ac.za

P. B. Pathare, M. S. Rahman (eds.), *Nondestructive Quality Assessment Techniques for Fresh Fruits and Vegetables*, https://doi.org/10.1007/978-981-19-5422-1_5

1 Introduction

Accurate assessment of the parameters associated with maturity, ripening, and quality of horticultural produce enhances the efficiency of managing safety, quality, and overall shelf life (Kader, 1997; Mditshwa et al., 2017). Physical and chemical characteristics of horticultural produce can be associated with the optimal harvest time, which determines whether the produce is sorted for fresh consumption or processed into a value-added product. These characteristics are also used to monitor changes during postharvest storage, which eventually determine the market value of the fresh produce (Kader, 2008).

Various techniques, either destructive or nondestructive, are normally used to determine quality parameters of horticultural produce (Mahajan et al., 2017; Su et al., 2017). Generally, nondestructive methods have a higher preference due to improved accuracy, rapidity, eco-friendliness, low costs, and ability to analyse every fruit in a given batch (Lin & Ying, 2009; Wedding et al., 2011). One of the topical nondestructive techniques is the application of visible to near infrared spectroscopy (Vis-NIRS) together with chemometric software for the assessment of internal parameters, such as soluble solutes and pH of intact fruits and vegetables (Magwaza et al., 2012). The application of Vis-NIRS has been demonstrated to detect the trueness and track the originality of various horticultural produce including grapes, wines, and other products (Aouadi et al., 2020). The Vis-NIRS has also been demonstrated as a technology for detecting chemical residues from normal treatments, such as pesticides that pose threat to consumer's health. Nazarloo et al. (2021) demonstrated the feasibility of using Vis-NIRS and multivariate analysis for detecting pesticide residue in tomatoes. The necessity of Vis-NIRS in the field of horticultural produce is enhanced by the increasing challenge for the industry to produce high-quality and safe products for consumption. Vis-NIRS is arguably the most suitable technology for noninvasive detection of produce parameters associated with consumer's preference including visual appearance, such as color, glossiness, surface texture, size, and absence of blemishes.

2 Basic Structures of Common Spectrometers

There are various designs of spectrometers. Their designs differ because they are intended to be used for various purposes and some spectrometers have to be portable for appropriateness and ease of handling during travel to various sites. The basic structures of a spectrometer include a probe, which is a part that bears a light source and a detector that records the radiation that is absorbed or reflected from the sample. The reflected radiation is sent via a network of wires to a computer or computer box that stores all the absorptions or reflections at different wavelengths of a selected spectrum. Spectrometers also contain a ventilation and a fan that cools its interior to avoid overheating. The fan is powered by an internal chargeable battery on portable

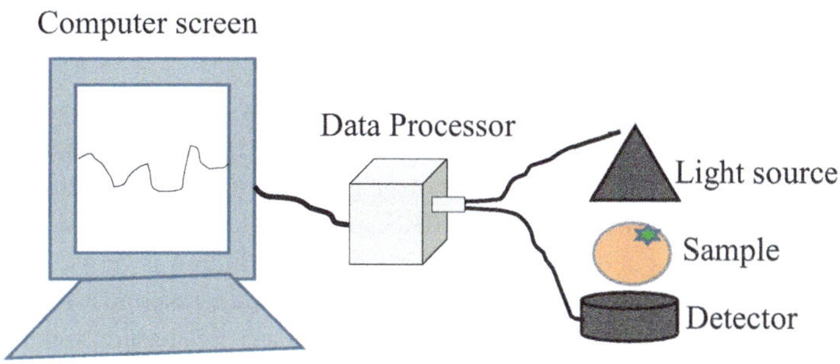

Fig. 1 The illustration of basic functional structures of common stationary spectrometers

spectrometers, while electric power is necessary on static spectrometers. Some spectrometers have an LED screen that shows graphs of absorptions against different wavelengths of a spectrum. Other spectrometers do not show graphs until the acquired spectral data is transferred to a data processor, and is processed into spectra. The spectral data is transferred to a computer that is programmed to read it and plot an appropriate data. All spectrometers have functional buttons that are used by a user for monitoring its functions. These are basic buttons to scan, store, or delete data collected from a sample. Figure 1 depicts an ordinary static spectrometer with its parts.

2.1 Relevance of Functional Components of Spectrometers

Spectrometers have different designs, tolerances, light source, light detector, and can function at various spectral ranges. These factors can result in a loss of correlation between spectra and wet chemistry data if they are not considered properly. Generally, the light-emitting and absorbing sides of a probe should be completely covered by a sample to eliminate interference from the surrounding environment, such as external radiation and irradiation. Some spectrometers change their functioning accuracy with a change in temperature and relative humidity of the surrounding environment (Cozzolino et al., 2011). This is associated with the environmental factors that affect the speed of internal parts, such as fans and sensors. It is also necessary that a spectrometer user matches the probable wavelengths where the analyzed biochemical compound can be absorbed. This alignment of spectral range to certain parameters is executed by adjusting the lens to sense the required section of a spectrum and ignore unnecessary wavelengths. Pigments such as carotenoids, chlorophylls, and anthocyanins are absorbed in the visible spectral range because of their bright-colored nature (Lichtenthaler & Buschmann, 2001). Water, carbohydrates, fats, and proteins are mostly absorbed in the NIR region because of their colorless nature (Williams & Norris, 1987). These phytochemical parameters must be measured where their absorption is found without allowing interference from external radiation.

2.2 Operation

The assessment of quality parameters by Vis-NIRS is performed by exposing a sample to a known radiation amount, and measuring the radiation reflection, absorption, or transmission that occurs when it is passing through the sample or gets reflected (Cozzolino et al., 2011). Vis-NIRS spectrometers distinguish the reaction of overtone vibrations occurring in the near infrared radiation regions, where overtones or combinations of fundamental stretching bands respond. The spectrum developed from the diffused radiation changes characteristics based on the chemical composition (absorption) and microstructures (scattering) it encounters while it is transferred through the sample (Rinnan et al., 2009). The detected change of radiation is normally plotted and saved as either reflectance or absorbance (log of reflectance, $1/R$) versus wavelength.

Specular reflection of radiation is normally associated with the shininess or coarse surfaces of a sample which causes indirect scattering and external diffuse reflection, respectively. The direct scattering and external reflection can only be associated with the surface of a sample. Some structural tissues such as cell wall interfaces, and suspended particles such as starch granules, chloroplasts, and mitochondria are the main causes of radiation scattering. This is based on their association with an abrupt change in the refractive index (McGlone et al., 1997; Mehinagic et al., 2004). Rinnan et al. (2009) categorized the types of scattering based on the diameter of particles found in the sample. According to the authors, Rayleigh scattering is caused by the smaller particles with a diameter smaller than the wavelength of electromagnetic radiation (<nm/10), while particles with bigger diameters cause Lormentz-Mie scattering. The types and intensity of spectra scattering in the biological samples are naturally complicated. Samples made up of tiny particles vary in size, number, and patterns in every sample even in a given batch. A sample containing major scattering organelles produces an abnormal spectrum that differs from all samples making a batch, and that sample is identified as an outlier during the calibration stage (McGlone et al., 1997). The size, shape, and homogeneity are among the factors that commonly cause radiation scattering that provides uniqueness to every sample in a batch. Other factors that cause scattering include interferences such as pores, openings, and capillaries randomly distributed within a biological sample (Magwaza et al., 2012).

The common scatterings of absorption spectra found in the near infrared region are overtone or combination bands of fundamental chemical compounds that causes vibrational and rotational transitions of radiation in plant material (Nicolai et al., 2007). Vis-NIR reflectance spectra of "Nules Clementine" mandarin, "Marsh" grapefruit, "Hass" avocado, and "Valencia" oranges are shown in Fig. 2. Noticeably, all the spectra depict very comparable absorption vibrations, that are significant at 490, 1000, 1234, 1470, and 1966 nm bands. The absorption bands that are found at 760, 970, and 1450 nm in spectra of biological samples are normally associated with the O-H bonds (i.e., typical characteristics of water) (Polessello & Giangiacomo, 1981). Water is the main compound that causes radiation reflection in the near

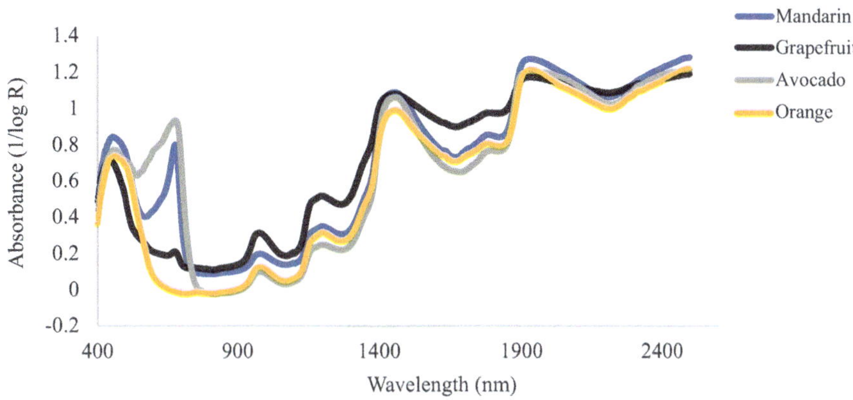

Fig. 2 The typical reflectance spectra of "Nules Clementine" mandarin, "Marsh" grapefruit, "Hass" avocado, and "Valencia" oranges attained using bench-top monochromatic Vis-NIR System (Model XDS spectrometer, FOSS NIR Systems Inc., Maryland, USA)

infrared radiation (NIR) regions of spectra from fresh horticultural produce (Cozzolino et al., 2011). Spectral regions associated with water reflection are normally removed to eliminate confusion and analyze only the necessary portions of the spectra. This practice increases the accuracy of the developed Vis-NIRS models to predict chemical parameters of the product instead of water. However, accuracy of each spectrum can also be reduced by wavelength-dependent scattering caused by environmental conditions (Cozzolino et al., 2011; Magwaza et al., 2014). Wavelength-dependent spectra associated with internal sample scattering effect and sample complexity are only a few factors of complexity that can not be avoided in spectroscopy. There are also unavoidable factors such as instrumental noise, instrument type, and matrix as well as environmental effects, such as ambient temperature and light (Dos Santos et al., 2013). It is ideal that a researcher fully explores the information in the spectra using multivariate analysis and spectral preprocessing techniques for extracting an informative region from raw spectra.

2.3 *Modeling*

The effectiveness of spectroscopy is highly dependent on the accuracy of the collected spectral data. No chemometric software or methods can obtain useful information from spectra that were collected inaccurately (Cozzolino et al., 2011). An appropriate experimental design of obtaining a spectrum from one sample is essential for good representative samples and a population. Samples with the widest variation of structural and physiological parameters need to be used for the collection of data and for developing models. The variability in the calibration data determines the robustness of developed models. The calibration set must have wide extremes in compositions, and samples should be within the limits of developed models (Peirs

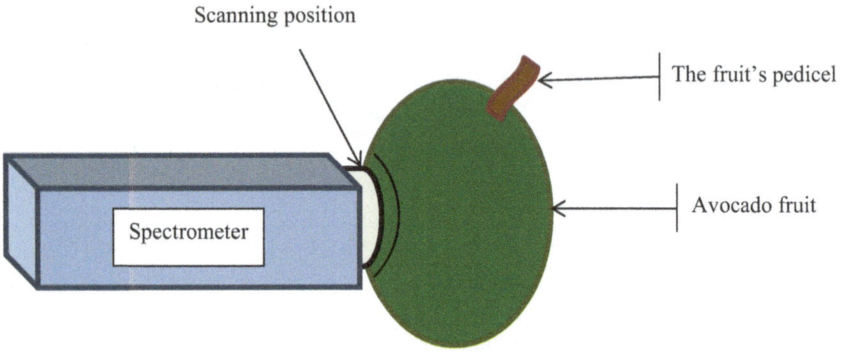

Fig. 3 The necessary illustration of the position where spectra were acquired from an avocado fruit for developing models to assess avocado maturity (Ncama et al., 2018)

et al., 2003; Wedding et al., 2013). During model calibration, data of physical and chemical parameters that is associated with spectra from a sample is normally acquired using standard techniques, such as wet chemistry or any other destructive methods. The assessment of physicochemical parameters normally possess a minor challenge to researchers in horticulture.

Cautiousness in the collection of spectral data from the sample is vital as a researcher needs to select the appropriate position. In most fruits, the ideal sampling position is normally the part that was exposed to sunlight when the fruit was on the tree (Fig. 3). A good practice for a study that does not have a huge sample number normally considers an average of spectra from different positions of a produce. The biochemical concentrations of compounds are known to differ within a sample (Guthrie et al., 2006; Peiris et al., 1999). The variation in biochemical concentration increases with the size of the fruit. Fruits such as papaya, melons, squash, and pineapples consist of higher variations compared to smaller fruit such as oranges and avocadoes (Guthrie & Walsh, 1997). Spectra can be collected at the same spot from all samples during the development of models to assess fruit that is not on conveyor belts. The position must be well indicated so that it can be recommended on the model information to increase the accuracy of models.

3 Important Considerations in Vis-Nirs Application on Fresh Fruit

The application of Vis-NIRS on fruits without big seeds is executed with cautions such as the avoidance of external radiation, keeping a spectrometer steady during scanning, and avoiding external moisture entering the space between the scanning probe and a sample. It is in drupe fruit that the user needs to be extra cautious. Vis-NIRS application on drupe fruit entails awareness about the stone that can interfere with radiation in assessing quality parameters of the mesocarp. This

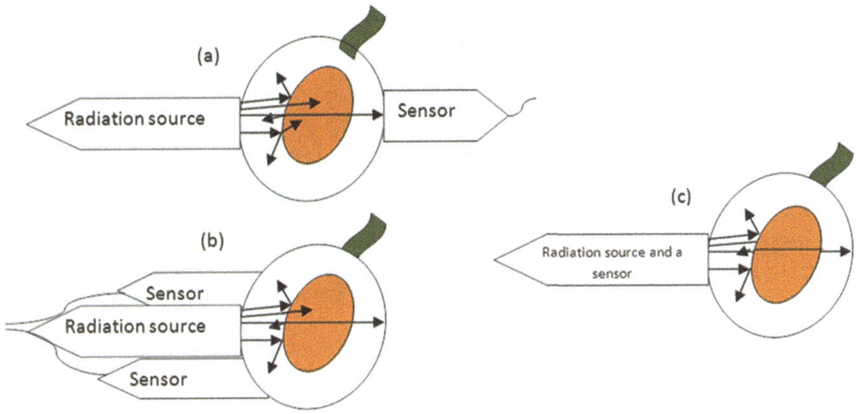

Fig. 4 The assumable radiation pathways that are possible inside stone fruit. (**a**) is the transmittance mode, (**b**) is the illustration of reflectance mode, and (**c**) is the interactance mode, and their ability to obtain the required information from only the mesocarp of the fruit

necessitates the avoidance of using spectrometers that operates in transmittance mode because the majority of radiation is likely to be deflected by the stone inside the fruit. Spectrometers working on reflectance and interactance modes are the most relevant unless the user is interested in assessing the quality or maturity of the stone. Operation in transmittance mode would be appropriate if the objective is to analyze the maturity, size, or hardness of the stone.

Operating in reflectance mode is the most appropriate method when spectral data is collected for assessing parameters on the fruit surface or in the mesocarp. The differences in the sensitivity and receiving the reflected radiation are illustrated in Fig. 4. Figure 4 illustrates a common assumption of radiation paths when a hypothetical spectrometer with transmittance (a), reflectance with a separate sensor (b), or reflectance with a sensing lens (c) is applied to a drupe fruit. Assuming that the number of streams illuminated by the spectrometer (n) is four, one, two, or one of the four arrays reach the sensor if a transmittance, reflectance, or interactance mode is used, respectively.

4 Important Considerations in Vis-NIRS Application on Fresh Vegetables

The application of Vis-NIRS on leafy vegetables is different because of their structural and shape characteristics in comparison with fruit phenology. Leafy vegetables have a relatively thin depth that is easily penetrated by radiation. A typical example of Vis-NIRS application on vegetables is similar to that of the handheld chlorophyll meter (TYS-A plant chlorophyll meter, China) that has a light source and the light detector on different sides of the sample (Fig. 5). The

Fig. 5 The handheld chlorophyll meter with a probe and light detector on different sides of a sample leaf

detector detects the amount of radiation that is not absorbed by the sample leaf and plots the difference between the emitted and the received radiation at specified wavelengths.

Spectroscopy has been successfully applied in assessing the freshness of lettuce leaves after irradiation of cold atmospheric pressure plasma (Abbasi et al., 2014). Mao et al. (2015) demonstrated the potency of Vis-NIRS as a nondestructive tool to measure total nitrogen in lettuce by applying spectroscopy and various chemometric software. Vis-NIRS was used for predicting the quality and safety attributes of spinach including ascorbic acid, nitrate, and soluble solutes concentration (Perez-Marin et al., 2019). Ye et al. (2008) showed the potential of Vis-NIRS in estimating adenosine-triphosphate (ATP) content per protoplast as an indicator of the freshness of spinach. The application of Vis-NIRS on vegetables does not only involve quality and safety but documentation as well. Wu et al. (2008) demonstrated the ability of Vis-NIRS in identifying the variety of Chinese cabbage seeds.

5 Chemometric Analysis of Collected Spectroscopic Data

A chemometric software is a software that is capable of extracting explainable information from multivariate data matrices such as the spectrum. The type and model of the software play a significant role in determining the level of accuracy and ability to simplify the reported information. The development of prediction models is the main purpose of chemometric software in the Vis-NIRS research field. The difference and grading of each software are based on the ease and accuracy in preprocessing the collected data, calibration, validation, and recalibration of prediction models.

5.1 Data Preprocessing

Preprocessing of spectral data is the first step in improving the accuracy of associating spectral and wet chemistry data during chemometric analysis. This process improves the accuracy of detecting the uniqueness of each spectrum or spectral region. A variety of preprocessing techniques are available for application to raw spectra for removing sections of spectra that have unexplainable scatterings and extracting analytical information from the entire spectra. Splitting the full visible and near infrared spectrum into smaller ranges and applying different preprocessing techniques to each range has been widely used in the research of Vis-NIRS applications. This wavelength-dependent technique is based on spectral wavelengths that are responsible for the absorption of specific biochemical compounds (Lichtenthaler & Buschmann, 2001; Williams & Norris, 1987). Certain biochemical compounds such as water and carbohydrates do not have specific absorption wavelengths (Polessello & Giangiacomo, 1981). Compounds with these characteristics normally respond to the entire spectrum rather than specific wavelengths. Only mathematical preprocessing methods can be applied to improve the process of obtaining information about the compounds that respond to the entire spectrum. Mathematical preprocessing methods that are commonly used can be categorized into two major groups: scatter correction methods and spectral derivatives (Stevens & Ramirez-Lopez, 2014). Scatter correction methods are applied to spectral data for removal or compressing spectra vibrations resulting from the instrumental error. Derivatives are applied to simplify spectra for improving its correlation with sample data from wet chemistry.

5.2 Calibrations

Modeling is the process of developing a Vis-NIRS model. The modeling process is divided into a calibration stage (developing a model) and validation (testing the model). A substitutive (leave-one-out) testing of all samples as independent validations is normally executed during the calibration stage to test the correlation of the predicted values to the biochemical values obtained from wet chemistry analysis. A batch of external samples with known biochemical data is used as a test set to analyze the model performance on external samples during the validation stage. Most researchers divide samples into 60% of the total number of the calibration, and the remaining 40% is assigned to the validation set. It is well known that increasing the calibration to validation sample ratio improves model accuracy when predicting external samples with unknown chemistry (Cen & He, 2007). Various types of models may be developed based on features of a chemometric tool and the skills of a researcher. However, the Vis-NIRS models commonly used can be categorized into quantitative and clustering models (Roggo et al., 2007).

5.3 Validations

The performance of Vis-NIRS models is typically based on its ability to assess the biochemistry of external samples. Models performance is statistically expressed by the value of the coefficient of determination (R^2; Eq. 1), the root mean square error of calibration (RMSEC; Eq. 2), and the root mean square error of prediction (RMSEP; Eq. 3) (Magwaza et al., 2012). An R^2—value is the correlation measure between values predicted by a Vis-NIRS model and values obtained using conventional destructive methods such as wet chemistry or other instruments. R^2—values range from 0.00 to 1.00 based on the proximity of predicted values to the biochemical values obtained via conventional methods. A good model should have a high R^2—value approaching 1.00. However, reaching 100% prediction accuracy ($R^2 = 1.00$) is practically invalid and indicates inaccurate analysis or a few number or variation among the calibration samples. Comparison of two analytical methods should be approaching 95% of the agreement ($R^2 = \pm 0.95$) to be statistically comparable. An RMSEC is known as the error of calibration during the correlation stage of a model. An RMSEP is the error of the developed model to predict parameter values of the validation set. An acceptable model must have a low RMSEC and RMSEP while having a small difference between the two root mean square errors (Gómez et al., 2006; Magwaza et al., 2014). The small difference results in a low ratio of prediction deviation when the model is used for predicting external samples. The residual predictive deviation (RPD; Eq. 4) is the value that expresses the abnormality of the predicted data from the reference data. The RPD value is a ratio of the standard deviation of reference data to the RMSEP (Magwaza et al., 2013). The RPD value was developed to measure the model robustness based on three model reliability categories: excellent models, with RPD greater than 2.0; fair models, with RPD between 1.4 and 2.0; non-reliable models with RPD less than 1.4 (Williams & Norris, 1987). However, the determination of those thresholds was not statistical and some researchers may use various values depending on their motivation.

Another statistical parameter that is used to measure the accuracy of a model is the biasness (Eq. 5). The average discrepancy between predicted and measured values is known as bias. The biasness expresses the average inequality in the contribution of each value towards the model development during calibration (Ncama et al., 2017). A model with the potential of high precision in the prediction of external samples is characterized by a low value of biasness. There are various other statistical items that a researcher can use to measure the effectiveness of a Vis-NIRS model including slope, intercept, and an average of the neighborhood or global Mahalanobis distances in addition to the mentioned statistical terms when necessary and possible.

$$R^2 = 1 - \sqrt{\frac{\sum (y_{\mathrm{cal}} - y_{\mathrm{act}})^2}{\sum (y_{\mathrm{cal}} - y_{\mathrm{mean}})^2}} \tag{1}$$

$$\mathrm{RMSEC} = \sqrt{\sum (y_{\mathrm{cal}} - y_{\mathrm{act}})^2 / n} \tag{2}$$

$$\mathrm{RMSEP} = \sqrt{\sum (y_{\mathrm{pred}} - y_{\mathrm{act}})^2 / n} \tag{3}$$

$$\mathrm{RPD} = \frac{\mathrm{SD}}{\mathrm{RMSEP}} \tag{4}$$

$$\mathrm{Bias} = \frac{1}{n} \sqrt{\sum (y_{\mathrm{pred}} - y_{\mathrm{act}})^2} \tag{5}$$

Classification or discrimination of horticultural produce forms the basis of Vis-NIRS relevance in quality management practices. The classification process is based on the pattern recognition technique which divides a batch of samples into groups of similar spectral characteristics (Roggo et al., 2007). The Vis-NIRS models developed for classification or discrimination operates by representing the Vis-NIRS spectral data by a few clusters. Each spectrum is projected as a dot in a scatter plot based on its characteristics. The credibility of the discrimination model is based on its ability to maximize simplification although it loses fine details compared to a quantitative model. It assigns spectral data of a sample with similar characteristics to clusters. The discrimination models can be grouped into unsupervised or supervised models. Unsupervised models can classify samples using spectral data without the analysis of biochemical data. Supervised models classify samples based on pre-stated categories of known biochemistry. Both categories are also categorized into linear and nonlinear models (Roggo et al., 2007). Common classification models include PCA, PLS-DA, FDA, SVM, and others. Notably, PCA and PLS-DA are the most common classification models in Vis-NIRS applications on fruits and vegetables.

5.4 *Models Recalibrations*

The robustness of a model is based on its ability to accurately classify or predict parameters of external samples that were not used for calibration and under different conditions. The model robustness usually diminishes if calibration models from an instrument are transferred to another instrument that functions differently. A model of an instrument can lose its robustness because of factors such as temperature fluctuations, electronic drift, and changes in wavelength or detector over time (Nicolai et al., 2007). Batches of a sample obtained from different seasons or locations can also result in poor performance of a model with poor robustness. Fruits from different trees, locations, seasons, or cultivars differ in their physicochemical attributes and that may be out of the range that was used during the calibration process. The difference is likely to result in poor performance of a Vis-NIRS model developed from similar samples on another batch with distinct origin. It is a common practice for researchers to test the robustness of calibration models on independent populations (Golic & Walsh, 2006). In some studies, full cross-validation is considered to be adequate to demonstrate that Vis-NIRS can be used to nondestructively evaluate a quality parameter of interest. The actual accuracy of the model must be assessed with an independent validation dataset for ensuring a model performance at wider ranges (Cozzolino et al., 2011; Guthrie et al., 2006).

Various studies have shown interest in the ability of Vis-NIRS models to maintain their robustness. McGlone et al. (1997) conducted a study to test model robustness on kiwifruit of different sizes, origins, and maturity stages. The authors obtained higher validation errors when fruit were all considered as a single batch. They only obtained accuracy when they separated the samples into batches of similar origins and characteristics. These results have been obtained from other fruits such as apples and melons (Guthrie et al., 2006; Peirs et al., 2003). Increasing the range or the number of samples during calibration is known to increase a model's ability for predicting external samples.

Validation of models using samples from various orchards and seasons depict their robustness. However, the performance of prediction models can be improved by the addition of samples from a season or an orchard it will be applied to. This process is called spiking (Magwaza et al., 2014). The addition of samples from similar conditions reduces the error that may occur because of unknown biochemical changes as a result of alternate bearing, environmental changes, or other factors during the produce development. The spiking techniques was ideally demonstrated by Magwaza et al. (2014) in the assessment of fruit from various orchard and it improved the model's performance. The authors concluded that while the ultimate goal of any prediction technique is to be universal, it is important to stress that there is variation between growing locations and seasons, and it is advisable that models are recalibrated to include the information of a new population.

6 Future of Spectroscopy Applications and Research in Fresh Produce

The success in the research and application of spectroscopy on fruit and vegetables can be widely found in the literature. Research has demonstrated the efficacy of Vis-NIRS research field from all angles. The development of models for predicting the quality of horticultural produce has been demonstrated on various fruit but fewer vegetable species. The effect of few vegetable studies can be associated with the fact that leafy vegetables possess similar characteristics such that if an effective model is developed for lettuce, it can be developed similarly for other leafy vegetables. Application of Vis-NIRS in assessing quality and safety of horticultural produce has been proven throughout the horticultural produce management from orchard management to the assessment of residual pathogens in postharvest storage.

Wang et al. (2021) demonstrated the use of Vis-NIRS as a tool for remote sensing the leaf nitrogen content in pear trees. The authors used a FieldSpec 3 portable field spectrometer (Analytical Spectral Devices, Boulder, CO, USA) that had a probe with an internal light source and a clip to hold the leaf against a black background. The spectroradiometer had a 2 nm sampling interval and a spectral resolution of 3 nm from 350 to 1000 nm, and 10 nm from 1000 to 2500 nm. To prevent outliers during in-field measurements, the spectrometer and the light source were preheated 30 min before the measurements were taken. Preprocessing and modeling of the collected spectral data were executed via ViewSpec Pro (Analytical Spectral Devices, Boulder, CO, USA) and Unscambler X 10.3 (Camo Software AS, Oslo, Norway), respectively. The best model ($R^2 = 0.82$) was established using the Moving Average Smoothing (MAS) preprocessing technique on data from 780 calibration and 230 validation samples. The leave-one-out validated model with RMSEC (0.16) and RMSEP (0.17) was considered ideal for relating the leaf spectra and N concentration using a partial least squares regression with full Vis-NIRS wavelength bands (350–2500 nm).

Alenazi et al. (2020) used the Vis-NIRS to determine the development and ripening stages of tomato fruit. The authors developed quantitative models to determine maturity and quality parameters of fresh tomato fruit at five distinctive maturity stages namely breaker, turning, pink, light-red, and red. The handheld near infrared enhanced spectrometer ((F-750 Data Viewer Software, Version: v1.1.0.51) at a wavelength range of 285–1200 nm was used to develop PLS models for nondestructive assessment of flesh firmness, lycopene, b-carotene, total phenolic content (TPC), and total flavonoids content (TFC). The Model Builder (Version: v1.1.0.105) was used to correlate the analytic data with reflectance (first and second derivatives). The PLS models developed were able to assess the parameters with distinct PCA models at different stages of development and ripening. Developed models had a high coefficient of determination (R^2): 0.83 for total phenolic content, 0.86 for lycopene, 0.79 for total flavonoid content, 0.71 for b-carotene, and 0.68 for flesh firmness.

Rahi et al. (2019) used the Vis-NIRS for nondestructive detection of *Escherichia coli* on lettuce. A green-wave spectrometer (StellarNet Inc., Florida, USA) with a highly sensitive CCD detector with 2048 element (signal to noise ratio: 400:1) was used with fiber optic coupled instruments for measurements in 350–1100 nm wavelength range with 1.6 nm resolution. The spectrometer was calibrated in 20 min intervals. The data were transferred to the lab top by the SpectraWiz software (StellarNet Inc., Florida, USA). The author executed the classification of lettuce samples with different concentrations of *E. coli* solution by means of Partial least

Table 1 Topical literature reviews of research in the application of NIR spectroscopy in fruits and vegetables in the last decade

Topic	Journal	Author
Application of NIR and Raman spectroscopy for quality and safety inspection of fruits and vegetables: A review	Transactions of the Chinese Society for Agricultural Machinery	Yibin (2013)
Contributions of Fourier-transform mid-infrared (FT-MIR) spectroscopy to the study of fruit and vegetables: A review	Postharvest Biology and Technology	Bureau et al. (2019)
Fruit quality evaluation using spectroscopy technology: a review	Sensors	Wang et al. (2015)
Monitoring strategies for quality control of agricultural products using visible and near infrared spectroscopy	Trends in Food Science and Technology	Cortés et al. (2019)
NIR spectroscopy applications for internal and external quality analysis of citrus fruit—a review	Food and Bioprocess Technology	Magwaza et al. (2012)
NIR spectroscopy as a suitable tool for the investigation of the horticultural field	Agronomy	Cattaneo and Stellari (2019)
Past, present, and future near infrared spectroscopy applications for fruits and vegetables	NIR news	Kawano (2016)
Visible-NIR "point" spectroscopy in postharvest fruit and vegetable assessment: The science behind three decades of commercial use	Postharvest Biology and Technology	Walsh et al. (2020)

Squares Discriminant Analysis (PLS-DA) with different preprocessing methods into the two different groups of "safe" and "unsafe" samples. The classification was assisted by Principal component analysis (PCA) that was conducted to recognize outliers in data and patterns or relationships between variables and groups. The best model was developed using standard normal variation (SNV) + second derivate (D_2) preprocessing techniques with the SECV of 0.26 and a good correlation (R^2 of 0.93) between spectral data and the presence of *E. coli*. The authors organized samples into the calibration and validation sets comprised of 75% and 25% of the total samples, respectively.

Research-wise, the application of visible to near infrared spectroscopy has been demonstrated as an efficient tool to detect the quality and safety of horticultural produce. The feasibility of in-line spectrometers has also been demonstrated successfully in various produce types including leafy vegetables, herbaceous fruit such as tomatoes and pepper, and pomology fruit such as citrus and avocado (Ignat et al., 2012; Ncama et al., 2017, 2018; Ye et al., 2008). The depth of research in studies demonstrating the application of spectroscopy in fresh horticultural produce can be read further in the ideal literature reviews from the last decade (Table 1).

The research section of Vis-NIRS spectroscopy application in fruits and vegetables has developed adequate information for this technology to be applicable on an industry level. The application of spectroscopy has been demonstrated in immobile and mobile samples for both sampling and in-line applications. The protocols of developed models are readily available for technicians to apply in different stages of horticultural produce including before harvesting, during harvesting, and postharvest handling. It is not easy to track Vis-NIRS success in the industry because it is not documented. But judging from the research success, the industry is not short of information and supplies of spectrometers and operating models.

7 Conclusion

The application of visible and near infrared spectroscopy in the assessment of the quality and safety of fresh fruits and vegetables has been thoroughly researched. Many models have been developed for predicting the quality parameters of fresh commodities. Prediction models for safety in terms of chemical residues is an ongoing research field. This field has a limitation that can be aligned with the fact that the type and concentration of chemicals applied by each farmer are always unique. Infections vary with specific commodities which results in different chemicals used. Spectrometers with easy calibration methods are necessary so that technicians at each farm can ideally calibrate and use the models that are applicable to their challenges.

References

Abbasi, H., Bani, M. A., Nazeri, M., & Mireei, S. A. (2014). Application of fluorescence spectroscopy in the freshness control of lettuce leaf after irradiation of cold atmospheric pressure plasma. In *The 2nd conference of plasma engineering and plasma physics (PEP 2014)* (pp. 51–57).

Alenazi, M. M., Shafiq, M., Alsadon, A. A., Alhelal, I. M., Alhamdan, A. M., Solieman, T. H., Ibrahim, A. A., Shady, M. R., & Saad, M. A. (2020). Non-destructive assessment of flesh firmness and dietary antioxidants of greenhouse-grown tomato (*Solanum lycopersicum* L.) at different fruit maturity stages. *Saudi Journal of Biological Sciences, 27*(10), 2839–2846.

Aouadi, B., Zaukuu, J. L. Z., Vitális, F., Bodor, Z., Fehér, O., Gillay, Z., Bazar, G., & Kovacs, Z. (2020). Historical evolution and food control achievements of near infrared spectroscopy, electronic nose, and electronic tongue—Critical overview. *Sensors, 20*, 5479.

Bureau, S., Cozzolino, D., & Clark, C. J. (2019). Contributions of Fourier-transform mid infrared (FT-MIR) spectroscopy to the study of fruit and vegetables: A review. *Postharvest Biology and Technology, 148*, 1 14.

Cattaneo, T. M., & Stellari, A. (2019). NIR spectroscopy as a suitable tool for the investigation of the horticultural field. *Agronomy, 9*, 503.

Cen, H., & He, Y. (2007). Theory and application of near infrared reflectance spectroscopy in determination of food quality. *Trends in Food Science & Technology, 18*(2), 72–83.

Cortés, V., Blasco, J., Aleixos, N., Cubero, S., & Talens, P. (2019). Monitoring strategies for quality control of agricultural products using visible and near-infrared spectroscopy: A review. *Trends in Food Science & Technology, 85*, 138–148.

Cozzolino, D., Cynkar, W. U., Shah, N., & Smith, P. (2011). Multivariate data analysis applied to spectroscopy: Potential application to juice and fruit quality. *Food Research International, 44*(7), 1888–1896.

Dos Santos, C. A. T., Lopo, M., Páscoa, R. N., & Lopes, J. A. (2013). A review on the applications of portable near infrared spectrometers in the agro-food industry. *Applied Spectroscopy, 67*(11), 1215–1233.

Golic, M., & Walsh, K. B. (2006). Robustness of calibration models based on near infrared spectroscopy for the in-line grading of stone fruit for total soluble solids content. *Analytica Chimica Acta, 555*(2), 286–291.

Gómez, A. H., He, Y., & Pereira, A. G. (2006). Non-destructive measurement of acidity, soluble solids and firmness of Satsuma mandarin using Vis/NIR-spectroscopy techniques. *Journal of Food Engineering, 77*(2), 313–319.

Guthrie, J. A., Liebenberg, C. J., & Walsh, K. B. (2006). NIR model development and robustness in prediction of melon fruit total soluble solids. *Australian Journal of Agricultural Research, 57*(4), 411–418.

Guthrie, J. A., & Walsh, K. B. (1997). Non-invasive assessment of pineapple and mango fruit quality using near infra-red spectroscopy. *Australian Journal of Experimental Agriculture, 37*(2), 253–263.

Ignat, T., Schmilovitch, Z., Fefoldi, J., Steiner, B., & Alkalai-Tuvia, S. (2012). Non-destructive measurement of ascorbic acid content in bell peppers by VIS-NIR and SWIR spectrometry. *Postharvest Biology and Technology, 74*, 91–99.

Kader, A. A. (1997). Fruit maturity, ripening, and quality relationships. *Acta Horticulturae, 485*, 203–208. https://doi.org/10.17660/ActaHortic.1999.485.27

Kader, A. A. (2008). Flavor quality of fruits and vegetables. *Journal of the Science of Food and Agriculture, 88*(11), 1863–1868.

Kawano, S. (2016). Past, present and future near infrared spectroscopy applications for fruit and vegetables. *NIR News, 27*(1), 7–9.

Lichtenthaler, H. K., & Buschmann, C. (2001). Chlorophylls and carotenoids: Measurement and characterization by UV-VIS spectroscopy. *Current Protocols in Food Analytical Chemistry, 1*(1), F4.3.

Lin, H., & Ying, Y. (2009). Theory and application of near infrared spectroscopy in assessment of fruit quality: A review. *Sensing and Instrumentation for Food Quality and Safety, 3*(2), 130–141.

Magwaza, L. S., Landahl, S., Cronje, P. J., Nieuwoudt, H. H., Mouazen, A. M., Nicolaï, B. M., Terry, L. A., & Opara, U. L. (2014). The use of Vis/NIRS and chemometric analysis to predict fruit defects and postharvest behaviour of 'Nules Clementine'mandarin fruit. *Food Chemistry, 163*, 267–274.

Magwaza, L. S., Opara, U. L., Nieuwoudt, H., Cronje, P. J., Saeys, W., & Nicolaï, B. (2012). NIR spectroscopy applications for internal and external quality analysis of citrus fruit—A review. *Food and Bioprocess Technology, 5*(2), 425–444.

Magwaza, L. S., Opara, U. L., Terry, L. A., Landahl, S., Cronje, P. J., Nieuwoudt, H. H., Hanssens, A., Saeys, W., & Nicolaï, B. M. (2013). Evaluation of Fourier transform-NIR spectroscopy for integrated external and internal quality assessment of Valencia oranges. *Journal of Food Composition and Analysis, 31*(1), 144–154.

Mahajan, P. V., Caleb, O. J., Gil, M. I., Izumi, H., Colelli, G., Watkins, C. B., & Zude, M. (2017). Quality and safety of fresh horticultural commodities: Recent advances and future perspectives. *Food Packaging and Shelf Life, 14*, 2–11.

Mao, H., Gao, H., Zhang, X., & Kumi, F. (2015). Nondestructive measurement of total nitrogen in lettuce by integrating spectroscopy and computer vision. *Scientia Horticulturae, 184*, 1–7.

McGlone, V. A., Abe, H., & Kawano, S. (1997). Kiwifruit firmness by near infrared light scattering. *Journal of Near Infrared Spectroscopy, 5*(2), 83–89.

Mditshwa, A., Magwaza, L. S., Tesfay, S. Z., & Mbili, N. (2017). Postharvest quality and composition of organically and conventionally produced fruits: A review. *Scientia Horticulturae, 216*(1), 148–159.

Mehinagic, E., Royer, G., Symoneaux, R., Bertrand, D., & Jourjon, F. (2004). Prediction of the sensory quality of apples by physical measurements. *Postharvest Biology and Technology, 34*(3), 257–269.

Nazarloo, A. S., Sharabiani, V. R., Gilandeh, Y. A., Taghinezhad, E., Szymanek, M., & Sprawka, M. (2021). Feasibility of using VIS/NIR spectroscopy and multivariate analysis for pesticide residue detection in tomatoes. *Processes, 9*(2), 196.

Ncama, K., Magwaza, L. S., Poblete-Echeverría, C. A., Nieuwoudt, H. H., Tesfay, S. Z., & Mditshwa, A. (2018). On-tree indexing of 'Hass' avocado fruit by non-destructive assessment of pulp dry matter and oil content. *Biosystems Engineering, 174*, 41–49.

Ncama, K., Opara, U. L., Tesfay, S. Z., Fawole, O. A., & Magwaza, L. S. (2017). Application of Vis/NIR spectroscopy for predicting sweetness and flavour parameters of 'Valencia' orange (*Citrus sinensis*) and 'star ruby' grapefruit (*Citrus x paradisi* Macfad). *Journal of Food Engineering, 193*, 86–94.

Nicolai, B. M., Beullens, K., Bobelyn, E., Peirs, A., Saeys, W., Theron, K. I., & Lammertyn, J. (2007). Nondestructive measurement of fruit and vegetable quality by means of NIR spectroscopy: A review. *Postharvest Biology and Technology, 46*(2), 99–118.

Peiris, K. H. S., Dull, G. G., Leffler, R. G., & Kays, S. J. (1999). Spatial variability of soluble solids or dry matter content within individual fruits, bulbs, or tubers: Implications for the development and use of NIR spectrometric techniques. *Horticultural Science, 34*(1), 114–118.

Peirs, A., Tirry, J., Verlinden, B., Darius, P., & Nicolaı□, B. M. (2003). Effect of biological variability on the robustness of NIR models for soluble solids content of apples. *Postharvest Biology and Technology, 28*(2), 269–280.

Perez-Marin, D., Torres, I., Entrenas, J. A., Vega, M., & Sánchez, M. T. (2019). Pre-harvest screening on-vine of spinach quality and safety using NIRS technology. *Spectrochimica Acta Part A: Molecular and Biomolecular Spectroscopy, 207*, 242–250.

Polesello, A., & Giangiacomo, R. (1981). Application of near infrared spectrophotometry to the nondestructive analysis of foods: A review of experimental results. *Critical Reviews in Food Science and Nutrition, 18*(3), 203–230.

Rahi, S., Mobli, H., Jamshidi, B., Azizi, A., & Sharifi, M. (2019). Visible/near-infrared spectroscopy as a novel technology for nondestructive detection of Escherichia coli ATCC 8739 in lettuce samples. In *Proceedings of the world conference on engineering and technology, 24–26 October 2019, Frankfurt, Germany* (pp. 24–26).

Rinnan, A., Van den Berg, F., & Engelsen, S. B. (2009). Review of the most common pre-processing techniques for near-infrared spectra. *Trends in Analytical Chemistry, 28*(10), 1201–1222.

Roggo, Y., Chalus, P., Maurer, L., Lema-Martinez, C., Edmond, A., & Jent, N. (2007). A review of near infrared spectroscopy and chemometrics in pharmaceutical technologies. *Journal of Pharmaceutical and Biomedical Analysis, 44*(3), 683–700.

Stevens, A., & Ramirez-Lopez, L. (2014). *An introduction to the prospectr package*. R Package Vignette, Report No.: R Package Version 0.1, 3.

Su, W. H., He, H. J., & Sun, D. W. (2017). Non-destructive and rapid evaluation of staple foods quality by using spectroscopic techniques: A review. *Critical Reviews in Food Science and Nutrition, 57*(5), 1039–1051.

Walsh, K. B., Blasco, J., Zude-Sasse, M., & Sun, X. (2020). Visible-NIR 'point' spectroscopy in postharvest fruit and vegetable assessment: The science behind three decades of commercial use. *Postharvest Biology and Technology, 168*, 111246.

Wang, H., Peng, J., Xie, C., Bao, Y., & He, Y. (2015). Fruit quality evaluation using spectroscopy technology: A review. *Sensors, 15*(5), 11889–11927.

Wang, J., Shi, X., Xu, Y., & Dong, C. (2021). Nitrogen management based on visible/near infrared spectroscopy in pear orchards. *Remote Sensing, 13*(5), 927.

Wedding, B. B., White, R. D., Grauf, S., Wright, C., Tilse, B., Hofman, P., & Gadek, P. A. (2011). Non-destructive prediction of 'Hass' avocado dry matter via FT-NIR spectroscopy. *Journal of the Science of Food and Agriculture, 91*(2), 233–238.

Wedding, B. B., Wright, C., Grauf, S., White, R. D., Tilse, B., & Gadek, P. (2013). Effects of seasonal variability on FT-NIR prediction of dry matter content for whole Hass avocado fruit. *Postharvest Biology and Technology, 75*(1), 9–16.

Williams, P., & Norris, K. (1987). *Near-infrared technology in the agricultural and food industries*. American Association of Cereal Chemists, Inc. https://www.cabdirect.org/cabdirect/abstract/19892442443.

Wu, D., Feng, L., He, Y., & Bao, Y. (2008). Variety identification of Chinese cabbage seeds using visible and near-infrared spectroscopy. *Transactions of the ASABE, 51*(6), 2193–2199.

Ye, X., Oshita, S., Jin, X., Makino, Y., & Kawagoe, Y. (2008). Potential of Vis/NIR spectroscopy in estimating ATP content per protoplast as an indicator of freshness of spinach. *IFAC Proceedings Volumes, 41*(2), 636–640.

Yibin, F. X. Y. (2013). Application of NIR and Raman spectroscopy for quality and safety inspection of fruits and vegetables: A review [J]. *Transactions of the Chinese Society of Agricultural Machinery, 8*, 027.

Fourier Transform Infrared Spectroscopy (FTIR) Technique for Food Analysis and Authentication

Asif Ahmad and Haris Ayub

Abstract Fourier Transform Infrared (FTIR) spectroscopy is a method of infrared absorption spectra and emanation of a solid, fluid, or gaseous material. It gathers and digitizes the interferogram followed by Fourier transformation and displays the spectrum. Infrared spectra obtained from the FTIR spectrometer lie in the mid-infrared region of wave number 4000–666 cm^{-1}. This technique is widely applied to foods, forensics, the environment, and pharmaceuticals. It is a rapid (i.e., less time-consuming), nondestructive, operator-friendly, and precise method. A highly accurate fingerprint of the material being analyzed is provided based on its chemical and physical state. FTIR in combination with other chemometric techniques has been successfully applied to analyze food products and their constituents, such as milk, meat, oil, fruits, and different carbohydrates. The determination of product quality and authenticity is also a major concern among food manufacturers and consumers. Authentication of different food products like honey, milk, oil, and tea is carried out successfully through the provision of the highly accurate fingerprint of the material under analysis. Single bonds are detected in the high wavenumber region of 2500–4000 cm^{-1}, double bonds in region 1500–2000 cm^{-1}, and triple bonds in the middle region of 2000–2500 cm^{-1}. This technique possesses a huge scope in the food industry and it is being immensely flourished worldwide over the past decade.

Keywords Fourier Transform Infrared (FTIR) · Spectroscopy · Food quality · Food authenticity · Food analysis · Beta-glucan

1 Introduction

The Infrared spectroscopic technique employs spectra in near-infrared and far-infrared regions for the quantitative and qualitative analysis of food products and other commodities (Rohman & Che Man, 2012a). A vast range of spectra which

A. Ahmad (✉) · H. Ayub
Department of Food Technology, PMAS-Arid Agriculture University, Rawalpindi, Pakistan
e-mail: asifahmad@uaar.edu.pk

P. B. Pathare, M. S. Rahman (eds.), *Nondestructive Quality Assessment Techniques for Fresh Fruits and Vegetables*, https://doi.org/10.1007/978-981-19-5422-1_6

falls within the region of 0.7–1000 μm can be applied to get useful information about food constituents but for chemical analysis mid-infrared region 2.5–25 μm (4000–400 cm^{-1}) is the most attractive. This information gave birth to the idea of commercial Fourier Transform Infrared (FTIR) Spectroscopy in the 1970s since then the science of FTIR has advanced in infrared spectroscopy (Doyle, 1992). However, the basic principle remained the same in earlier and modern models of the equipment that makes use of an infrared source to irradiate the sample. The absorption of these radiations excites the molecules into vibrational motion by the deposition of energy (Duygu et al., 2009). The interferogram collected is translated into a spectrum through a mathematical model called Fourier transformation Modern FTIR systems are equipped with chemometric software that increases the application of this equipment to obtain useful information on the qualitative and quantitative parameters (Rahman, 2012; Rohman & Che Man, 2012a). Several elements like a non-homogenous sample, temperature, refractive index, nonlinear detector, and reflection losses may affect FTIR results. These elements can be minimized by frequent calibration checks, using more samples for calibration, using peak areas rather than heights, and maintaining the absorbance between 0.2 and 0.7 absorbance units (Van de Voort & Ismail, 1991). A wide range of food and non-food items can be analyzed using the FTIR technique. At the industrial and research level, it applies to four major areas, i.e., identification of compounds, elucidation of structures, quantitative analysis of species, and measurement of molecular properties (Gaffney et al., 2002). Valuable information can be achieved about carbohydrates, lipids, proteins, moisture, and other components in food products using this technique (Sivakesava & Irudayaraj, 2000; Van de Voort & Ismail, 1991). Furthermore, this technique has a great potential to evaluate adulteration in different food products including jams, purees, oils, coffee, meat, and milk products (Syahariza et al., 2005). Recent research in this decade elucidated its potential as a reliable technique for the identification of microorganisms, especially food spoilage and food pathogenic bacteria down to the strain level (Duygu et al., 2009; Wenning et al., 2014). Some researchers showed its application in the pharmaceutical industry, astronomical measurements, and waste management (Baravkar et al., 2011; Smidt & Schwanninger, 2005). All of these applications depicted the perspective of FTIR spectroscopy as a fast, efficient, free of chemicals, and relatively low-cost technique, which requires no or less sample preparation and small sample size (Triebel et al., 2007; Vongsvivut et al., 2014). As compared to other conventional instruments, FTIR reproduces better, has high energy throughput, and has better accuracy from the laser source (Baravkar et al., 2011; Syahariza et al., 2005). This technique is equally good for aqueous, powder, or dehydrated samples (Duygu et al., 2009). Major advantages associated with this technique are fast, reliable, environmentally friendly, and has no waste generation (Vongsvivut et al., 2012). The modern models of FTIR also offer improved signal-to-noise ratio, detection of all frequencies simultaneously, wavelength accuracy, and superior spectral resolution (Sivakesava & Irudayaraj, 2000; Van de Voort & Ismail, 1991).

Today in the modern era, a large variety of analytical techniques are available for food analysis in research laboratories and industry. The stakeholders are always interested in accuracy and precision while making use of analytical techniques for the assessment of quality, nutritive value determination, and detect adulteration (Franca & Oliveira, 2011). For this purpose, they make use of the traditional high-performance liquid chromatography, mass spectroscopy, gas chromatography, atomic absorption spectroscopy, X-ray fluorescence spectroscopy, polymerase chain reaction (PCR), and gas chromatography-mass spectrometry. However, these techniques may have certain limitations (i.e., the maintenance costs, the time required for measurement, and sampling difficulty) and these limitations justify to develop of new rapid and efficient analysis techniques (Cifuentes, 2012; Franca & Oliveira, 2011; Kizil et al., 2002; Sezer et al., 2017). These conventional analysis techniques may be replaced by more efficient techniques like near-infrared spectroscopy, laser-induced breakdown spectroscopy, Raman spectroscopy, and mid-Fourier transform infrared spectroscopy (Sezer et al., 2017). Among all these techniques FTIR spectroscopy has turned out to be a powerful tool for analyzing foods and their components (Kačuráková & Wilson, 2001).

Food product authentication is a major concern in today's world for food processors, consumers, and regulatory bodies (Černá et al., 2003; Syahariza et al., 2005). For a long time, various methods of authenticating food are applicable worldwide, such as gas chromatography, high-pressure liquid chromatography, DNA-based methods (PCR), ELISA, nuclear magnetic resonance, differential scanning calorimetry, specific natural isotope fractionation (SNIF), electronic nose x-ray diffraction, mass spectrometry, fluorescence spectroscopy, and visible spectroscopy (Gouilleux et al., 2018). Most of the conventional authentication methods are time-consuming and expensive. There is always a need for rapid, accurate, and efficient methods for food authentication. Raman spectroscopy, laser-induced breakdown spectroscopy, near-infrared spectroscopy, and FTIR spectroscopy have a great potential to achieve desired objectives for the future industry and regulatory authorities to make decisions on the authentication of foods. All types of food products (e.g., milk, meat, oils, and nuts) and ingredients are efficiently analyzed and authenticated through these techniques (Grassi et al., 2018; Sezer et al., 2017; Yang et al., 2005; Zhang et al., 2017).

The first part of this chapter presents the advantages, basic principles, and operation of the FTIR technique. The second part of this chapter explains the application of the FTIR technique in analyzing and authenticating different foods and their constituents. It also reviews the analysis of bacteria involved in food and authentication of halal products. The gathered information can be beneficial for the food industry, researchers, students, data scientists, academics, and food business personnel.

2 Fourier Transform Infrared Spectroscopy (FTIR)

Early infrared spectrometers are referred to as dispersive instruments as they use monochromators to scatter radiations into their spectral constituents while modern infrared spectrometers make use of interferometers for the distribution of source radiations (Gaffney et al., 2002). Modern Fourier transforms infrared

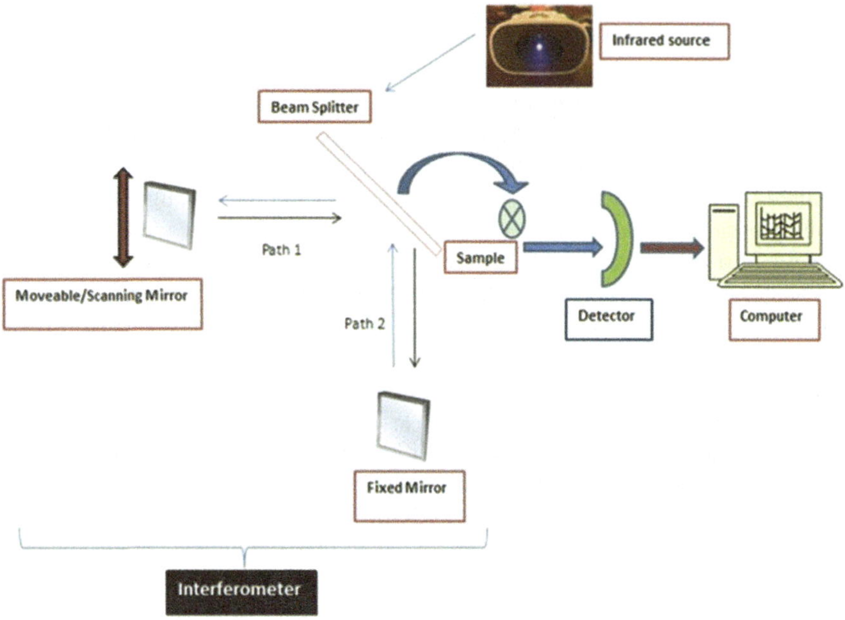

Fig. 1 Working of FTIR spectroscope

spectrophotometers provide a much rapid infrared spectrum as compared to traditional spectrophotometers (Hilal et al., 2017). A resolution of 4 cm^{-1} can be obtained at mid-infrared spectrum on dispersive spectroscopy in 15 min while the modern Fourier transform infrared spectrometer can get the same spectra in just 1 s (Griffiths & De Haseth, 2007). The FTIR spectra reflect the molecular structure as well as the molecular environment. In the FTIR technique, infrared radiations from the infrared source irradiate the sample and the absorption of these radiations deposit the quanta of energy into vibrational modes hence stimulating the vibrational motions. Thus, the exposure of a molecule to the radiations emitted by the infrared source absorbs only at frequencies that correspond to its mode of vibration in the electromagnetic spectrum region between visible and microwaves (Duygu et al., 2009). A flow diagram of the FTIR system is shown in Fig. 1.

2.1 *Principle*

Fourier transform spectroscopy records the radiant power data as a function of time. The interferometer changes the electromagnetic radiation frequency to slow the oscillating signal. A Computer then mathematically separates the sum of slower oscillating signals into individual oscillation signals and further computes the fluctuations in the frequencies of the radiations observed followed by a continuous recording of this data. Fourier transform is a mathematical method that converts the time domain spectrum into frequency domain spectrum (Baravkar et al., 2011). The transformation of the interferogram is carried out by instrument software. The intensity of infrared radiations before and after passing through the sample is calculated to measure the infrared spectrum. Moreover, the spectrum is plotted with Y-axis units as absorbance/transmittance and the X-axis as wavenumber units. FTIR absorbance spectrum follows Beer's law which relates the absorbance with the concentration as shown in the equation below:

$$A_\lambda = l\varepsilon_\lambda c$$

where A_λ is Absorbance, l is path length, ε is Absorptivity, c is concentration

However, the transmittance is not directly proportional to concentration and is shown in the below equation:

$$\%T = I_S/I_R$$

where T is transmittance, I_S is intensity of the beam after passing sample, I_R is intensity of the beam before passing sample.

2.2 *Equipment and Their Components*

2.2.1 Infrared [IR] Source

Generally, infrared sources are hot body type, which is made up of a strong material warmed by an electric current. The power of the radiant and the wavenumber of peak emission increases as the temperature rises. Common infrared sources used are silicon carbide rods commonly known as Globar, Nernst glowers, and tungsten glowers. Most of the equipment working in the mid-infrared region make use of Globar in the form of a silicon carbide heated rod, with 2–5 cm in length and 0.5–1 cm in width. It usually operates at a temperature of 1300 K. As an electric current passes through the silicon carbide rods (infrared source), it becomes hot thus emitting a huge amount of infrared radiation. Nernst glowers are quite different compared to globar and are made of refractory oxide mixtures such as cerium, zirconium, and thorium and they can achieve more temperature than Globar although they are not capable of producing infrared radiations above 2000 cm^{-1}. The range of temperature a Nernst glower can achieve is around 1500–2000 K but it falls off quickly above 2000 cm^{-1}. Thus, these glowers are a good source for spectral regions below 2000 cm^{-1} although they are not effective for the whole mid-infrared region. Quartz-tungsten-halogen (QTH) lamp contains additional inert gas and halogen forming tungsten oxy-halide extending the shelf life of the lamp (Gaffney et al., 2002). Nernst glower is shown in Fig. 2.

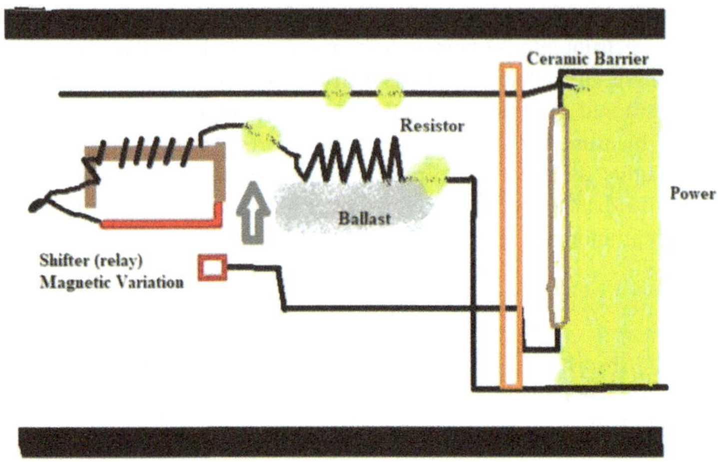

Fig. 2 Components of Nernst glower

2.2.2 Interferometer

The Interferometer was first invented by Albert Abraham Michelson and it was also known as the Michelson interferometer. The Interferometer is the heart of any FTIR spectrometer. It consists of a beam splitter, a compensating plate, and two mirrors (stationary & movable) (Doyle, 1992). Two beams reflect from the mirrors back to the beam splitter where they are reunited, focused, and sent to the detector. The path difference because the two traveling beams create a phase difference between them. If the paths of the two traveling beams are the same then the signal will be maximum with both beams being in phase while the path difference between two beams can generate the smallest signal at the detector causing destructive interference (Larkin, 2017). Entrance apertures of >1 cm in diameter are used in the interferometers while monochromators need narrow slits for achieving good spectral resolution (Griffiths & De Haseth, 2007).

2.2.3 Beam Splitter

A Beam splitter is composed of a different material transmitting half of the radiations while reflecting the remaining half. The infrared radiations from the infrared source after striking the beam splitter are converted into two beams. One beam gets transmitted to the fixed mirror through the beam splitter while the second beam reflects the movable mirror through a beam splitter. Both of these mirrors reflect the radiations to the beam splitter where the two beams interfere to get an interferogram (Baravkar et al., 2011).

2.2.4 Mirrors

A moving mirror is an extremely reactive flat surface allowing the mirror to move with high speed. The mirror moves after every millisecond and a few millimeters away from the beam splitter. The fixed mirror is also a highly reflective flat surface (Larkin, 2017). Ideally, 50% of the light is refracted towards a fixed mirror while 50% is transmitted towards the moving mirror. Light reflects from the mirrors to the beam splitter and focused on the sample. Moreover, if the distance from the splitter to the fixed mirror is not the same as the distance from the splitter to the moveable mirror then there will be constructive and destructive interference depending upon the distance between the two mirrors.

2.2.5 Laser

Helium-Neon laser is employed by many instruments as a calibration standard of internal wavelength. It makes a record of optical path differences in the interferometer precisely (Larkin, 2017). The precise constant velocity of the moving mirror is timed through the use of accurate laser wavelength. The laser beam intensity in the interferometer is measured at two points. Due to the enhancement and cancellation of He-Ne beam paths, the intensity of the moving mirror rise and fall at these two points. The relative phase of the sine wave tells the direction of the moving mirror while the number of fringes in the wave tells how far the mirror has moved (Willey, 1976).

2.2.6 Detectors

Infrared [IR] detectors compute the strength of infrared radiations from an infrared source converting them into electrical signals which are further processed by generating spectra. Photosensitivity, detectivity, and noise equivalent power are the major parameters determining the performance of a detector. Detectivity is defined as the photosensitivity per unit area of the detector. Infrared detectors are either thermal or photonic types. Thermal detectors use infrared radiations as heat and detect the temperature changes of absorbing material. Thermal detectors are mostly used in the MIR region but these are equally good in the NIR region. The most common of these is the pyroelectric detector providing good linearity over a wide spectral region. Photodetectors use infrared radiations as light and are more sensitive as compared to thermal detectors. These detectors are exclusively made to compute the distinctive interferogram signal coming from the interferometer (Gaffney et al., 2002).

2.2.7 Computer System and Software

Information received from the detector in the form of a digitized signal is forwarded to a computer where the signal is transformed into Fourier series. It converts it into final infrared spectra which are used by the user for further elucidation and operation (Larkin, 2017). The Fourier transformation of an interferogram using a computer is a rapid way to obtain the composite spectrum of the source, instrument, and sample. It simultaneously measures the wavelengths rather than sequential measurement (Doyle, 1992).

2.2.8 Background Spectrum

A background spectral measurement is a procedure performed where the beam run without the sample. This measurement can be matched with the sample in the beam to determine the % transmittance. This technique removes all the instrumental characteristics of a spectrum originating as noise. Hence, all the spectral features obtained while running the sample can be purely due to the sample. Computer software subtracts the background spectrum from the sample spectrum using the Fourier transform technique to generate a reliable spectrum originating solely from the sample (Hilal et al., 2017).

2.2.9 Parameters Optimization

Fourier transform infrared spectrometers are furnished with powerful hardware and software to get accurate and precise results for a specific sample. For the data collection, certain processing parameters are set by these software packages which should be adjusted for a specific sample and application. The parameters are the spectral resolution, several scans, choosing the mirror velocity, digital filtering, and the application of phase correction to the interferogram (Gaffney et al., 2002).

2.3 Operation of FTIR

FTIR spectroscopy comprises an infrared radiation source, an interferometer (fixed and moving mirror, a beam splitter, and laser), the sample chamber, and a detector (Fig. 1). It uses a Michelson interferometer to analyze the infrared radiations passing through a sample. The infrared radiations are made accurately parallel by the mirror and the infrared beam is divided at the beam splitter. The stationary mirror passes half of the beams while the moving mirror refracts the other half. After reflecting from the moveable and stationary mirror, the beams reunite at the beam splitter and the radiation is focused on the detector (Baravkar et al., 2011). The path differences between stationary and movable mirrors bring back together the beams experiencing a constructive and destructive interference called an interferogram (Hilal et al., 2017; Van de Voort & Ismail, 1991). The radiations only from the two mirrors can arise in phase at the beam splitter, cell, and detector when the distance between the beam splitter and stationary mirror is the same as that of the beam splitter and moving mirror. With the changing position of the moving mirror remoteness between mirror and splitter also changes hence the radiations of the stable wavelength reach in phase only to the cell and detector. With the change in the distance, the wavelength of the radiation beams either arrive in phase or out of phase. By controlling the mirror motion rate a series of simultaneous signals arrive at the detector which measures the interferogram signal for all frequencies and carries the information to the computer (Baravkar et al., 2011; Hilal et al., 2017).

A beam is superimposed to provide a background for the instrumental operation. After the subtraction of the background spectrum from the sample spectrum through Fourier transformation software, the desired spectrum is obtained (Hilal et al., 2017). A Computer using a Fourier transform method with mathematical algorithms converts the output into a plot of absorption/transmission against wavenumber (Munajad et al., 2018; Schmitt & Flemming, 1998). It usually takes 2 min to record a plot starting with the sample insertion (Baravkar et al., 2011).

2.4 *Modeling Techniques Used for FTIR*

Different modeling techniques are used to analyze and authenticate food products using FTIR. FTIR is considered powerful as a quantitative tool because of its ability to carry out multicomponent analysis. Most of the FTIR multicomponent analysis procedures are based on K and P matrix methods. They use the known concentration of mixtures and compute the concentration of an unidentified blend. Analytical methods based on spectroscopy possess some limitations when it comes to similar or overlapping bands of some compounds which led to the use of chemometric techniques like principal component analysis, partial least square regression, and discriminant analysis (Franca & Oliveira, 2011). Several studies show that FTIR-based methods can be applied in the food industry to detect and authenticate products successfully.

PLS (Partial Least Square) is one of the modeling techniques commonly used in combination with FTIR. It is a form of factor analysis. It is very useful in studying the complete spectrum of the mixture and determining the spectrums of interest through correlation (Van de Voort, 1992). PLS does not need to establish a direct relationship between absorbance and concentration at specified frequencies, it is rather developed by compacting the spectral data for training set into mathematical spectra. When the spectrum of an unidentified sample is analyzed the spectrum is renovated by PLS from loading spectra and their volumes of each are then used to forecast the unidentified concentration. It possesses the advantages of both the CLS and Inverse Least Square but does not possess the limitations they have (Ismail et al., 1997).

One of the most crucial modeling techniques used for the analysis and authentication of food products based on inverse calibration and data reduction is principal component analysis (PCA). It is a type of factor analysis in which the spectral and concentration data is incorporated into the model in one step. It condenses the multidimensional data into major dominant components while conserving the applicable variation. The natural groups are recognized by PCA with the first principal component showing the major variation followed by the second and third components. Moreover, score plots in PCA are used to look for similarities or differentiate the compounds (Davis & Mauer, 2010). Likewise, the closer the samples more similarities they possess. CLS (Classical Least Square) is another such multivariate least square technique used in FTIR spectroscopy. A linear arrangement of single constituent reference spectra of separate composites is tailored to the measured spectrum such that the sum of squares of residuals at each frequency is minimized (Bacsik et al., 2004). Quantitative results are provided by the amount of each absorber required for the best fit. However, this method does possess certain limitations as well.

Hierarchical cluster analysis (HCA) is also used to detect similarities between the spectra of samples. It recognizes them based on the aggression algorithms and distance between spectra. Furthermore, discriminant analysis (DA) is another statistical tool to categorize substances into clusters. It determines the similarity of values from an unidentified sample to an identified sample. Moreover, soft independent modeling by class analogy (SIMCA) is another model used for the spectral organization of data and is mainly applied in the classification of bacteria. The sample is first analyzed through PCA before SIMCA modeling so that the important components are retained. It requires a training data set comprising samples with a set of classes and qualities (Davis & Mauer, 2010).

3 Food Analysis Using FTIR Spectroscopy

Proximate analysis of food is the most common analytical requirement of the food industry whether it is used for processed products, raw materials, or finished products. Proteins, fats, carbohydrates, moisture, and the other minor constituents, like vitamins and minerals are the core of food systems and these constituents contribute to the obtained spectrum. The distinctive absorption bands are related to these main constituents. To determine the carbohydrates using FTIR, the presence of hydroxyl (–OH) bonds elucidate their molecular structure, whereas for proteins presence of amide I and amide II bands explain useful information. The presence of fat bands depicting ester linkage and C-H stretch at specific wavelengths provide fundamental information (Van de Voort & Ismail, 1991).

The infrared spectrum obtained from the FTIR spectrometer lies usually in the mid-infrared region between 4000–666 cm^{-1}. The transition energies which correspond to the changes in vibrational energy states of most functional groups within the food lie in the mid-infrared region, i.e., 4000–400 cm^{-1}. The absorption band appearance in this region can be applied to check the presence of the particular functional groups in the molecule. Single bonds (O–H, C–H, N–H) can be detected in the high wavenumber region of 2500–4000 cm^{-1}, double bonds are detected in the wavenumber region 1500–2000 cm^{-1} while triple bonds can be detected in the middle wavenumber region 2000–2500 cm^{-1} (Hilal et al., 2017).

3.1 Carbohydrates

Natural polysaccharides or food additives polysaccharides like pectin, starch, carrageenan, and galactan are well distinguished through FTIR in the region of 1200–800 cm^{-1}. These polysaccharides constitute the plant cell wall along with proteins, lignin, and inorganic compounds (Kyomugasho et al., 2015). The shape of the spectrum of a polysaccharide is mainly determined by the composition of polysaccharides and their side-chain components. The information about the major polysaccharides obtained from the FTIR spectrum appears in the region of 1200–800 cm^{-1} (Kacurakova et al., 2000). The anomeric region, i.e., 950–750 cm^{-1} is the most commonly used analysis region in carbohydrates providing vital information about α and β anomers of polysaccharide monomers. The conformers (alpha and beta) of mannose, glucose, and galactose are illustrated using bands at 870–840 and 890 cm^{-1} (Kačuráková & Wilson, 2001). Moreover, the detection of these polysaccharides may be affected by their water absorption but the studies show that though the spectra are affected by hydration but still no information about these polysaccharides was lost (Čopíková et al., 2006).

The effect of additives (agar, alginate, lecithin, and glycerol) on the potato puree prepared from commercial powder is also studied through FTIR spectroscopy. The presence of various starch-based food additives in processing brings variable characteristics to the product and can be analyzed through the FTIR technique. In this regard, spectra may be recorded in the region of 349–4000 cm^{-1}, the obtained peaks depict how these starch-based food additives affect water holding capacity by interacting with the hydroxyl group and hydrogen bonds. Similar investigations can be confirmed for other food additives such as glycerol and lecithin that may induce a greater effect on the structure because of their small molecules (Dankar et al., 2018). This technique is also valuable to detect other structural groups of exopolysaccharides (EPS) that is added to food products as food additives. A good example is the development of milk and yogurt-based products by the use of EPS-producing cultures in these products. All the cocultured exopolysaccharides in these products show peaks in a range of 3395.1–589.1 cm^{-1}. The stretching vibration of the –OH group showed a band at 3395.1 cm^{-1} while the band at 2924.2 cm^{-1} indicates the C–H stretching vibration. The carbohydrates presence is depicted by the presence of broad stretches of C–O–C, C–O at 1000–1200 cm^{-1} (Ahmed, Wang, Anjum, Ahmad, & Khan, 2013).

3.1.1 Starch

Starch is present in plants as α-(1 → 4) linked glucan. The gelation and retrogradation are the two major factors that affect the nature and characteristics of these starches during the processing of food. It is possible to study these changes in characteristics through FTIR. The characteristics of infrared bands for starch gels can be detected at 1046 and 1019 cm^{-1}. The intensity and shape changes of peaks between 995 and 1020 cm^{-1} indicate the onset of retrogradation. The stretching region (1300–800 cm^{-1}) of C–C and C–O bonds are very sensitive to retrogradation. The retrogradation rates vary considerably for both amylose and amylopectin (Kačuráková & Wilson, 2001). Irradiated starches are also characterized and classified through this technique (Kizil et al., 2002). Moreover, the abundance of resistant starches in refrigerated tortillas is because of the changes in the starch structure imposed by increased crystallinity. The bands appear at 1047 cm^{-1} and disappear at 1022 cm^{-1} which depict the retrogradation of starch. The band at 1047 cm^{-1} is because of the weakening intermolecular hydrogen bonds. The peaks observed show that the storage of tortillas has a great effect on the resistant starch formation (Flores-Morales et al., 2012). FTIR spectroscopy is also helpful in elucidating structural and conformational information of protein in conjunction with polysaccharides in bread and other baking products (Sivam et al., 2013).

3.1.2 Pectin

The most extensively studied polysaccharides through infrared spectroscopy are the pectin. These constituents are available mainly in the cell walls of higher plants. Pectins are made up of (1 → 4) linked α-D-galacturonan and residues of (1 → 2)-α-L-rhamnopyranoseyl (Kacurakova et al., 2000). FTIR read the functional groups of pectin and its derivatives in the region of 1900–1500 cm^{-1}. Pectate interacts with the potassium and magnesium ions even when there is no gel formation. Cadmium, copper, zinc, and lead form complexes with pectinates at 59% esterification while lead and copper form complexes even with 93% esterified pectinate (Kačuráková & Wilson, 2001). Esterification also reduces the negative charges and ion-binding sites. The complexes of ions are merely constant in the zones of the gel joint. The gels are formed by calcium (Ca^{+2}) and strontium (Sr^{+2}) with pectate and pectinates having 23.8% esterification. The spectra show a complex formation of nickel, zinc, copper, and lead with pectinates up to 59.1% esterification. In prominently esterified pectinate, i.e., 93.4%, copper (Cu^{+2}) to a greater extent and lead and zinc to a lesser extent interact strongly with the pectinate chain. Copper complex even forms a gel when carboxylate-free groups are just 6% because of their high stability (Wellner et al., 1998). Moreover, this technique is also used to analyze the degree of methyl esterification in pectin thus providing information about how the setting characteristics of pectin (Kyomugasho et al., 2015).

3.1.3 Cellulose

Cellulose is a β-(1 → 4) linked glucan present in plants, algae, and bacteria. The identification of the types of cellulose, the degree of crystallinity, and the effect of the source on the crystallinity are studied by the Fourier transform infrared spectroscopy. FTIR data about celluloses from various sources indicate that the intensity of the bands is limited while the infrared band position is the same in the bacterial cellulose and cotton linter cellulose (Kačuráková & Wilson, 2001). Hemicellulose polysaccharides are often composed of xylans, glucomannans, galactoglucomannans, and xyloglucans. Among these molecules β-(1 → 6), (1 → 3) linked galactan bands appear in the range of 1078–1072 cm^{-1}, β-(1 → 4) linked mannans in 1066–1064 cm^{-1}, xyloglucan and β-glucan at 1041 cm^{-1}, arabinan at 1039 cm^{-1}, rhamnose at 1043 cm^{-1} while galactose showed the strongest band at 1070 cm^{-1}, mannose at 1070 cm^{-1}, and glucose at 1035 cm^{-1}. The position of these distinctive bands identifies the structures and composition of polysaccharides (Kacurakova et al., 2000).

Protein, pectin, lignin, and phenolics may be distinctively detected in the complex samples of xylan. The presence of both phenolic and proteins in a sample might overlap in the region 1700–1500 cm^{-1} but their band intensities vary. Xylan structures are different depending upon the source. Xylan from cereal is a dietary fiber that is often used as gum. They are composed of β-(1 → 4)-D-xylopyranosyl residues while their branches are made up of L-arabinose and D-glucuronic acid. Xylan is bound to lignin through its uronic acid constituent. Xylan with (1 → 4) spectrum is dominated by the band at 1047 cm^{-1} while in xylan (1 → 3) two overlapping bands exist at 1066 and 1030 cm^{-1} (Kačuráková et al., 1999). The (1 → 3) linkage in xylans affects the infrared spectral pattern while the gelling mechanism of xylan is greatly affected by the position and type of glycosidic linkage. Hydration may relate to the existence or non-existence of 1150 and 1000 cm^{-1} band in infrared spectra in xylooligosaccharides and Xylo-polysaccharides (Kačuráková et al., 1998).

3.1.4 Dextran and Pullulan

Dextrans are produced by L. mesenteroides, Lactobacilli, and Weisella species while the dextran produced by L. mesenteroides and Weissellacibaria has 95% α-(1 → 6) and 5% α-(1 → 3) linkage (Ahmed & Ahmad, 2017). Pullulan and dextrans are used in food, the cosmetic, and pharmaceutical industry. The mobility of the dextran chain increase because of the presence of (1 → 6) glycosidic linkage but pullulan consists of both (1 → 4) and (1 → 6) linkages which differentiate both. Highly merged bands are observed in the spectral region (1175–975 cm^{-1}) of irradiated dextran and pullulan. The band at 1150 cm^{-1} is assigned to the exocyclic stretching –CO vibrations which were previously assigned to –COC. The varying frequency position of this band is because of the different repeating units and the OH group orientation of different polysaccharides. The dextran spectra contain a less distinct 1080 cm^{-1} band as compared to pullulan. The band 1024 cm^{-1} in pullulan appears in the spectra of dextran at 1018 cm^{-1} while the band at 996 cm^{-1} is not present in the dextran spectra. Hence, these bands differentiate the pullulan and dextran based on their presence or absence (Shingel, 2002).

3.1.5 Alginate

Phycocolloids are the polysaccharides present in seaweeds and are used as thickeners in the food industry. Brown seaweeds' major structural polysaccharide is alginate while that of red seaweed is carrageenan (Pereira et al., 2009). Alginic acid is a heteropolysaccharide present in brown seaweeds. Alginic acid is made up of β-D-mannuronic acid and α-L-guluronic acid (Papageorgiou et al., 2010). In brown seaweed alginate, the value of mannuronic acid is higher than guluronic acid blocks. The quality of raw seaweed can be checked nondestructively and rapidly through this technique (Gómez-Ordóñez & Rupérez, 2011). Bands around 700 and 800 cm^{-1} are assigned to the guluronic and mannuronic acids while the wavenumber ranging from 3000 to 3600 cm^{-1} shows the stretching O–H bonds while 2920–2850 cm^{-1} shows the stretching of aliphatic C–H. Furthermore, enzyme tannase is entrapped in the alginate beads it can be helpful in the controlled release of enzymes or in improving tea and juice quality (Larosa et al., 2018).

3.1.6 Guar Gum, Chitosan, and Agar

Guar gum has a wide application in the food industry as a stabilizer and thickener. It is derived from cluster bean and is composed of β-1 → 4-linked mannose and α-1 → 6-linked galactose units. Partly hydrolyzed guar gum has gained consideration as water-soluble dietary fibers and its production is controlled through mannase and pectinase enzymes. There is no alteration in the functional group of partially hydrolyzed guar gum compared with the native gum. Sharp absorption bands around 1648 cm^{-1} also indicate the increased solubility of semi-hydrolyzed guar gum (Mudgil et al., 2012). Chitin is also a polysaccharide widely present in nature and its films are promising due to its good antimicrobial and barrier properties. The effect of moisture on the neutralization of amorphous chitosan films shows that chitosan does not present bands in the range of 2500–1750 cm^{-1}. The shift of the amide band to 1640 cm^{-1} and the vibration of carboxylate ions at 1560 and 1410 cm^{-1} distinguish the spectrum of chitosan acetate. Moreover, at the water activity of 0.44 and 0.55, the spontaneous neutralization of acetate chitosan films takes place which is an alternative neutralization method without film damage. These films also maintain their original form at $a_w = 0.11$ (Mauricio-Sánchez et al., 2018). Agars and crystalline cellulose structure are also studied through FTIR spectroscopy which shows that the agar and its derivatives show characteristic bands in the region between 800 and 700 cm^{-1} (Kačuráková & Wilson, 2001).

3.1.7 β-Glucan

FTIR is a valuable tool used to study the glycans and carboxymethyl glucan hence obtaining fingerprints of different bonds (Ahmed & Ahmad, 2017). The stretching (υ) vibrations of carboxylate anion in carboxymethyl glucan are observed at 1600 and 1421 cm^{-1} (Kačuráková & Wilson, 2001). Along with barley and oat, other plants such as rye, mushrooms, apple pomace, and tomato fibers are also rich in β-glucan (Ahmad & Khalid, 2018). FTIR also provides structural information of β-glucans thus a valuable technique to distinguish between yeast and fungi β-glucans (Ahmad & Anjum, 2010). The importance of oat cereal and its use as healthy human food throughout the world is well known (Ahmad et al., 2010). Similarly, dietary fibers from barley such as β-glucan reduce the glycemic and cholesterol responses and can be used as a functional or nutraceutical ingredient in food products (Ahmad et al., 2009; Ahmad, Munir, et al., 2012). It has also got emulsification, gelation, and stabilizing properties (Grumezescu, 2016). The chemical characterization revealed that barley β-glucan is composed of β-[1-4]-linked glucose units (Din et al., 2009). If the sequence of (1 → 4)-linkage is longer then the glucan is less soluble (Ahmad, Anjum, et al., 2012). Most of the beta-glucan portion lies in the endosperm cell wall while its quantity and extraction are influenced by the cultivars. Dehulled barley and oats contain 3–7% β-glucan while wheat contains <0.5%. Seventy-five percent is present in the cell wall while 25% is in the aleurone layer. Beta-glucan extracted from the endosperm of barley also improves the quality parameters of bread. FTIR technique serves as an important tool to identify β-glucan in various products and even distinguish the sources of these glucan molecules based on the spectra obtained from the FTIR technique. (Ahmad et al., 2008). Similarly, β-glucan extracted from the oat grains is incorporated into different products. The comparison between oat and barley β-glucan gum pellets indicates that the absorption patterns are almost similar but band intensities vary. β-glucan from different sources show different bands at different frequencies (Table 1).

Carbohydrates in β-glucan gum pellets show a peak at 1040 cm^{-1} for C–O bond, 2940 cm^{-1} for C–H stretching, and 3400 cm^{-1} for –OH stretching (Fig. 3). The stretching of C–O in the region of 1200–950 cm^{-1} indicate the carbohydrate nature of β-glucan. Band in the region 1000–1200 cm^{-1} depicts polysaccharide nature while the region 1500–1700 cm^{-1} depicts residual proteins attached with β-glucan molecules (Ahmad & Anjum, 2010). The absorption peak at 1156–1165 cm^{-1} depicts the linearity in the structure of β-D-glucan (Ahmad et al., 2010). Bands at 1156, 1076, and 894 cm^{-1} characterize the (1 → 3) and (1 → 4)-linked β -D-glucan while the absence of a band at 834 cm^{-1} region shows that there are no alpha linkages present in the samples (Ahmad & Anjum, 2010). The stretching of protein groups (CN and NH) give bands at 1566 and 1658 cm^{-1} (Ahmad et al., 2010). Absorption peaks for lipids appeared in the 2840–3000 cm^{-1} region while the major peak at 2918–2924 cm^{-1} is because of the CH_2 stretching. The bands in the region of 1640 and 3415 cm^{-1} indicate the presence of water. The water spectra (1600–1650 cm^{-1}) overlap with the protein spectra. A major portion of β-glucan

Table 1 Frequencies of FTIR bands of β-glucan from oat and yeast

Absorption bands (Frequency cm^{-1})	Assignment	Comments
810	C-O-S bond	Sulfated β-glucan (oat)
894	δ (C-1-H)	β-Glucan (oat)
1076	ν (CC) (CO)	β (1 → 3) glucan (oat)
1074	ν (CC) (CO)	Glucopyranose (oat)
1156–65	ν (CC) (CO) (COC)	β–glucan (oat)
1103–09		β-(1 → 3) glucan
1465–1468	δ CH_2	Sugars (yeast)
1560	ν CN	Amide II (yeast)
1566	ν CN	Amide II (oat)
1640	ν NH	Amide I (yeast)
1658	ν NH	Amide I (oat)
2853, 2922	ν CH_2	Sugars (yeast)
2855	ν CH symmetric	Lipids (oat)
2918–24	ν CH asymmetric	Lipids (oat)
3415, 3447	ν OH	Water (oat)
3350–3380	ν OH	Water (yeast)

ν Stretching, *δ* Bending
Sources: Ahmad & Anjum, 2010; Chang et al., 2006; Novák et al., 2012; Shin et al., 2005

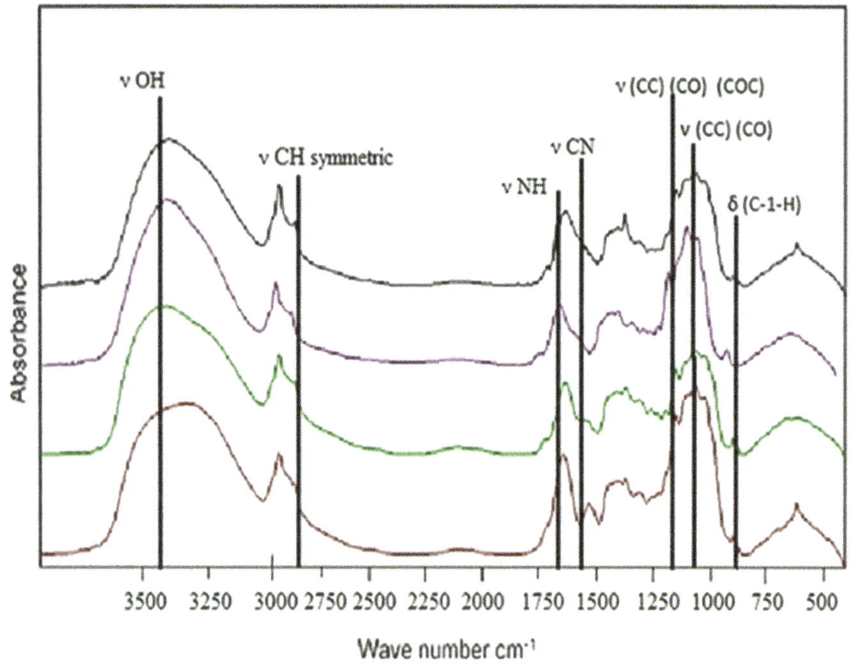

Fig. 3 FTIR spectra of β-glucan gum pellets

gum pellets is composed of carbohydrates with some fat, protein, and water impurities (Ahmad & Anjum, 2010).

3.2 Meat and Meat Products

The information about the quality of meat throughout the production process is beneficial both for the consumers and manufacturers. FTIR together with chemometric methods quickly calculates the bacterial count from the surface of meat while it also quantifies the pork in beef meatballs. It also studies the collagen and gelatin from skin and bones along with the denaturation process of myofibrillar and connective tissues. It shows that with the increase in temperature α-helix structure decreases while β-sheet structure increases while these changes are more noticeable in myofibrils (Damez & Clerjon, 2013). As the freezing and thawing of chicken greatly affect its quality by forming crystals, protein degradation, and lipid oxidation hence it is important to differentiate fresh meat from frozen one. Based on this information, the FTIR technique can now predict the storage conditions of the meat products at which these are kept during storage. Hence, the frozen minced meat when analyzed using FTIR, revealed the differences in the refrigerated and frozen samples at 1660–1628 cm^{-1}. The amide I band indicates the alterations in the backbone of protein and the pattern of hydrogen bonding. The use of improved pattern recognizing algorithms could improve the performance of FTIR spectroscopy in identifying frozen and thawed meat (Grunert et al., 2016).

The composition of meat fat is the major determining factor of meat quality. Mid-infrared Fourier transform spectroscopy productively forecast fatty acids in the beef, chicken, and pork fats. As per the regulation of the European Union (EU), it is compulsory to specify the components of saturated fats along with total fat content. FTIR in combination with chemometric methods quantifies the saturated and unsaturated fats efficiently (Lucarini et al., 2018). Fish are rich in omega-3 EPA (eicosapentaenoic acid) and DHA (docosahexaenoic acid). The quality of fish meat is affected by the change in the lipid and protein portions. Fatty acid composition, total fat, and nutritional parameters of fish are studied by FTIR in combination with partial least square where the strong bands at 3360 cm^{-1} are assigned to the OH stretching of water, 3270 cm^{-1} to N–H stretching of proteins, 1550 cm^{-1} to the bending and stretching of N–H and C–N protein respectively and 1652 cm^{-1} to C=O stretching of proteins while weak bands at 2958 cm^{-1} are assigned to methyl asymmetric stretching of lipids, 1239 cm^{-1} to asymmetric stretching of phospholipids, and 1124 cm^{-1} to the C–O stretching vibrations of ester (Hernández-Martínez et al., 2013).

3.3 Fruits and Vegetables

Natural polysaccharides or food additives polysaccharides like pectin, starch, and galactan are well distinguished through FTIR in the region of 1200–800 cm^{-1}. Mid-infrared Fourier transform spectroscopy in combination with multivariate analysis study the contents of capsicum. Capsicum has an antioxidant activity while capsaicinoids possess an antibacterial effect. The absorption peak 3600–3000 cm^{-1} are assigned to hydroxyl stretching of water while peaks at 2950–2800 cm^{-1} were allocated to C–H stretching of methylene and methyl groups. The peak at 1750–1710 cm^{-1} is allocated to the stretching of the carbonyl group while the peak at 1650 cm^{-1} to amide and OH stretching. The bands at 1200–950 cm^{-1} represented the stretching of C–O and C–C (Domínguez-Martínez et al., 2014). Tomato cultivars are also classified and determined through ATR-FTIR based on sugar and acid along with the prediction of its quality parameters (Bobelyn et al., 2006). The absorbance peaks of tomato sample at 3359 and 1677 cm^{-1} are the same as that of other fruits and vegetables while the major difference is in the regions of 1500–1000 and 3300–2750 cm^{-1} which depict the presence of sugars and organic acids. The absorption range, i.e., 1153–904 cm^{-1} shows the stretching of C–O and C–C along with O–C–H, C–C–H, and C–O–H bend in the region of 1474–1199 cm^{-1} are indicators for the presence of carbohydrates. The revealing C–H bending at 1583 cm^{-1} relates to the presence of organic acids. Soluble solids are depicted by the peaks 1067, 1082, and 1593 cm^{-1} while the presence of glucose and fructose in tomatoes is depicted at 1034 and 1067 cm^{-1}. Absorption peaks at 1718, 1593, 1369, and 1227 cm^{-1} are strong indicators of acidity and reveal the presence of citric acid (Ścibisz et al., 2011).

Apple wax, vitamin C in mixtures, and polymethoxylated flavones in orange peel oil are also analyzed through this technique. Polymethoxylated flavones are anticancer and anti-inflammatory. The low intensive C=O vibration at 1724 cm^{-1} depicts the low amount of non-PMF constituents (Bernal et al., 2011). In apricot slurries, the absorption zones 3700–2800 and 1800–1470 cm^{-1} are assigned to water while sugars and organic acids are present in the zone (1500–900 cm^{-1}). Band in 1500–1200 cm^{-1} region depicts the deformation of $-CH_2$, C–C–H, and H–C–O while bands in the range 1200–950 cm^{-1} show the C–C and C–O stretching (Bureau et al., 2009). Moreover, the contents of bioactive compounds in kiwi fruit are also studied through FTIR showing that kiwi fruits are composed of different components such as polyphenols, tannins, ascorbic acid, phenolic, and flavonoids. The wavelength numbers (866, 1026, 1237, 1451, 1542, and 1619 cm^{-1}) are shown by the gallic acid while catechin spectra at wavelength numbers (831, 1040, 1112, 1144, 1285, 1478, 1512, and 1611 cm^{-1}) are assigned to –CH alkenes, COH alcohols, CO alcohols, CH alkane, C=C aromatic ring, and C=C alkenes (Park et al., 2011).

3.4 Dairy and Dairy Products

UHT-treated milk endures significant changes in the physiochemical properties during processing and storage leading towards sedimentation or gelation. These changes being high in skim and whole milk at high temperature are affiliated with the protein-fat and protein-protein interaction. Milk proteins exhibit three absorption bands in the IR spectrum, i.e., amide I (1600–1695 cm^{-1}), amide II (1520–1560 cm^{-1}), and amide III (1230–1300 cm^{-1}) while the carbohydrates appear in the range of 800–1300 cm^{-1}. Moreover, water bands predominate around 1600 cm^{-1} and 3000 cm^{-1}. An increase in the sedimentation of milk is associated with the increasing intermolecular β-sheets. The presence of lipid absorbance in regions of 3000–2800 cm^{-1} and 1800–1700 cm^{-1} indicate the participation of protein-lipid interactions in sedimentation. Water bands predominated in the region 3500–3000 cm^{-1} and 1730–1600 cm^{-1} while the changes in the carbohydrates and protein are less intensive (Grewal et al., 2017). The microwave treatment of β-lactoglobulin in D_2O solution shows that the increasing temperature increases the protein aggregation more as compared to conventional heated solutions. The unfolding of β-lactoglobulin is also enhanced at 70–90 °C (Gomaa et al., 2013). Cholesterol in milk products is also determined in the region of 3200–2800 cm^{-1} in nearly 5 min. Vibrations in the region 2940–2962 cm^{-1} and 2872–2875 cm^{-1} are associated with the stretching of the CH bond in the methyl group. The peaks in region 3100–2990 cm^{-1} are related to the stretching of C-H bonds in cyclic alkanes (Paradkar & Irudayaraj, 2002).

Fourier transform infrared spectroscopy is also used to analyze the fresh milk over the range of 5000–900 cm^{-1} to predict the yield of cheese. The region 1582–930 cm^{-1} consists of peaks for carbohydrates and organic acids. It predicts the major nutritional contents and their retention in the cheese (Ferragina et al., 2013). Ripening of cheese is a complex procedure including events like metabolism of lipid, protein, lactose, citrate, and lactate. The analysis of water-soluble fractions shows a high amount of amino acids like leucine, glutamic acid, asparagine, etc., while the changes are more intense in the range of 1800–900 cm^{-1} (Subramanian et al., 2011). Italian cheeses are also classified based on their ripening time and their technique of manufacturing. Major differences in the cheese spectra are observed in the regions 3700–3000 and 1700–1000 cm^{-1} while it also inspects the bands of fats, lactose, lactic acid, and proteins in the milk products (Lerma-García, Gori, et al., 2010).

The study of chemical composition, fatty acid profile, and trans-fat of margarine show that the major variations are in the region 3690–2942 cm^{-1} due to OH stretching and cis-trans double bonds while the region 2911–2876 cm^{-1} is associated with the vibration of methyl asymmetrical stretching. Different parameters of margarine composition are obtained in 10 min while the same analysis takes hours during conventional analysis while the concentration of adulterants such as sodium bicarbonate, sodium citrate, and cheese whey in milk is also measured through FTIR. It is an effective technique to monitor adulterants in dairy products possessing a low cost of operation and higher analytical performance (Cassoli et al., 2011; Hernández-Martínez et al., 2010).

3.5 Oil and Fats

Physicochemical properties of corn and mustard oil to assess their compositional quality, as well as the effect of frying on various characteristics of oil, can be best studied through FTIR. This technique is also effective in the elucidation of changes in oils during repeating frying operation and shed light on oxidative and thermal reactions which alter the physiochemical, nutritional, and sensory properties. Though all these changes in the fried oil depend upon the frying conditions, oil quality, and can also be detected through conventional chemical analysis FTIR provides a rapid solution to get the overall quality of frying oils. As fats and oils are simple molecular systems and mainly possess an uncomplicated spectrum between 800 and 3500 cm^{-1} (Ismail et al., 1997). In oil samples at room temperature, the specific band at 3473 cm^{-1} is associated with stretching of O-H, stretching of C-H is revealed at 3006 cm^{-1} and bands between 2854 and 2925 cm^{-1} appear due to double bond vibration. At boiling point, the band 1745.5 cm^{-1} is assigned to double bond stretching while 1654–1658 cm^{-1} band to C=C stretching. The fresh and used oils show similar spectra though the heights in specific bands indicate the difference in quality (Zahir et al., 2017). Whereas, the oil absorption by the starchy fried products is also a huge concern for the consumers as it affects the health and its content in fried starchy products are determined through ATR-FTIR spectroscopy showing that the huge amount of oil is absorbed by the starch but the major quantity of it is located near the surface. Soybean oil exhibits $-CH_3$ characteristics in the range 2995–2886 cm^{-1} while ester group at 1793–1693 cm^{-1} (Chen et al., 2018). Moreover, the analysis of the peanut butter and mayonnaise show that the fat and water peaks are quite prominent in mayonnaise while the water peak in the peanut butter spectrum is missing which may be because of less water content (van de Voort et al., 1993).

Virgin coconut oil has a variable effect on HDL cholesterol, total cholesterol, LDL, and triglycerides levels. It differentiates from olive oil in the frequency region of 1111 and 962 cm^{-1} while with palm oil at 1120–1105 and 965–960 cm^{-1}. The virgin coconut oil if mixed with palm and olive oil show bands in the region of frequency 1120–1105 and 965–960 cm^{-1} is effectively analyzed through FTIR spectroscopy in combination with ATR and PLS (Rohman et al., 2010). Whereas, the polyunsaturated fatty acids are quantitatively analyzed in oils showing the band at 3013 cm^{-1} which represent the unsaturated fatty acids while bands in the range of 3000–2800 cm^{-1} attribute to the C–H stretching. Strong peaks at 1145 and 1098 cm^{-1} are assigned to bonds found in esters and ethers (Vongsvivut et al., 2014). Likewise, the analysis of cinnamon oil shows a peak at 1727 cm^{-1} which is assigned to the aldehyde of saturated fat while peaks at 1679 and 1626 cm^{-1} represent the stretching of aldehyde carbonyl C O. Peak at 1573 cm^{-1} represent aromatic ring C C, 1450 cm^{-1} for alcohol COH, 1248 cm^{-1} to aromatic acid ester C O C. The volatile components of oil display high diversity and complexity (Li et al., 2013).

Oils from different marine sources including marine microorganisms are rich in n-3 polyunsaturated fatty acids (eicosapentaenoic acid and docosahexaenoic acid). The bands at 3013 cm^{-1} are allocated to C–H stretching of cis-alkene which represent unsaturated fatty acids while bands in region 3000–2800 cm^{-1} represent the C–H stretching of methyl and methylene pillars of lipids. The peak at 1743 cm^{-1} is because of the C=O stretching of ester groups from lipids while the intense bands at 720 cm^{-1} attribute to methylene deformation in cis-substituted olefins (Vongsvivut et al., 2014). Moreover, encapsulated fish oil is also analyzed through FTIR spectroscopy revealing bands at 3290 cm^{-1} assigned to O–H stretching, C–H stretching at 3015 cm^{-1} and C–H stretching of methyl and methylene pillars of lipids at 3000–2800 cm^{-1}. Overlapping bands at 1460–1370 cm^{-1} represent the deformed modes of methyl and methylene groups from proteins as well as lipids. The band intensity tends to increase at 2923, 1744, 1146 cm^{-1} with the increase in oil quantity (Vongsvivut et al., 2012).

3.6 *Limitations of FTIR*

One of the greatest powers of FTIR is its capability to acquire spectra from a wide range of solids. However, at times certain solids and liquids need sample preparation to acquire a spectrum of good quality (Franca & Oliveira, 2011). For instance, one of the common methods of sample preparation for solids is to grind them into powder form and then dissolve them in the solid medium which in most cases is KBr. Moreover, the liquid transmission needs an easier sample preparation as compared to solids though both the medium of samples can face unavoidable reproducibility concerns. However, this problem has been somewhat minimized recently through Attenuated Total Reflectance (ATR) and will get better over time because of the invention of new supporting methods and techniques. Moreover, FTIR is more applied to determine the change in the molecular structure during processing and treatments. It is a challenge to use it to determine the actual compound present in the sample (e.g., how much sugar). However, it can be done by using standards that are tough to find and calibrate.

In FTIR, the results can also be misinterpreted because of the overlapping spectra produced by complex samples especially a mixture of bacteria. Therefore, in certain cases a bacterial purification or separation step becomes necessary. This overlapping can be resolved by the Fourier deconvolution as it increases the peak height and reduced the bandwidth. Moreover, it may well need hard data gathering, calibration, and skill in chemometric analysis of spectra. Furthermore, few protocols are being developed which could help one understand and apply this technique better. Hence, reproducible, rapid, and accurate methods need to be developed to make the process smoother. Also, for the analysis of a compound in IR spectroscopy dipole moment is required thus the compounds having no dipole moment will not absorb IR rays and would not be analyzed. Likewise, during the analysis of a mixture of compounds two compounds may absorb IR rays at the same wavelength which causes the masking of peaks. All these limitations can be reduced by developing more and more protocols considering the major limitations. FTIR is more applied to determine the change in the molecular structure during processing, source, and treatments. It is a challenge to use it to determine the actual compound present in the sample (e.g., how much sugar). However, it could be done by using standards, which are not easy to find and calibrate.

3.7 Comparison with Other Similar Methods

Food is a complex system that makes food analysis an intricate process. Various techniques are being used to analyze food while the choice of an analytical technique depends upon the type of analysis and the properties of food to be analyzed. Ultraviolet and visible spectroscopy is one of the commonly used techniques to analyze food. Ultraviolet wavelength ranges from 200–350 nm while visible wavelength range from 350 to 700 nm. The sample solution used must be clear and uniform while the sample solution is normally water or an aqueous buffer. Moreover, for good results, the cuvette should be quartz or silicate glass. Different anthocyanins, flavonoids, phenolic, and carotenoid content in food can be measured through this technique (Tiwari et al., 2013). Another analytical technique used in food analysis is fluorescence spectroscopy but it is more sensitive and uses a different optical system than UV. Moreover, two-wavelength selectors are required so they can study the conformational changes in protein as well as the content of vitamins (Lakowicz, 2013). Both these techniques differ from FTIR as UV spectroscopy uses a different range of wavelength and source while it is mainly a quantitative technique whereas FTIR is qualitative as well as a quantitative technique. Moreover, the limitations of fluorescence include loss of photostability and not all molecules being fluorescent which is not an issue when it comes to FTIR.

Atomic spectroscopy is another spectroscopy technique used mainly for the measurement of trace amounts of minerals in food which limits its functioning to elements. On the other hand, it requires more samples, is less sensitive, and can analyze only solutions. Furthermore, Raman spectroscopy is a vibrational technique that complements infrared spectroscopy specially FTIR. It is mainly used for the analysis of organic compounds. It can also measure the changes in chemical bonding, e.g., when a substrate is added to an enzyme (Li-Chan et al., 2010). As compared to the infrared it possesses weak water scattering hence a low quantity of organic molecules in an aqueous solution is allowed.

Nuclear magnetic resonance spectroscopy (NMR) studies a wide range of food constituents and provides their structural data. It can analyze samples in solution as well as solid state. Insufficiently concentrated samples, low sensitivity, difficulty in handling mixtures, and magnetic field drift are some of the limitations of this technique. Additionally, the X-ray diffraction technique is also used to identify the structures of carbohydrates, lipids, and proteins. X-ray fluorescence detects the small number of metal ions in food. All these and some more techniques can be used for the analysis of different food components and products. Moreover, they are all different from each other in one way or the other and possess certain limitations. What makes FTIR different from these techniques is its properties such as rapid, little or no sample preparation, nondestructive, sensitive, multiple sample environment, qualitative as well as quantitative and mechanical simplicity. Moreover, it can be modified according to the sample and is internally calibrated. Additionally, different variety of food components and products can be analyzed using FTIR such as butter, sugar, milk, meat, oils, and juices which makes it superior to the other methods.

3.8 *Bacteria in Food Industry*

The composition of bacterial cells is analyzed by FTIR spectroscopy. In combination with different chemometrics, it is possible to detect, classify, quantify, and differentiate the bacterial cultures in food (Davis & Mauer, 2010). Bacteria are often used in the food industry as starter cultures but they can also cause food spoilage and food poisoning. FTIR identifies these strains by comparing the obtained identification spectra with the reference spectra (Wenning et al., 2014). For the identification of bacteria five main regions of absorbance in infrared spectra should be analyzed. The fatty acid region is 2800–3000 cm^{-1}, proteins and peptides (1500–1700 cm^{-1}), a mixed region of peptides, phosphate compounds and fatty acid (1200–1500 cm^{-1}), carbohydrates (900–1200 cm^{-1}), and 700–900 cm^{-1} is called as fingerprint region which contains absorbances that are the attributes to specific bacteria. This technique is also used to differentiate between the fresh, semi-fresh, and spoiled beef fillets. The comparison provides information about the biochemical changes during spoilage. The peak at 1640 and 1550 cm^{-1} represent the moisture and amide respectively while peaks at 1460, 1240, and 1175 cm^{-1} represent the fat presence (Argyri et al., 2010). Raw meat deteriorates rapidly even if stored under refrigeration. The major peak is observed at 1639 cm^{-1} indicating the presence of water while the peak at 1550 cm^{-1} represents the absorbance of amide II. The wavenumbers from 1408 to 1370 and 1320–1305 cm^{-1} are linked to the amides and amines and are strongly associated with the spoilage of meat (Vasconcelos et al., 2014).

Metabolites formed by the different microorganisms cause the spoilage of meat. Salmonella, Pseudomonas, and Campylobacter species are a major threat to meat quality which requires a rapid detection method to identify and control them. The biochemical changes in the meat are effectively measured by FTIR spectroscopy which enhances microbial spoilage detection. The most intensive peaks in fresh meat are of amide bands at 1640 and 1550 cm^{-1} suggesting the decrease in protein content during spoilage. These peaks start to increase after 16 h because of the production of more free amino acids by aerobic microorganisms (Ellis et al., 2002). Likewise, beef is a widely consumed food and is also very susceptible to microbial spoilage. FTIR spectroscopy allows the rapid detection of microbial spoilage in fresh meat. The absorption band at 1513 cm^{-1} represents C–N stretching and N–H bending while 1413 cm^{-1} refers to C–H stretch, 1382 cm^{-1} to NO_2, and 1112 cm^{-1} to C–N stretch (Ellis et al., 2004).

Escherichia coli and Pseudomonas in the bottled drinking water are also identified through Fourier transform infrared spectroscopy using multivariate analysis. The presence of these coliforms in drinking water indicates insufficient treatment or posttreatment impurity. The comparison of inoculated and non-inoculated drinking water showed that this technique also discriminates against bacteria based on the differences in biochemical and phenotypic features (Al-Qadiri, Al-Holy, et al., 2006). It is also applicable in the characterization of different microorganisms, characterization of E. coli and Alicyclobacillus strains in the juice while it also detects the effect of chlorine on bacterial injury in water. The absorption bands vary from bacteria to bacteria and product to product (Table 2). Listeria monocytogenes is also a major threat to the safety of food products because they can grow in a refrigeration environment. Different lactic acid bacteria produce bacteriocins which are antibacterial peptides hence hindering the growth of other bacteria. The susceptibility of L. monocytogenes to these bacteriocins is also evaluated through FTIR (Oust et al., 2006).

Milk contains a high amount of water and exists at about a neutral pH range (6.5–6.7) which makes it more prone to microbial attack. Temperature between 2 °C and 30 °C is also ideal for microbial growth. FTIR technique rapidly quantifies the microbial load in pasteurized milk (Nicolaou & Goodacre, 2008). This technique also discloses the entrance of Staphylococcus aureus in dairy production through milk contamination caused by cow infections (Kümmel et al., 2016). The fortune of smear starters during the cheese ripening studied through FTIR shows that the starter cultures do not outcompete the residential bacteria which may be due to the starter culture application, competitiveness of resident microorganisms, or both these factors (Goerges et al., 2008).

Lactic acid bacteria are a group of microorganisms that highly relate to fermented food and human health. They form lactic acid by fermenting carbohydrates. Lactic acid bacteria include genera like Lactococcus, Pediococcus, Leuconostoc, Streptococcus, Bifidobacterium and Lactobacillus, etc. It is difficult to identify the LAB because of their rich and closely related species. Hence, FTIR is an effective technique to identify and characterize these bacterial strains rapidly in the spectral region of 1450–700 cm^{-1} (Wenning et al., 2010). Moreover, the effect of exopolysaccharide-producing strains and non-producers is also studied through FTIR. Exopolysaccharides are the macromolecules produced by different microbes. Lactobacillus kefiranofaciens is a unique lactic acid bacterium, which produce kefiran. Different cocultures improve product quality and its nutritional value. The strong peaks at 1647.31 cm^{-1} represent the polysaccharides while peaks at 3395.07, 2924, 1558.07, 1245, and 1061 cm^{-1} represent sugar residues, stretching of C H in sugar ring, N H bending, sulfate groups, and C O stretching, respectively. This study shows that the coculturing technique effectively produces exopolysaccharides along

Table 2 Different absorption ranges of the microbial constituents

Study	Absorption bands (Frequency cm^{-1})	Assignment	Comments
Characterization of microorganisms	1085	ν (P=O) symmetric	Phospholipids
	1000–1200	ν (C-O-C)	Carbohydrates
	1457	δ (CH_2)	Lipids, proteins
	1670	ν (C=N)	RNA, DNA bases
	1708	ν (C=O), hydrogen bonded	RNA, DNA
	1741	ν (C=O)	Lipids
Detection and discrimination of L. monocytogenes from other strains	2852–2957	ν (CH_3) asymmetric	Fatty acids
	1250	ν (P=O) asymmetric	Phosphate
	1455	δ (CH_3)	Proteins
	1735–1745	ν (C=O)	Fatty acids/ esters
	3000	ν (CH_2, CH_3)	Fatty acids
Detection and identification of Pseudomonas aeruginosa and Escherichia coli in bottled drinking water	950–1050	ν (C-O-C)	Polysaccharides
	1398	δ (CH_2) symmetric	Protein
	1455	δ (CH_3) asymmetric	Protein
	1540	δ (NH)	Amide II
	1650	ν (C=O)	Amide I
	1740	ν (C=O)	Lipids
	2929,2960	ν (CH_2, CH_3) asymmetric	Fatty acids
Effect of chlorine-induced bacterial injury in water	950–1100	ν (C-O-C)	Polysaccharides
	1242	ν (P=O) asymmetric	Nucleic acid
	1398	δ (CH_3) symmetric	Nucleic acid
	1455	δ (CH_2) symmetric, δ (CH_3) asymmetric	Protein
	1540	ν (C-N), δ (N-H), δ (CH_2, CH_3)	Amide II
	1650	ν (C=O)	Protein/amide I
Differentiation and detection of microorganisms	1047	ν (COC)	Carbohydrates
	1732	ν (C=O)	Esters
	2856, 2928	ν (CH)	Fatty acid

(continued)

Table 2 (continued)

Study	Absorption bands (Frequency cm^{-1})	Assignment	Comments
	3372	ν (OH)	Water
Detection and identification of bacterial strains in juice	900–1200	ν (COC)	Polysaccharides
	1080, 1242	ν (P=O)asymmetric, symmetric	Nucleic acid
	1398, 1455	δ (CH_3) asymmetric, δ (CH_2) symmetric	Proteins
	1540	δ (NH)	Amide II
	1650	ν (C=O)	Amide I
Detection and evaluation of beef and chicken breast fillet spoilage	1025–1140	ν (CN)	Amines
	1175, 1240, 1460	C=O ester	Fat
	1240, 1315	ν (CN) ν (CO) δ (NH) δ (O=CN)	Amide III
	1550	ν (CN) δ (NH)	Amide II
	1640	ν (OH)	Water

δ bending, ν stretching
Al-Qadiri, Al-Holy, et al., 2006; Al-Qadiri, Lin, et al., 2006; Al-Qadiri et al., 2008; Argyri et al., 2010; Irudayaraj et al., 2002; Lin et al., 2004; Yu & Irudayaraj, 2005

with protein and sulfur-containing substances (Ahmed, Wang, Anjum, Ahmad, Ahmad, & Raza, 2013).

Polyhydroxyalkanoates (PHA) are synthesized by the prokaryotic microorganisms while polyhydroxy butyrate (PHB) is the common PHA having application at the industrial level. It is an alternative to plastics hence its industrial fermentation is under focus. FTIR is applied to quantitatively investigate the polyhydroxy butyrate content of bacterial cells through the combined use of the Partial least square (PLS) technique. The bands at 1738–1728 cm^{-1} are due to ester carbonyl stretch, at 1450 cm^{-1} due to methyl deformation, and 1654 cm^{-1} show the amide carbonyl stretching. The study shows the amorphous nature of polyhydroxy butyrate while it also predicts the degree of PHB crystallinity (Kansiz et al., 2000). Furthermore, the strains of yeast are also identified through this technique and it identifies the S. cerevisiae strains with 92% success while that of D. hansenii with 91% (Wenning et al., 2002). This technique in combination with PLS utilizes the spectral range of 2000–1000 cm^{-1} to determine the microbial quality of minimally processed pomegranate arils (Adiani et al., 2018). Microorganisms in the biofilms are also studied efficiently within the region 4000–500 cm^{-1} because it holds characteristic bands for microbial characterization (Schmitt & Flemming, 1998). The FTIR study of the fermentation of shrimp shells to produce chitin and chitosan through the *Bacillus subtilis* starter cultures show that this bacteria is an efficient starter culture (Sini et al., 2007).

Microorganisms are also identified and differentiated through FTIR spectroscopic imaging which is a novel method having received great consideration. It possesses high molecular sensitivity in combination with spatial resolution down to the micrometer level while it also examines samples under native conditions and applies to a wide range of products (Steiner & Koch, 2009). This imaging technique also diagnoses pathogenic bacteria rapidly and cost-effectively. Different bacteria are accurately identified through the investigation of the levels of taxonomic resolution. The strains from the same species are successfully differentiated through FTIR microspectroscopic imaging (Lasch et al., 2018). It also measures the micro-colonies reducing the incubation time and accelerating the identification process. Moreover, the old colonies are better identified through conventional spectroscopy as compared to microspectroscopy (Wenning & Scherer, 2013). Furthermore, different wheat cultivars are also analyzed through this technique while the images ranging from bran to the endosperm center are obtained showing delicate differences between the cell walls of different varieties (Barron et al., 2005).

3.9 Comparison with Other Methods

Bacteria are present in large numbers in food products among which some are beneficial while some are harmful to humans. Hence, it becomes critical to analyze them in different products. Various analytical techniques are being used to analyze the bacteria for a long time. However, with each passing year, new techniques are being developed which are more efficient than the previous ones. Culturing, enumeration, and isolation of colonies for further identification is the basis of the traditional way to detect and identify the bacteria in food (Hakovirta, 2008). Culture media technique has been a traditional choice for either fresh or processed food. The microscopic colony count technique is a simple method to morphologically identify and count bacteria at initial levels. However, it has certain disadvantages as both viable and nonviable cells are enumerated while some cells might not get stained properly. Plate count is another traditional method of bacterial identification in which samples are blended, diluted, plated, and incubated at an appropriate temperature. It identifies the colony-forming units in food products. Dye reduction is another traditional technique used to detect the microbes present in food. The dye reduction time is inversely proportional to the number of microbes present in the food sample. However, this technique is not efficient for food samples that contain reductive enzymes.

Some other comparatively older bacterial identification techniques are flow cytometry, ATP bioluminescence assay, epifluorescent filter technique, ELISA, ELFA, and PCR. However, these techniques are more rapid than purely traditional techniques. All these techniques are used to enumerate bacteria in various food items such as milk, wine, water, and juices (Van Alfen, 2014). Moreover, the traditional techniques require a huge amount of time, glassware, laboratory work, and culture media preparation. Likewise, these techniques are rather insensitive, nonspecific visual appearance, and are limited to certain bacteria with known growth requirements. On the other hand, FTIR is less time-consuming, reliable, economic, more sensitive, no glassware required, less laboratory work, and no media preparation (Davis & Mauer, 2010; Vongsvivut et al., 2012). Moreover, it is more applied to detect the changes at the molecular level which makes it more efficient than the other traditional techniques. It can analyze a huge variety of bacteria present in different food products.

4 Food Authentication Using FTIR Spectroscopy

Adulteration is the addition of any non-authentic substance or removal of authentic substance without the consumer knowledge for economic gain (Zhao et al., 2014). Authentication is the process that verifies the compliance of the food with its label description. Fraud cases (68%) in 2016 were reported in different products with variable fat content (meat 27%, fish 13%, fats 11%, milk 10%, nuts 4%, and animal by-products 3%). In the USA, 1500 lbs of chicken were found contaminated with nitrofurazone (a veterinary drug). In China and India, milk, edible oils, meat, nuts, etc., are highly susceptible to adulterations (Valdés et al., 2018).

4.1 Oils and Fat Authentication

Fats and oils are simple molecular systems and mainly possess uncomplicated spectrum between 800 and 3500 cm^{-1} (Ismail et al., 1997). Olive oil is available in diversified forms as extra-virgin oil (EVO), virgin, refined, and pomace (de la Mata et al., 2012). Extra-virgin olive oil is adulterated with canola, hazelnut and sunflower oil as the cost of EVO is high as compared to others hence more profit is gained through adulteration (Maggio et al., 2010). Olive oil being costly is often adulterated with less expensive oils like palm oil. Both the oils possess differences at frequency 1402 and 1030 cm^{-1} (Rohman & Che Man, 2010). Products of olive oil contain a high amount of oleic acid and are adulterated with cheap oils such as canola and palm oil. They possess similar spectra though both the oils show a slight difference at 3005 cm^{-1} (Rohman et al., 2014). Moreover, sesame oil present in olive oil can be judged using frequencies 1207–1018, 1517–1222, and 3050–2927 cm^{-1}. The presence of high amounts of stearic and linoleic acids in these oils increase with the increase in the amount of sesame oil and can be detected using the FTIR technique (Rohman & Che Man, 2012b).

FTIR coupled with the multivariate analysis is used to classify vegetable oils according to their origin (Lerma-García, Ramis-Ramos, et al., 2010). Major peaks at 2937, 2856, 1749, 1454, and 1166 cm^{-1} represent CH stretching, CH bending, and

CO stretching. These play a vital role in the discriminant analysis (Yang et al., 2005). Vegetable oils are also adulterated with used frying oil showing less visible absorption bands in the region 4000–1650 cm^{-1} while additional bands in the range below 1650 cm^{-1} during FTIR study (Zhang et al., 2012). Different vegetable oils such as corn oil, canola oil, soybean, and sunflower oil are adulterated with lard whereas lard and vegetable oils show minor differences at 1117 and 1097 cm^{-1} corresponding to CH bending and deformation. An increase in the lard concentration will increase the peak height at 1097 cm^{-1} (Rohman et al., 2011).

Virgin coconut oil is often adulterated with cheap canola oil. This adulteration can be detected in the region of 3007 and 1654 cm^{-1}. In region 1120–1090 cm^{-1} canola exhibited two peaks as compared to coconut oil (Che Man & Rohman, 2013). Avocado oil is a newcomer in the oils market and is costly hence adulterated with different oils to gain profit and deceive the consumer. FTIR in combination with the SIMCA model rapidly identify and quantify avocado oil and its adulterants (Quiñones-Islas et al., 2013). Furthermore, cod liver oil offer many nutritional advantages as it contains EPA and DHA. It is also adulterated with lard which is determined through FTIR in region 1500–1030 cm^{-1} (Rohman & Che Man, 2009).

4.2 *Honey Authentication Using FTIR*

Honey is composed of fructose, glucose, maltose, sucrose, water, minerals, and high sugars. It is often adulterated with corn syrup, cane sugar, jaggery, and beet sugar but this adulteration is difficult to determine because of the variation in the origin, species, and environment (Mishra et al., 2010). This adulteration affects both the consumers and authentic honey producers hence the development of reliable and quantitative techniques for detection is important (Li et al., 2017). Adulteration of honey with inverted cane sugar is determined through ATR-FTIR which provides a benefit over traditional techniques like wet chemistry, enzymatic technique, and chromatographic procedures which are time-consuming, destructive, and costly (Irudayaraj et al., 2003). FTIR spectroscopy coupled with PLS is expected to replace the traditional, time to consume, and high-cost analysis of honey (Wang et al., 2010). Sugars are observed in the range of 800–1500 cm^{-1}. As the adulterated honey presents similar spectra of different components so principal component regression and partial least squares are applied to develop models for effective prediction of different variables. The bands in region 1470–1149 cm^{-1} represent bending C–O–H, O–C–H, and C–C–H while the bands at 1028 and 1053 cm^{-1} are the most intensive ones representing C–O and C–C stretching (Rios-Corripio et al., 2012).

4.3 Milk and Milk Products

Melamine consumption through adulterated milk in China during the year 2008 posed adverse effects on the consumers and the economy. Mid-infrared spectroscopy is effective in melamine detection coupled with the partial least square method (Domingo et al., 2014). Melamine is a synthetic nitrogenous chemical used to produce synthetic resins. It is added to foods to deceive the consumer about the protein content of the product but is a harmful chemical if over-consumed. The band at 806 cm^{-1} indicates the bending of the triazine ring in melamine while this band is not present in milk. The exclusive fingerprint of each compound is detected in the fingerprint region, i.e., 1300–910 cm^{-1} (Jawaid et al., 2013).

Milk proteins exhibit three absorption bands in the IR spectrum, i.e., amide I (1600–1695 cm^{-1}), amide II (1520–1560 cm^{-1}), and amide III (1230–1300 cm^{-1}). Milk consumption has increased in Asia over the years hence to meet the supply-demand vendors adulterate cow milk with soymilk. Soymilk contains the same proteins as cow milk excluding sulfur-containing amino acids. FTIR coupled with the chemometric analysis detects the presence of soymilk in cow milk. In soy milk amide I, amide II, and amide III are present in a high amount as compared to cow milk. Soy milk also depicts stronger absorption in the region 1639–1613 cm^{-1} because of the occurrence of secondary structures (Jaiswal et al., 2015). Adulteration with sodium bicarbonate, cheese whey, and sodium citrate in raw milk while chicken fat in butter is also determined through FTIR. Butter consists of more than 500 fatty acids including saturated, monounsaturated, and polyunsaturated fatty acids. The absorption band at 966 cm^{-1} is the marker band for determining the trans-fat associated with –HC=CH. The band at 3006 cm^{-1} indicates the high amount of unsaturated fats as possessed by the chicken fat particularly linoleic acid (Cassoli et al., 2011; Nurrulhidayah et al., 2013).

4.4 FTIR Authentication in Other Products

One of the most popular drinks throughout the world is tea while its color is very important for its market price and quality. Talcum powder is often added to the tea to hide the quality defects. FTIR spectroscopy discriminates the pure and adulterated tea samples showing absorbance peaks at wavenumbers 1401, 1617, and 3117 cm^{-1} are similar while peaks at 1016, 1182, 1249, 1340, 1631, and 2296 cm^{-1} distinguish talcum adulterated tea samples from pure tea sample (Li et al., 2016). The adulteration of sibutramine in green tea, green coffee, and herbal tea is also determined through this rapid technique. Sibutramine is a serotonin-norepinephrine reuptake inhibitor often added to herbal diets and supplements to enhance weight loss. The bands observed in regions 2698 and 2700–2600 cm^{-1} are specifically used for sibutramine detection (Cebi et al., 2017). FTIR also finds its application in the detection of adulterated chocolate with lard when coupled with ATR. The region of 4000–650 cm^{-1} is used to predict the lard and chocolate components. The band at 2924 cm^{-1} represents methylene asymmetrical stretching while the band at 2853 cm^{-1} represents methylene symmetrical stretching (Che Man et al., 2005).

4.5 Comparison with Other Methods

There has been a rise in the demand for food authentication because of the increasing awareness in the public considering food safety. There are different traditional methods to authenticate food being used such as sensory analysis (taste, color, and smell) and physiochemical testing. Moreover, in past few years certain other methods have also been developed such as chromatographic techniques, isotopic techniques, PCR, ELISA, mass spectrometry, and elemental profiling (Danezis et al., 2016). Physiochemical and organoleptic testing methods are comparatively simple and less expensive though they are often time-consuming and require different equipment and glassware. Gas chromatography is rapid, require small sample, and is sensitive; however, it is limited to volatile samples and some samples require extensive preparation which is not the case with FTIR spectroscopy. Moreover, ELISA can test a large number of samples, reproducible, and cost-effective; however, in some cases, it may be destructive to certain samples.

Polymerase chain reaction (PCR) is highly specific, sensitive, easy to use, and reproducible while on the other hand, it may at times give false-positive. Furthermore, mass spectrometry is fast and requires a small sample size though it does not directly give the structural information and nonvolatile compounds are difficult to handle. HPLC is another technique used to analyze food products. It is efficient, automated with less contamination however it is a costly technique, less sensitive, and complex which makes it less appealing when it comes to analyzing food products regularly. All these techniques aside, FTIR has its pros and cons. It is rapid, highly sensitive, high resolution, less expensive, provides rich information, and reproducible (Davis & Mauer, 2010; Vongsvivut et al., 2012). However, the limitations of FTIR include the inability to detect atoms or monoatomic ions, aqueous solutions are hard to analyze (i.e., water is a strong infrared absorber), inability to recognize molecules comprised of duplicate atoms, and complex blends give complex spectra.

5 Conclusion and Future Progress

Fourier transform infrared spectroscopy is a fast, nondestructive, and reliable technique requiring minimal or no sample preparation as compared to other techniques. It has the potential to analyze and authenticate a wide range of food products at macro and micro levels, though more research work is required to develop new protocols for the further acceptance of the technique in the scientific community. The use of FTIR imaging capacity enables us to examine the samples at the microscopic level and hence increase its application in the discipline of microbiology. Using this instrument, we can now analyze complex substances with ease and efficiency. Furthermore, it could be of great choice in the field of food science, food analysis, and product authentication with the further development of new protocols. Overall, it is a dynamic technique to keep a check on the quality and quantity of products and their constituents. Fourier transform infrared spectroscopy despite being an efficient tool to analyze and authenticate products has certain limitations in the application of this technique. Little protocols are being developed which could have helped one understand and apply this technique better. Reproducible, rapid, and accurate methods need to be developed to make the process smoother. Furthermore, the hardware and software of the technique should be improved to make it more cost-effective.

References

Adiani, V., Gupta, S., Ambolikar, R., & Variyar, P. S. (2018). Development of rapid method to assess microbial quality of minimally processed pomegranate arils using FTIR. *Sensors and Actuators B: Chemical, 260*, 800–807.

Ahmad, A., & Anjum, F. M. (2010). *Persepctive of βglucan: Extraction and utilization*. VDM Verlag.

Ahmad, A., Anjum, F. M., Zahoor, T., Chatha, Z. A., & Nawaz, H. (2008). Effect of barley β-glucan on sensory characteristics of bread. *Pakistan Journal of Agricultural Sciences, 45*, 88–94.

Ahmad, A., Munir, B., Abrar, M., Bashir, S., Adnan, M., & Tabassum, T. (2012). Perspective of β-glucan as functional ingredient for food industry. *Journal of Nutrition & Food Sciences, 2*, 133.

Ahmad, A., Anjum, F. M., Zahoor, T., Nawaz, H., & Ahmed, Z. (2010). Extraction and characterization of β-d-glucan from oat for industrial utilization. *International Journal of Biological Macromolecules, 46*, 304–309.

Ahmad, A., Anjum, F. M., Zahoor, T., Nawaz, H., & Dilshad, S. M. R. (2012). Beta glucan: A valuable functional ingredient in foods. *Critical Reviews in Food Science and Nutrition, 52*, 201–212.

Ahmad, A., Anjum, F. M., Zahoor, T., Nawaz, H., & Din, A. (2009). Physicochemical and functional properties of barley β-glucan as affected by different extraction procedures. *International Journal of Food Science & Technology, 44*, 181–187.

Ahmad, A., & Khalid, N. (2018). Dietary fibers in modern food production: A special perspective with β-glucans. In A. Grumezescu & A. M. Holban (Eds.), *Biopolymers for food design* (pp. 123–154). Elsevier.

Ahmed, Z., & Ahmad, A. (2017). Biopolymer produced by the lactic acid bacteria: Production and practical application. In A. M. Grumezescu & A. M. Holban (Eds.), *Microbial production of food ingredients and additives* (pp. 217–257). Elsevier.

Ahmed, Z., Wang, Y., Anjum, N., Ahmad, A., & Khan, S. T. (2013). Characterization of exopolysaccharide produced by lactobacillus kefiranofaciens ZW3 isolated from Tibet kefir– part II. *Food Hydrocolloids, 30*, 343–350.

Ahmed, Z., Wang, Y., Anjum, N., Ahmad, H., Ahmad, A., & Raza, M. (2013). Characterization of new exopolysaccharides produced by coculturing of L. kefiranofaciens with yoghurt strains. *International Journal of Biological Macromolecules, 59*, 377–383.

Al-Qadiri, H. M., Al-Alami, N. I., Al-Holy, M. A., & Rasco, B. A. (2008). Using Fourier transform infrared (FT-IR) absorbance spectroscopy and multivariate analysis to study the effect of chlorine-induced bacterial injury in water. *Journal of Agricultural and Food Chemistry, 56*, 8992–8997.

Al-Qadiri, H. M., Al-Holy, M. A., Lin, M., Alami, N. I., Cavinato, A. G., & Rasco, B. A. (2006). Rapid detection and identification of Pseudomonas aeruginosa and Escherichia coli as pure and mixed cultures in bottled drinking water using Fourier transform infrared spectroscopy and multivariate analysis. *Journal of Agricultural and Food Chemistry, 54*, 5749–5754.

Al-Qadiri, H. M., Lin, M., Cavinato, A. G., & Rasco, B. A. (2006). Fourier transform infrared spectroscopy, detection and identification of Escherichia coli O157: H7 and Alicyclobacillus strains in apple juice. *International Journal of Food Microbiology, 111*, 73–80.

Argyri, A. A., Panagou, E. Z., Tarantilis, P. A., Polysiou, M., & Nychas, G.-J. E. (2010). Rapid qualitative and quantitative detection of beef fillets spoilage based on Fourier transform infrared spectroscopy data and artificial neural networks. *Sensors and Actuators B: Chemical, 145*, 146–154.

Bacsik, Z., Mink, J., & Keresztury, G. (2004). FTIR spectroscopy of the atmosphere. I. Principles and methods. *Applied Spectroscopy Reviews, 39*, 295–363.

Baravkar, A. A., Kale, R. N., & Sawant, S. D. (2011). FT-IR spectroscopy: Principle, technique and mathematics. *International Journal of Pharma and Bio Sciences, 2*, 513–519.

Barron, C., Parker, M. L., Mills, E. N. C., Rouau, X., & Wilson, R. H. (2005). FTIR imaging of wheat endosperm cell walls in situ reveals compositional and architectural heterogeneity related to grain hardness. *Planta, 220*, 667–677.

Bernal, J., Mendiola, J. A., Ibáñez, E., & Cifuentes, A. (2011). Advanced analysis of nutraceuticals. *Journal of Pharmaceutical and Biomedical Analysis, 55*, 758–774.

Bobelyn, E., Hertog, M. L. A. T. M., Schenk, A., Vermeir, S., Lammertyn, J., Nicolaï, B. M., & Beullens, K. (2006). Systems to characterise internal quality of fruit and vegetables. *IV International Conference on Managing Quality in Chains—The Integrated View on Fruits and Vegetables Quality, 712*, 59–66.

Bureau, S., Ruiz, D., Reich, M., Gouble, B., Bertrand, D., Audergon, J.-M., & Renard, C. M. G. C. (2009). Application of ATR-FTIR for a rapid and simultaneous determination of sugars and organic acids in apricot fruit. *Food Chemistry, 115*, 1133–1140.

Cassoli, L. D., Sartori, B., & Machado, P. F. (2011). The use of the Fourier transform infrared spectroscopy to determine adulterants in raw milk. *Revista Brasileira de Zootecnia, 40*, 2591–2596.

Cebi, N., Yilmaz, M. T., & Sagdic, O. (2017). A rapid ATR-FTIR spectroscopic method for detection of sibutramine adulteration in tea and coffee based on hierarchical cluster and principal component analyses. *Food Chemistry, 229*, 517–526.

Černá, M., Barros, A. S., Nunes, A., Rocha, S. M., Delgadillo, I., Čopíková, J., & Coimbra, M. A. (2003). Use of FT-IR spectroscopy as a tool for the analysis of polysaccharide food additives. *Carbohydrate Polymers, 51*, 383–389.

Chang, Y. J., Lee, S., Yoo, M. A., & Lee, H. G. (2006). Structural and biological characterization of sulfated-derivatized oat β-glucan. *Journal of Agricultural and Food Chemistry, 54*, 3815–3818.

Che Man, Y. B., Syahariza, Z. A., Mirghani, M. E. S., Jinap, S., & Bakar, J. (2005). Analysis of potential lard adulteration in chocolate and chocolate products using Fourier transform infrared spectroscopy. *Food Chemistry, 90*, 815–819.

Che Man, Y. B., & Rohman, A. (2013). Analysis of canola oil in virgin coconut oil using FTIR spectroscopy and chemometrics. *Journal of Food and Pharmaceutical Sciences, 1*, 5–9.

Chen, L., Tian, Y., Sun, B., Cai, C., Ma, R., & Jin, Z. (2018). Measurement and characterization of external oil in the fried waxy maize starch granules using ATR-FTIR and XRD. *Food Chemistry, 242*, 131–138.

Cifuentes, A. (2012). Food analysis: Present, future, and foodomics. *ISRN Analytical Chemistry, 2012*, 801607.

Čopíková, J., Barros, A. S., Šmídová, I., Černá, M., Teixeira, D. H., Delgadillo, I., Synytsya, A., & Coimbra, M. A. (2006). Influence of hydration of food additive polysaccharides on FT-IR spectra distinction. *Carbohydrate Polymers, 63*, 355–359.

Damez, J.-L., & Clerjon, S. (2013). Quantifying and predicting meat and meat products quality attributes using electromagnetic waves: An overview. *Meat Science, 95*, 879–896.

Danezis, G. P., Tsagkaris, A. S., Camin, F., Brusic, V., & Georgiou, C. A. (2016). Food authentication: Techniques, trends & emerging approaches. *Trends in Analytical Chemistry, 85*, 123–132.

Dankar, I., Haddarah, A., EL Omar, F., Pujolà, M., & Sepulcre, F. (2018). Characterization of food additive-potato starch complexes by FTIR and X-ray diffraction. *Food Chemistry, 260*, 7–12.

Davis, R., & Mauer, L. J. (2010). Fourier transform infrared (FT-IR) spectroscopy: A rapid tool for detection and analysis of foodborne pathogenic bacteria. In A. Mendez-Vilasa (Ed.), *Current research, technology, education topics in applied microbiology, and microbial biotechnology* (Vol. 2, pp. 1582–1594). Formatex Research Center.

de la Mata, P., Dominguez-Vidal, A., Bosque-Sendra, J. M., Ruiz-Medina, A., Cuadros-Rodríguez, L., & Ayora-Cañada, M. J. (2012). Olive oil assessment in edible oil blends by means of ATR-FTIR and chemometrics. *Food Control, 23*, 449–455.

Din, A., Anjum, F. M., Zahoor, T., & Nawaz, H. (2009). Extraction and utilization of barley β-glucan for the preparation of functional beverage. *International Journal of Agriculture and Biology, 11*, 737–740.

Domingo, E., Tirelli, A. A., Nunes, C. A., Guerreiro, M. C., & Pinto, S. M. (2014). Melamine detection in milk using vibrational spectroscopy and chemometrics analysis: A review. *Food Research International, 60*, 131–139.

Domínguez-Martínez, I., Meza-Márquez, O. G., Osorio-Revilla, G., Proal-Nájera, J., & Gallardo-Velázquez, T. (2014). Determination of capsaicin, ascorbic acid, total phenolic compounds and antioxidant activity of Capsicum annuum L. var. serrano by mid infrared spectroscopy (mid-FTIR) and chemometric analysis. *Journal of Korean Society for Applied Biological Chemistry, 57*, 133–142.

Doyle, W. M. (1992). Principles and applications of Fourier transform infrared (FTIR) process analysis. *Process Control and Quality, 2*, 11–41.

Duygu, D. Y., Baykal, T., Açikgöz, İ., & Yildiz, K. (2009). Fourier transform infrared (FT-IR) spectroscopy for biological studies. *Gazi University Journal of Science, 22*, 117–121.

Ellis, D. I., Broadhurst, D., & Goodacre, R. (2004). Rapid and quantitative detection of the microbial spoilage of beef by Fourier transform infrared spectroscopy and machine learning. *Analytica Chimica Acta, 514*, 193–201.

Ellis, D. I., Broadhurst, D., Kell, D. B., Rowland, J. J., & Goodacre, R. (2002). Rapid and quantitative detection of the microbial spoilage of meat by Fourier transform infrared spectroscopy and machine learning. *Applied and Environmental Microbiology, 68*, 2822–2828.

Ferragina, A., Cipolat-Gotet, C., Cecchinato, A., & Bittante, G. (2013). The use of Fourier-transform infrared spectroscopy to predict cheese yield and nutrient recovery or whey loss traits from unprocessed bovine milk samples. *Journal of Dairy Science, 96*, 7980–7990.

Flores-Morales, A., Jiménez-Estrada, M., & Mora-Escobedo, R. (2012). Determination of the structural changes by FT-IR, Raman, and CP/MAS 13C NMR spectroscopy on retrograded starch of maize tortillas. *Carbohydrate Polymers, 87*, 61–68.

Franca, A. S., & Oliveira, L. S. (2011). Potential uses of fourier transform infrared spectroscopy (FTIR) in food processing and engineering. In B. C. Siegler (Ed.), *Food engineering* (pp. 211–257). Nova Science Publishers Inc..

Gaffney, Jeffrey S, Nancy A Marley, and Darin E Jones. 2002. Fourier transform infrared (FTIR) spectroscopy. In *Characterization of materials* (pp. 1–33). Wiley.

Goerges, S., Mounier, J., Rea, M. C., Gelsomino, R., Heise, V., Beduhn, R., Cogan, T. M., Vancanneyt, M., & Scherer, S. (2008). Commercial ripening starter microorganisms inoculated into cheese milk do not successfully establish themselves in the resident microbial ripening consortia of a south German red smear cheese. *Applied and Environmental Microbiology, 74*, 2210–2217.

Gomaa, A. I., Sedman, J., & Ismail, A. A. (2013). An investigation of the effect of microwave treatment on the structure and unfolding pathways of β-lactoglobulin using FTIR spectroscopy with the application of two-dimensional correlation spectroscopy (2D-COS). *Vibrational Spectroscopy, 65*, 101–109.

Gómez-Ordóñez, E., & Rupérez, P. (2011). FTIR-ATR spectroscopy as a tool for polysaccharide identification in edible brown and red seaweeds. *Food Hydrocolloids, 25*, 1514–1520.

Gouilleux, B., Marchand, J., Charrier, B., Remaud, G. S., & Giraudeau, P. (2018). High-throughput authentication of edible oils with benchtop ultrafast 2D NMR. *Food Chemistry, 244*, 153–158.

Grassi, S., Casiraghi, E., & Alamprese, C. (2018). Handheld NIR device: A non-targeted approach to assess authenticity of fish fillets and patties. *Food Chemistry, 243*, 382–388.

Grewal, M. K., Chandrapala, J., Donkor, O., Apostolopoulos, V., Stojanovska, L., & Vasiljevic, T. (2017). Fourier transform infrared spectroscopy analysis of physicochemical changes in UHT milk during accelerated storage. *International Dairy Journal, 66*, 99–107.

Griffiths, P. R., & De Haseth, J. A. (2007). *Fourier transform infrared spectrometry*. Wiley.

Grumezescu, A. (2016). *Nutraceuticals*. Academic Press.

Grunert, T., Stephan, R., Ehling-Schulz, M., & Johler, S. (2016). Fourier transform infrared spectroscopy enables rapid differentiation of fresh and frozen/thawed chicken. *Food Control, 60*, 361–364.

Hakovirta, J. (2008). *Modern techniques in detection, identification and quantification of bacteria and peptides from foods*. University of Helsinki.

Hernández-Martínez, M., Gallardo-Velázquez, T., & Osorio-Revilla, G. (2010). Rapid characterization and identification of fatty acids in margarines using horizontal attenuate total reflectance Fourier transform infrared spectroscopy (HATR-FTIR). *European Food Research and Technology, 231*, 321–329.

Hernández-Martínez, M., Gallardo-Velázquez, T., Osorio-Revilla, G., Almaraz-Abarca, N., Ponce-Mendoza, A., & Vásquez-Murrieta, M. S. (2013). Prediction of total fat, fatty acid composition and nutritional parameters in fish fillets using MID-FTIR spectroscopy and chemometrics. *LWT-Food Science and Technology, 52*, 12–20.

Hilal, N., Ismail, A. F., Matsuura, T., & Oatley-Radcliffe, D. (2017). *Membrane characterization*. Elsevier.

Irudayaraj, J., Xu, R., & Tewari, J. (2003). Rapid determination of invert cane sugar adulteration in honey using FTIR spectroscopy and multivariate analysis. *Journal of Food Science, 68*, 2040–2045.

Irudayaraj, J., Yang, H., & Sakhamuri, S. (2002). Differentiation and detection of microorganisms using Fourier transform infrared photoacoustic spectroscopy. *Journal of Molecular Structure, 606*, 181–188.

Ismail, A. A., van de Voort, F. R., & Sedman, J. (1997). Fourier transform infrared spectroscopy: principles and applications. *Techniques and Instrumentation in Analytical Chemistry, 18*, 93–139.

Jaiswal, P., Jha, S. N., Borah, A., Gautam, A., Grewal, M. K., & Jindal, G. (2015). Detection and quantification of soymilk in cow–buffalo milk using attenuated total reflectance Fourier transform infrared spectroscopy (ATR–FTIR). *Food Chemistry, 168*, 41–47.

Jawaid, S., Talpur, F. N., Sherazi, S. T. H., Nizamani, S. M., & Khaskheli, A. A. (2013). Rapid detection of melamine adulteration in dairy milk by SB-ATR–Fourier transform infrared spectroscopy. *Food Chemistry, 141*, 3066–3071.

Kacurakova, M., Capek, P., Sasinkova, V., Wellner, N., & Ebringerova, A. (2000). FT-IR study of plant cell wall model compounds: Pectic polysaccharides and hemicelluloses. *Carbohydrate Polymers, 43*, 195–203.

Kačuráková, M., Wellner, N., Ebringerová, A., Hromádková, Z., Wilson, R. H., & Belton, P. S. (1999). Characterisation of xylan-type polysaccharides and associated cell wall components by FT-IR and FT-Raman spectroscopies. *Food Hydrocolloids, 13*, 35–41.

Kačuráková, M., & Wilson, R. H. (2001). Developments in mid-infrared FT-IR spectroscopy of selected carbohydrates. *Carbohydrate Polymers, 44*, 291–303.

Kačuráková, M., Belton, P. S., Wilson, R. H., Hirsch, J., & Ebringerová, A. (1998). Hydration properties of xylan-type structures: An FTIR study of xylooligosaccharides. *Journal of the Science of Food and Agriculture, 77*, 38–44.

Kansiz, M., Billman-Jacobe, H., & McNaughton, D. (2000). Quantitative determination of the biodegradable polymer poly (β-hydroxybutyrate) in a recombinant Escherichia coli strain by use of mid-infrared spectroscopy and multivariative statistics. *Applied and Environmental Microbiology, 66*, 3415–3420.

Kizil, R., Irudayaraj, J., & Seetharaman, K. (2002). Characterization of irradiated starches by using FT-Raman and FTIR spectroscopy. *Journal of Agricultural and Food Chemistry, 50*, 3912–3918.

Kümmel, J., Stessl, B., Gonano, M., Walcher, G., Bereuter, O., Fricker, M., Grunert, T., Wagner, M., & Ehling-Schulz, M. (2016). Staphylococcus aureus entrance into the dairy chain: Tracking S. aureus from dairy cow to cheese. *Frontiers in Microbiology, 7*, 1603.

Kyomugasho, C., Christiaens, S., Shpigelman, A., Van Loey, A. M., & Hendrickx, M. E. (2015). FT-IR spectroscopy, a reliable method for routine analysis of the degree of methylesterification of pectin in different fruit-and vegetable-based matrices. *Food Chemistry, 176*, 82–90.

Lakowicz, J. R. (2013). *Principles of fluorescence spectroscopy*. Springer Science & Business Media.

Larkin, P. (2017). *Infrared and Raman spectroscopy: Principles and spectral interpretation*. Elsevier.

Larosa, C., Salerno, M., de Lima, J. S., Meri, R. M., da Silva, M. F., de Carvalho, L. B., & Converti, A. (2018). Characterisation of bare and tannase-loaded calcium alginate beads by microscopic, thermogravimetric, FTIR and XRD analyses. *International Journal of Biological Macromolecules, 115*, 900–906.

Lasch, P., Stämmler, M., Zhang, M., Baranska, M., Bosch, A., & Majzner, K. (2018). FT-IR hyperspectral imaging and artificial neural network analysis for identification of pathogenic bacteria. *Analytical Chemistry, 90*, 8896–8904.

Lerma-García, M. J., Gori, A., Cerretani, L., Simó-Alfonso, E. F., & Caboni, M. F. (2010). Classification of pecorino cheeses produced in Italy according to their ripening time and manufacturing technique using Fourier transform infrared spectroscopy. *Journal of Dairy Science, 93*, 4490–4496.

Lerma-García, M. J., Ramis-Ramos, G., Herrero-Martínez, J. M., & Simó-Alfonso, E. F. (2010). Authentication of extra virgin olive oils by Fourier-transform infrared spectroscopy. *Food Chemistry, 118*, 78–83.

Li, S., Zhang, X., Yang, S., Donglin, S., Ma, Q., Wen, R., & Li, J. (2017). Qualitative and quantitative detection of honey adulterated with high-fructose corn syrup and maltose syrup by using near-infrared spectroscopy. *Food Chemistry, 218*, 231–236.

Li, X., Zhang, Y., & He, Y. (2016). Rapid detection of talcum powder in tea using FT-IR spectroscopy coupled with chemometrics. *Scientific Reports, 6*, 30313.

Li, Y.-q., Kong, D.-x., & Wu, H. (2013). Analysis and evaluation of essential oil components of cinnamon barks using GC–MS and FTIR spectroscopy. *Industrial Crops and Products, 41*, 269–278.

Li-Chan, E., Chalmers, J. M., & Griffiths, P. R. (2010). *Applications of vibrational spectroscopy in food science*. Wiley.

Lin, M., Al-Holy, M., Al-Qadiri, H., Kang, D.-H., Cavinato, A. G., Huang, Y., & Rasco, B. A. (2004). Discrimination of intact and injured listeria monocytogenes by Fourier transform infrared spectroscopy and principal component analysis. *Journal of Agricultural and Food Chemistry, 52*, 5769–5772.

Lucarini, M., Durazzo, A., Sánchez del Pulgar, J., Gabrielli, P., & Lombardi-Boccia, G. (2018). Determination of fatty acid content in meat and meat products: The FTIR-ATR approach. *Food Chemistry, 267*, 223–230.

Maggio, R. M., Cerretani, L., Chiavaro, E., Kaufman, T. S., & Bendini, A. (2010). A novel chemometric strategy for the estimation of extra virgin olive oil adulteration with edible oils. *Food Control, 21*, 890–895.

Mauricio-Sánchez, R. A., Ricardo, S., Gabriel Luna-Bárcenas, J., & Mendoza-Galván, A. (2018). FTIR spectroscopy studies on the spontaneous neutralization of chitosan acetate films by moisture conditioning. *Vibrational Spectroscopy, 94*, 1–6.

Mishra, S., Kamboj, U., Kaur, H., & Kapur, P. (2010). Detection of jaggery syrup in honey using near-infrared spectroscopy. *International Journal of Food Sciences and Nutrition, 61*, 306–315.

Mudgil, D., Barak, S., & Khatkar, B. S. (2012). X-ray diffraction, IR spectroscopy and thermal characterization of partially hydrolyzed guar gum. *International Journal of Biological Macromolecules, 50*, 1035–1039.

Munajad, A., Subroto, C., & Suwarno, S. (2018). Fourier transform infrared (FTIR) spectroscopy analysis of transformer paper in mineral oil-paper composite insulation under accelerated thermal aging. *Energies, 11*, 364.

Nicolaou, N., & Goodacre, R. (2008). Rapid and quantitative detection of the microbial spoilage in milk using Fourier transform infrared spectroscopy and chemometrics. *Analyst, 133*, 1424–1431.

Novák, M., Synytsya, A., Gedeon, O., Slepička, P., Procházka, V., Synytsya, A., Blahovec, J., Hejlová, A., & Čopíková, J. (2012). Yeast β (1–3),(1–6)-d-glucan films: preparation and characterization of some structural and physical properties. *Carbohydrate Polymers, 87*, 2496–2504.

Nurrulhidayah, A. F., Rohman, A., Amin, I., Shuhaimi, M., & Khatib, A. (2013). Analysis of chicken fat as adulterant in butter using Fourier transform infrared spectroscopy and chemometrics. *Grasas y Aceites, 64*, 349–355.

Oust, A., Møretrø, T., Naterstad, K., Sockalingum, G. D., Adt, I., Manfait, M., & Kohler, A. (2006). Fourier transform infrared and Raman spectroscopy for characterization of listeria monocytogenes strains. *Applied and Environmental Microbiology, 72*, 228–232.

Papageorgiou, S. K., Kouvelos, E. P., Favvas, E. P., Sapalidis, A. A., Romanos, G. E., & Katsaros, F. K. (2010). Metal–carboxylate interactions in metal–alginate complexes studied with FTIR spectroscopy. *Carbohydrate Research, 345*, 469–473.

Paradkar, M. M., & Irudayaraj, J. (2002). Determination of cholesterol in dairy products using infrared techniques: 1. FTIR spectroscopy. *International Journal of Dairy Technology, 55*, 127–132.

Park, Y.-S., Leontowicz, H., Leontowicz, M., Namiesnik, J., Suhaj, M. M. C. M. O., Cvikrová, M., Olga, M., Weisz, M., & Gorinstein, S. (2011). Comparison of the contents of bioactive compounds and the level of antioxidant activity in different kiwifruit cultivars. *Journal of Food Composition and Analysis, 24*, 963–970.

Pereira, L., Amado, A. M., Critchley, A. T., Van de Velde, F., & Ribeiro-Claro, P. J. A. (2009). Identification of selected seaweed polysaccharides (phycocolloids) by vibrational spectroscopy (FTIR-ATR and FT-Raman). *Food Hydrocolloids, 23*, 1903–1909.

Quiñones-Islas, N., Meza-Márquez, O. G., Osorio-Revilla, G., & Gallardo-Velazquez, T. (2013). Detection of adulterants in avocado oil by mid-FTIR spectroscopy and multivariate analysis. *Food Research International, 51*, 148–154.

Rahman, A. (2012). Application of Fourier transform infrared spectroscopy for quality control of pharmaceutical products: A review. *Indonesian Journal of Pharmacy, 23*, 1–8.

Rios-Corripio, M. A., Rojas-López, M., & Delgado-Macuil, R. (2012). Analysis of adulteration in honey with standard sugar solutions and syrups using attenuated total reflectance-Fourier transform infrared spectroscopy and multivariate methods. *CyTA Journal of Food, 10*, 119–122.

Rohman, A., & Che Man, Y. B. (2010). Fourier transform infrared (FTIR) spectroscopy for analysis of extra virgin olive oil adulterated with palm oil. *Food Research International, 43*, 886–892.

Rohman, A., & Che Man, Y. B. (2012a). Analysis of pig derivatives for halal authentication studies. *Food Reviews International, 28*, 97–112.

Rohman, A., & Che Man, Y. B. (2012b). Authentication of extra virgin olive oil from sesame oil using FTIR spectroscopy and gas chromatography. *International Journal of Food Properties, 15*, 1309–1318.

Rohman, A., Che Man, Y. B., Hashim, P., & Ismail, A. (2011). FTIR spectroscopy combined with chemometrics for analysis of lard adulteration in some vegetable oils Espectroscopia FTIR combinada con quimiometría para el análisis de adulteración con grasa de cerdo de aceites vegetales. *CyTA Journal of Food, 9*, 96–101.

Rohman, A., Che Man, Y. B., & Yusof, F. M. (2014). The use of FTIR spectroscopy and chemometrics for rapid authentication of extra virgin olive oil. *Journal of the American Oil Chemists' Society, 91*, 207–213.

Rohman, A., & Che Man, Y. B. (2009). Analysis of cod-liver oil adulteration using Fourier transform infrared (FTIR) spectroscopy. *Journal of the American Oil Chemists' Society, 86*, 1149.

Rohman, A., Che Man, Y. B., Ismail, A., & Hashim, P. (2010). Application of FTIR spectroscopy for the determination of virgin coconut oil in binary mixtures with olive oil and palm oil. *Journal of the American Oil Chemists' Society, 87*, 601–606.

Schmitt, J., & Flemming, H.-C. (1998). FTIR-spectroscopy in microbial and material analysis. *International Biodeterioration & Biodegradation, 41*, 1–11.

Ścibisz, I., Reich, M., Bureau, S., Gouble, B., Causse, M., Bertrand, D., & Renard, C. M. G. C. (2011). Mid-infrared spectroscopy as a tool for rapid determination of internal quality parameters in tomato. *Food Chemistry, 125*, 1390–1397.

Sezer, B., Bilge, G., & Boyaci, I. H. (2017). Capabilities and limitations of LIBS in food analysis. *TrAC Trends in Analytical Chemistry, 97*, 345–353.

Shin, M. S., Lee, S., Lee, K. Y., & Lee, H. G. (2005). Structural and biological characterization of aminated-derivatized oat β-glucan. *Journal of Agricultural and Food Chemistry, 53*, 5554–5558.

Shingel, K. I. (2002). Determination of structural peculiarities of dexran, pullulan and γ-irradiated pullulan by Fourier-transform IR spectroscopy. *Carbohydrate Research, 337*, 1445–1451.

Sini, T. K., Santhosh, S., & Mathew, P. T. (2007). Study on the production of chitin and chitosan from shrimp shell by using Bacillus subtilis fermentation. *Carbohydrate Research, 342*, 2423–2429.

Sivakesava, S., & Irudayaraj, J. (2000). Analysis of potato chips using FTIR photoacoustic spectroscopy. *Journal of the Science of Food and Agriculture, 80*, 1805–1810.

Sivam, A. S., Sun-Waterhouse, D., Perera, C. O., & Waterhouse, G. I. N. (2013). Application of FT-IR and Raman spectroscopy for the study of biopolymers in breads fortified with fibre and polyphenols. *Food Research International, 50*, 574–585.

Smidt, E., & Schwanninger, M. (2005). Characterization of waste materials using FTIR spectroscopy: Process monitoring and quality assessment. *Spectroscopy Letters, 38*, 247–270.

Steiner, G., & Koch, E. (2009). Trends in Fourier transform infrared spectroscopic imaging. *Analytical and Bioanalytical Chemistry, 394*, 671–678.

Subramanian, A., Alvarez, V. B., Harper, W. J., & Rodriguez-Saona, L. E. (2011). Monitoring amino acids, organic acids, and ripening changes in Cheddar cheese using Fourier-transform infrared spectroscopy. *International Dairy Journal, 21*, 434–440.

Syahariza, Z. A., Che Man, Y. B., Selamat, J., & Bakar, J. (2005). Detection of lard adulteration in cake formulation by Fourier transform infrared (FTIR) spectroscopy. *Food Chemistry, 92*, 365–371.

Tiwari, B. K., Brunton, N. P., & Brennan, C. (2013). *Handbook of plant food phytochemicals: Sources, stability and extraction*. Wiley.

Triebel, S., Sproll, C., Reusch, H., Godelmann, R., & Lachenmeier, D. W. (2007). Rapid analysis of taurine in energy drinks using amino acid analyzer and Fourier transform infrared (FTIR) spectroscopy as basis for toxicological evaluation. *Amino Acids, 33*, 451–457.

Valdés, A., Beltrán, A., Mellinas, C., Jiménez, A., & Garrigós, M. C. (2018). Analytical methods combined with multivariate analysis for authentication of animal and vegetable food products with high fat content. *Trends in Food Science & Technology, 77*, 120–130.

Van Alfen, N. K. (2014). *Encyclopedia of agriculture and food systems*. Elsevier.

Van de Voort, F. R. (1992). Fourier transform infrared spectroscopy applied to food analysis. *Food Research International, 25*, 397–403.

Van de Voort, F. R., & Ismail, A. A. (1991). Proximate analysis of foods by mid-FTIR spectroscopy. *Trends in Food Science & Technology, 2*, 13–17.

van de Voort, F. R., Sedman, J., & Ismail, A. A. (1993). A rapid FTIR quality-control method for determining fat and moisture in high-fat products. *Food Chemistry, 48*, 213–221.

Vasconcelos, H., Saraiva, C., & de Almeida, J. M. M. M. (2014). Evaluation of the spoilage of raw chicken breast fillets using Fourier Transform Infrared Spectroscopy in tandem with chemometrics. *Food and Bioprocess Technology, 7*, 2330–2341.

Vongsvivut, J., Heraud, P., Zhang, W., Kralovec, J. A., McNaughton, D., & Barrow, C. J. (2012). Quantitative determination of fatty acid compositions in micro-encapsulated fish-oil supplements using Fourier transform infrared (FTIR) spectroscopy. *Food Chemistry, 135*, 603–609.

Vongsvivut, J., Miller, M. R., McNaughton, D., Heraud, P., & Barrow, C. J. (2014). Rapid discrimination and determination of polyunsaturated fatty acid composition in marine oils by FTIR spectroscopy and multivariate data analysis. *Food and Bioprocess Technology, 7*, 2410–2422.

Wang, J., Kliks, M. M., Jun, S., Jackson, M., & Li, Q. X. (2010). Rapid analysis of glucose, fructose, sucrose, and maltose in honeys from different geographic regions using Fourier transform infrared spectroscopy and multivariate analysis. *Journal of Food Science, 75*, C208–CC14.

Wellner, N., Kačuráková, M., Malovíková, A., Wilson, R. H., & Belton, P. S. (1998). FT-IR study of pectate and pectinate gels formed by divalent cations. *Carbohydrate Research, 308*, 123–131.

Wenning, M., Breitenwieser, F., Konrad, R., Huber, I., Busch, U., & Scherer, S. (2014). Identification and differentiation of food-related bacteria: A comparison of FTIR spectroscopy and MALDI-TOF mass spectrometry. *Journal of Microbiological Methods, 103*, 44–52.

Wenning, M., Büchl, N. R., & Scherer, S. (2010). Species and strain identification of lactic acid bacteria using FTIR spectroscopy and artificial neural networks. *Journal of Biophotonics, 3*, 493–505.

Wenning, M., & Scherer, S. (2013). Identification of microorganisms by FTIR spectroscopy: Perspectives and limitations of the method. *Applied Microbiology and Biotechnology, 97*, 7111–7120.

Wenning, M., Seiler, H., & Scherer, S. (2002). Fourier-transform infrared microspectroscopy, a novel and rapid tool for identification of yeasts. *Applied and Environmental Microbiology, 68*, 4717–4721.

Willey, R. R. (1976). Fourier transform infrared spectrophotometer for transmittance and diffuse reflectance measurements. *Applied Spectroscopy, 30*, 593–601.

Yang, H., Irudayaraj, J., & Paradkar, M. M. (2005). Discriminant analysis of edible oils and fats by FTIR, FT-NIR and FT-Raman spectroscopy. *Food Chemistry, 93*, 25–32.

Yu, C., & Irudayaraj, J. (2005). Spectroscopic characterization of microorganisms by Fourier transform infrared microspectroscopy. *Biopolymers, 77*, 368–377.

Zahir, E., Saeed, R., Hameed, M. A., & Yousuf, A. (2017). Study of physicochemical properties of edible oil and evaluation of frying oil quality by Fourier transform-infrared (FT-IR) Spectroscopy. *Arabian Journal of Chemistry, 10*, S3870–S3S76.

Zhang, L., Yuan, Z., Li, P., Wang, X., Mao, J., Zhang, Q., & Chundi, H. (2017). Targeted multivariate adulteration detection based on fatty acid profiles and Monte Carlo one-class partial least squares. *Chemometrics and Intelligent Laboratory Systems, 169*, 94–99.

Zhang, Q., Liu, C., Sun, Z., Hu, X., Shen, Q., & Wu, J. (2012). Authentication of edible vegetable oils adulterated with used frying oil by Fourier Transform Infrared Spectroscopy. *Food Chemistry, 132*, 1607–1613.

Zhao, M., Downey, G., & O'Donnell, C. P. (2014). Detection of adulteration in fresh and frozen beefburger products by beef offal using mid-infrared ATR spectroscopy and multivariate data analysis. *Meat Science, 96*, 1003–1011.

Multi/Hyper Spectral Imaging for Mango

Maimunah Mohd Ali and Norhashila Hashim

Abstract Mango is a tropical fruit of great economic importance due to its incredible taste and high nutritional content. Quality is an important parameter in monitoring the postharvest handling and storage of the fruit. In line with agricultural advancement, multispectral and hyperspectral imaging techniques have been widely developed and used to provide rapid and nondestructive detection of mango quality. In this chapter, the fundamental and system components of multi/hyper spectral imaging are introduced and discussed. In addition, the applications of multi/hyper spectral imaging for the mango quality evaluation of mango are elucidated. Some of the perspectives of the challenges and future trends of using multi/hyper spectral imaging are addressed and evaluated.

Keywords Mango · Fruit quality evaluation · Safety inspection · Hyperspectral imaging · Multispectral imaging

1 Introduction

Mango is a tropical fruit that comprises an abundant amount of fiber, minerals, and vitamins, which can be either eaten fresh or processed into various products. The top five mango producers worldwide in 2018 were India (21.8 million tonnes), China (4.8 million tonnes), Thailand (3.8 million tonnes), Indonesia (3.1 million tonnes), and Pakistan (2.3 million tonnes) (FAOSTAT, 2021). Similar to other fresh produce, mango is easily perishable, which requires extensive preservation methods such as

M. Mohd Ali
Department of Biological and Agricultural Engineering, Faculty of Engineering, Universiti Putra Malaysia, Serdang, Selangor, Malaysia

N. Hashim (✉)
Department of Biological and Agricultural Engineering, Faculty of Engineering, Universiti Putra Malaysia, Serdang, Selangor, Malaysia

SMART Farming Technology Research Centre, Faculty of Engineering, Universiti Putra Malaysia, Serdang, Selangor, Malaysia
e-mail: norhashila@upm.edu.my

P. B. Pathare, M. S. Rahman (eds.), *Nondestructive Quality Assessment Techniques for Fresh Fruits and Vegetables*, https://doi.org/10.1007/978-981-19-5422-1_7

drying, coating, canning, and freezing in order to prolong the shelf life (Pu & Sun, 2016). Generally, mango is evaluated in terms of maturity stage, fruit size, quality attributes (e.g., color, flavor, and texture), and postharvest characteristics (Wang et al., 2015). These criteria are crucial in delivering good-quality mangoes to the consumers through the supply chain. There are several methods that have been used to determine mango quality either destructive or nondestructive. These methods have been developed to meet the expectations of the consumers when purchasing. Nevertheless, most of the quality of mangoes is still determined using conventional methods, and these are time-consuming, low precision, and subjective.

Over the last decades, the potential of nondestructive technologies has emerged extensively with the rapid expansion of the economy and enhancement of living standards. Among all the emerging technologies, multi/hyper spectral imaging techniques have drawn huge attention for quality detection of mangoes that combines the employment of spectroscopy and imaging methods. In this sense, the spectroscopy method delivers optical information related to the quality attributes of the fruits whereas the imaging method acquires the images in terms of spatial information (Fu & Chen, 2019). Multi/hyper spectral imaging techniques have proven to be feasible for the quantitative and qualitative evaluation of mangoes, especially considering the fact that the potential applications of these methods are real-time monitoring detection. In practical applications, multi/hyper spectral imaging techniques have been utilized to detect quality changes of mangoes, such as maturity estimation (Gutiérrez et al., 2019b), firmness (Xu et al., 2018), total soluble solids (Rungpichayapichet et al., 2017), and chilling injury (Hashim et al., 2018). To fulfill the real-time inspection of quality attributes of mangoes, the transformation from fundamental work is required to rapid applications of multi/hyper spectral imaging. This might be advantageous to the existing multi/hyper spectral imaging devices since the use of customized filters is required for specific applications (Zhang et al., 2018). In a multi/hyper spectral imaging system, light signifies a key role in terms of detailed information regarding an object. In a sense, light carries information within the range of human vision, which is useful to facilitate quality assessment. The evolution of imaging technology and computerized systems has led the multi/hyper spectral imaging devices for rapid, precise, and nondestructive analysis of food quality detection with minimum human interference.

Although multi/hyper spectral imaging techniques could perform quality detection of mangoes, the determination of chemical attributes of the fruit is quite challenging (Ma et al., 2019). To effectively implement multi/hyper spectral imaging, proper development of this technique should be carefully considered to ensure the quality and safety inspection of fruits. Apart from that, the measurement of spatial and spectral distribution based on the multi/hyper spectral images offers new information relevant to the physical and chemical attributes of the fruit. In this sense, multi/hyper spectral imaging is advantageous over the conventional approaches, while routine analyses are susceptible to sample destruction. The application of multi/hyper spectral imaging coupled with multivariate algorithms and chemometrics makes it possible for the detection of the analytical composition of mangoes. Thus, multi/hyper spectral imaging techniques can be successfully employed for the qualitative and quantitative measurement of the characteristics of mangoes. It is a promising technology as the trends and potential applications are evolving.

2 Quality Assessment of Mangoes

Mango quality comprises many attributes such as sweetness, acidity, texture, aroma, shelf life, color, size, and nutritional composition. Mango contains more than 80% of water during fruit growth, which relies on environmental conditions (Léchaudel & Joas, 2007). Apart from that, the flesh of the mango depends on the balance between soluble sugars and organic acids. The predominant chemical composition in the fruit consists of citric acid, malic acids, sucrose, fructose, and glucose, which are essential for the maturation of mango fruits (Theanjumpol et al., 2013). For practical purposes, mango quality is defined as the mathematical function of a sequence of quality traits by assigning an overall score based on the sensory evaluation for decision-making during purchase. Another quality attribute of mango is the shelf life that varies with postharvest storage conditions, ripening stages, and temperature (Ibarra-Garza et al., 2015). Six ripening stages (RS) of mangoes stored at room temperature for several days can be observed in Fig. 1. Nevertheless, mango shelf life can be affected by environmental factors, which are driven by the water fluxes and carbohydrate formation during fruit growth (Durán Zuazo et al., 2019).

The quality attributes of mango consist of two different components, which are external and internal quality attributes. The external quality of mango depends on the visual appearance of the fruit, such as shape, size, color, and surface defects (Nicolaï et al., 2014). Meanwhile, internal quality attributes of mango are related to the taste, aroma, internal defects, and firmness, which provide the organoleptic satisfaction of the consumers (Cortés et al., 2016). Several elusive quality traits involving health-promoting composition, authentication, freshness, safety aspect, and chemical residues of mango may also influence the consumer's perception, which is subjective and abstract. Since the internal quality attributes cannot be determined during purchase, it is noted that the overall quality of mango is often evaluated in such a

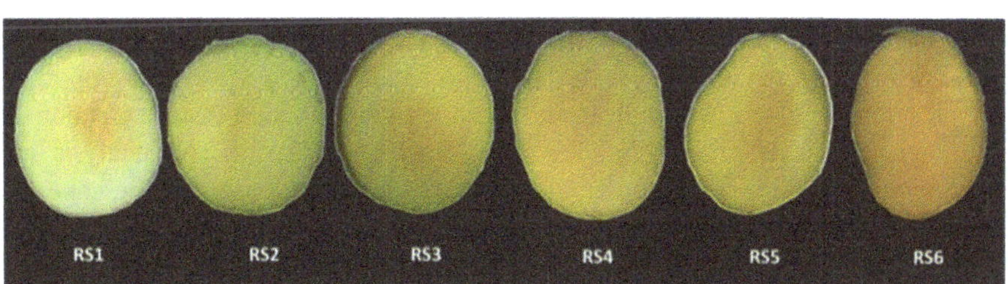

Fig. 1 Mango samples at six ripening stages (RS) stored at room temperature for 10 days (Ibarra-Garza et al., 2015)

way that the color changes, size, and shape of the fruit would be the main indicator (Jha et al., 2010). Usually, the fruits should fulfill certain requirements, which are easily recognized by the consumers using the senses of sight and touch to determine the decision of purchase (Mahmud et al., 2020). The internal quality is perceived in a destructive way by the senses of taste and smell either by sensory evaluation or consumer acceptability (Handa & Singh, 2016). However, this process could take a long time since there is no significant approach that can be observed between consumer acceptability and instrumental analysis using traditional inspection. The identification of mango quality using nondestructive methods has led to the evaluation of various types of food commodities, which is useful to replace analytical and routine analysis.

In this regard, comprehensive quality evaluation of mango combines the visual inspection and intrinsic traits of the fruit. Conventionally, mango quality detection is heavily dependent on manual handling, which is quite laborious and time-consuming. In addition, several internal quality detection methods have also relied on routine chemical analysis and destructive methods (Pu et al., 2015). To improve the efficiency and real-time quality monitoring of mango, many efforts have been proposed to find successful solutions for nondestructive and rapid determination of fruit quality. Nondestructive methods have massive advantages, which do not influence the characteristics of the mango. These are suitable to be used for fruit grading as well as postharvest storage prior to packaging and sales. Thus, utilizing nondestructive methods in quality detection of mango have become an urgent need in order to employ the robustness of the algorithms as well as the rapid optical detection system.

3 Fundamental and System Components

Multi/hyper spectral imaging aims to integrate both spectral and spatial information, which derives from spectroscopy and machine vision systems (Nicolaï et al., 2014). This technique is based on optical imaging that focuses on the interaction between the incident light and the object that offers more detailed and descriptive measurements. Generally, multi/hyper spectral imaging is used to generate chemical mapping to demonstrate the distribution of quality attributes involving the selected wavelength bands (Ropodi et al., 2016). The success of multi/hyper spectral imaging depends on the efficiency of the technique, which needs to select key important wavelengths for a specific application. When the configuration instrumentation of multi/hyper spectral imaging is developed, image processing and analysis are performed by a computer in order to generate information based on the captured images. With a well-designed multi/hyper spectral imaging system, many tedious image processing procedures can be reduced to meet the real-time needs encountered in modern food applications.

3.1 Hyperspectral Imaging

Hyperspectral imaging involves the interaction of spectral wavelengths from 400 to 1100 nm with molecular overtones as well as a mixture of molecular vibrations due to compound bending (Liu et al., 2017). Hyperspectral imaging exploits both imaging and spectroscopic techniques, which can determine spatial and spectral information simultaneously. Each hyperspectral image is defined as a three-dimensional (3D) hypercube comprised of spatial dimensions and wavebands (Wang et al., 2018). The unique spectral structure of a pixel in a hyperspectral image can be used to determine the chemical attributes of the mango. For the quality detection of mango, visible and near-infrared regions are the most commonly used system for hyperspectral imaging, which offers great flexibility. Hyperspectral imaging has been employed in several sensing modes by obtaining a hypercube through line-scanning, area-scanning, point-scanning, and single-shot methods (Qin et al., 2017). Most of the hyperspectral imaging applications use line-scanning method in which the spectral information is retrieved continuously by moving the sample linearly to the scanning line. The area-scanning method is implemented by scanning the image across the whole wavelength region, whereas point-scanning method is suitable for obtaining high spectral resolution image. For single-shot method, the scanning speed is the fastest due to the ability to obtain simultaneously both spatial and spectral information with one exposure (Ma et al., 2019). In terms of light illumination, hyperspectral imaging can be conducted in transmittance, reflectance, fluorescence, and scattering modes (Pu et al., 2015). Line-scanning and area-scanning are widely used in the quality evaluation of food products, whereas point-scanning and single-shot methods are suitable for real-time monitoring detection due to the rapid imaging speed.

A typical hyperspectral imaging system consists of a light source, imaging unit, spectrometer, and a computer equipped with the software for spectral analysis (Fig. 2). The core part of hyperspectral imaging is the imaging unit, which comprises a zoom lens and a highly sensitive spectrograph with charge-coupled device (CCD) or complementary metal-oxide-semiconductor (CMOS) camera device (Hussain et al., 2018). The spectrograph acts as a dispersion of light into continuous spectral wavelength ranged in the form of optical devices including diffraction grating, prism, and tunable filter (Pu et al., 2015). The light that is captured by the imaging unit, carries information associated with the physical and chemical properties of the sample via the light beam onto the region arrays of the device. Normally, the reliability of the imaging unit could also influence the quality features of the hyperspectral image. Specifically, most of the light beam penetrates into the sample, whereas the rest is reflected off the surface of the sample (Zhao et al., 2020). In this sense, different quality parameters are observed from the data acquisition of the sample based on the configuration and position of the lighting unit.

The light source is another important component in a hyperspectral imaging system that acts as an optical probe for the detection of physical and chemical attributes of mango. For mango fruit which has round and oval shapes, the indirect

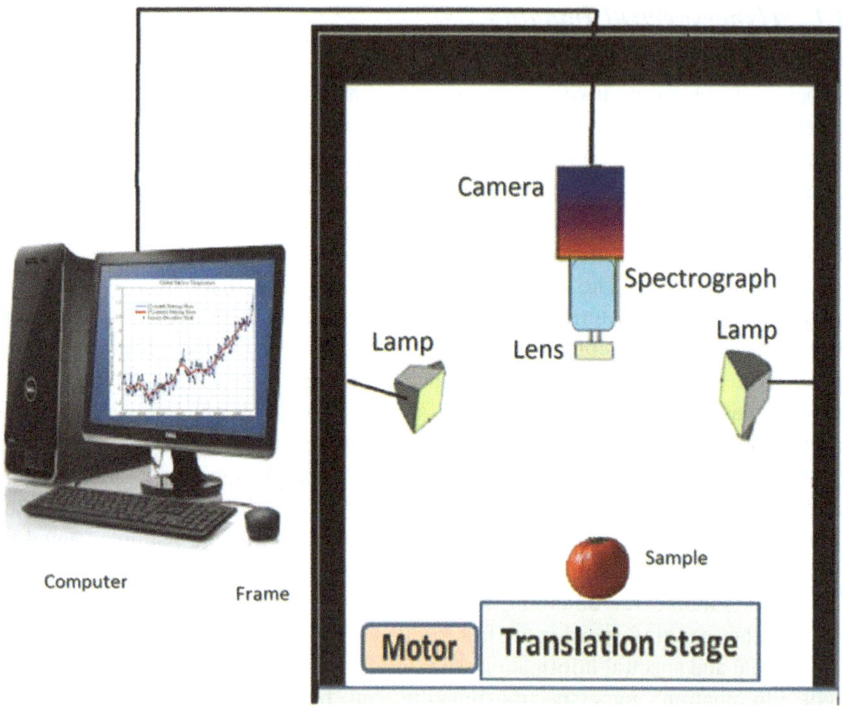

Fig. 2 The main components of a hyperspectral imaging system (Hussain et al., 2018)

light source was placed around the edge of a hemispherical diffuser in order to provide homogeneous illumination conditions (Lu et al., 2020). Halogen lamps are extensively used in hyperspectral imaging systems as the illumination source, especially for food quality detection. Despite being widely used in most of the hyperspectral imaging applications due to the smooth spectrum in the visible to infrared wavelength region, halogen lamps are sensitive to vibration and generate a high level of heat output. For this reason, many studies have shifted to light-emitting diodes (LEDs) due to the reduced heat output, high durability, and low energy consumption (Armstrong et al., 2019). Further, the LEDs can be exploited in different structured arrays depending on the specific application. However, halogen lamps are still irreplaceable in hyperspectral imaging systems because of their large advantage in implementing a wide range of wavelengths compared to LEDs, which cover only narrow wavelengths. On the other hand, the function of a spectrometer is to diffuse the light into a single-wavelength light at different frequencies via the imaging unit (Fu & Chen, 2019). There are three different types of spectrometers that are available for hyperspectral imaging applications including whisk broom, tunable filter, and push broom (Pathan et al., 2020). The three configurations of spectrometers vary in terms of scanning modes according to the specific needs of hyperspectral imaging.

In addition, the spatial characteristics of hyperspectral imaging can be observed in different spectral dimensions by means of a single or a combination of wavebands (Siche et al., 2016). Basically, the consecutive acquisition of hyperspectral images is obtained within a specific range of spectral wavelengths or by a full spectrum depending on the pixel requirements of the image (Du et al., 2020). Due to the variation in terms of physical and chemical properties, the light hitting the sample is scattered, absorbed, or reflected using hyperspectral imaging obtained from different patterns in specific wavelength regions. The captured hyperspectral image of a sample represents a series of sub-images at contiguous, discrete, and wide wavebands over the reflectance spectrum (ElMasry & Nakauchi, 2016). Before exploring different types of image processing performed for hyperspectral images, it is crucial to understand that hyperspectral imaging relies on light irradiance, sensitivity, and exposure time of the device. As an alternative to a typical visible camera, a hyperspectral imaging device is able to separate the electromagnetic spectrum into a series of finite wavebands. As a result, hyperspectral images could perceptively provide much more detailed information compared to other imaging devices.

3.2 *Multispectral Imaging*

Multispectral imaging is a technology that provides a 3D hypercube comprising spatial information in full wavelength bands (Wang et al., 2018). There are similar characteristics between hyperspectral and multispectral imaging in which both techniques simultaneously obtain spectral and spatial information (Li et al., 2020). The distinct difference between multi/hyper spectral imaging techniques relies on the number of spectral wavebands in the hypercube (Bhargava & Bansal, 2020). Generally, there are less than 30 noncontiguous and irregular spectral wavebands that significantly could influence the effective application of multispectral imaging (Raju & Sazonov, 2020). Due to the discrete wavebands, multispectral images have higher spatial resolution than hyperspectral images. The instrumentation of multispectral imaging in terms of complexity and cost is lower than hyperspectral imaging despite the fact that the system cannot describe fine details in the spectral distribution based on the multispectral image.

Each spatial location of the multispectral image contains information of a set of spectral wavebands that implement the purpose of fruit detection. Moreover, the multispectral imaging technique can be adopted in applications involving filtering devices, such as tunable and optical filters, which may be described by many bandwidths and spacing (Sendin et al., 2018). The filtering devices are used to allow rapid mapping of pigments for selecting feature wavelengths based on the food quality attributes. In general, a multispectral imaging system comprises of a charge-coupled device (CCD) camera equipped with interference filters in order to exploit image data for modeling food quality attributes (Su & Sun, 2018). A typical multispectral imaging system is shown in Fig. 3. Regarding the quality detection of

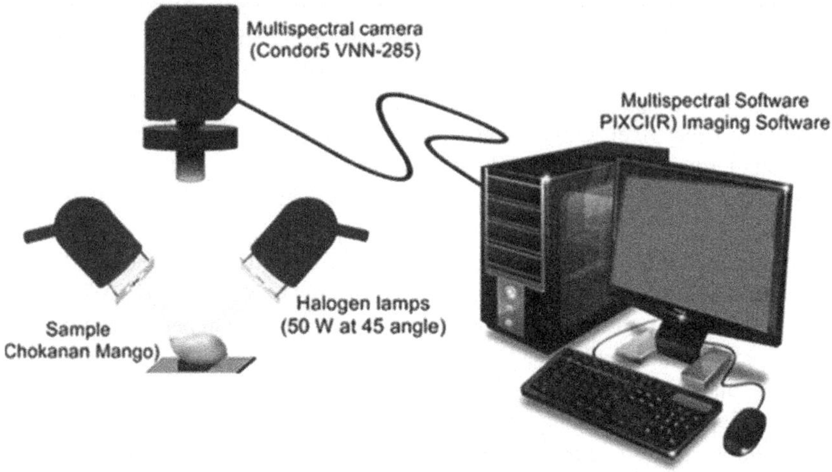

Fig. 3 A typical diagram of multispectral imaging system (Hashim et al., 2018)

mango, the most widely used multispectral imaging technique is developed to obtain spectral data in the visible and near-infrared region ranging from 380 to 2500 nm (Spyrelli et al., 2020). These regions are feasible for providing complex features of the fruit correlated with the absorption of pigments as well as molecular bonds (Jia et al., 2017). Since multispectral imaging has the ability to acquire image datasets within discontinuous spectral ranges, 3D multispectral data can be transformed into two-dimensional (2D) images by establishing multivariate modeling (Di Nisio et al., 2020). Nonetheless, multispectral imaging that is composed of narrow-band images could significantly obtain the feature wavelength of each pixel in the desired sample.

Multispectral imaging explores the optical property of different samples with distinctive spectral signatures. Each pixel in a multispectral image possesses spectral information regarding the surface and its constituents (Khan et al., 2018). In this case, the spectral resolution of multispectral images provides detailed and a fine measurement based on the observation of the spectrum at each pixel intensity. Compared to hyperspectral imaging, multispectral imaging is more easily developed in various configurations according to the spectral device (i.e., structure and illumination source during the image acquisition process). With a specific configuration, separate grayscale images can be obtained along with the raw spectral image acquired from a selected bandwidth (Elmasry et al., 2019). Furthermore, the multispectral imaging device is able to filter the light reaching the surface of the camera lens via specific band-pass filters. It allows the storage of multispectral images at various wavelength bands (Tsakanikas et al., 2015). For multispectral imaging, images could be obtained in the form of transmittance, reflectance, or interactance modes according to the light source and the camera detector (Qin et al., 2013). Among the three types of modes, the reflectance configuration is the most commonly used in a multispectral imaging system.

4 Applications of Multi/Hyper Spectral Imaging

Being a fast detection and noncontact device, multi/hyper spectral imaging has found vast utilization, especially for agricultural and food products. The expansion of multi/hyper spectral imaging for quality detection of mangoes is a vital goal and it is the required fundamental work in the fruit industry. With the introduction of multi/hyper spectral imaging technologies, recent applications of these techniques for the past 10 years are discussed related to the maturity estimation, shelf life, and quality changes of mangoes. Table 1 illustrates the applications of the multi/hyper spectral imaging system for the quality inspection of mango fruit.

The chemical and physical attributes of mango are highly related to the quality evaluation of the fruit, which includes maturity estimation, moisture content, soluble solid content, textural properties, dry matter, fat and protein content, carbohydrates, and other physicochemical properties. The multi/hyper spectral imaging technology has great potential in real-time monitoring of quality changes of mango as compared to other conventional analysis methods. Makino et al. (2013) investigated the potential of hyperspectral imaging for the detection of soluble solid content (SSC) of mangoes stored at 27 °C. Spectral reflectance ranging from 380 to 1000 nm was used to predict the SSC using an artificial neural network (ANN) model with a correlation coefficient of 0.79. Vélez Rivera et al. (2014) evaluated the maturity estimation of mangoes at different ripeness levels ranging from 650 to 1100 nm using a hyperspectral imaging system. Eleven features can be selected using the k-nearest neighbors (kNN) model with a correct classification rate of 98%. In another approach, Pu and Sun (2015) determined the prediction of moisture content of mango slices using a hyperspectral imaging system. The two-term thin layer and the Page drying models were found to be feasible to define the drying process of mango slices with a coefficient of determination (R^2) of 0.98. The Page model was used as the optimal model to evaluate the drying kinetics in mango, which was corresponded to moisture prediction from three waveband selections of hyperspectral imaging.

Pu and Sun (2016) examined the feasibility of hyperspectral imaging in the moisture prediction of mango slices with four different shapes, namely round, square, triangle, and rectangle. The moisture content distribution based on the hyperspectral map of mango slices of different shapes before and during microwave-vacuum drying is shown in Fig. 4. It was observed that the mango slices with a moisture content mean value of 89% signified an even moisture content distribution. Seven wavelengths were selected to develop a partial least squares (PLS) model in order to predict the moisture content of mango slices with an R^2 of 0.99. In a similar manner, Pu and Sun (2017) reported the application of hyperspectral imaging coupled with chemometrics to predict moisture in mangoes using three drying methods (microwave-vacuum, hot air, and a combination of microwave-vacuum and hot air). These findings demonstrated that drying uniformity was improved in color retention when combined microwave-vacuum and hot air drying were used. Furthermore, Wendel et al. (2018) proposed the application of

Table 1 The applications of multi/hyper spectral imaging system for the quality inspection of mango

Quality parameter	Data analysis	Result	References
Soluble solid content detection	ANN	Correlation coefficient of 0.79	Makino et al. (2013)
Maturity estimation	kNN	Correct classification rate of 98%	Vélez Rivera et al. (2014)
Moisture content prediction	Two-term thin layer and the page drying models	R^2 of 0.98	Pu and Sun (2015)
Moisture prediction	PLS	Moisture content mean value of 89%	Pu and Sun (2016)
Moisture prediction	Drying models	Drying uniformity was improved in color retention	Pu and Sun (2017)
Dry matter content prediction	Regression analysis	Low classification error of 0.29%	Wendel et al. (2018)
Physicochemical attributes determination	PLS	R^2 higher than 0.84	Xu et al. (2018)
Physicochemical attributes determination	PLS	R^2 higher than 0.80	Rungpichayapichet et al. (2017)
Yield estimation	Pixel identification model	Prediction accuracy of 0.75	Gutiérrez et al. (2019a)
Maturity classification	Spectral reflectance data point analysis	Overall classification accuracy of 98%	Mithun et al. (2017)
Moisture prediction	PCA	R^2 of 0.96	Jödicke et al. (2020)
Chilling injury	PCA	Successfully classified the samples according to the severity of chilling injury	Hashim et al. (2018)
Yield estimation	ANN	Strong correlation of 0.70	Rahman et al. (2018)
Maturity estimation	SVM	R^2 of 0.74	Gutiérrez et al. (2019b)

PLS Partial least squares, *ANN* artificial neural network, *kNN* k-nearest neighbors, R^2 coefficient of determination, *PCA* principal component analysis, *SVM* support vector machines

hyperspectral imaging ranging from 411 to 867 nm for the prediction of dry matter of mangoes in orchards. For dry matter prediction, hyperspectral data was evaluated using different orientations to allow efficient repeatability using the regression method and obtained a low classification error of 0.29%. Xu et al. (2018) determined the quality attributes of mangoes using hyperspectral imaging using reflectance spectra ranging from 900 to 1700 nm. Based on the findings, it was revealed that PLS models were successfully used to represent the predictive performance of quality changes of mangoes with R^2 higher than 0.84 for all the physicochemical properties of the fruit.

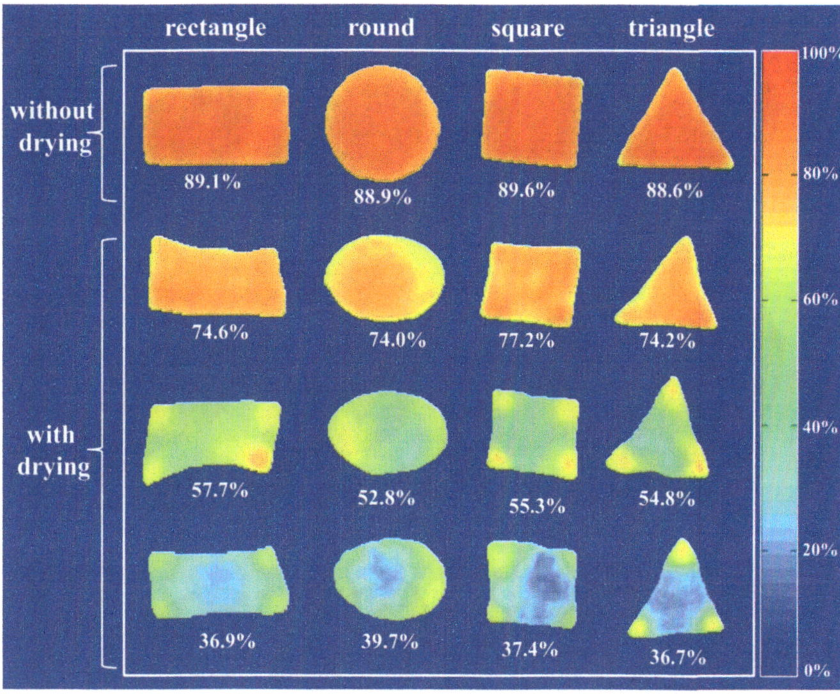

Fig. 4 Moisture content distribution based on the hyperspectral map of mango slices of different shapes (Pu & Sun, 2016)

Rungpichayapichet et al. (2017) assessed the physicochemical properties of mangoes using hyperspectral imaging in the wavelength region of 450–998 nm. The prediction maps of the physicochemical properties of mangoes during storage based on the hyperspectral prediction model are shown in Fig. 5. The PLS model was used to predict the physicochemical properties of mangoes based on the spatial distribution of hyperspectral images, which achieved R^2 higher than 0.80. Gutiérrez et al. (2019a) discussed the potential of hyperspectral imaging system to determine the yield estimation of mangoes from an unmanned ground vehicle. The hyperspectral images were acquired for mango pixel identification to achieve the best models for counting fruit yield. The finding suggested that hyperspectral imaging was feasible for yield estimation of mangoes with a prediction accuracy of 0.75. Mithun et al. (2017) investigated the classification of mangoes according to maturity levels using a multispectral imaging sensor. This finding showed that multispectral imaging was feasible for the mango classification with an overall classification accuracy of 98% using 36 spectral reflectance data points.

A recent study conducted by Jödicke et al. (2020) suggested the potential of multispectral imaging for the moisture prediction of mangoes during drying. It was revealed that the multispectral images were compared with the spectral reflectance in relation to the moisture content of the fruit, which can be used for quality detection in

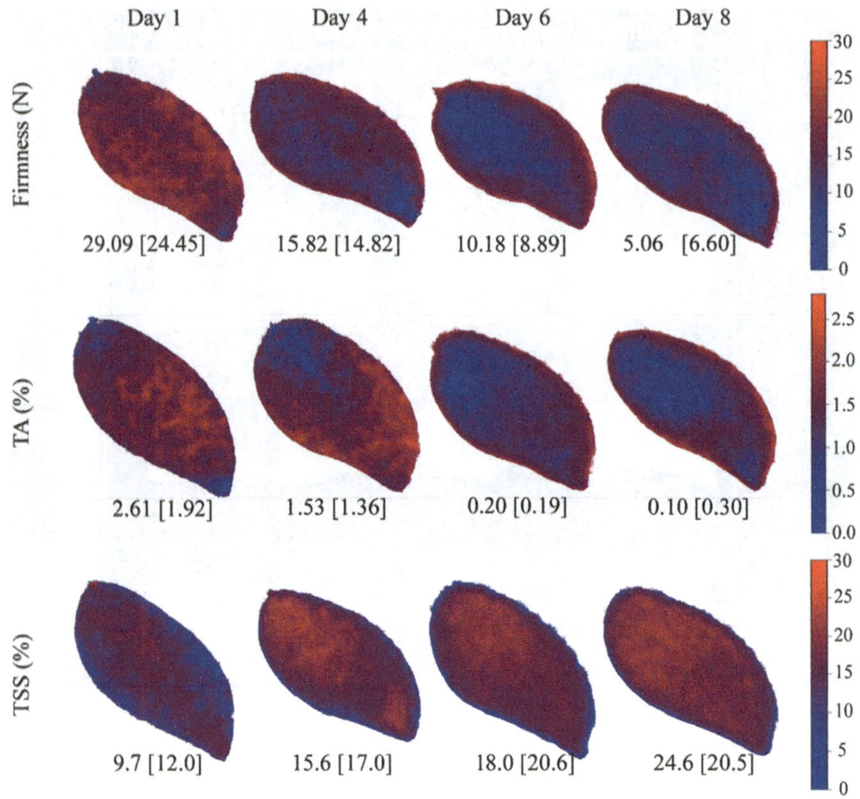

Fig. 5 The prediction maps of the physicochemical properties of mangoes during storage days based on a hyperspectral prediction model (Rungpichayapichet et al., 2017)

fruit drying. Hashim et al. (2018) investigated the chilling injury in mangoes stored at 4 °C using multispectral imaging. Two multivariate analyses including least square-support vector machine (LS-SVM) and principal component analysis (PCA) successfully classified the mango samples according to the severity of the chilling injury. In addition, Rahman et al. (2018) evaluated the feasibility of multispectral imaging for yield estimation of mangoes coupled with the ANN model. Their results demonstrated that the ANN model achieved a strong correlation of 0.70 for mango yield collected from two growing seasons. Likewise, Gutiérrez et al. (2019b) developed filter selection using hyperspectral imaging to estimate the maturity of mangoes under field conditions. Figure 6 demonstrates the reflectance plots of mango spectra obtained from hyperspectral imaging coupled with an unmanned ground vehicle by averaging the pixels corresponding to each fruit sample. The support vector machines (SVM) method was used to simulate the device using several filter combinations, which obtained R^2 of 0.74 with full hyperspectral datasets.

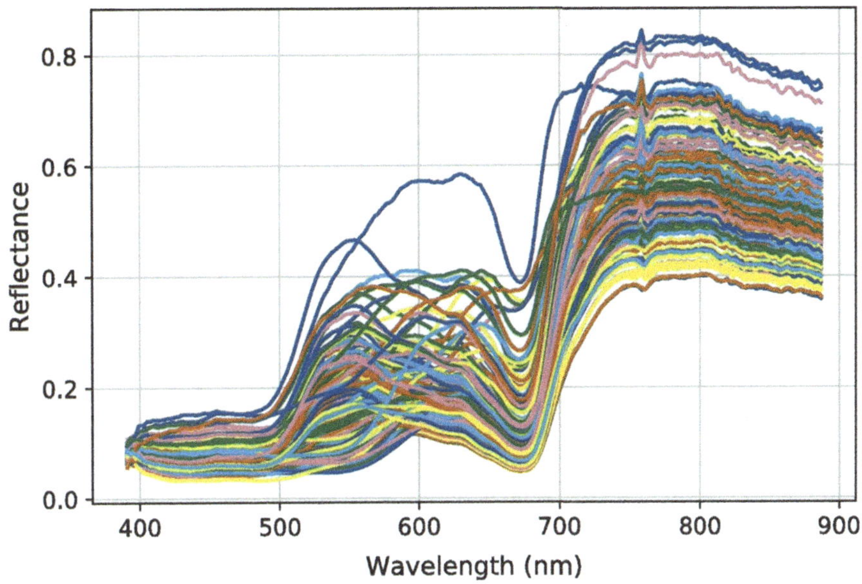

Fig. 6 The reflectance plot of mango spectra obtained from hyperspectral imaging coupled with an unmanned ground vehicle (Gutiérrez et al., 2019b)

Haff et al. (2013) developed an algorithm by incorporating Gaussian blur, thresholding, and particle count analysis based on hyperspectral images to detect fruit fly infestation in mango. An optimal combination for the detection of infestation was obtained using adjustable parameters of false positive and false negative results. The false positive and false negative results produced 11% and 1% error rates in the classification infested and healthy samples. In addition, hyperspectral imaging was also employed in mango drying to monitor the moisture distribution of the fruit (Ma et al., 2019). The moisture content distribution of each pixel of hyperspectral image can be visually determined and predicted using established models. When it comes to the inspection of the internal quality of fruit, multi/hyper spectral imaging techniques are conducted using different image acquisition methods (e.g., reflectance, transmittance, and fluorescence) to enhance the sensing capability of the system rather than using a single imaging mode. Thus, it can be denoted that multi/hyper spectral imaging techniques have huge potential to be a promising and alternative approach for a wide spectrum of industrial and real-time applications.

5 Challenges and Future Trends

Multi/hyper spectral imaging has been widely utilized due to its prominent benefits in terms of precise and simultaneous analysis for the qualitative detection of mangoes. By combining the implementation of spectroscopy and imaging technologies, multi/hyper spectral imaging could deliver abundant information associated with mango quality. For this reason, the successful implementation of multi/hyper spectral imaging has progressed towards the utilization of chemometrics to assist in the massive data collection and robustness of predictive modeling of the spectra mode (Wang et al., 2015). A special focus should lie on the complex operation of multi/hyper spectral imaging systems since the existing application requires improvement related to the acquisition speed and stability of the model for newly developed spectrographs. In future applications, more effective algorithms should be developed for spectral data processing and reference databases involving online detection and in situ operation. It is important to ensure the selection of wavelength range for the implementation of new algorithms involving applications of multi/hyper spectral imaging (Wang et al., 2018).

Despite the foregoing studies related to multi/hyper spectral imaging systems, there are still barriers that need to be monitored. The issue regarding the limited ability of software and hardware of multi/hyper spectral imaging systems requires some improvement, especially in terms of data processing and analysis. Other difficulties arise from the highly dimensional nature of spectral data, which also involves redundant spectral and spatial data information (Liu et al., 2017). In addition, employing algorithms for utilizing spectral information sometimes contributes to key information being missed due to the poor extraction of spatial information. Those challenges are mainly concerned with a multi/hyper spectral imaging system along with the model development and implementation of algorithms. With regard to the quality detection of mangoes, future studies should explore data fusion to provide robust and more superior prediction performance. Therefore, the selection of wavelength bands is the most efficient way to simplify calibration and predictive modeling in order to extract useful information.

Considering the fact that multi/hyper spectral imaging is not recommended for simultaneous detection in online applications, robust model transfer algorithms are needed to tackle this issue (i.e., chemical-based analytical techniques for quality detection of mangoes). In addition, various multivariate analyses are commonly used to eliminate the hidden data and multicollinearity in multi/hyper spectral imaging. Since multi/hyper spectral images typically contain unnecessary redundant data, multivariate analysis is beneficial to extract hidden and valuable information (Cheng & Sun, 2015). Multi/hyper spectral imaging can be executed in different nodes that have the advantage of delivering specific sample information (Pu et al., 2015). It is suggested that spectroscopic and image distortion should be reduced by developing multi/hyper spectral imaging devices with higher spectral and spatial resolution. The selection of a light source and environment conditions is essential in order to eliminate specular reflection during image acquisition (Fu & Chen, 2019). Since the 3D data cube of the hyperspectral image comprises abundant information regarding the physical and chemical attributes of the fruit, multiple mathematical algorithms can be developed in order to explore the feasibility of the multi/hyper spectral imaging technique.

With the booming innovations in the quality monitoring of various food products, the development of multi/hyper spectral imaging techniques is very promising to achieve high-quality mango fruits. More intensive work should be geared towards the technological advancements of data transfer from the laboratory. It should be set to real-time and online implementations. Multi/hyper spectral imaging techniques are being rapidly developed not only for laboratory purpose but also extended for commercial use for wholesale fruit distribution (Su & Sun, 2018). For this, the design complexity of multi/hyper spectral imaging should be exceptionally fast and robust to control the processing time in the real-time detection of quality attributes of the fruit. Identifying solutions for real-time problems implicate challenges in both software and hardware requirements. It is needed to detect characteristics that are invisible to the human eye (ElMasry & Nakauchi, 2016). Moreover, the integrated sensing node of multi/hyper spectral imaging is faster than existing optic fiber-based configurations with regard to spatial resolution in data acquisition, mainly for simultaneous detection of internal and external quality attributes (Lu et al., 2017). Hence, the continuous demand for multi/hyper spectral imaging provides a valuable approach and an effective sensing technology for the quality assessment of mangoes as well as other food and agricultural products.

6 Conclusion

Getting the full benefit of nondestructive spectral imaging implementation, multi/hyper spectral imaging is a potential device for qualitative and quantitative measurement of the properties of mangoes. The striking advantage of this approach underlies the rapid and chemical-free detection along with abundant information that provides various quality attributes of mangoes to be evaluated. An increasing number of multi/hyper spectral imaging applications have arisen in order to meet the demands of modern technologies in the mango industries. With this, a huge contribution from multi/hyper spectral imaging techniques is necessary to move the developed applications forward (i.e., laboratory scale to actual field applications). With continuous technical inventions in mango quality detection, multi/hyper spectral imaging aids in the development of low-cost and real-time monitoring of multiple quality features. It has been anticipated that the existing multi/hyper spectral imaging systems are established simultaneously with artificial intelligence and chemometrics applications. These combined systems can be used for the quality evaluation of fruit. The improvements in imaging software and hardware of multi/hyper spectral imaging have addressed many related issues regarding mango quality. For example, two-dimensional spatial resolution between spectral characteristics and quality attributes of the fruit. Multi/hyper spectral imaging techniques as a logical advancement as compared to traditional spectroscopic approaches have paved the way towards mango quality assessment. This technique can be useful for quality monitoring of other agricultural and food products.

Acknowledgments The authors would like to acknowledge the Department of Biological and Agricultural Engineering, Faculty of Engineering, Universiti Putra Malaysia for providing the support and technical facilities.

Conflict of Interest The authors have declared no conflict of interest.

References

Armstrong, P. R., McClung, A. M., Maghirang, E. B., Chen, M. H., Brabec, D. L., Yaptenco, K. F., Famoso, A. N., & Addison, C. K. (2019). Detection of chalk in single kernels of long-grain milled rice using imaging and visible/near-infrared instruments. *Cereal Chemistry, 96*(6), 1103–1111. https://doi.org/10.1002/cche.10220

Bhargava, A., & Bansal, A. (2020). Quality evaluation of mono & bi-colored apples with computer vision and multispectral imaging. *Multimedia Tools and Applications, 79*(11–12), 7857–7874. https://doi.org/10.1007/s11042-019-08564-3

Cheng, J. H., & Sun, D. W. (2015). Recent applications of spectroscopic and hyperspectral imaging techniques with chemometric analysis for rapid inspection of microbial spoilage in muscle foods. *Comprehensive Reviews in Food Science and Food Safety, 14*(4), 478–490. https://doi.org/10.1111/1541-4337.12141

Cortés, V., Ortiz, C., Aleixos, N., Blasco, J., Cubero, S., & Talens, P. (2016). A new internal quality index for mango and its prediction by external visible and near-infrared reflection spectroscopy. *Postharvest Biology and Technology, 118*, 148–158. https://doi.org/10.1016/j.postharvbio.2016.04.011

Di Nisio, A., Adamo, F., Acciani, G., & Attivissimo, F. (2020). Fast detection of olive trees affected by xylella fastidiosa from uavs using multispectral imaging. *Sensors (Basel), 20*(17), 1–23. https://doi.org/10.3390/s20174915

Du, Z., Zeng, X., Li, X., Ding, X., Cao, J., & Jiang, W. (2020). Recent advances in imaging techniques for bruise detection in fruits and vegetables. *Trends in Food Science and Technology, 99*, 133–141. https://doi.org/10.1016/j.tifs.2020.02.024

Durán Zuazo, V. H., Rodríguez Pleguezuelo, C. R., Gálvez Ruiz, B., Gutiérrez Gordillo, S., & García-Tejero, I. F. (2019). Water use and fruit yield of mango (Mangifera indica L.) grown in a subtropical Mediterranean climate. *International Journal of Fruit Science, 19*(2), 136–150. https://doi.org/10.1080/15538362.2018.1493960

Elmasry, G., Mandour, N., Al-Rejaie, S., Belin, E., & Rousseau, D. (2019). Recent applications of multispectral imaging in seed phenotyping and quality monitoring—An overview. *Sensors (Basel), 19*(1090), 1–32. https://doi.org/10.3390/s19051090

ElMasry, G. M., & Nakauchi, S. (2016). Image analysis operations applied to hyperspectral images for non-invasive sensing of food quality—A comprehensive review. *Biosystems Engineering, 142*, 53–82. https://doi.org/10.1016/j.biosystemseng.2015.11.009

FAOSTAT. (2021). *Food and Agriculture Organization of the United Nations, Crops data*. http://www.fao.org/faostat/en/#data/QC

Fu, X., & Chen, J. (2019). A review of hyperspectral imaging for chicken meat safety and quality evaluation: Application, hardware, and software. *Comprehensive Reviews in Food Science and Food Safety, 18*(2), 535–547. https://doi.org/10.1111/1541-4337.12428

Gutiérrez, S., Wendel, A., & Underwood, J. (2019a). Ground based hyperspectral imaging for extensive mango yield estimation. *Computers and Electronics in Agriculture, 157*, 126–135. https://doi.org/10.1016/j.compag.2018.12.041

Gutiérrez, S., Wendel, A., & Underwood, J. (2019b). Spectral filter design based on in-field hyperspectral imaging and machine learning for mango ripeness estimation. *Computers and Electronics in Agriculture, 164*, 1–12. https://doi.org/10.1016/j.compag.2019.104890

Haff, R. P., Saranwong, S., Thanapase, W., Janhiran, A., Kasemsumran, S., & Kawano, S. (2013). Automatic image analysis and spot classification for detection of fruit fly infestation in hyperspectral images of mangoes. *Postharvest Biology and Technology, 86*, 23–28. https://doi.org/10.1016/j.postharvbio.2013.06.003

Handa, P., & Singh, B. (2016). Electronic nose and their application in food industries. *Food Science Research Journal, 7*(2), 314–318. https://doi.org/10.15740/has/fsrj/7.2/314-318

Hashim, N., Onwude, D. I., & Osman, M. S. (2018). Evaluation of chilling injury in mangoes using multispectral imaging. *Journal of Food Science, 83*(5), 1271–1279. https://doi.org/10.1111/1750-3841.14127

Hussain, A., Pu, H., & Sun, D. W. (2018). Innovative nondestructive imaging techniques for ripening and maturity of fruits—A review of recent applications. *Trends in Food Science and Technology, 72*, 144–152. https://doi.org/10.1016/j.tifs.2017.12.010

Ibarra-Garza, I. P., Ramos-Parra, P. A., Hernández-Brenes, C., & Jacobo-Velázquez, D. A. (2015). Effects of postharvest ripening on the nutraceutical and physicochemical properties of mango (Mangifera indica L. cv Keitt). *Postharvest Biology and Technology, 103*, 45–54. https://doi.org/10.1016/j.postharvbio.2015.02.014

Jha, S. N., Narsaiah, K., Sharma, A. D., Singh, M., Bansal, S., & Kumar, R. (2010). Quality parameters of mango and potential of non-destructive techniques for their measurement—A review. *Journal of Food Science and Technology, 47*(1), 1–14. https://doi.org/10.1007/s13197-010-0004-6

Jia, B., Yoon, S. C., Zhuang, H., Wang, W., & Li, C. (2017). Prediction of pH of fresh chicken breast fillets by VNIR hyperspectral imaging. *Journal of Food Engineering, 208*, 57–65. https://doi.org/10.1016/j.jfoodeng.2017.03.023

Jödicke, K., Zirkler, R., Eckhard, T., Hofacker, W., & Jödicke, B. (2020). High end quality measuring in mango drying through multi-spectral imaging systems. *ChemEngineering, 4*(8), 1–13. https://doi.org/10.3390/chemengineering4010008

Khan, M. J., Khan, H. S., Yousaf, A., Khurshid, K., & Abbas, A. (2018). Modern trends in hyperspectral image analysis: A review. *IEEE Access, 6*, 14118–14129. https://doi.org/10.1109/ACCESS.2018.2812999

Léchaudel, M., & Joas, J. (2007). An overview of preharvest factors influencing mango fruit growth, quality and postharvest behaviour. *Brazilian Journal of Plant Physiology, 19*(4), 287–298. https://doi.org/10.1590/S1677-04202007000400004

Li, M., Huang, M., Zhu, Q., Zhang, M., Guo, Y., & Qin, J. (2020). Pickled and dried mustard foreign matter detection using multispectral imaging system based on single shot method. *Journal of Food Engineering, 285*, 1–12. https://doi.org/10.1016/j.jfoodeng.2020.110106

Liu, Y., Pu, H., & Sun, D. W. (2017). Hyperspectral imaging technique for evaluating food quality and safety during various processes: A review of recent applications. *Trends in Food Science and Technology, 69*, 25–35. https://doi.org/10.1016/j.tifs.2017.08.013

Lu, Y., Huang, Y., & Lu, R. (2017). Innovative hyperspectral imaging-based techniques for quality evaluation of fruits and vegetables: A review. *Applied Sciences, 7*(189), 1–36. https://doi.org/10.3390/app7020189

Lu, Y., Saeys, W., Kim, M., Peng, Y., & Lu, R. (2020). Hyperspectral imaging technology for quality and safety evaluation of horticultural products: A review and celebration of the past 20-year progress. *Postharvest Biology and Technology, 170*, 1–19. https://doi.org/10.1016/j.postharvbio.2020.111318

Ma, J., Sun, D. W., Pu, H., Cheng, J. H., & Wei, Q. (2019). Advanced techniques for hyperspectral imaging in the food industry: Principles and recent applications. *Annual Review of Food Science and Technology, 10*, 197–220. https://doi.org/10.1146/annurev-food-032818-121155

Mahmud, M. M. C., Shellie, R. A., & Keast, R. (2020). Unravelling the relationship between aroma compounds and consumer acceptance: Coffee as an example. *Comprehensive Reviews in Food Science and Food Safety, 19*, 2380–2420. https://doi.org/10.1111/1541-4337.12595

Makino, Y., Isami, A., Suhara, T., Oshita, S., Tsukada, M., Ishiyama, R., Serizawa, M., Kuroki, S., Kawagoe, Y., Purwanto, Y. A., Ahmad, U., & Mardjan, S. (2013). Non-destructive analysis of internal and external qualities of mango fruits during storage by hyperspectral imaging. *Acta Horticulturae, 1011*(2002), 443–450. https://doi.org/10.17660/actahortic.2013.1011.56

Mithun, B. S., Mondal, M., Vishwakarma, H., Shinde, S., & Kimbahune, S. (2017). Detection of artificially ripened mango using spectrometric analysis. *Sensing for Agriculture and Food Quality and Safety IX, 10217*, 1–9. https://doi.org/10.1117/12.2262457

Nicolaï, B. M., Defraeye, T., De Ketelaere, B., Herremans, E., Hertog, M. L. A. T. M., Saeys, W., Torricelli, A., Vandendriessche, T., & Verboven, P. (2014). Nondestructive measurement of fruit and vegetable quality. *Annual Review of Food Science and Technology, 5*(1), 285–312. https://doi.org/10.1146/annurev-food-030713-092410

Pathan, M., Patel, N., Yagnik, H., & Shah, M. (2020). Artificial cognition for applications in smart agriculture: A comprehensive review. *Artificial Intelligence in Agriculture, 4*, 81–95. https://doi.org/10.1016/j.aiia.2020.06.001

Pu, Y.-Y., Feng, Y.-Z., & Sun, D.-W. (2015). Recent progress of hyperspectral imaging on quality and safety inspection of fruits and vegetables: A review. *Comprehensive Reviews in Food Science and Food Safety, 14*(2), 176–188. https://doi.org/10.1111/1541-4337.12123

Pu, Y. Y., & Sun, D. W. (2015). Vis-NIR hyperspectral imaging in visualizing moisture distribution of mango slices during microwave-vacuum drying. *Food Chemistry, 188*, 271–278. https://doi.org/10.1016/j.foodchem.2015.04.120

Pu, Y. Y., & Sun, D. W. (2016). Prediction of moisture content uniformity of microwave-vacuum dried mangoes as affected by different shapes using NIR hyperspectral imaging. *Innovative Food Science and Emerging Technologies, 33*, 348–356. https://doi.org/10.1016/j.ifset.2015.11.003

Pu, Y. Y., & Sun, D. W. (2017). Combined hot-air and microwave-vacuum drying for improving drying uniformity of mango slices based on hyperspectral imaging visualisation of moisture content distribution. *Biosystems Engineering, 156*, 108–119. https://doi.org/10.1016/j.biosystemseng.2017.01.006

Qin, J., Chao, K., Kim, M. S., Lu, R., & Burks, T. F. (2013). Hyperspectral and multispectral imaging for evaluating food safety and quality. *Journal of Food Engineering, 118*(2), 157–171. https://doi.org/10.1016/j.jfoodeng.2013.04.001

Qin, J., Kim, M. S., Chao, K., Chan, D. E., Delwiche, S. R., & Cho, B. K. (2017). Line-scan hyperspectral imaging techniques for food safety and quality applications. *Applied Sciences, 7*, 1–22. https://doi.org/10.3390/app7020125

Rahman, M. M., Robson, A., & Bristow, M. (2018). Exploring the potential of high resolution worldview-3 imagery for estimating yield of mango. *Remote Sensing, 10*(1866), 1–18. https://doi.org/10.3390/rs10121866

Raju, V. B., & Sazonov, E. (2020). Detection of oil-containing dressing on salad leaves using multispectral imaging. *IEEE Access, 8*, 86196–86206. https://doi.org/10.1109/ACCESS.2020.2992326

Ropodi, A. I., Panagou, E. Z., & Nychas, G. J. E. (2016). Data mining derived from food analyses using non-invasive/non-destructive analytical techniques; determination of food authenticity, quality & safety in tandem with computer science disciplines. *Trends in Food Science and Technology, 50*, 11–25. https://doi.org/10.1016/j.tifs.2016.01.011

Rungpichayapichet, P., Nagle, M., Yuwanbun, P., Khuwijitjaru, P., Mahayothee, B., & Müller, J. (2017). Prediction mapping of physicochemical properties in mango by hyperspectral imaging. *Biosystems Engineering, 159*, 109–120. https://doi.org/10.1016/j.biosystemseng.2017.04.006

Sendin, K., Manley, M., & Williams, P. J. (2018). Classification of white maize defects with multispectral imaging. *Food Chemistry, 243*, 311–318. https://doi.org/10.1016/j.foodchem.2017.09.133

Siche, R., Vejarano, R., Aredo, V., Velasquez, L., Saldaña, E., & Quevedo, R. (2016). Evaluation of food quality and safety with hyperspectral imaging (HSI). *Food Engineering Reviews, 8*(3), 306–322. https://doi.org/10.1007/s12393-015-9137-8

Spyrelli, E. D., Doulgeraki, A. I., Argyri, A. A., Tassou, C. C., Panagou, E. Z., & George-John, E. N. (2020). Implementation of multispectral imaging (MSI) for microbiological quality assessment of poultry products. *Microorganisms, 8*(552), 1–14. https://doi.org/10.3390/microorganisms8040552

Su, W. H., & Sun, D. W. (2018). Multispectral imaging for plant food quality analysis and visualization. *Comprehensive Reviews in Food Science and Food Safety, 17*, 220–239. https://doi.org/10.1111/1541-4337.12317

Theanjumpol, P., Self, G., Rittiron, R., Pankasemsuk, T., & Sardsud, V. (2013). Selecting variables for near infrared spectroscopy (NIRS) evaluation of mango fruit quality. *Journal of Agricultural Science, 5*(7), 146–159. https://doi.org/10.5539/jas.v5n7p146

Tsakanikas, P., Pavlidis, D., & Nychas, G. J. (2015). High throughput multispectral image processing with applications in food science. *PLoS One, 10*(10), 1–15. https://doi.org/10.1371/journal.pone.0140122

Vélez Rivera, N., Gómez-Sanchis, J., Chanona-Pérez, J., Carrasco, J. J., Millán-Giraldo, M., Lorente, D., Cubero, S., & Blasco, J. (2014). Early detection of mechanical damage in mango using NIR hyperspectral images and machine learning. *Biosystems Engineering, 122*, 91–98. https://doi.org/10.1016/j.biosystemseng.2014.03.009

Wang, H., Peng, J., Xie, C., Bao, Y., & He, Y. (2015). Fruit quality evaluation using spectroscopy technology: A review. *Sensors (Basel), 15*(5), 11889–11927. https://doi.org/10.3390/s150511889

Wang, K., Pu, H., & Sun, D. W. (2018). Emerging spectroscopic and spectral imaging techniques for the rapid detection of microorganisms: An overview. *Comprehensive Reviews in Food Science and Food Safety, 17*(2), 256–273. https://doi.org/10.1111/1541-4337.12323

Wendel, A., Underwood, J., & Walsh, K. (2018). Maturity estimation of mangoes using hyperspectral imaging from a ground based mobile platform. *Computers and Electronics in Agriculture, 155*, 298–313. https://doi.org/10.1016/j.compag.2018.10.021

Xu, D., Wang, H., Ji, H., Zhang, X., Wang, Y., Zhang, Z., & Zheng, H. (2018). Hyperspectral imaging for evaluating impact damage to mango according to changes in quality attributes. *Sensors (Basel), 18*(3920), 1–14. https://doi.org/10.3390/s18113920

Zhang, B., Liu, L., Gu, B., Zhou, J., Huang, J., & Tian, G. (2018). From hyperspectral imaging to multispectral imaging: Portability and stability of HIS-MIS algorithms for common defect detection. *Postharvest Biology and Technology, 137*, 95–105. https://doi.org/10.1016/j.postharvbio.2017.11.004

Zhao, Y., Zhang, C., Zhu, S., Li, Y., He, Y., & Liu, F. (2020). Shape induced reflectance correction for non-destructive determination and visualization of soluble solids content in winter jujubes using hyperspectral imaging in two different spectral ranges. *Postharvest Biology and Technology, 161*, 1–11. https://doi.org/10.1016/j.postharvbio.2019.111080

Raman Spectroscopy for Fresh Fruits and Vegetables

Rasool Khodabakhshian

Abstract The need for accurate and reliable techniques for biological materials analysis at the molecular level has been consistently increasing over the past decades. This trend has been linked to recent food quality control. Raman spectroscopy plays an important role in the investigation of the structure of food and food quality control. Nowadays, along with many advances in research equipment design, Raman spectroscopy has become simpler, more accessible, and more cost-effective. Of course, despite the many advances that have been made, interpreting the Raman spectra is still a major challenge and requires special skills. Raman spectroscopy is used to observe vibrational, rotational, and other low-frequency states in a system. This type of spectroscopy typically provides a special structural fingerprint that is used to identify different molecules. In general, Raman spectra can provide some minor changes that allow the isolation of specific local variations of certain phytochemicals. This chapter provides the theoretical background needed for a full spectra analysis, a guide to the methods by which Raman spectroscopy has been used successfully, and leading to more sophisticated but less commonly available Raman spectrometer techniques. The historical background and various aspects of Raman spectroscopy theory including basic Raman scattering theory, instrumentation, and statistical data analysis was considered in this chapter. It also provides brief overviews of the application of Raman spectroscopy in the analysis and control of fruits and vegetables.

Keywords Raman spectroscopy · Agricultural products · Instrumentation · Sampling · Data analysis

---

R. Khodabakhshian (✉)
Department of Biosystems Engineering, Ferdowsi University of Mashhad, Mashhad, Iran
e-mail: khodabakhshian@um.ac.ir

P. B. Pathare, M. S. Rahman (eds.), *Nondestructive Quality Assessment Techniques for Fresh Fruits and Vegetables*, https://doi.org/10.1007/978-981-19-5422-1_8

1 Introduction

1.1 Raman Spectroscopy in General

Raman spectroscopy is a technique to detect vibrations in molecules, which is based on the inelastic scattering of light (called Raman scattering). Raman spectroscopy is an exciting alternative measurement technique that provides detailed information on the chemical and physical properties of biological materials, such as agricultural products and food ingredients. Raman spectroscopy plays an important role in the investigation of the structure of biological materials and food quality. Nowadays, along with many advances in research equipment design, Raman spectroscopy has become simpler, more accessible, and more cost-effective. Raman spectroscopy is used to observe vibrational, rotational, and other low-frequency states in a system. This type of spectroscopy typically provides a special structural fingerprint that is used to identify different molecules or amount of a substance in a sample. Samples can be tested under a full range of physical conditions; for example, solids, liquids, or vapors, in hot or cold conditions, in bulk, as fine particles or as surface layers. In general, Raman spectra can provide some minor changes that allow the isolation of specific local variations of certain phytochemicals. Due to these reasons, Raman spectroscopy holds great promise as an instrument for online or in-line applications in the food industry. This chapter considers the historical background and various aspects of Raman spectroscopy theory including basic Raman scattering theory, instrumentation, and statistical data analysis. It also provides brief overviews of the application of Raman spectroscopy in the analysis and control of fruits and vegetables.

1.2 Historical Background

The inelastic scattering of light in contact with substance was not reported until 1928, while this important physical phenomenon was first predicted by Smekal (1923). Chandrasekhara Venkata Raman also observed this phenomenon experimentally in 1928 for the first time after studying the sunlight passing through organic solutions, and in 1930 he was awarded the Nobel Prize in Physics for this important discovery and his other studies in this field (Raman & Krishnan, 1928). The development of this physical work was done by George Placzek between 1930 and 1934. Also, John William Strutt, third Baron Rayleigh in 1899, while proposing a new hypothesis, was able to justify the elastic scattering of light (Raman & Krishnan, 1928; Smekal, 1923).

This theory of light scattering was actually an answer to why the sky was blue. At that time, light scattering studies were also seriously pursued in countries, such as Russia, France, India, the United States, and Germany. In the early twentieth century, people such as Raman (Chandrasekhara Venkata Raman) and Krishnan

(Kariamanikkam Srinivasa Krishnan) in India and Landsberg (Grigory Landsberg) and Mandelstam (Leonid Mandelstam) in Russia pioneered this field of study. While studying the change in frequency of light scattered in different physical conditions, these people achieved results that they had not actually planned for before. Landsberg and Mandelstam also studied the scattering of light in quartz and several other crystals to find scattered beams that have changed frequency relative to incoming light (Butler et al., 2016; Downes & Elfick, 2010).

At the same time, Raman and Krishnan were studying changes in light due to Compton in India. With the publication of three papers in 1928, they recorded the change in the frequency of light scattered while colliding with substance, while the reports of Raman and Krishnan were only slightly earlier than those of Russian scientists. Since then, the phenomenon has been referred to as Raman spectroscopy. However, it was not until 1953 that the first commercial Raman spectrometer was developed after the development of the monochromator, which used mercury with a radiation wavelength of 435.8 nm. Therefore, prior to the 1960s, this configuration could be used to identify over 40,000 compounds. Though, the source used made the application hard, especially for photosensitive and colored samples that strongly absorbed this radiation (Sala et al., 1984). Thus, the use of lasers as a light source for the Raman spectrometer, which was invented in 1960 and introduced by Brazilian physicists Sérgio Porto and Wood in 1962, brings another great advance to this type of spectroscopy, allowing its application in a faster and more reproducible manner (Porto & Wood, 1962).

In 1966, Delhaye and Migeon proposed that the Raman scattering sample volume is autonomous and therefore applies to fine particle analysis (Delhaye & Migeon, 1966). With this concept, in 1974, the first Raman micro-spectrometers were developed and commercialized, providing Raman point and imaging analysis. Raman imaging spectroscopy combines a microscope with a Raman spectrometer (Louden, 1989; Turrell & Dhamelincourt, 1966). This new technology allowed the use of Raman spectroscopy for surface mapping. Another important step since 1987 has been the use of charge-coupled device (CCD) detectors (Turrell & Dhamelincourt, 1966). Since Raman scattering is an inherently weak signal, this problem was solved by using CCD detectors, which improved Raman sensitivity and further expanded the range of applications. Since the 1950s, several variants of Raman spectroscopy have also been developed, such as Coherent Anti-Stokes Raman Spectroscopy (CARS), Surface-Enhanced Raman Spectroscopy (SERS), Resonant Raman Spectroscopy (RRS), Tip Enhanced Raman Spectroscopy (TRES), and Stimulated Raman Spectroscopy (SRS), which has produced significant improvements in several aspects (speed and/or resolution) compared to the spontaneous Raman spectrum. Raman spectra can now also be acquired using Fourier Transform (FT) Spectroscopy. Some of these techniques that were used in agricultural products and food analysis are stated in Sect. 3. Nowadays, extensive studies are performed on the distribution of light while dealing with substance, and the high volume of studies and numerous scientific articles published about this discovery indicate the special importance of this issue.

2 Theoretical Progress

2.1 The Basis of Raman Spectroscopy

When light interacts with matter, the photons which make up the light may be absorbed, transmitted, or scattered. While the first two phenomena are related to middle infrared spectra (IR), the latter one is responsible for Raman spectra. In Raman spectroscopy, photons of monochromatic light (single wavelength light) are scattered in different directions after interacting with the sample. In fact, scattered photons of the sample are important in Raman spectroscopy. Most of the photons that interact with the molecule are scattered elastically. This type of scattering is called Rayleigh scattering, in which the scattered photons from the sample have the same energy or wavelength as the photons that hit the sample (Downes & Elfick, 2010; Nemecek et al., 2013).

As mentioned in the introduction section, Chandrasekhara Venkata Raman discovered the phenomenon called Raman in 1928. In this phenomenon, the energy or wavelength of the beam scattered by the molecules is different from the initial beam wavelength that strikes the sample. This type of scattering of light beams is called inelastic scattering. About 1 in 10 million photons scatter inelastically after interacting with matter. The amount of difference in energy or wavelength of inelastically scattered light also depends on the molecular structure of the compound. In fact, Raman spectroscopy is based on the analysis of these differences and with the aim of determining the molecular structure of different compounds. Changing the wavelength or energy of the initial radiation provides very important information about the molecular movements within the system. In Raman scattering, the photon interacts with the material and after scattering, its wavelength shifts to longer or shorter wavelengths (Fig. 1a). In this type of scattering, the transmission of beams to longer wavelengths is called Stock Raman scattering. Also, the transmission to shorter wavelengths is called anti-Stock Raman scattering (Downes & Elfick, 2010; Nemecek et al., 2013).

It has been reported that the ratio of anti-stock to stock intensity increases with increasing temperature. In fact, the initial photon interacts with the electron cloud of functional groups and excites the electrons to a virtual state. The electron then returns from the virtual state to an excited vibrational or rotational state (Fig. 1b). This phenomenon causes the photon to lose some of its energy and become obvious as a scattering of Stock Raman. The energy lost is directly related to the chemical identity of the agent group, the molecular structure attached to it, the type of atoms of the molecule, and the environment around it. Therefore, the Raman spectra of each molecule are specific and can be used as a "fingerprint" in the chemical identification of molecular compounds in a liquid, on a surface, or in air (Nemecek et al., 2013).

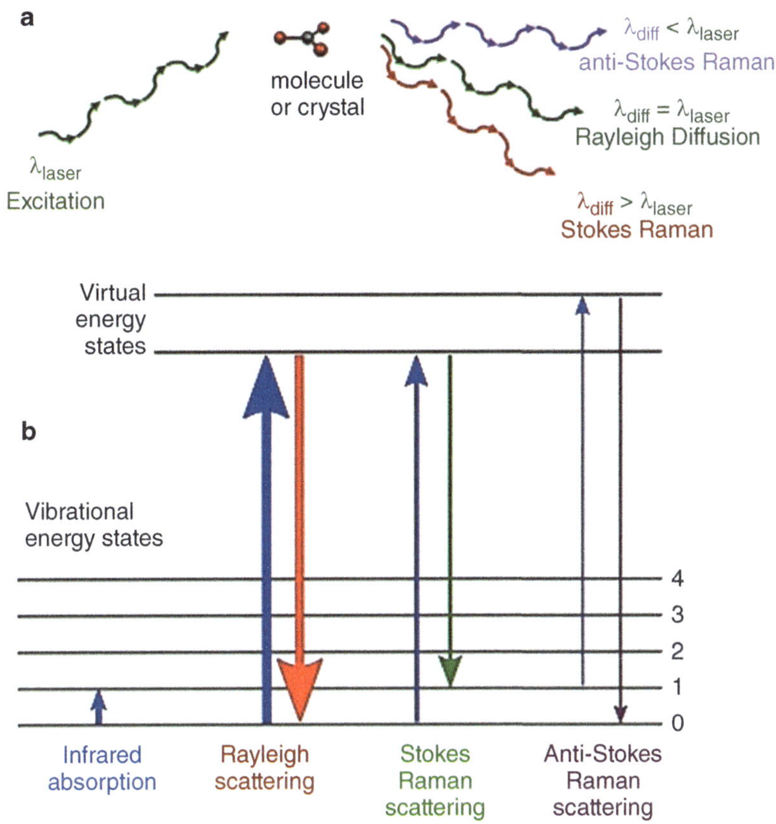

Fig. 1 (**a**) The frequency ($\nu = 1/\lambda$) difference between the incident and the scattered light characteristics of the molecule vibration (https://www.smacgigworld.com/blog/principle-raman-spectroscopy.php). (**b**) Schematic illustration of energy levels in the Raman spectrum (https://en.wikipedia.org/wiki/Raman_spectroscopy#/media/File:Raman_energy_levels.svg)

2.2 *Raman Effect*

The degree of Raman effect is directly related to the polarizability of the electrons of the molecule. The Raman effect is actually the interaction between the electron cloud of the sample and the external electric field of the incoming monochromatic light beams. This creates an induced dipole moment whose formation depends on the polarizability of the sample. Since laser light does not excite the molecule, no real transitions between energy levels occur in Raman studies. Hence, Raman signal is obtained from the collision of a light beam (photons) with the intermolecular oscillations (phonons) of the sample.

The information obtained in Raman spectroscopy leads to the determination of the structure, qualitative measurement, and quantification of molecules. It also studies the effect of many different physical parameters, such as temperature, pressure, and stress on interatomic and intermolecular fluctuations. The Raman scattering spectrum and the infrared absorption spectrum of a molecule have many similarities with each other, which are due to the similarities between the two techniques. Also, despite their similarity, the two methods are fundamentally different from each other and they are commonly used as complementary methods.

In infrared absorption, the amount of energy absorbed by the input photon corresponds to the energy difference between the initial and final rotational-vibrational states, while in the Raman scattering the amount of energy of the input photon is not the same as the output (usually more or less). Raman's dependence on the polarizability of electric dipole-electric dipole also distinguishes it from infrared spectroscopy, which is dependent only on electric dipole moment (atomic polar tensor). These differences indicate that transitions between rotational-vibrational states may not be active in infrared absorption but can be studied using Raman spectroscopy. There is also the opposite of this phenomenon so that infrared absorption spectroscopy is used in cases where Raman spectroscopy is not applicable to the study of molecules (Louden, 1989).

Therefore, transmissions that are highly intense in the Raman spectrum often have poor infrared absorption, and vice versa. In other words, vibration is active in infrared spectroscopy when a change in the dipole moment of the molecule occurs. Vibration is also active in Raman spectroscopy, which also changes the polarization of the molecule. For example, molecules with identical nuclei such as N_2, H_2, and O_2 are active in the Raman study but are not active in infrared spectroscopy. In the CO_2 molecule, symmetrical vibrational motion is active in Raman and not active in infrared. Conversely, asymmetric vibrational motion is not active in Raman but is active in infrared. Some vibrations are also active in both infrared and Raman. Table 1 shows the differences and similarities between Raman spectroscopy and infrared absorption (Louden, 1989).

Table 1 Comparison of Raman and IR spectroscopy (Krstić & Ražić, 2018)

Raman spectroscopy	IR spectroscopy
Analysis of scattered light of the vibrating molecules	Analysis of absorption of the vibrating molecules
Vibration is Raman active if it causes a change in the polarizability	Vibration is IR active if a change in the dipole moment during the vibration occurs
Molecules may not have a dipole moment	Chemical bond must have the characteristics of an electric dipole
Water can be used as a solvent	Water cannot be used as a solvent due to intense absorption
No need for specific sample preparation	Requires specific sample preparation
Provides information on the covalent characteristics of a molecule	Provides information on the ion characteristics of a molecule
Expensive instrumentation costs are relatively high	Relatively inexpensive instrumentation

2.3 Raman Shift

Raman shift displacements are typically reported as wave numbers, the unit of which is the inverse of the wavelength unit, and its value is directly related to energy changes. The following formula can be used to convert between spectral wavelength and shift wavenumbers in the Raman spectrum:

$$\Delta\nu = \left(\frac{1}{\lambda_0} - \frac{1}{\lambda_1}\right), \tag{1}$$

where $\Delta\nu$ is the Raman shift expressed in wavenumber, λ_0 is the excitation wavelength, and λ_1 is the Raman spectrum wavelength. Most commonly, the unit chosen for expressing wavenumber in Raman spectra is inverse centimeters (cm^{-1}). Since wavelength is often expressed in units of nanometers (nm), the formula above can scale for this unit conversion explicitly, giving:

$$\Delta\nu\left(cm^{-1}\right) = \left(\frac{1}{\lambda_{0(nm)}} - \frac{1}{\lambda_1(nm)}\right) \times \frac{10^7 nm}{(cm)}, \tag{2}$$

3 Raman Instrumentation and Sampling

3.1 Raman Techniques

Each Raman device consists of four main components, including a laser light source, a sample lighting system and light-collecting lenses, a wavelength selector (filter or spectrometer), and a detector. Figure 2 shows a schematic diagram of the typical arrangement of these components. Currently, various systems with this general format have been marketed. The main types of different devices are described by the wavelength region for the excitation sources, the type of wavelength selector, and the detection system. Most devices have a similar pattern light and collection system. The following sections provide a brief overview of the Raman techniques commonly used in the analysis of agricultural products and foods with their advantages and are also shown in Table 2.

3.1.1 Dispersive Raman Spectroscopy

Dispersive Raman Spectroscopy is the most common and fundamental technique with intense monochromatic radiation normally produced by continuous wave gas lasers such as argon (488.0 and 514.5 nm), krypton ions (647.1 nm), or diodes

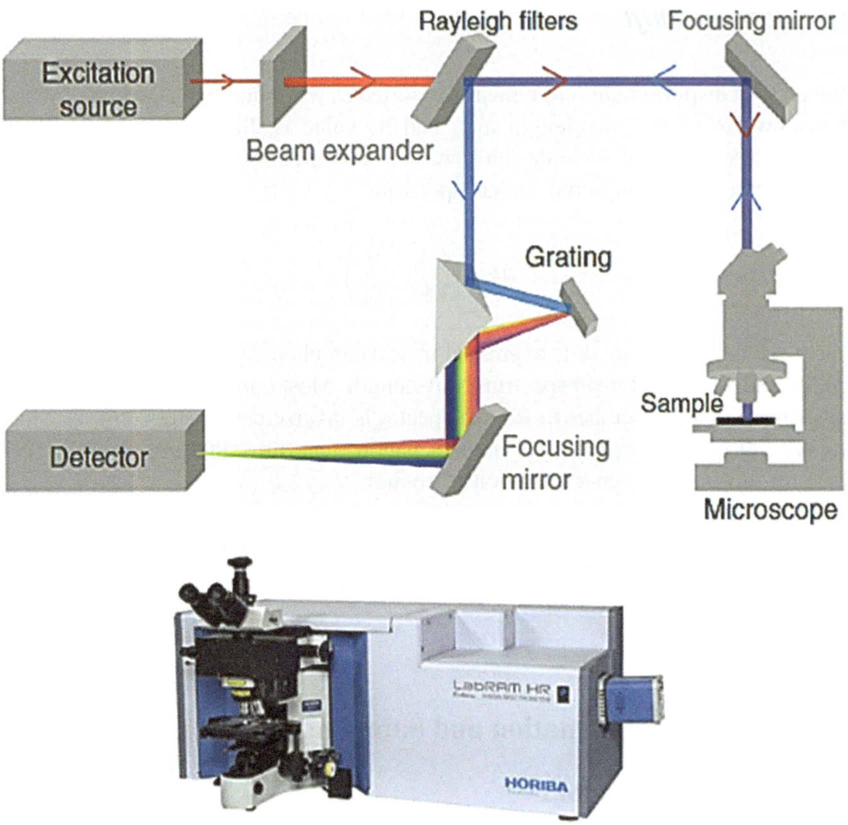

Fig. 2 (**a**) Internal components of Raman device. (**b**) An overview of a Raman device equipped with a microscope (Butler et al., 2016)

Table 2 Advantages of Raman techniques (Borio et al., 2012)

Techniques	Advantages
Dispersive Raman spectroscopy	Is suitable for aqueous samples, samples with elevated temperature, black samples
	Suppresses fluorescence at 780 or 830 nm
Fourier transform Raman spectroscopy	Reduces fluorescence
	Eases operation with FT-IR spectrometer
	Possesses high spectral resolution
Surface-enhanced Raman spectroscopy	Possesses high molecule specificity, sensitivity, and resolution
Spatially offset Raman spectroscopy	Has more effective illumination
	Suppresses fluorescence
	Facilitates analysis of a variety of samples

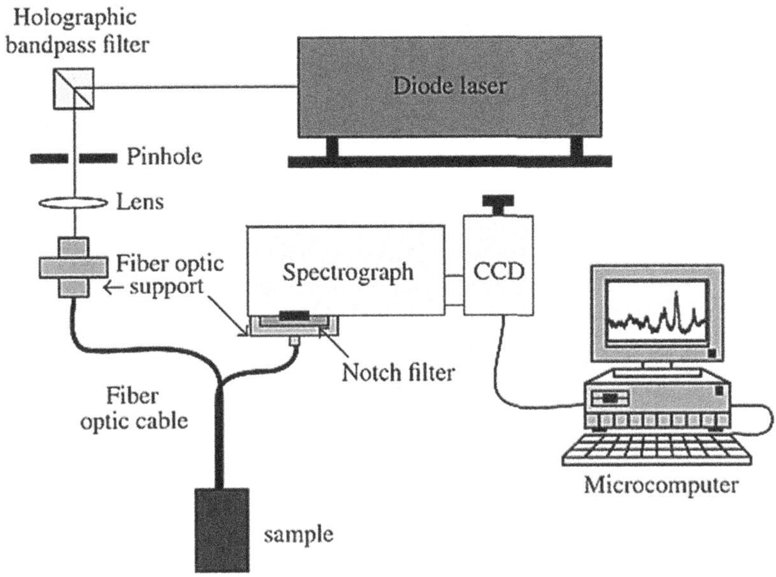

Fig. 3 Schematic diagram of dispersive Raman spectrometer (Borio et al., 2012)

(780 or 830 nm.) by combining a charge-coupled device (CCD) array detector with a single grating spectrograph. A schematic of a dispersive Raman spectrometer is shown in Fig. 3. Dispersive Raman spectroscopy has some simple advantages: (1) it is suitable for water phase samples; (2) it can also be used for high-temperature samples, even above 1000 °C; (3) allows you to analyze black samples; and (4) with 780 or 830 nm lasers, it can help suppress fluorescence from samples.

3.1.2 Fourier Transform Raman Spectroscopy

Fourier Transform Raman Spectroscopy systems, which use near IR excitation sources, were first marketed in 1987. With its ease of use, Raman spectroscopy has become a common feature in many industrial laboratories. Commercial systems use the 1064 nm excitation from Nd:YAG (Neodymium doped yttrium aluminum garnet) lasers with an NIR interferometer coupled to either a liquid nitrogen-cooled Ge or an InGaAs detector. A schematic diagram of the FT-Raman spectrometer is shown in Fig. 4. FT-Raman has three main advantages over dispersive Raman systems: (1) reducing the laser-induced fluorescence that a number of samples exhibit; (2) It is as easy to operate as a Fourier Transform Infrared (FT-IR) spectrometer; and (3) it shows high spectral resolution with good wavelength precision.

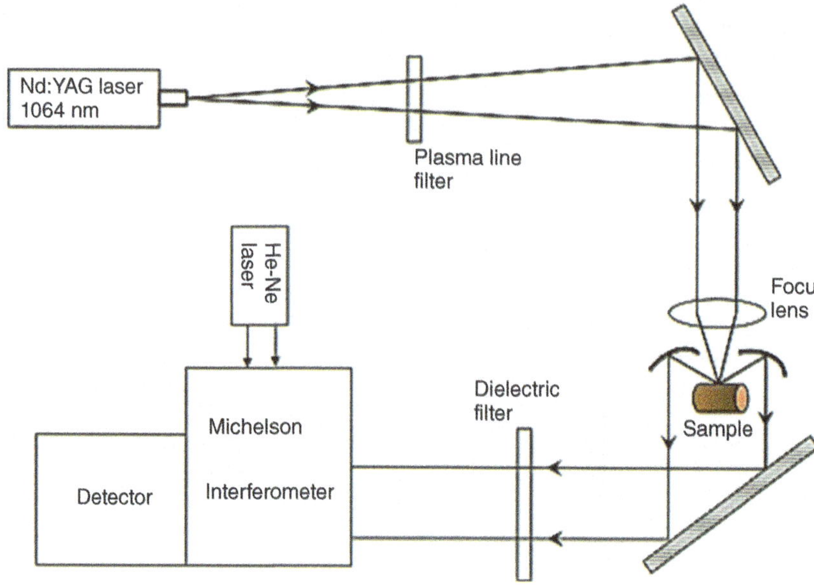

Fig. 4 Schematic diagram of Fourier Transform Raman Spectroscopy (Kizil & Irudayaraj, 2018)

3.1.3 Surface-Enhanced Raman Spectroscopy (SERS)

SERS has attracted a lot of attention in recent years due to its superior sensitivity compared to spontaneous Raman scattering. Using the two main classes of enhancement media, solid substrates and colloidal nanoparticles, total enhancement of the observed Raman signal can be up to 10^{14} times that of the unobserved signal. The improvement of the signal is based on two effects: the chemical mechanism and electromagnetic mechanism, the latter being dominant. This technique has been studied in several biological applications, including in vivo studies. Comprehensive reviews on SERS can be found in the literature, among them those of Guerrini and Graham (2012), McNay et al. (2011), Pieczonka and Aroca (2008), and Schlücker (2014). Figure 5 shows a schematic diagram of the SERS system.

3.1.4 Spatially Offset Raman Spectroscopy (SORS)

Spatially offset Raman spectroscopy is based on previous research on the Raman photon migration process. Unlike conventional Raman spectroscopy, shown in Fig. 6, Raman spectra for SORS are collected from spatially displaced areas of light on the sample surface. Due to the wide lateral scattering of photons emanating from greater depths, SORS has different signals from the sample layers at different depths. With a large spatial shift, greater suppression of the surface layer signal is achieved in SORS. SORS relies on the separation of the illuminated and collected

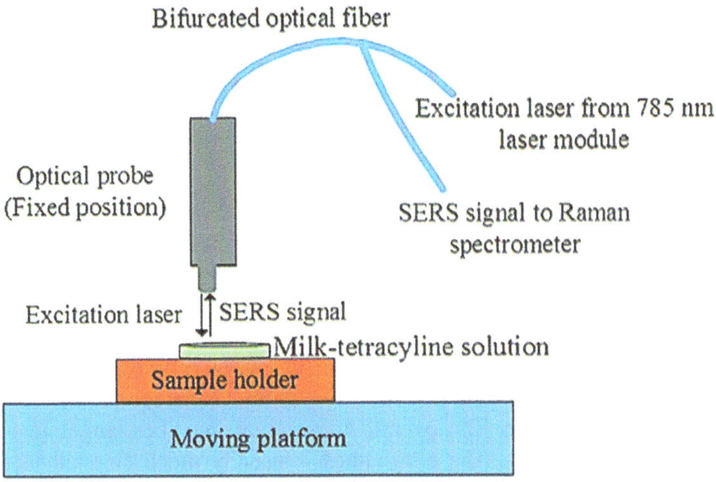

Fig. 5 Schematic diagram of the portable SERS system (Dhakal et al., 2018)

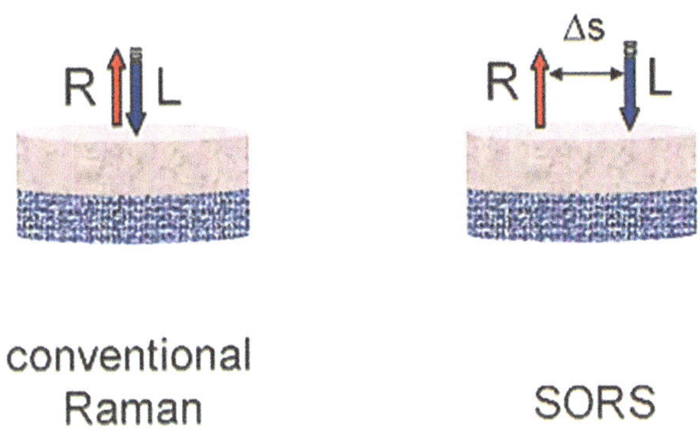

Fig. 6 Schematic diagram of conventional Raman and SORS approach (Matousek, 2018)

areas of the sample surface. The greater the separation (spatial shift, Δs), the deeper the Raman signal can be retrieved. This is because the lateral migration of photons is statistically skewed to deeper areas, which favors photon paths that stay away from the air-sample interface with loss. This is because if a photon migrates near this interface, it is more likely to cross the line and fade from the pattern.

3.2 *Raman Sampling*

A clear difference between IR and Raman spectroscopy can be seen in the sampling methods. In IR spectroscopy, sampling methods for conventional measurements are relatively simple. In contrast, Raman sampling techniques are complex and versatile, and individual workers use different sampling techniques designed to meet their needs. Some of these methods are described below.

3.2.1 Fiber Optics

Fiber optics are ideal light pipes for transmitting optical signals in and out of the sample. Using fiber optics to connect the sample to a remote location, the instrument can be located in the laboratory or in another clean place. Optical fibers can bend the light ray in any direction and transport optical signals to unhealthy or dangerous environments such as high-temperature processing lines and nuclear facilities.

Fiber optics consists of a main material with a relatively high refractive index, surrounded by a cladding with a low refractive index. Radiation is fully reflected in the core-cladding interface. The overall diameter of the fiber ranges from a few micrometers to millimeters. The core radius is usually much larger than the clad radius. For Raman spectroscopy, a single fiber can be used to carry the laser beam, stimulate the sample, and one or more fibers can be used to return the Raman scattered light back to the spectrometer.

Figure 7 shows a schematic of an FT-Raman system optimized for fiber optic interfaces. Excitation radiation from an Nd:YAG laser is focused onto a single, center fiber of a bifurcated fiber bundle by lens L1. The radiation scattered in Raman is transmitted to FT-IR through 18 fibers around the central fiber. Low-hydroxy

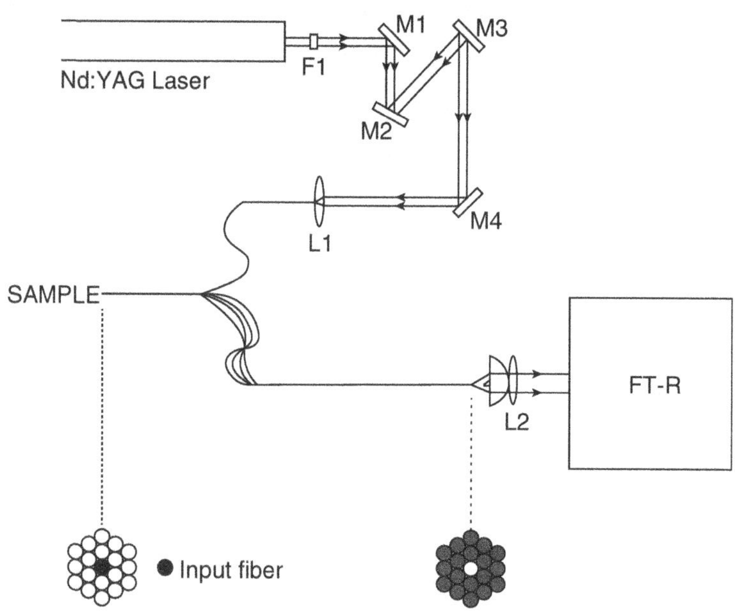

Fig. 7 Schematic diagram of FT-Raman instrumentation fitted with a bifurcated fiber optic interface (De Bakker & Fredericks, 1995)

quartz fibers with a core diameter of 200 μm are used for the probe. This type of two-way fiber bundle is ideal for combining excitement and Raman scattering, as the intense laser intensity propagates through a single fiber, while the weak Raman signal propagates through 18 input fibers (De Bakker & Fredericks, 1995).

3.2.2 Colorless Compounds

If the sample is colorless, the Raman spectrum can easily be obtained by irradiating the sample with a laser beam whose wavelength is in the visible range (normal Raman scattering).

Gases: The sample gas normally is contained in glass tubing of diameter 1–2 cm and thickness of 1 mm. The gas can be sealed in a small capillary tube whose diameter is slightly larger than that of the laser beam (~1 mm). For weak Raman scatter, an external resonant setup is used to increase the intensity of Raman by passing the laser beam more than once through the sample (Fig. 8a).

Liquids: Liquid samples can be sealed in ampoules, tubes, or capillaries, depending on the amount of sample available (Fig. 8b). For microquantities (~10 μL), capillaries with a diameter of 0.5–0.1 mm and a length of ~1 mm were used. The use of large cylinder cell cells as shown in Fig. 8c reduces local heating and allows a more accurate determination of the deportation ratio. If the solution contains solid particles, a loud noise can be made.

Solids: Depending on the sample size, powder samples can be packed in ampoules or capillaries, and their Raman spectra are measured in the same way as for liquid samples. The same KBr pellet method used in IR spectroscopy is also useful. First, 200 mg of KBr powder is compressed to provide support.

3.2.3 Colored Compounds

For color samples that absorb the energy of the laser beam, decomposition can occur by local heating. In this case, there are several methods available besides reducing laser power. These include: (1) changing the laser wavelength, (2) concentrating the laser beam into the sample, (3) avoiding absorption by the sample by diluting the sample density in the pellets or solution, (4) cooling the sample, (5) rotate the sample, and (6) rotate or vibrate the laser beam in a fixed pattern.

3.2.4 Special Cells

In addition to the cells already described, there are still many cells that are suitable for special applications. These include (1) Thermostated Cells, (2) High-Temperature Cells, (3) Low-Temperature Cells, and (4) UV Resonance Cells.

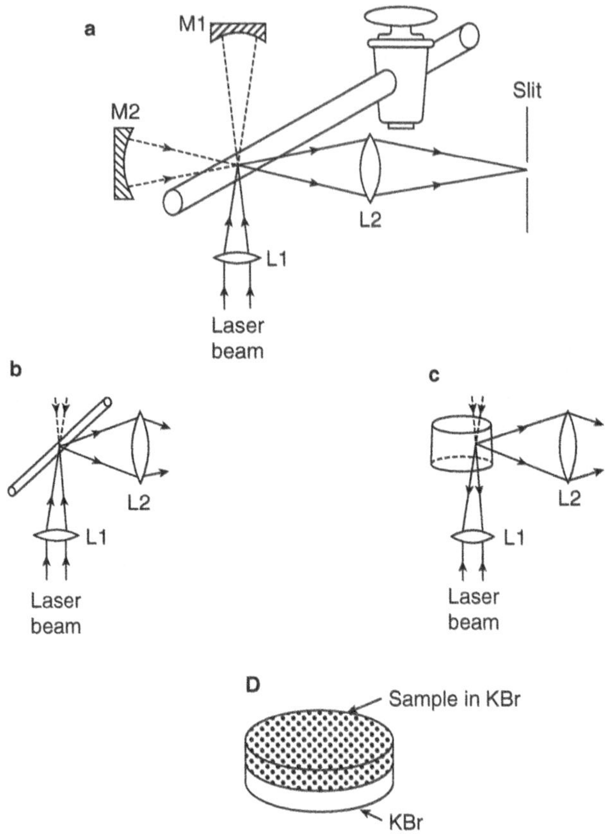

Fig. 8 (**a**) Gas cell with external resonating mirrors. (**b**) Capillary cell for liquids. (**c**) Cylindrical cell for liquids. (**d**) KBr pellet for solid samples (Ferraro et al., 2003)

4 Statistical Data Analysis

4.1 Calibration and Data Processing

Until now we have considered the components of the spectrometer and the presentation of the sample. Before continuing to research and manipulate data, a question that is regularly asked is "How do you know if your device is working correctly and consistently?" FT-Raman instruments are calibrated with an internal laser, which is used to determine the precise position of the moving mirror on the interferometer. The intensity of the interphagram is thus known as a function of the mirror location (distance in cm) and is converted into the fast Fourier variable distance or wavenumber (cm^{-1}) in the spectral domain.

For dispersive instruments, the wavenumber or wavelength readings on the instrument should not be considered a nominal value. It is recommended to calibrate the instrument regularly. The time it takes to calibrate depends on the accuracy desired for a particular experiment. Dispersive spectrometers are typically frequency calibrated using one of the following methods.

4.1.1 Internal Standards

If 1 cm^{-1} accuracy is required, internal standards can be used. This can be the frequency of solvent bands or bands of added, noninteractive solutions. The measured compound band is compared to the internal standard frequency. However, care must be taken to ensure that the chemical interaction between the substance under investigation and the reference itself does not cause a large band shift. In addition to its simplicity, this method has a distinct advantage over other methods in that the frequency determined by the position of the band according to the internal standard is necessarily independent of temperature.

4.1.2 Indene

If additional precision (on the order of 0.5 cm^{-1}) is desired, indene can be used (Hendra & Loader, 1968). Indene has also been used as a frequency calibrator for IR spectrophotometers. Before use, it must be purified by vacuum distillation and stored in a sealed capillary or NMR tube.

4.1.3 Laser Plasma Lines

The principal plasma lines of the argon-ion laser, some of which may be used for calibration, are listed by many researchers, such as Hendra and Loader (1968), Beck et al. (1978) and Zaidel et al. (1961). In order to observe these lines, the laser beam must be detuned and the scattered radiation must be collected from the Kimax melting point tube. This method gives a calibration of better than 1 cm^{-1} accuracy.

4.1.4 Neon Emission Lines

If the neon lamp is available, the Ne emitting lines can be used to obtain high-frequency calibration over a wide frequency range. The Ne frequencies that are useful for calibrating the Raman spectra obtained by excitation with a He-Ne and a Kr-ion laser are listed by Hendra and Loader (1968). For a complete listing of neon emission lines, see Strommen and Nakamoto (1984).

4.2 Data Analysis

Spectroscopic studies can quickly generate large data sets that require computer processing to obtain biochemical information. You can extract useful images, spectral biomarkers, and patterns and classify samples based on spectral fingerprints, depending on your specific spectral acquisition and experimental purpose. As a rule, data analysis can be divided into three different stages: (1) preprocessing of data set, (2) feature extraction, and (3) classification. A list of available analysis software is shown in Table 3.

Table 3 Data analysis software for Raman spectral data sets (Butler et al., 2016)

Software	Website	License
CytoSpec	http://www.cytospec.com/ftir.php	Commercial
ImageLab	http://www.imagelab.at/en_home.html	Commercial
MATLAB	–	Commercial
Biodata Toolbox	http://www.mathworks.com/matlabcentral/fileexchange/22068-biodata-toolbox	Open source
Extended Multiplicative Signal Correction (EMSC) Toolbox	http://www.models.life.ku.dk/emsctoolbox	–
IRootLab	https://code.google.com/p/irootlab/	Open source
Multivariate Image Analysis (MIA) Toolbox	http://www.eigenvector.com/software/mia_toolbox.htm	Open source
Multivariate Curve Resolution–Alternating Least Squares (MCR-ALS) Toolbox	http://www.cid.csic.es/homes/rtaqam/tmp/WEB_MCR/welcome.htm	Commercial
PLS Toolbox	http://www.eigenvector.com/software/pls_toolbox.htm	Open source
Raman Processing Program	http://cares.wayne.edu/rp/	Commercial
Origin for Spectroscopy	http://www.originlab.com/index.aspx?go=Solutions/Applications/Spectroscopy	Commercial
PeakFit	https://systatsoftware.com/products/peakfit/	Commercial
Python	–	Open source
PyChem	http://pychem.sourceforge.net/	Open source
PyVib2	http://pyvib2.sourceforge.net/	Open source
R	–	
HyperSpec	http://hyperspec.r-forge.r-project.org/	Open source
The Unscrambler X	http://www.camo.com/	Commercial

4.2.1 Preprocessing

Immediately after acquiring the spectrum, the quality of the spectral dataset should be evaluated and proactive treatment should be applied to reduce small fluctuations and improve study accuracy. At this point, the spectrum should be corrected for cosmic rays and the quality of the spectrum can be visually evaluated. There are some situations where visual inspection of the spectrum shows obvious outliers, such as significant spectral contamination, fluorescence, or a very low signal-to-noise ratio. In such cases, explicit outliers can be removed from the data set. Other spectra containing outliers in the data set can be identified using Q-tests or other outlier-detect algorithms; this can be done in addition to or beyond visual inspection, to reduce user bias (Khodabakhshian & Abbaspour-Fard, 2020).

Raman spectra are particularly prone to noise, and the data may require noise reduction to improve spectral quality. The first approach would be to modify the spectral acquisition parameters to improve the quality of the Raman signal, for example, by using increased integration time and higher laser power. Sample preparation can be adjusted by prefocusing the sample or by photobleaching. If these methods are inadequate, the spectrum can be significantly maintained after acquisition to improve the signal-to-noise ratio. Principal Component Analysis (PCA) is a powerful Raman preprocessing technique that can efficiently reduce spectra to a number of principal components (PCs) that represent large spectral dispersion. This technique can only be used to reconstruct spectra using a prominent PC thus eliminating background noise and preserving prominent spectral data. Other noise reduction techniques include Savitzky–Golay (SG) smoothing, minimal noise fraction transform, and high-frequency noise filtering techniques such as wavelet denoising (WDN). However, as smoothing of the spectra reduces the apparent noise, we note that these processes also degrade the spectrum features, and we recommend the limited and careful use of perfumes (Khodabakhshian, 2019; Khodabakhshian & Abbaspour-Fard, 2020).

Fluorescence and background noise of the sample and thermal fluctuations of the CCD can have a significant impact on the spectral baseline and require baseline correction. Polynomial baseline fitting attempts to estimate unknown background. These are usually based on user-defined polynomial points and can effectively eliminate sloped or oscillatory baselines. Caution should always be exercised with any underlying deduction procedure as they may take unannounced samples. Alternatively, the first- or second-order differential in conjunction with SG smooth can be used to mathematically eliminate scalar baseline offsets or slope contributions while resolving overlapping peaks. Unlike polynomial fittings that cause spectra with conventional morphology, the obtained spectra are transformed and have no regular Raman appearance.

After baseline correction, the spectra may also need to be normalized to correct for sample and experiment variables, such as thickness and density. Vector normalization and min-max normalization are two popular methods that can be applied to spectra after any baseline correction algorithm without significantly affecting the vernacular properties. Amide I peak normalization is also commonly used in IR and Raman studies. However, due to the shift of the typical amide bands, or for experiments where you are measuring a change in protein structure, this technique is not appropriate after the distinction, since all values are made 1 (Khodabakhshian, 2019).

Finally, optional data reduction step can be added in the preprocess to optimize statistical analysis. As highlighted above, large spectral data sets can often cause significant computational loads, as there are many absorption intensities in the same spectrum. This load can be reduced by cutting the spectrum to shorter wave number ranges, especially if the range is concentrated on individual Raman peaks. Alternatively, data reduction algorithms such as partial least squares regression (PLS) and PCA can be used to reduce individual spectra to several key components, which are widely implemented as preprocess and feature extraction steps.

It has been shown that the choice of preprocessing steps and the order in which they are performed have a major impact on the results of the spectral analysis, making it difficult to develop a uniform approach. Whenever possible, we recommend using estimated baseline correction and vector normalization. This requires high SNR data but is especially effective in diagnostic studies. In addition, tools for noise and data reduction are available as needed.

4.2.2 Feature Extraction

In both research and diagnostic studies, it is important to turn individual spectacles into appropriate variables that provide biological information. Feature extraction methods range from very simple ones, such as determining band regions, to very complex computational tasks. Feature extraction can be divided into two different ways: feature construction and feature selection. These methods have been extensively reviewed by Trevisan et al. (2012). In summary, the construction of characteristics can be defined as the creation of new characteristics in a data set that can infer otherwise obscured information, for example, the linear methods mentioned above are PCA and PLS. It is exceptional for diagnosis, biomarker extraction, and pattern recognition in homogeneous data sets. Feature selection extrapolates existing features from a dataset, such as specific wavenumbers, that can be used to identify spectral biomarkers and/or input into diagnostic frameworks. Techniques such as genetic algorithms, multidimensional curve resolution, and successive projection algorithms have proven to be particularly popular as feature extraction techniques because only informative variables are included in the resulting model.

4.2.3 Classification

Classification of samples based on their spectrum is often desirable in both imaging and diagnostic studies, as spectra can be classified based on previous user input (supervised classification) or spectrum variance alone (unsupervised). Unsupervised classification usually relies on clustering techniques of which hierarchical cluster analysis, k-means clustering, and fuzzy C-means clustering are three common options for this approach. PCA is another unsupervised approach commonly used to extract principal components that represent the largest variance in a dataset. For imaging studies, this approach does not require prior knowledge of the sample in question and produces pseudo-spectral images rich in research information. The loadings or spectral features used to calculate scores or weight icons can provide important information about the molecular distribution in a sample (Khodabakhshian, 2019; Khodabakhshian & Abbaspour-Fard, 2020).

For diagnostic analysis, supervised classification is most commonly used because the desired results depend on the class labels provided by the user. This could be baseline histopathology, cell type, or other class of measurement. These class assignments are taken into account when implementing classification techniques using parts of the dataset called the "training dataset." Then a classifier can categorize a separate "test data set" accordingly. It is a common approach to validate classification results using an independent data set. Linear discriminant classifiers (LDCs), artificial neural networks, and support vector machines (SVMs) are particularly common machine learning techniques that efficiently classify spectral data, although other techniques are available (Khodabakhshian, 2019; Khodabakhshian & Abbaspour-Fard, 2020). A semi-supervised classification is an alternative option when some, but not all, data classes can be determined.

5 Application of Raman Spectroscopy in Fresh Fruits and Vegetables

On the basis of Raman spectroscopy characteristics, such as rapid intrinsic analysis, minimal sample preparation, and compatibility with aqueous solutions, this technique has become a useful tool for at-line or online analysis in various fields, including food science, agriculture and biomass, biological and medical, environmental, biochemical, geological, and material science. Given that fresh fruits and vegetables are an important factor in people's daily lives, this section introduces current applications that support the use of Raman spectroscopy in fresh fruits and vegetables, most of which were published in structural analysis, safety control, classification, and quantification of fruits and vegetables between 2000 and 2021.

5.1 Structural Analysis

Veraverbeke et al. used FT-Raman spectroscopy for evaluating natural intact wax layers on the surface of whole fruits, but due to the high fluorescence of apple tissue, no continuous interference was obtained (Veraverbeke et al., 2005). López-Casado et al. have succeeded in the biomechanical behavior analysis of the cuticle of isolated tomato (*Solanum lycopersicum L.*) fruits using Raman spectroscopy (López-Casado et al., 2007). FT-Raman spectroscopy was illustrated as a useful tool in the structural analysis of commercial citrus and sugar beet pectin by Synytsya et al. and was showed that the combination of FT-Raman and FT-IR spectroscopic methods provided a more complete characterization of samples of pectin than any method (Synytsya et al., 2003). A good agreement with HPLC and ultraviolet-visible (UV-Vis) spectroscopies in quercetin concentrations determined by Raman spectroscopy was found by Numata and Tanaka with a relative error of about 1% (Numata & Tanaka, 2011). A spectroscopic description of *Rosa aff. Rubiginosa* flesh and seeds using Raman spectroscopy to elucidate the significant chemical aspects of this fruit was provided by Da Silva et al. (2008). In a study by Pudney et al., three different types of carotenoids found in tomatoes, lycopene, β-carotene, and lutein were investigated and the physical state of the carotenoids was recognized by this technique, which provided necessary knowledge that Raman spectroscopy could be fruitfully used in nutrient bioavailability trials (Pudney et al., 2011).

5.2 *Safety Control*

The safety of fruits and vegetables can be carried out by Raman spectroscopic methods. Yang and Irudayaraj pioneered the FT-Raman approach for the detection and classification of foodborne microorganisms on the whole apple surface (Yang & Irudayaraj, 2003). Successful differentiation of five different strains of Escherichia coli and 100% accuracy in separating pathogens from non-pathogens has demonstrated that FT-Raman spectroscopy is an excellent tool for the investigation of microbial contamination on food surfaces and classification of microbial cultures. The FT-Raman spectroscopy was used to detect microbial disease by Bonora et al. (Zhang et al., 2006). The investigation of the structural properties of "Hayward" kiwifruit affected by elephantiasis was accomplished using a FT-Raman method. Pesticides are a key component in preserving fruits and vegetables, but excessive concentrations pose a risk to human health. However, rapid chemical analysis of pesticide residues is not available. The Raman spectra of clean fruits and pesticides were measured by Zhang et al. and it was found that FT-Raman method can be applied to measure effective components in fruits and vegetables, as well as pesticides left on the surface (Zhang et al., 2006). Based on its high sensitivity and ability to identify the molecular structure of pesticides, the SERS sampling device was developed by Shende et al. for detection of pesticides on the surface of fruits and in juices extracted from fruits (Shende et al., 2004). Liu and Liu developed a simple, rapid, and environmentally friendly technique for micro-Raman spectroscopy determination of pesticide residues on the surface of fruits (Liu & Liu, 2010). Overall results indicate that micro-Raman spectroscopy is a potential tool for determining fruit surface and pesticide residues for fruit quality and safety control. A fast and accurate method for detection of fenthion on fruit and vegetable peel was developed by Weng et al. using dynamic surface-enhanced Raman spectroscopy and random forests with variable selection (Weng et al., 2018). They found that dynamic surface-enhanced Raman spectroscopy and random forests with variable selection can achieve the fast, ultrasensitive, and accurate analysis of fenthion residue on fruit and vegetable peel. Also in research, the freshness assessment of citrus fruits using Raman spectroscopy was done recently (Nekvapil et al., 2018). It was demonstrated that the freshness assessment could be achieved using a portable Raman instrument.

5.3 Classification

A very small portion of rotten fruits and vegetables spoils the whole batch. Good separation from bad things, better than usual, is the key to maintaining good fruit and vegetable quality. Nowadays many researchers found that associated with chemometrics, spectroscopy-based techniques such as Raman spectroscopy become a powerful tool in extracting chemical information from complex matrices (Khodabakhshian, 2019; Khodabakhshian & Abbaspour-Fard, 2020; Yang & Ying, 2011). Muik et al. combined Fourier-Raman spectroscopy and partial least squares (PLS) to classify olives based on the fatty acid content (FFA) of olives and achieved an accuracy of 80% (Muik et al., 2004). In a subsequent study, they used FT-Raman and pattern recognition to distinguish olives of different qualities, including healthy olives, frostbite olives, olives collected from the ground, fermented olives, and olive samples with diseases (Muik et al., 2004). The results showed that the Raman spectra of the samples were correctly clustered using soft independent modeling of class analogy (SIMCA) with discriminant power of 95% for sound, 93% for frostbite, 96% for ground, 92% for fermented olives, and 100% for damaged olives. The above research has shown that Raman spectroscopy works directly on olives and therefore makes it possible to check olives based on their quality, which could help optimize the production of high-quality virgin olive oil. Changes that occur during the development and maturation of olive fruits, such as increasing the content of carotenoids and phenolic compounds during olive growth and decreasing them in the ripening stage, can be successfully controlled by Raman spectroscopy, a rapid tool for the classification of olives depends on their stage of maturity.

Recently in another study, the identification and characterization of Aspergillus specics of fruit rot fungi using Raman spectroscopy was investigated (Saif et al., 2021). The results found by Raman spectroscopy could be a useful tool, sensitive, fast, economical, accurate, not require sample preparation, and successfully used to identify fungi. The potential of FT-Raman spectroscopy as a rapid and noninvasive technique to distinguish four different maturity stages of pomegranate fruits (Immature stage (S1), Fairly Half-ripe stage (S2), Half-ripe stage (S3), and Full ripe stage (S4)) on the basis of tannin changes was investigated by Khodabakhshian (2019). The analysis of the Raman signal changes that happen during pomegranate maturity and its relationship with the maturity degree of the pomegranate fruits was studied. It concluded that Raman spectroscopy has the potential for the future development of a

Raman-based nondestructive approach for subsurface detection of tannin as an indicator of pomegranate maturity. In a later study (Khodabakhshian & Abbaspour-Fard, 2020), the feasibility of utilizing Fourier transform Raman spectroscopy, combined with supervised and unsupervised pattern recognition methods were considered, to distinguish the maturity stage of pomegranate "Ashraf variety" during four distinct maturity stages between 88 and 143 days after full bloom. Principal component analysis (PCA) as a method of unsupervised pattern recognition was performed to test the possibility of clustering the pomegranate samples into four groups. Two supervised pattern recognition techniques, partial least squares discriminant analysis (PLSDA) and soft independent modeling of class analogy (SIMCA), were also used as powerful supervised pattern recognition methods for classifying samples. The results showed that in all sample groups, the Raman spectra of the samples were correctly clustered using PCA. The accuracy of the SIMCA classification for differentiation of four pomegranate groups was 82%. In addition, the overall differential performance of the PLSDA classes was approximately 96%, and 95% for the calibration and validation sample sets, respectively.

5.4 *Quantification*

Raman spectroscopy is commonly applied for the detection of carotenoids in biomaterials, such as fruits and vegetables, as carotenoids are known to contribute to the beneficial effects of fruits and vegetables. In a research by Darvin et al., Raman spectroscopy was applied to determine carotenoid antioxidant substances in fruits and vegetables (Darvin et al., 2005). Schulz et al. developed a NIR-FT-Raman spectroscopy method to measure lycopene and used Raman mapping techniques to deepen his knowledge (Schulz et al., 2005a). The comparison of the levels of lycopene and α-carotene in tomato fruits and related crops using FT-Raman and NIR spectroscopy was studied by Baranska et al. (2006). The Raman spectrum of tomato products exhibited distinct key bands of the studied carotenoids, which provided the basis for quantitative measurements of lycopene and β-carotene. The model of FT-Raman could predict better lycopene and α-carotene with $R^2 = 0.91$ and 0.89, standard error of cross-validation [SECV] = 74.34 and 0.34, respectively than NIR prediction model ($R^2 = 0.85$ and 0.80, SECV = 91.19 and 0.41 for lycopene and α-carotene, respectively). Similar research was conducted by Bicanic et al., who evaluated the uniform trans-lycopene content from fresh tomatoes by micro-Raman spectroscopy (Bicanic et al., 2010a). The strongest typical Raman band occurred at 1520 cm^{-1}, but the concentration of trans-lycopene in the homogeneity showed a weak correlation ($R^2 = 0.52$), which was probably due to the fact that Raman measurements were not done after 1 week during a storage time of 1–20 °C. Resonance Raman spectroscopy (RRS) can amplify the signal of samples for better quantitative analysis. Bicanic et al. used RRS to estimate the concentration of β-carotene in mango homogenates and found a good correlation between the intensity of Raman signals and the concentration with $R^2 = 0.9618$ (Bicanic et al.,

2010b). In addition to carotenoids, a feasibility study performed to predict the level of piperine and essential oil in different samples of ground pepper by NIRFT-Raman spectroscopy showed a high correlation between Raman spectroscopy data and chromatography with $R^2 = 0.82–0.84$ for piperine and $R^2 > 0.95$ for essential oil (Schulz et al., 2005b). Also, the feasibility of using Raman spectroscopy for the quantification of soluble solids and individual sugars in apples was studied by Monago-Marana et al. (2021). Six different commercially available apple varieties were measured by Raman spectroscopy: (1) intact apples with skin, (2) apples without skin, and (3) juices obtained from apples. The results showed that a large area Raman probe can be used to measure Raman signals at an apple depth of 8 mm. A multivariate calibration model was established to assess how well it can be used to estimate the quality parameters SSC (%), total sugar, glucose, fructose, and sucrose using Raman spectra. References number and year format are mixed up?

6 Conclusion

In this chapter, we have attempted to introduce the reader to the basics of Raman spectroscopy, delving too deep into the theory and details of the practice to generate any further interest. Also, this chapter provides the theoretical background needed for a full spectra analysis, a guide to the methods by which Raman spectroscopy has been used successfully, and leading to more sophisticated but less commonly available Raman spectrometer techniques. The historical background and various aspects of Raman spectroscopy theory including basic Raman scattering theory, instrumentation, and statistical data analysis were considered in this chapter. It also provides brief overviews of the application of Raman spectroscopy in the analysis and control of fruits and vegetables.

References

Baranska, M., Schultze, W., & Schulz, H. (2006). Determination of lycopene and α-carotene content in tomato fruits and related products: Comparison of FT-Raman, ATR-IR, and NIR spectroscopy. *Analytical Chemistry, 78*, 8456–8461.

Beck, R., Englisch, W., & Gurs, K. (1978). *Tables of laser lines in gases and vapors* (2nd ed.). Springer-Verlag.

Bicanic, D., Dimitrovski, D., Luterotti, S., Marković, K., van Twisk, C., Buijnsters, J. G., & Dóka, O. (2010a). Correlation of trans-lycopene measurements by the HPLC method with the optothermal and photoacoustic signals and the color readings of fresh tomato homogenates. *Food Biophysics, 5*, 24–33.

Bicanic, D., Dimitrovski, D., Luterotti, S., Twisk, C., Buijnsters, J. G., & Dóka, O. (2010b). Estimating rapidly and precisely the concentration of beta carotene in mango homogenates by measuring the amplitude of optothermal signals, chromaticity indices and the intensities of Raman peaks. *Food Chemistry, 121*, 832–838.

Borio, V. G., Vinha, R., Nicolau, R. A., Oliveira, H. P., Lima, C. J., & Silveira, L. (2012). Quantitative evaluation of acetaminophen in oral solutions by dispersive Raman spectroscopy for quality control. *Spectroscopy, 25*(1), 3956–3972.

Butler, H. J., Ashton, L., Bird, B., Cinque, G., Curtis, K., Dorney, J., & Martin, F. L. (2016). Using Raman spectroscopy to characterize biological materials. *Nature Protocols, 11*(4), 664–687.

Da Silva, C. E., Vandenabeele, P., Edwards, H. G. M., & De Oliveira, L. F. C. (2008). NIR-FTRaman spectroscopic analytical characterization of the fruits, seeds, and phytotherapeutic oils from rosehips. *Analytical and Bioanalytical Chemistry, 392*, 1489–1496.

Darvin, E., Gersonde, I., Meinke, M., Sterry, W., & Lademann, J. (2005). Non-invasive in vivo determination of the carotenoids beta-carotene and lycopene concentrations in the human skin using the Raman spectroscopic method. *Journal of Physics D: Applied Physics, 38*, 2696.

De Bakker, C. J., & Fredericks, P. M. (1995). Determination of petroleum properties by fiber-optic Fourier transform Raman spectrometry and partial least-squares analysis. *Applied Spectroscopy, 49*, 1766.

Delhaye, M., & Migeon, M. (1966). *Comptes Rendus de l'Académie des Sciences, 262*, 702.

Dhakal, S., Chao, K., Huang, Q., Kim, M., Schmidt, W., Qin, J., & Broadhurst, C. L. (2018). A simple surface-enhanced Raman spectroscopic method for on-site screening of tetracycline residue in whole milk. *Sensors (Basel), 18*(2), 424.

Downes, A., & Elfick, A. (2010). Raman spectroscopy and related techniques in biomedicine. *Sensors (Basel), 10*(3), 1871–1889.

Ferraro, J. R., Nakamoto, K., & Brown, C. W. (2003). Instrumentation and experimental techniques. In J. R. Ferraro, K. Nakamoto, & C. W. Brown (Eds.), *Introductory Raman spectroscopy* (2nd ed., pp. 95–146). Academic Press.

Guerrini, L., & Graham, D. (2012). Molecularly-mediated assemblies of plasmonic nanoparticles for Surface-Enhanced Raman Spectroscopy applications. *Chemical Society Reviews, 41*, 7085.

Hendra, P. J., & Loader, E. J. (1968). Laser Raman spectra of sorbed species. *Nature, 217*(5129), 637–638. https://doi.org/10.1038/217637a0

Khodabakhshian, R. (2019). Feasibility of using Raman spectroscopy for detection of tannin changes in pomegranate fruits during maturity. *Scientia Horticulturae, 257*, 108670.

Khodabakhshian, R., & Abbaspour-Fard, M. H. (2020). Pattern recognition-based Raman spectroscopy for non-destructive detection of pomegranates during maturity. *Spectrochimica Acta Part A: Molecular and Biomolecular Spectroscopy, 231*, 118127.

Kizil, R., & Irudayaraj, J. (2018). Spectroscopic technique: Fourier Transform Raman (FT-Raman) spectroscopy. In D.-W. Sun (Ed.), *Modern techniques for food authentication* (2nd ed., pp. 193–217). Academic Press.

Krstić, M., & Ražić, S. S. (2018). Analytical approaches to the characterization of solid drug delivery systems with porous adsorbent carriers. *Current Medicinal Chemistry, 25*(1), 3956–3972.

López-Casado, G., Matas, A. J., Domínguez, E., Cuartero, J., & Heredia, A. (2007). Biomechanics of isolated tomato (Solanum lycopersicum L.) fruit cuticles: The role of the cutin matrix and polysaccharides. *Journal of Experimental Botany, 58*(14), 3875–3883.

Liu, Y., & Liu, T. (2010). Determination of pesticide residues on the surface of fruits using micro-Raman spectroscopy. *IFIP Advances in Information and Communication Technology, 347*, 427–434.

Louden, J. D. (1989). Raman microscopy. In D. J. Gardiner & P. R. Graves (Eds.), *Practical Raman spectroscopy* (pp. 119–151). Springer-Verlag.

Matousek, P. (2018). Spatially offset Raman spectroscopy for non-invasive analysis of turbid samples. *TrAC Trends in Analytical Chemistry, 103*, 209–214.

McNay, G., Eustace, D., Smith, W. E., Faulds, K., & Graham, D. (2011). Surface-enhanced Raman scattering (SERS) and surface-enhanced resonance Raman scattering (SERRS): A review of applications. *Applied Spectroscopy, 65*, 825.

Monago-Maraña, O., Afseth, N. K., Knutsen, S. H., Wubshet, S. G., & Wold, J. P. (2021). Quantification of soluble solids and individual sugars in apples by Raman spectroscopy: A feasibility study. *Postharvest Biology and Technology, 180*, 111620. https://doi.org/10.1016/j.postharvbio.2021.111620
Muik, B., Lendl, B., Molina-Díaz, A., Ortega-Calderón, D., & Ayora-Cañada, M. J. (2004). Discrimination of olives according to fruit quality using Fourier transform Raman spectroscopy and pattern recognition techniques. *Journal of Agricultural and Food Chemistry, 52*, 6055–6060.
Nekvapil, F., Brezestean, I., Barchewitz, D., Glamuzina, B., Chiş, V., & Cintă Pinzaru, S. (2018). Citrus fruits freshness assessment using Raman spectroscopy. *Food Chemistry, 242*, 560–567.
Nemecek, D., Stepanek, J., & Thomas, G. J., Jr. (2013). Raman spectroscopy of proteins and nucleoproteins. *Current Protocols in Protein Science, 17*, 8.
Numata, Y., & Tanaka, H. (2011). Quantitative analysis of quercetin using Raman spectroscopy. *Food Chemistry, 126*(2), 751–755.
Pieczonka, N. P. W., & Aroca, R. F. (2008). Single molecule analysis by surfaced-enhanced Raman scattering. *Chemical Society Reviews, 37*, 946.
Porto, S. P. S., & Wood, D. L. (1962). Ruby optical maser as a Raman source. *Journal of the Optical Society of America, 52*, 251.
Pudney, P. D. A., Gambelli, L., & Gidley, M. J. (2011). Confocal Raman microspectroscopic study of the molecular status of carotenoids in tomato fruits and foods. *Applied Spectroscopy, 65*(2), 127–134.
Raman, C. V., & Krishnan, K. S. (1928). A new type of secondary radiation. *Nature, 121*, 501.
Saif, F. A., Yaseen, S. A., Alameen, A. S., Mane, S. B., & Undre, P. B. (2021). Identification and characterization of Aspergillus species of fruit rot fungi using microscopy, FT-IR, Raman and UV–Vis spectroscopy. *Spectrochimica Acta Part A: Molecular and Biomolecular Spectroscopy, 246*, 119010.
Sala, O., Bassi, D., Santos, P. S., Hase, Y., Forneris, R. I. M. G., Temperini, M. L. A., & Kawano, Y. (1984). *Espectroscopia Raman: Princípios e Aplicações*. Fundação Valeparaibana de Ensino.
Schlücker, S. (2014). Surface-enhanced Raman spectroscopy: Concepts and chemical applications. *Angewandte Chemie International Edition, 53*, 4756.
Schulz, H., Baranska, M., & Baranski, R. (2005a). Potential of NIR-FT-Raman spectroscopy in natural carotenoid analysis. *Biopolymers, 77*(4), 212–221.
Schulz, H., Baranska, M., Quilitzsch, R., Schutze, W., & Losing, G. (2005b). Characterization of peppercorn, pepper oil, and pepper oleoresin by vibrational spectroscopy methods. *Journal of Agricultural and Food Chemistry, 53*, 3358–3363.
Shende, C., Gift, A., Inscore, F., Maksymiuk, P., & Farquharson, S. (2004). Inspection of pesticide residues on food by surface-enhanced Raman spectroscopy. *Proceedings of SPIE, 5271*, 28–34.
Smekal, A. (1923). Zur Quantentheorie der Dispersion. *Naturwissenschaften, 43*, 873.
Strommen, D. P., & Nakamoto, K. (1984). *Laboratory Raman spectroscopy*. Wiley.
Synytsya, A., Čopíková, J., Matějka, P., & Machovič, V. (2003). Fourier transform Raman and infrared spectroscopy of pectins. *Carbohydrate Polymers, 54*(1), 97–106.
Trevisan, J., Angelov, P. P., Carmichael, P. L., Scott, A. D., & Martin, F. L. (2012). Extracting biological information with computational analysis of Fourier-transform infrared (FTIR) biospectroscopy datasets: Current practices to future perspectives. *Analyst, 137*, 3202–3215.
Turrell, G., & Dhamelincourt, P. (1966). Micro-Raman spectroscopy. In J. J. Laserna (Ed.), *Modern techniques in Raman spectroscopy* (pp. 109–142). Wiley.
Veraverbeke, E. A., Lammertyn, J., Nicola, B. M., & Irudayaraj, J. (2005). Spectroscopic evaluation of the surface quality of apple. *Journal of Agricultural and Food Chemistry, 53*(4), 1046–1051.
Weng, S., Qiu, M., Dong, R., Wang, F., Huang, L., Zhang, D., & Zhao, J. (2018). Fast detection of fenthion on fruit and vegetable peel using dynamic surface-enhanced Raman spectroscopy and random forests with variable selection. *Spectrochimica Acta Part A: Molecular and Biomolecular Spectroscopy, 200*, 20–25.

Yang, D., & Ying, Y. (2011). Applications of Raman spectroscopy in agricultural products and food analysis: A review. *Applied Spectroscopy Reviews, 46*(7), 539–560.
Yang, H., & Irudayaraj, J. (2003). Rapid detection of foodborne microorganisms on food surface using Fourier transform Raman spectroscopy. *Journal of Molecular Structure, 646*, 35–43.
Zaidel, A. N., Prokofev, V. K., & Raiskii, S. M. (1961). *Tables of spectrum lines*. Pergamon Press.
Zhang, P. X., Zhou, X. F., Cheng, A. Y. S., & Fang, Y. (2006). Raman spectra from pesticides on the surface of fruits. *Journal of Physics, 28*, 7–11.

NIR Spectroscopy for Internal and External Quality Measurement and Analysis of Thick Rind Fruits

Umezuruike Linus Opara, Ebrahiema Arendse, and Lembe Samukelo Magwaza

Abstract The global horticultural industry is continually faced with new technological challenges to be more internationally competitive and to meet the growing demand for quality-assured agricultural products. To meet these challenges, a global trend in postharvest research has resulted in the development of rapid detection and prediction of external and internal product quality attributes. Among the various non-destructive technologies, near-infrared spectroscopy (NIRS) is the most cost-effective and rapid method for qualitative and quantitative analysis of a wide range of fresh and processed horticultural products. NIRS offers tremendous prospects for application in measuring and predicting the organoleptic and nutritional quality of fruits. Unlike deciduous fruit such as apples and pears with thin rind/skin/peel, thick rind fruit has been reported to interfere with the prediction of internal quality parameters by near-infrared spectroscopy, primary due to the inadequate penetration depth of NIR radiation. This chapter provides a brief overview of the application of near-infrared spectroscopy to evaluate various internal and external quality parameters of thick rind fruit, including the measurement and prediction of internal and external disorders and diseases. A brief overview of the role of spectral acquisition, wavelength selection, chemometrics, and preprocessing techniques play in the model development of thick rind fruit. Finally, the prospects of NIRS technology in evaluating the quality attributes of fruits with thick rind are highlighted.

U. L. Opara (✉)
SARChI Postharvest Technology Research Laboratory, Africa Institute for Postharvest Technology, Faculty of AgriSciences, Stellenbosch University, Stellenbosch, South Africa

UNESCO International Centre for Biotechnology, Nsukka, Nigeria
e-mail: opara@sun.ac.za

E. Arendse
SARChI Postharvest Technology Research Laboratory, Africa Institute for Postharvest Technology, Faculty of AgriSciences, Stellenbosch University, Stellenbosch, South Africa

L. S. Magwaza
Department of Crop and Horticultural Science, School of Agricultural, Earth and Environmental Science, University of KwaZulu-Natal, Pietermaritzburg, South Africa

P. B. Pathare, M. S. Rahman (eds.), *Nondestructive Quality Assessment Techniques for Fresh Fruits and Vegetables*, https://doi.org/10.1007/978-981-19-5422-1_9

1 Introduction

This rise in global consumption of fresh fruit and vegetables is largely driven by increased consumer demand for a regular supply of good quality, safe, convenient, and nutritious fresh food products. Consumer satisfaction with product quality is the primary aim of the production, handling, storage, distribution, and processing of fresh produce (Opara, 2009). Quality assessment of fruit and vegetables is a combination of appearance, taste, aroma, and textural properties (Kader & Rolle, 2004). These characteristics are indices that govern the marketability of fruit, and these measurements allow for comparison against industrial standards by ensuring that the product meets the limits of acceptability by the consumer (Kader, 2002). Consumers primarily judge the initial purchase of fresh fruits, ornamentals, and vegetables based on the external appearance (i.e., firmness, color, shape, size, absence of blemishes). Their subsequent and repeated purchase of the same commodity is often based on consumer satisfaction, which is based on sensory properties (i.e., sweetness, acidity, flavor, and aroma) (Chen & Opara, 2013; Kader, 2000).

Currently, quality assessment of fruits or vegetables may be evaluated either subjectively using a trained panel and/or destructive methods based on instrumental analysis. Subjective and destructive methods have several drawbacks compared to non-destructive methods, they are time-consuming, intensive, and often require specialized preparation. Fruit quality attributes may show significant variation due to differences in cultivars, fruit maturity stage, and growing location. Thus quality attributes determined using destructive methods would only represent a specific batch being tested (Magwaza et al., 2014a; Peirs et al., 2003). Currently, destructive techniques are incapable of online or inline grading due to the large biological variability of the quality attributes of fruit and vegetables (Nicolaï et al. 2007). The increased growth by consumers for quality-assured fresh fruits and vegetables has prompted the agro-industry to perform research and develop fast, accurate, and cost-effective non-destructive approaches that can be employed for online or inline grading.

Near-infrared spectroscopy (NIRS) is a non-destructive tool that operates within the electromagnetic spectrum of 800–2500 nm. When NIR light interacts with materials it causes vibration of functional molecular groups such as carbon-hydrogen (C-H), oxygen-hydrogen (O-H), and nitrogen-hydrogen (N-H) (Williams et al., 2019). The absorption of energy from these molecular groups can be detected and measured spectroscopically. A simplified illustration of the basic steps used for developing NIR calibration models is presented in Fig. 1.

For fruit and vegetables, NIRS has been one of the most successful non-destructive techniques used for the assessment of external and internal quality parameters (Walsh et al., 2020, 2020, 2021). The most successful application of NIRS in agricultural produce includes the prediction of composition in terms of constituents such as water, carbohydrates, fat, protein, cellulose, polysaccharides, and various other constituents which have an abundance of O-H, C-H, and N-H bonds (Williams et al., 2019). Furthermore, NIRS has extensively been applied in

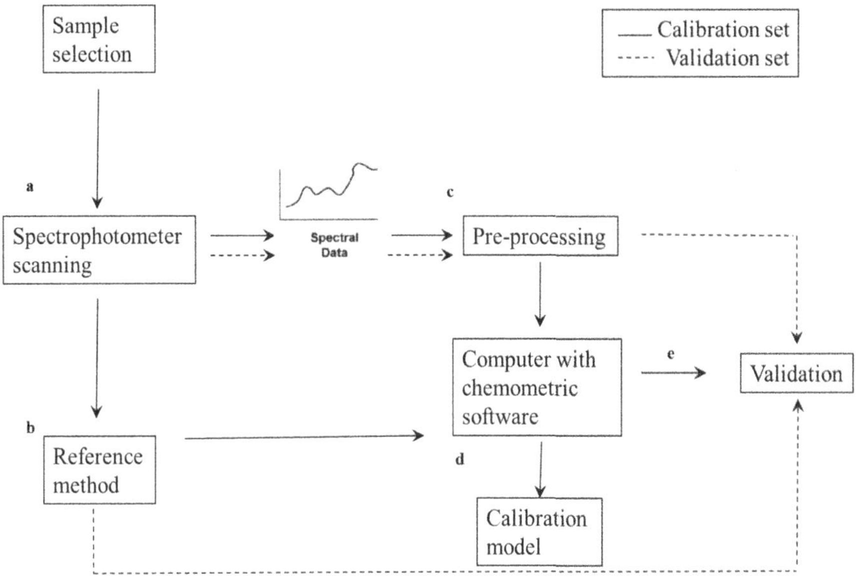

Fig. 1 Block diagram illustrating the basic steps for developing NIR calibration models (**a**) the sample is irradiated with NIR radiation, (**b**) analytical methods also known as reference method is used to obtain the dependant variable to be calibrated, (**c**) the acquired spectral data is subjected to preprocessing methods using chemometric software, (**d**) the combination of reference and spectral data are used to develop calibration model, (**e**) calibration models are validated to test the model performance. Okere et al. (2021)

the food and beverage industry for authentication and detection of adulteration in processed horticultural products (Arendse et al., 2020; Lohumi et al., 2015). Therefore, non-destructive measurement and prediction methods, such as NIRS, offer tremendous prospects for application in measuring and predicting the organoleptic and nutritional quality of horticultural produce.

However, the successful application of NIRS has mainly been restricted to evaluating the quality attributes of thin rind fruit with homogenous pulp (Arendse et al., 2018a; De Oliveira et al., 2014). Several investigators have reported that thick rind fruit interferes with the NIRS measurement of internal quality attributes (De Oliveira et al., 2014; Guthrie et al., 2006; Li et al., 2020). This was buttressed by Arendse et al. (2018a) who reviewed the application of various non-destructive methods to measure the internal quality attributes of different thick rind fruit and observed that NIRS faces several challenges for internal quality assessment. To measure the internal quality of fruits and vegetables, NIR radiation needs to adequately penetrate the internal tissue, a subject not often discussed in the literature. Furthermore, limited application of NIRS has been reported to monitor and detect external physiological rind disorders (Magwaza et al., 2014; Magwaza et al., 2014a, 2014b. Therefore, this chapter evaluates the application of near-infrared spectroscopy for the measurement and prediction of external and internal quality attributes of thick rind fruit.

2 Application of NIRS in Quality Analysis of Thick Rind Fruit

NIRS technology has been used to monitor and predict various internal and external physicochemical properties, including the detection of physiological disorders and diseases in fruits such as oranges, mangoes, mandarins, pomegranates, melons, and watermelons have been limited due to the inadequate penetration depth of NIR radiation within the tissue compared to other thin skin/peel deciduous fruits such as apple and pears. This section discusses various aspects that may affect the application of NIRS in the prediction of internal and external quality attributes of thick rind fruit.

2.1 Data Acquisition Modes

Generally, there are five types of spectral data acquisition modes that can be applied depending on the application, these include transmittance, interactance, transflectance, diffuse transmittance, and diffuse reflectance (Williams et al., 2019). However, for thick rind fruit, three types are frequently applied during spectral acquisition these include reflectance, transmittance, and interactance (Magwaza et al., 2012). In reflectance mode, the light source and detector are on the same side of the scanned sample and the radiation is diffusely reflected back to the detector (Fig. 2a). Generally, capturing spectra in reflectance mode mainly provides information about the external attributes related to quality evaluation (Krivoshiev et al., 2000). Therefore, calibrations developed using reflectance are highly susceptible to variations due to the fruit's surface properties. In transmittance mode, the light source is positioned opposite to the detector (Fig. 2b). Generally, transmittance is the most frequently used spectral acquisition method to predict the internal quality attributes of thick rind fruit. In interactance mode, the measurements are made with special fiber optic sensors in which the light source and detector are positioned parallel to each other in such a way that light due to specular reflection cannot directly enter the detector (Fig. 2c). Thus, the fibers collect and transmit the light back to the sample and diffusely reflect it to the detector, which measures the interactance (Williams et al., 2019).

Researchers over the years have applied NIRS to evaluate the quality attributes of thick rind fruits. However, the spectral acquisition approach may differ among researchers. For instance, McGlone et al. (2003) assessed different spectral data acquisition modes (reflectance, interactance, and transmittance) in predicting the

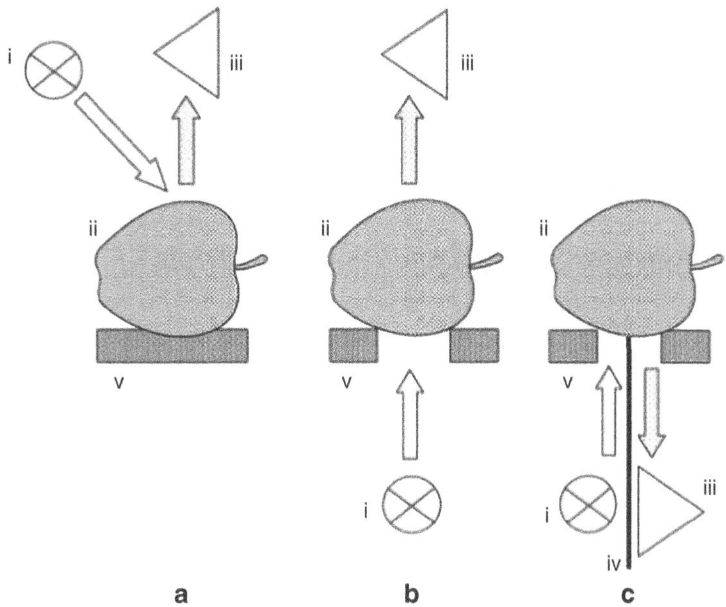

Fig. 2 Setup of three types of spectral acquisition mode: (**a**) reflectance, (**b**) transmittance, and (**c**) interactance spectra, with (i) the light source, (ii) fruit, (iii) monochromator/detector, (iv) light barrier, and (v) support. Nicolaï et al. (2007)

soluble solid content (SSC) of Satsuma Mandarin (*Citrus reticulata* cv. "Miyagawa") by visible to near-infrared spectroscopy (vis-NIRS). The authors observed that spectral acquisition using transmission mode provided the most accurate SSC predictions. The authors suggested that SSC measurements using reflectance mode should not be considered due to the limited penetration depth of reflectance or interactance measurements. A similar argument was reported in a research paper by Krivoshiev et al. (2000) who reported that transmittance can be used to penetrate the peel and the flesh and sufficiently predict the internal quality. Several other investigators observed that the application of transmission measurements provided an adequate prediction of quality attributes of thick rind fruit (Cayuela, 2008; Maniwara et al., 2014; Masithoh et al., 2016). For successful measurements to occur in transmission, most often the light intensity needs to be very high and fruit may require longer exposure time which could result in damage to the fruit surface and altering its spectral properties. Contrary to the above results, Khodabakhshian et al. (2019) studied reflectance, interactance, and transmittance measurement using vis-NIRS in evaluating the firmness of pomegranate fruit. The authors reported high prediction accuracy in reflectance mode and recommended acquiring spectral data in reflectance mode. They observed that performing measurements in transmittance resulted in reduced model performance and that the spectral data obtained in transmittance mode could be a result of fruit size and the amount of light penetrating the fruit. Li et al. (2020) evaluated the SSC in oranges and mandarins using diffuse reflectance, they observed that the peel thickness of the fruit got in the way of obtaining spectral information of the flesh layer. Several other authors also reported that diffuse reflectance is appropriate in predicting sugar and acidity levels in thick rind fruit, such as oranges, pomegranates, and melons (Arendse et al., 2018; Magwaza et al., 2013; Zeb et al., 2021).

2.2 *Wavelength Selection and Penetration Depth of NIRS*

The visible and NIR regions of the electromagnetic spectrum ranges are 400–750 nm and 750–2500 nm, respectively. The wavelength region of 750–1100 nm is often referred to as the short-wavelength range, while the 1100–2500 nm region is considered the NIR region proper (Walsh et al., 2020). These wavelength ranges have been used to evaluate the sweetness and the acidic taste of a variety of thick rind fruits (Table 1). However, among researchers the selection of wavelength ranges may differ for the quality evaluation of fresh fruits, this is due to inherent variability among fruit species, cultivars, and growing conditions. McGlone et al. (2003), used the wavelength range of 750–1100 nm to develop models to predict SSC in mandarin fruit. In a similar study on mandarin fruit, Guthrie et al. (2005) and Guthrie et al. (2005) used vis-NIRS to evaluate the SSC and dry matter content (DMC) using the spectral wavelength range between 720 and 950 nm. Tian et al. (2009) reported that the wavelength region of 650–1000 nm can be used to measure SSC in watermelon. For "Navel" oranges, Liu et al. (2010) reported that the wavelength range of 350–1800 nm can be used for model development to predict SSC within Khodabakhshian et al. (2017) determined that a wavelength range from 400 to 1100 nm can be used to develop models to assess the internal quality parameters of the pomegranate fruit. Zeb et al. (2021) reported that selected wavelength ranges between 729–975 and 500–1050 nm can be used to develop SSC prediction models in melon fruit. The literature suggests that the short-wavelength spectral range (650–1100 nm) can be used for the evaluation of the sugar content of thick rind fruit. The main reason is that absorbers within the region of 650–1100 nm are all characterized by low intensity therefore it is possible to use a thicker sample (Williams et al., 2019). In the higher wavelengths, above 1100 nm the absorbers are much stronger, and the sample thickness would need to be thin for transmittance to work. However, NIRS has also been successfully used in the wavelength region of 800–2400 nm to evaluate the internal quality of thick rind fruit (Table 1). Magwaza et al. (2013) observed that a spectral region of 900–1990 nm can be used to develop models for physicochemical properties, such as mass, color index, and total soluble solids (TSS) content. Olarewaju et al. (2016) reported that the wavelength of 800–2400 nm can be used for the development of calibration models for DMC and moisture content in avocados. According to Arendse et al. (2018), the wavelength region of 1064–1333 and 1640–1835 nm can be used for model development to measure total phenolic and total anthocyanin content.

Table 1 Applications of NIR spectroscopy to measure internal quality attributes of selected fruits with thick rind

Commodity	Measured parameter	Wavelength range (nm)	Statistical analysis	Prediction accuracy	References
Avocado	Dry matter	500–1000	PLSR, MLR	$R^2 = 0.88$, RMSEP = 1.30	Clark et al. (2004)
	Oil content	800–2400	PLSR	$R^2 = 0.58$, RMSEP = 5.44	Olarewaju et al. (2016)
	Dry matter			$R^2 = 0.84$, RMSEP = 2.49	
	Moisture content			$R^2 = 0.85$, RMSEP = 2.49	
	Dry matter	720–975	PLSR	$R^2 = 0.71$, SEP = 1.37	Subedi and Walsh (2020)
Mandarin	SSC	720–950	MPLS	$R^2 = 0.50–0.80$, SEP = 0.60–0.80, RPD = 120–3.40	Guthrie et al. (2005)
	SSC	720–950	MPLS	$R^2 = 0.60–0.90$, SEP = 0.40–0.50	Guthrie et al. (2005)
	TSS	400–1040	PLSR	$R^2 = 0.85$, RMSEP = 0.65	Xudong et al. (2009)
	TA			$R^2 = 0.71$, RMSEP = 1.37	
	TSS		Local PLSR, MPLS	$R^2 = 0.39–0.57$, SEP = 0.71–0.84, RPD = 127–151	Torres et al. (2019)
	TA			$R^2 = 0.65–0.76$, SEP = 0.13–0.18, RPD = 1.92–2.08	
	TSS:TA ratio			$R^2 = 0.68–0.79$, SEP = 1.13–1.75, RPD = 148–2.14	
	TSS		PLSR	$R^2 = 0.78–0.85$, RMSEP = 0.59–0.65	Li et al. (2020)
Mango	SSC	1000–2500	PLSR, PCR	$R^2 = 0.66$ to 0.91, RPD = 0.45 to 1.80	Munawar et al. (2016)
	TA	1000–2500	PLSR, PCR	$R^2 = 0.94$ to 0.98, RPD = 3.21 to 4.87	
	TSS	720–744, 880–910	PLSR	$R^2 = 0.91$, RMSEP = 0.51	Li et al. (2021)
	Dry matter	684–990	PLSR	RMSEP = 0.79%	Mishra and Passos (2021)

(continued)

Table 1 (continued)

Commodity	Measured parameter	Wavelength range (nm)	Statistical analysis	Prediction accuracy	References
Oranges	SSC	578–1850	PLSR	$R^2 = 0.91$, RMSEP = 0.51	Cayuela (2008)
	TA			$R^2 = 0.56$, RMSEP = 0.33	
	pH			$R^2 = 0.45$, RMSEP = 0.49	
	SSC	780–2500	PLSR, PCA	$R^2 = 0.83$, RMSEP = 0.58	Magwaza et al. (2013)
	TA			$R^2 = 0.22$, RMSEP = 0.15	
	Vitamin C			$R^2 = 0.66$, RMSEP = 8.01	
	TSS		PLSR, PCA	$R^2 = 0.85$, RMSEP = 0.60	Li et al. (2020)
Watermelon	SSC	650–950	PLSR	$R^2 = 0.90$, RMSEP = 0.76	Tian et al. (2009)
	SSC	687–920	PLSR, SMLR	$R^2 = 0.66$, RMSEP = 0.39	Jie et al. (2014)
	TSS	900–1700	PLSR	$R^2 = 0.70$, RMSEP = 1.4	Tamburini et al. (2017)
Melon	SSC	630–1030	PLSR	$R^2 = 0.30$–0.93, RMSEP = 0.7–1.3	Greensill et al. (2001)
	SSC	306–1130	PLSR	$R^2 = 0.28$, RMSEP = 1.0	Guthrie et al. (2006)
	SSC	300–1150	PLSR, PCA	$R^2 = 0.59$–0.88, RMSECV = 0.4–0.81	Long and Walsh (2006)
	SSC	729–975, 500–1050	MLR	$R^2 = 0.93$, RMSE = 1.63	Zeb et al. (2021)
			SVM	$R^2 = 0.78$, RMSE = 1.58	
			ANN	$R^2 = 0.79$, RMSE = 1.63	

TSS total soluble solids, *SSC* soluble solid content, *TA* titratable acidity, *PLS* partial least squares regression, *PCA* principal component analysis, *PCR* principal component regression, *MLR* multiple linear regression, *SMLR* stepwise multivariate linear regression, *SVM* support vector machine, *ANN* artificial neural network, R^2 coefficient of determination, *RMSEP* root mean square error of prediction, *RMSE* root mean square error, *RMSECV* root mean square error of cross-validation, *RPD* residual predictive deviation

The light penetration depth of NIRS is related to the optical path length and optical density of the fruit. The optical properties of the fruit can differ as a result of fruit shape, size, and thickness of the rind; hence, the spectrum obtained from the fruit flesh will vary depending on the changes in rind optical density (Krivoshiev et al., 2000). Several researchers have studied the effects of wavelength range and light penetration depth on thin and thick rind fruit. Guthrie and Walsh (1997) investigated the penetration depth of NIRS for evaluating the internal quality of mango. The authors observed that all spectral information was derived within the first 5 mm from the fruit surface. Lammertyn et al. (2000) reported that the light penetration depth was higher (4 mm) in the short-wavelength range (700–900 nm) compared to the NIR range of 900–1900 nm for apples. Similarly, Fraser et al. (2001) observed that the light penetration depth in the apple was much larger in the wavelength range of 700–900 nm range compared to the 1400–1600 nm range. Fraser et al. (2003) assessed the penetration depth of NIRS using an illuminated laser at a wavelength of 808 nm. The authors observed that the thick rind of mandarins resulted in a reduction in light penetration up to 4 mm compared to the flesh of the fruit (14 mm). Xu et al. (2018) explored the light penetration depth and different wavelength ranges of observed an effective wavelength of provided light penetration up to 21 mm within the fruit. For the detection of internal disorders in mangoes, Raghavendra et al. (2021) studied the wavelength selection and the penetration depth of vis-NIRS. The authors observed that penetration depth was higher in the short-wavelength range of 673–1100 nm compared to the NIR region of 1100–1900 nm. From the review of the foregoing literature, it is evident that fruit with irregular or thick rind interferes with NIRS-based measurements for evaluating internal quality and detecting internal disorders of thick rind fruit. Therefore, to remove the interference of unwanted spectra of the peel and improve the model performance, multivariate techniques and preprocessing techniques are often applied to the acquired spectra.

2.3 *Chemometrics and Preprocessing Techniques*

Nearly 90% of the composition of fruits and vegetables constitutes water. The NIR spectrum is mainly dominated by water thus causing the near-infrared spectrum to be highly convoluted due to the broadband and large set of overtones (Williams et al., 2019). Therefore, multivariate statistical tools (also known as chemometrics) are often used to acquire relevant information from the NIR spectrum. In brief, calibration models are developed using various chemometric algorithms such as partial least squares regression (PLS or PLSR), multivariate linear regression (MLR), artificial neural networks (ANN), and genetic algorithm (GA) (Blanco & Villarroya, 2002; Cozzolino, 2014). Other applications require classification techniques that are often used instead of quantification of different quality attributes. The most common classification techniques include principal component regression (PCR) also referred to as principal component analysis (PCA), linear discriminant analysis (LDA), and partial least squares-discriminant analysis (PLS-DA). Further details on the application of these chemometric tools to the measurement of postharvest quality attributes have been reviewed by Cozzolino et al. (2011).

Today, PLS or PLSR is arguably the most widely applied multivariate analysis tool used in the prediction of fruit quality attributes. Researchers have explored PLS or PLSR and different regression techniques on a variety of thick rind fruit to develop calibration models. For example, Clark et al. (2004) assessed the DMC of intact avocado using Fourier transform near-infrared spectroscopy (FT-NIRS) and reported similar results for PLSR and MLR algorithms. Lu et al. (2006) compared both PLSR and PCR for model development of SSC in citrus fruit and observed that PLSR provided better prediction than PCR. For the prediction of maturity index in mangoes using NIRS, Jha et al. (2014) observed that PLSR models provided better prediction statistics compared to MLR. In a study on oranges, Torres et al. (2019) used a portable NIRS instrument in reflectance. The researchers compared Local PLS and modified partial least squares (MPLS) in predicting SSC and TA content. The models developed for SSC was shown for Local PLSR was superior ($R^2 = 0.80$, RPD = 2.23) to MPLS models ($R^2 = 0.75$, RPD = 1.84). Generally, PLSR has been shown by researchers to be a better method for model development than the other regression techniques (Cayuela, 2008; Clark et al., 2004; Nicolaï et al., 2007). The application of various multivariate regression techniques is often combined with spectral preprocessing or pretreatment methods.

The capturing of spectra creates noise problems due to the physical characteristics of the fruit, such as the shape, size, the thickness of the peel (Blanco & Villarroya, 2002). During spectral acquisition, undesirable scatter effects (both baseline shift and noise) occur within the broad wavelength regions. Therefore, to remove these undesirable scatter effects, preprocessing or pretreatment of the spectral data is often required before subsequent multivariate regression, or classification (Rinnan et al., 2009). The most common used spectral pretreatments include normalization, derivatives (usually first and second), the multiplicative signal correction also known as multiplicative scatter correction (MSC), the standard vector normalization (SNV), smoothing using moving average and detrending or a combination thereof (Blanco & Villarroya, 2002). Further details and the purpose of each preprocessing method have extensively been reviewed by several authors (Blanco & Villarroya, 2002; Nicolaï et al., 2007; Rinnan et al., 2009).

The effects of spectral pretreatment on the model's performance on thick rind fruit have been investigated over the years by many researchers. Derivatives are used to correct the light scattering effects by removing baseline shifts and superposed peaks. In Satsuma mandarins, Gómez et al. (2006), applied derivatives (first and second) as a spectral pretreatment to reduce the effects of nonlinearity introduced by light scatter. Working on watermelon in transmittance mode, Hai-qing et al. (2007) applied derivatives (first and second) and smoothing (Savitzky-Golay) and reported that both spectra pretreatment methods improved the performance of the calibration models. However, derivatives as spectral pretreatment cannot correct path-length variations within the spectra. Thus, MSC pretreatment is often used to correct path-length variations. For the prediction of quality attributes in "Valencia" oranges, Magwaza et al. (2013) applied both derivatives (first and second) and MSC, the authors observed that the use of MSC pretreatment resulted in the development of better calibration models. Similarly, Liu et al. (2010), applied SNV and MSC as pretreatments to the spectral data acquired from "Navel" oranges. The authors observed that the MSC pretreatment method provided better calibration models compared to SNV. For pomegranates, Arendse et al. (2018) observed that pretreatment MSC combined with derivatives provided superior model performance. In oranges, Li et al. (2020) applied mean normalization as a spectral pretreatment to correct spectral variability caused by the optical thickness of the fruit skin and improved the model's performance. Contrary, Lu et al. (2006) reported that the spectral pretreatments of constant, SNV and MSC provided similar prediction statistics and reasoned that the application of spectral pretreatments was not effective in improving the performance of the model.

3 Measurement of Internal Quality Attributes of Fruit with Thick Rind

Quality parameters such as TSS, SSC, and TA have been associated with flavor and sensory properties. The molecule structure of these quality parameters contains C–H, O–H, C–O, and C–C bonds; therefore NIRS can detect and predict their concentration. The prediction of various internal quality attributes for thick rind fruits has been studied by several researchers over the years. NIRS has been reported to interfere with the measurement of internal quality due to the optical thickness of the rind. Dull et al. (1989) performed a comparative study by evaluating the SSC content in whole and cut cantaloupe fruit. They observed a lower prediction accuracy ($R^2 = 0.60$ and SEC = 1.67 °Brix) for intact cantaloupe compared to cantaloupe slices ($R^2 = 0.97$ and SEC = 0.56 °Brix). Guthrie and Walsh (1997) evaluated the SSC content of pineapple and observed that the thick irregular rough surface provided unsatisfactory prediction results compared to the smooth skin of mango fruit. Guthrie et al. (2006) studied the TSS content of intact melons and reported a lower correlation coefficient compared to other fruits, such as mandarins. The authors attributed the low prediction accuracy to the poor penetration depth of NIR radiation through the rind of the fruit and high SSC heterogeneity within the fruit. Similarly, Flores et al. (2008) studied the SSC in two types of fruit, melons and watermelons using NIR reflectance spectroscopy. The authors compared both sliced and whole fruit and observed that sliced fruit provided better prediction results for SSC, the poor prediction statistics were attributed to the heterogeneous SSC distribution within the fruit. De Oliveira et al. (2014) conducted a comparative study to predict the internal quality attributes of three different types of fruit species (apricot, tomatoes, and passion fruit). The authors observed that NIRS in diffuse reflectance mode can be used to evaluate the apricot's internal quality ($R^2 = 0.93$, RMSEP = 0.69 °Brix). However, the authors observed that NIRS were not suitable for evaluating thick rind fruits such as passion fruit ($R^2 = 0.63$, RMSEP = 1.66 °Brix), and/or fruits with heterogeneous internal structures (tomato) ($R^2 = 0.52$, RMSEP = 0.53 °Brix).

Although studies have reported difficulty in predicting internal quality parameters in thick rind fruit, NIRS has been successfully used to evaluate quality attributes such as TSS/SCC and TA in a variety of thick rind fruits (Table 1). For mangoes (cv. Kent), Munawar et al. (2016) used FT-NIRS within the spectral wavelength range of 1000–2500 nm and developed PLRr models to measure SSC ($R^2 = 0.66$–0.91) and TA ($R^2 = 0.94$–0.98). In a study to predict the maturity status of mango (cv. Palmer) using SSC and DMC, Neto et al. (2017) developed PLSr calibration models within the short NIR wavelength range (699–981 nm) using a portable vis-NIR instrument. The authors developed models for SSC ($R^2 = 0.87$, RPD = 2.77) and DMC ($R^2 = 0.84$, RPD = 2.51) and according to Nicolaï et al. (2007), RPD values above 2.50 are considered good models. In a more recent study, Li et al. (2021) applied the wavelength selection method using NIR to predict TSS in mangoes (cv. Golek). The authors applied PLSR and reported high prediction statistics for TSS ($R^2 = 0.89$, RMSEP = 1.24) when spectral data was collected in reflectance mode.

For citrus fruits, such as oranges and mandarins, Cayuela (2008) used NIRS in diffuse reflectance to develop calibration models to measure the TSS ($R^2 = 0.91$, RMSEP = 0.51), but observed poor prediction statistics for TA ($R^2 = 0.56$, RMSEP = 0.33). Xudong et al. (2009) used NIRS in the region of 400–1040 nm to develop calibration models to measure the TSS ($R^2 = 0.85$). In a similar study, Magwaza et al. (2013) evaluated NIRS in diffuse reflectance mode and reported greater prediction accuracy for TSS ($R^2 = 0.56$, RMSEP = 0.33 °Brix) compared to the low prediction statistics observed for pH ($R^2 = 0.45$, RMSEP = 0.49) and TA ($R^2 = 0.22$, RMSEP = 0.15) (Table 1). Li et al. (2020) evaluated different positions (i.e., stem, equator, and navel) using vis-NIRS evaluation. The authors observed that the SSC was heterogeneous due to the rind thickness at different positions. When combining the models for different positions, the authors reported optimal prediction performance ($R^2 = 0.85$, RMSEP = 0.60 °Brix) compared to separate local models. Melons and watermelons are another fruit that has a thick skin/rind, and several authors have successfully applied NIRS to measure their internal properties. Tian et al. (2009) applied PLSR within the short-wavelength region (650–950 nm) and provided a satisfactory prediction ($R^2 = 0.90$, RMSEP = 0.76) of SSC in watermelon. In a similar study on watermelon, Tamburini et al. (2017) observed that PLS regression within the spectral region of 900–1700 nm is appropriate for predicting the TSS content ($R^2 = 0.70$, RMSEP = 1.4) with reasonable accuracy (Table 1). Zeb et al. (2021) reported that diffuse reflectance and shortwave range is appropriate for developing calibration models for measuring the SSC using MLS ($R^2 = 0.93$, RMSE = 1.63). Other internal parameters that have been successfully predicted in thick rind fruit include, TSS:TA ratio, BrimA, DMC, vitamin C content, total phenolic content, and total anthocyanin content (Arendse et al., 2018; Jha et al., 2014; Mishra & Passos, 2021; Ncama et al., 2018).

4 Measurement of External Quality Attributes of Thick Rind Fruits

In NIRS, physical properties like fruit mass are often measured indirectly through the textural properties of the fruit (Williams et al., 2019) or quantification of the water content (Louw & Theron, 2010). Several researchers have evaluated the usage of NIRS for predicting the fruit mass of thick rind fruit (Table 2). Cayuela and Weiland (2010) performed a comparative study by investigating the fruit mass of oranges using two different NIR instruments. The authors observed that type of instrument and wavelength range affected the prediction accuracy of the models. Sánchez et al. (2013) reported that the application of MPLS can be used to predict the fruit mass of oranges with a coefficient of determination of 0.79 and SEP of 27.94 g. Magwaza et al. (2013) reported that their successful prediction of fruit mass for oranges ($R^2 = 0.96$, RPD = 3.53) was based on the fruit size and shape since both are closely associated with fruit mass. Arendse et al. (2018) on the other hand reported that the irregular shape and thickness of pomegranates resulted in poor prediction of fruit mass.

For fruit firmness, several investigators have successfully applied NIRS on thick rind fruit (Table 2). For mandarins, Gómez et al. (2006) applied NIRS within the region of 350–2500 nm and developed PLS calibration models for fruit firmness ($R^2 = 0.94$, RMSEP = 0.33 N). Rungpichayapichet et al. (2016) used the vis-NIRS to develop robust PLS calibration models to measure fruit firmness in mangoes. The authors applied diffuse reflectance over a wavelength region of 700–1100 nm and reported an R^2 of 0.82 and SEP of 4.22 N. In a similar study on mangoes in the vis-NIR region, Mishra et al. (2020) developed calibration models that provided good prediction statistics ($R^2 = 0.75$ and RMSEP = 5.92 N) for fruit firmness. Several other investigators have successfully evaluated fruit firmness using NIRS on a variety of thick rind fruit (Arendse et al., 2018; Khodabakhshian et al., 2019; Lu et al., 2015).

The measurement of color properties in fruits and vegetables can be used as an indirect measurement of various color pigments (Pathare et al., 2013). Sun et al. (2009) developed PLS calibration models in the vis-NIR region (400–1140 nm) for the prediction of the color index ($R^2 = 0.93$ and RMSEP = 0.65) in mandarins. Similarly, Magwaza et al. (2014b) developed PLS calibration models for predicting the external color parameters (L*, C*, h°) in "Nules Clementine" mandarins (Table 2). Model development occurred within the region of 350 and 2400 nm and the authors reported a prediction accuracy for L* ($R^2 = 0.93$ and RPD = 2.73), C* ($R^2 = 0.89$ and RPD = 2.40), and h° ($R^2 = 0.96$ and RPD = 3.41). In another study on citrus, Magwaza et al. (2013) successfully developed calibration models and predicted the color index ($R^2 = 0.83$, RPD = 2.08) in the NIR region of 800–2500 nm. For color properties in mango fruit peel, Watanawan et al. (2014) applied NIRS in reflectance mode using a short-wavelength range of 700–1100 nm. The authors developed regression models for color component a* ($R^2 = 0.84$) and b* ($R^2 = 0.72$). Arendse et al. (2018) observed that PLSR applied within the spectral

Table 2 Applications of NIR spectroscopy to measure external quality attributes in selected fruits with thick rind

Commodity	Measured parameter	Wavelength range (nm)	Statistical analysis	Prediction accuracy	References
Mandarin	Firmness	350–2500	PLSR	R^2 = 0.94 RMSEP = 0.33	Gómez et al. (2006)
	Color index	400–1140	PLSR	R^2 = 0.93 RMSEP = 0.65	Sun et al. (2009)
	Color (Chroma)	350–2500	PLSR, PCA	R^2 = 0.87 RMSEP = 2.63, RPD = 2.09	Magwaza et al. (2014b)
	Color (hue)			R^2 = 0.93 RMSEP = 2.62, RPD = 3.32	
Orange	Mass	780–2500	PLSR	R^2 = 0.97, RMSEP = 10.45 g, RPD = 2.57	Magwaza et al. (2013)
	Color index			R^2 = 0.83, RMSEP = 0.82, RPD = 2.08	
	Mass	1100–1690	LOCAL	R^2 = 0.79, SEP = 27.94 g	Sánchez et al. (2013)
	Color (a*)	500–1690	LOCAL	R^2 = 0.74, SEP = 0.91	
	Firmness	1600–2400	MPLS	R^2 = 0.50, SEP = 10.58 N	
Mango	Color (a*)	700–1100	PLSR	R^2 = 0.84	Watanawan et al. (2014)
	Color (b*)	700–1100	PLSR	R^2 = 0.72	
	Firmness	700–1100	PLSR	R^2 = 0.99	
	Firmness	400–1130	PLSR	R^2 = 0.75, RMSEP = 5.92 N	Mishra et al. (2020)
	Firmness	700–1100	PLSR	R^2 = 0.82, RMSEP = 4.22 N	Rungpichayapichet et al. (2016)

PLSR partial least squares regression, *PCA* principal component analysis, *MPLS* modified partial least squares, R^2 coefficient of determination, *RMSEP* root mean square error of prediction, *RMSE* root mean square error, *RPD* residual predictive deviation region of 900–1700 nm is appropriate for predicting the color components a* (R^2 = 0.90 and RPD = 3.34), C* (R^2 = 0.83 and RPD = 2.43), and h° (R^2 = 0.83 and RPD = 2.50) in pomegranates. The results of these studies show that NIRS combined with chemometrics has the potential to evaluate the external quality attributes of thick rind fruit.

5 NIRS Measurement and Prediction of Internal and External Disorders and Diseases Affecting Thick Rind Fruit

A review of the literature suggests that NIRS can be applied to quantitatively and qualitatively assess various internal disorders, diseases, and rind disorders within thick rind fruit. Lorente et al. (2015) investigated vis-NIRS (650–1700 nm) for the detection and prediction of fungal infection within citrus fruit and reported a classification accuracy of 97.8%. For detection of carob moth infestation within pomegranate fruit, Khodabakhshian et al. (2016) explored the application of vis-NIRS (400–1100 nm). The authors applied PLS-DA and reported a classification accuracy of 88%, 90%, and 86% for different fruit maturity stages. For the detection of internal disorders in mangoes after storage, Mogollón et al. (2020) used a portable hand-held vis-NIRS instrument in the interactance mode. The authors applied linear discriminate analysis and supporting vector machine and reported that internal disorders (jelly seed and black flesh) were classified up to an accuracy of 76%. Other successful applications of NIRS for the detection of internal disorders or pests include internal browning in mango (Gabriëls et al., 2020), moldy core in apples (Tian et al., 2020), and insect damage in pomegranates (Jamshidi et al., 2019).

For the assessment of external disorders in thick rind fruit, Magwaza et al. (2014a) evaluated vis-NIRS for predicting the rind breakdown (RBD) in "Nules Clementine" mandarin. The authors examined the effects of fruit canopy position, bagged fruit, and storage duration on the susceptibility of mandarins to rind breakdown. Based on the spectral characteristics, the PCA technique was successfully used to differentiate between RBD based on different fruit canopy positions. In a later study, Magwaza et al. (2014b) reported that the rind carbohydrates (i.e., glucose, sucrose, fructose) and dry matter are associated with RBD and can be predicted (RPD for fructose model = 2.50) using vis-NIRS. In a similar study on "Marsh" grapefruit, Ncama et al. (2018) applied vis-NIRS to predict the development of rind pitting disorder during storage. The authors reported that rind pitting can be predicted (R^2 = 0.78, RMSEP = 1.41, RPD = 2.03) and that PCA can be used to differentiate between healthy and fruit affected with rind pitting disorder. For the evaluation of other surface defects, NIRS has successfully been applied to detect bruising on avocado (Wedding et al., 2019), husk scald on pomegranates (Arendse et al., 2018c), and chilling injury on kiwifruit (Wang et al., 2020).

6 Prospects for NIRS for Evaluating Quality Attributes of Thick Rind Fruit

Over the years, NIRS has arguably become the most widely used non-destructive tool that can be applied for quality evaluation to a broad range of applications.. Tremendous advances have been made regarding NIR instrumentation accessories and the development of more suitable chemometric software packages. This chapter highlighted the application of NIRS to predict the external and internal quality attributes of thick rind fruit offers considerable potential for the measurement and prediction of physiological disorders and diseases.

A NIR instrument can only record what it sees, the development of calibrations and future analysis depends mainly on the quality of the acquired spectra. This is particularly important in the case of fruits that differ in optical density due to size, shape, and peel thickness. Therefore, selection of an appropriate NIR instrument for quality analysis of fruits with thick rinds, one needs to consider the spectral acquisition mode, wavelength range, and spectral resolution. The majority of FT-NIR and FT-IR instruments are regarded as being superior to those of conventional dispersive NIR instruments due to their superior wavelength accuracy and spectral resolution. Considering that the spectra of intact fruit and vegetables are commonly influenced by physical properties, such as fruit shape, size, and surface gloss, the selection of suitable regression techniques and preprocessing or pretreatment methods is an important step in improving model development. Another important factor that has not been discussed in this chapter includes model robustness. Robustness is an important index in evaluating the performance stability of calibration models. Robustness often includes a variation of certain operational parameters and/or environmental factors to ensure the developed models are insensitive to unknown variables. Therefore, adding robustness through biological variability (i.e., season, growing locations, cultivar, and fruit maturity) and different environmental conditions within a study can improve the model performance and prediction accuracy. Further information on model robustness has been studied and reviewed by various authors (Magwaza et al., 2012, 2014b; Nicolaï et al., 2007; Peirs et al., 2003). Considering that different parts of the fruit have an inherent variability in sweetness, acidity, and high-water content, future NIR instrumentation design needs to consider the heterogeneous nature of the fruit, by improving the NIR light intensity source, which would provide more adequate light penetration depth, faster integration time, and increased spectrometer aperture/detector size for improving the internal spectral acquisition. Through the improvement of NIR instruments and multivariate software packages, the problems associated with thick rind fruit are gradually being eliminated.

7 Conclusion

Recent advances in NIR systems have contributed to the implementation of these technologies for quality control in the fresh and processed food industries. In the horticultural industry, thick rind fruit such as banana, watermelon, pomegranate, and citrus present considerable challenges for light penetration into fruit tissue thereby impairing the quality of prediction. Near-infrared spectroscopy has been successfully applied on a wide range of fruit with rind to quantify and predict both external and internal quality, including coproducts such as juice, seed, and seed oil. Despite the considerable progress made, various factors related to postharvest handling techniques and inherent product variability within and across batches limit the applicability of quality prediction models. Ongoing advancements in data science, machine learning, and artificial intelligence offer new prospects for developing robust NIRS models for rapid and accurate prediction of product quality.

Acknowledgements This work is based on the research supported wholly by the National Research Foundation of South Africa (Grant 593 Numbers: 64813) and UNESCO ICB, Nsukka. The opinions, findings, and conclusions or recommendations expressed are those of the authors alone, and the NRF accepts no liability whatsoever in this regard.

References

Arendse, E., Fawole, O. A., Magwaza, L. S., Nieuwoudt, H., & Opara, U. L. (2018). Fourier transform near infrared diffuse reflectance spectroscopy and two spectral acquisition modes for evaluation of external and internal quality of intact pomegranate fruit. *Postharvest Biology and Technology, 138*, 91–98.

Arendse, E., Fawole, O. A., Magwaza, L. S., & Opara, U. L. (2018a). Non-destructive prediction of internal and external quality attributes of fruit with thick rind: A review. *Journal of Food Engineering, 217*, 11–23.

Arendse, E., Fawole, O. A., Magwaza, L. S., & Opara, U. L. (2018c). Evaluation of biochemical markers associated with the development of husk scald and the use of diffuse reflectance NIR spectroscopy to predict husk scald in pomegranate fruit. *Scientia Horticulturae, 232*, 240–249.

Arendse, E., Nieuwoudt, H. H., Nturambirwe, J. F. I., Fawole, O. A., Magwaza, L. S., & Opara, U. L. (2020). Recent advancements on vibrational spectroscopic techniques for the detection of authenticity and adulteration in horticultural products with a specific focus on oils, juices, and powders. *Food and Bioprocess Technology, 14*, 1–22.

Blanco, M., & Villarroya, I. (2002). NIR spectroscopy: A rapid-response analytical tool. *Trends in Analytical Chemistry, 21*, 240–250.

Cayuela, J. A. (2008). Vis–NIR soluble solids prediction in intact oranges (*Citrus sinensis* L.) cv. Valencia late by reflectance. *Postharvest Biology and Technology, 47*, 75–80.

Cayuela, J. A., & Weiland, C. (2010). Intact orange quality prediction with two portable NIR spectrometers. *Postharvest Biology and Technology, 58*, 113–120.

Chen, L., & Opara, U. L. (2013). Texture measurement approaches in fresh and processed foods—A review. *Food Research International, 51*, 823–835.

Clark, C. J., McGlone, V. A., Requejo, C., White, A., & Woolf, A. B. (2004). Dry matter determination in 'Hass' avocado by NIR spectroscopy. *Postharvest Biology and Technology, 29*, 300–307.

Cozzolino, D. (2014). An overview of the use of infrared spectroscopy and chemometrics in authenticity and traceability of cereals. *Food Research International, 60*, 262–265.

Cozzolino, D., Cyankar, W. U., Shah, N., & Smith, P. (2011). Multivariate data analysis applied to spectroscopy: Potential application to juice and fruit quality. *Food Research International, 44*, 1888–1896.

De Oliveira, G. A., Bureau, S., Renard, C. M. G. C., Pereira-Netto, A. B., & de Castilhos, F. (2014). Comparison of NIRS approach for prediction of internal quality traits in three fruit species. *Food Chemistry, 143*, 223–230.

Dull, G. G., Birth, G. S., Smittle, D. A., & Leffler, R. G. (1989). Near infrared analysis of soluble solids of intact cantaloupe. *Journal of Food Science, 54*, 393–395.

Flores, K., Sánchez, M. T., Pérez-Martin, D. C., López, M. D., Guerrero, J. E., & Garrido-Varo, A. (2008). Prediction of soluble solid content of intact and cut melons and watermelons using near infrared spectroscopy. *Journal of Near Infrared Spectroscopy, 16*, 91–98.

Fraser, D. G., Jordan, R. B., Kűnnemeyer, R., & McGlone, V. A. (2003). Light distribution inside mandarin fruit during internal quality assessment by NIR spectroscopy. *Postharvest Biology and Technology, 27*, 185–196.

Fraser, D. G., McGlone, V. A., Jordan, R. B., & Kűnnemeyer, R. (2001). NIR (near infra-red) light penetration into an apple. *Postharvest Biology and Technology, 22*, 191–194.

Gabriëls, S. H. E. J., Mishra, P., Mensink, M. J. G., Spoelstra, P., & Woltering, E. J. (2020). Non-destructive measurement of internal browning in mangoes using visible and near-infrared spectroscopy supported by artificial neural network analysis. *Postharvest Biology and Technology, 166*, 111206.

Gómez, A. H., He, Y., & Pereira, A. G. (2006). Non-destructive measurement of acidity, soluble solids and firmness of Satsuma mandarin using Vis–NIR spectroscopy techniques. *Journal of Food Engineering, 77*, 313–319.

Greensill, C. V., Wolf, P. J., Spiegelman, C. H., & Walsh, K. B. (2001). Calibration transfer between PDA-based NIR spectrometers in the NIR assessment of melon soluble solids content. *Applied Spectroscopy, 55*, 647–653.

Guthrie, J. A., Liebenberg, C. J., & Walsh, K. B. (2006). NIR model development and robustness in prediction of melon fruit total soluble solids. *Australian Journal of Agricultural Research, 57*, 1–8.

Guthrie, J. A., Reid, D. J., & Walsh, K. B. (2005). Assessment of internal quality attributes of mandarin fruit. 2. NIR calibration model robustness. *Australian Journal of Agricultural Research, 56*, 417–426.

Guthrie, J. A., & Walsh, K. B. (1997). Non-invasive assessment of pineapple and mango fruit quality using near infra-red spectroscopy. *Australian Journal of Experimental Agriculture, 37*, 253–263.

Guthrie, J. A., Walsh, K. B., Reid, D. J., & Liebenberg, C. J. (2005). Assessment of internal quality attributes of mandarin fruit. 1. NIR calibration model development. *Australian Journal of Agricultural Research, 56*, 405–416.

Hai-qing, T., Yi-bin, Y., Hui-shan, L., Xia-ping, F., & Hai-yan, Y. (2007). Measurement of soluble solids content in watermelon by Vis/NIR diffuse transmittance technique. *Journal of Zhejiang University. Science. B, 8*, 105–110.

Jamshidi, B., Mohajerani, E., Frazmand, H., Mahmoudi, A., & Hemmati, A. (2019). Pattern recognition-based optical technique for non-destructive detection of Ectomyelois ceratoniae infestation in pomegranates during hidden activity of the larvae. Spectrochim. Spectrochimica Acta. *Part A: Molecular and Biomolecular Spectroscopy, 206*, 552–557.

Jha, S. N., Narsaiah, K., Jaiswal, P., Bhardwaj, R., Gupta, M., Kumar, R., & Sharma, R. (2014). Nondestructive prediction of maturity of mango using near infrared spectroscopy. *Journal of Food Engineering, 124*, 152–157.

Jie, D., Xie, L., Rao, X., & Ying, Y. (2014). Using visible and near infrared diffuse transmittance technique to predict soluble solids content of watermelon in an on-line detection system. *Postharvest Biology and Technology, 90*, 1–6.

Kader, A. A. (2000). Quality of horticultural products. *Acta Horticulturae, 517*, 17–18.

Kader, A. A. (2002). Standardization and inspection. In *Postharvest technology of horticultural crops* (3rd ed., pp. 290–298). University of. California, Agriculture and Natural Resources.

Kader, A. A., & Rolle, R. S. (2004). The role of post-harvest management in assuring the quality and safety of horticultural produce. *FAO Agricultural Service Bulletin, 152*, 2–5.

Khodabakhshian, R., Emadi, B., Khojastehpour, M., & Golzarian, M. R. (2016). Carob moth, Ectomyelois ceratoniae, detection in pomegranate using visible/near infrared spectroscopy. *Computers and Electronics in Agriculture, 129*, 9–14.

Khodabakhshian, R., Emadi, B., Khojastehpour, M., & Golzarian, M. R. (2019). A comparative study of reflectance and transmittance modes of Vis/NIR spectroscopy used in determining internal quality attributes in pomegranate fruits. *Journal of Food Measurement and Characterization, 13*, 3130–3139.

Khodabakhshian, R., Emadi, B., Khojastehpour, M., Golzarian, M. R., & Sazgarnia, A. (2017). Non-destructive evaluation of maturity and quality parameters of pomegranate fruit by visible/near infrared spectroscopy. *International Journal of Food Properties, 20*, 41–52.

Krivoshiev, G. P., Chalucova, R. P., & Moukarev, M. I. (2000). A possibility for elimination of the interference from the peel in nondestructive determination of the internal quality of fruit and vegetables by Vis–NIR spectroscopy. *LWT - Food Science and Technology, 33*, 344–353.

Lammertyn, J., Peirs, J., De Baerdemaeker, J., & Nicolaï, B. M. (2000). Light penetration properties of NIR radiation in fruit with respect to non-destructive quality assessment. *Postharvest Biology and Technology, 18*, 121–132.

Li, L., Jang, X., Li, B., & Liu, Y. (2021). Wavelength selection method for near-infrared spectroscopy based on standard-sample calibration transfer of mango and apple. *Computers and Electronics in Agriculture, 190*, 106448.

Li, Y., Jin, G., Jiang, X., Yi, S., & Tian, X. (2020). Non-destructive determination of soluble solids content using a multi-region combination model in hybrid citrus. *Infrared Physics and Technology, 104*, 103138.

Liu, Y., Sun, X., & Ouyang, A. (2010). Non-destructive measurements of soluble solid content of navel orange fruit by visible-NIR spectrometric technique with PLS and PCA-BPNN. *LWT - Food Science and Technology, 43*, 602–607.

Lohumi, S., Lee, S., Lee, H., & Cho, B. K. (2015). A review of vibrational spectroscopic techniques for the detection of food authenticity and adulteration. *Trends in Food Science and Technology, 46*, 85–98.

Long, R. L., & Walsh, K. B. (2006). Limitations to the measurement of intact melon total soluble solids using near infrared spectroscopy. *Australian Journal of Agricultural Research, 57*, 403–410.

Lorente, D., Escandell-Montero, P., Cubero, S., Gómez-Sanchis, J., & Blasco, J. (2015). Visible–NIR reflectance spectroscopy and manifold learning methods applied to the detection of fungal infections on citrus fruit. *Journal of Food Engineering, 163*, 17–24.

Louw, E. D., & Theron, K. I. (2010). Robust prediction models for quality parameters in Japanese plums (Prunus salicina L.) using NIR spectroscopy. *Postharvest Biology and Technology, 58*(3), 176–184.

Lu, H., Xu, H., Ying, Y., Fu, X., Yu, H., & Tian, H. (2006). Application Fourier transform near infrared spectrometer in rapid estimation of soluble solids content of intact citrus fruits. *Journal of Zhejiang University. Science, 7*, 794–799.

Lu, J., Qi, S., Liu, R., Zhou, E., Li, W., Song, S., & Han, D. (2015). Nondestructive determination of soluble solids and firmness in mix-cultivar melon using near-infrared CCD spectroscopy. *Journal of Innovative Optical Health Sciences, 8*, 1550032.

Magwaza, L. S., Landahl, S., Cronje, P. J. R., Nieuwoudt, H. H., Mouazen, A. M., Nicolaï, B. M., Terry, L. A., & Opara, U. L. (2014). The use of Vis/NIRS and chemometric analysis to predict fruit defects and postharvest behaviour of 'Nules Clementine' mandarin fruit. *Food Chemistry, 163*, 267–274.

Magwaza, L. S., Opara, U. L., Cronje, P. J. R., Landahl, S., Nieuwoudt, H. H., Mouazen, A. M., Nicolaï, B. M., & Terry, L. A. (2014a). Assessment of rind quality of 'Nules Clementine' mandarin fruit during postharvest storage: 1. Vis/NIRS PCA models and relationship with canopy position. *Scientia Horticulturae, 165*, 410–420.

Magwaza, L. S., Opara, U. L., Cronje, P. J. R., Landahl, S., Nieuwoudt, H. H., Mouazen, A. M., Nicolaï, B. M., & Terry, L. A. (2014b). Assessment of rind quality of 'Nules Clementine' mandarin fruit during postharvest storage: 2. Robust Vis/NIRS PLS models for prediction of physico-chemical attributes. *Scientia Horticulturae, 165*, 421–432.

Magwaza, L. S., Opara, U. L., Nieuwoudt, H., Cronje, P. J. R., Saeys, W., & Nicolaï, B. (2012). NIR spectroscopy applications for internal and external quality analysis of citrus fruit-a review. *Food and Bioprocess Technology, 5*, 425–444.

Magwaza, L. S., Opara, U. L., Terry, L. A., Landahl, S., Cronje, P. J. R., Nieuwoudt, H. H., Hanssens, A., Saeys, W., & Nicolaï, B. M. (2013). Evaluation of Fourier transform-NIR spectroscopy for integrated external and internal quality assessment of 'Valencia' oranges. *Journal of Food Composition and Analysis, 31*, 144–154.

Maniwara, P., Nakano, K., Boonyakiat, D., Ohashi, S., Hiroi, M., & Tohyama, T. (2014). The use of visible and near infrared spectroscopy for evaluating passion fruit postharvest quality. *Journal of Food Engineering, 143*, 33–43.

Masithoh, R. E., Haff, R., & Kawano, S. (2016). Determination of soluble solids content and titratable acidity of intact fruit and juice of Satsuma mandarin using a hand-held near infrared instrument in transmittance mode. *Journal of Near Infrared Spectroscopy, 24*, 83–88.

McGlone, V. A., Fraser, D. G., Jordan, R. B., & Kunnemeyer, R. (2003). Internal quality assessment of mandarin fruit by Vis–NIR spectroscopy. *Journal of Near Infrared Spectroscopy, 11*, 323–332.

Mishra, P., & Passos, D. (2021). A synergistic use of chemometrics and deep learning improved the predictive performance of near-infrared spectroscopy models for dry matter prediction in mango fruit. *Chemometrics and Intelligent Laboratory Systems, 212*, 104287.

Mishra, P., Woltering, E., & Harchioui, N. E. (2020). Improved prediction of 'Kent' mango firmness during ripening by near-infrared spectroscopy supported by interval partial least square regression. *Infrared Physics and Technology, 110*, 10345.

Mogollón, R., Contreras, C., Neta, M. L. D., Marques, E. J. N., Zoffoli, J. P., & de Freitas, S. T. (2020). Non-destructive prediction and detection of internal physiological disorders in 'Keitt' mango using a hand-held Vis-NIR spectrometer. *Postharvest Biology and Technology, 167*, 111251.

Munawar, A. A., Von Hörsten, D., Wegener, J. K., Pawelzik, E., & Mörlein, D. (2016). Rapid and non-destructive prediction of mango quality attributes using Fourier transform near infrared spectroscopy and chemometrics. *Engineering in Agriculture, Environment and Food, 9*, 208–215.

Ncama, K., Tesfay, S. Z., Fawole, O. A., Opara, U. L., & Magwaza, L. S. (2018). Non-destructive prediction of 'marsh' grapefruit susceptibility to postharvest rind pitting disorder using reflectance Vis/NIR spectroscopy. *Scientia Horticulturae, 231*, 265–271.

Neto, J. P. S., de Assis, M. W. D., Casagrande, I. P., Júnior, L. C. C., & Teixeira, G. H. A. (2017). Determination of 'palmer' mango maturity indices using portable near infrared (VIS-NIR) spectrometer. *Postharvest Biology and Technology, 130*, 75–80.

Nicolaï, B. M., Beullens, K., Bobelyn, E., Peirs, A., Saeys, W., & Theron, I. K. (2007). Non-destructive measurement of fruit and vegetable quality by means of NIR spectroscopy: A review. *Postharvest Biology and Technology, 46*, 99–118.

Okere, E. E., Arendse, E., Nieuwoudt, H., Fawole, O. A., Perold, W. J., & Opara, U. L. (2021). Non-invasive methods for predicting the quality of processed horticultural food products, with emphasis on dried powders, juices and oils: A review. *Food, 10*, 3061.

Olarewaju, O. O., Bertling, I., & Magwaza, L. S. (2016). Non-destructive evaluation of avocado fruit maturity using near infrared spectroscopy and PLS regression models. *Scientia Horticulturae, 199*, 229–236.

Opara, U. L. (2009). Quality management: An industrial approach to produce handling. In J. Wojciech, R. L. Florkowski, B. B. Shewfelt, & E. P. Stanley (Eds.), *Postharvest handling. A systems approach* (pp. 153–204). Elsevier.

Pathare, P. B., Opara, U. L., & Al-Said, F. A. J. (2013). Colour measurement and analysis in fresh and processed foods: A review. *Food and Bioprocess Technology, 6*, 36–60.

Peirs, A., Tirry, J., Verlinden, B., Darius, P., & Nicolaï, B. M. (2003). Effect of biological variability on the robustness of NIR-models for soluble solids content of apples. *Postharvest Biology and Technology, 28*, 269–280.

Raghavendra, A., Guru, D. S., & Rao, M. K. (2021). Mango internal defect detection based on optimal wavelength selection method using NIR spectroscopy. *Artificial Intelligence in Agriculture, 5*, 43–51.

Rinnan, A., Van den Berg, F., & Engelsen, S. B. (2009). Review of the most common pre-processing techniques for near-infrared spectra. *Trends in Analytical Chemistry, 28*, 1201–1222.

Rungpichayapichet, P., Mahayothee, B., Nagle, M., Khuwijitjaru, P., & Müller, J. (2016). Robust NIRS models for non-destructive prediction of postharvest fruit ripeness and quality in mango. *Postharvest Biology and Technology, 111*, 31–40.

Sánchez, M. T., Haba, M. J. D. L., Serrano, I., & Perez-Marin, D. C. (2013). Application of NIRS for non-destructive measurement of quality parameters in intact oranges during on-tree ripening and at harvest. *Food Analytical Methods, 6*, 826–837.

Subedi, P. P., & Walsh, K. B. (2020). Assessment of avocado fruit dry matter content using portable near infrared spectroscopy: Method and instrumentation optimisation. *Postharvest Biology and Technology, 161*, 111078.

Sun, X., Zhang, H., & Liu, Y. (2009). Nondestructive assessment of quality of 'Nanfeng' mandarin fruit by a portable near infrared spectroscopy. *International Journal of Agricultural and Biological Engineering, 2*, 65–71.

Tamburini, E., Costa, S., Rugiero, I., Pedrini, P., & Marchetti, M. G. (2017). Quantification of lycopene, β-carotene, and total soluble solids in intact red-flesh watermelon (*Citrullus lanatus*) using on-line near-infrared spectroscopy. *Sensors, 17*, 746.

Tian, H., Ying, Y., Xu, H., Lu, H., & Xie, L. (2009). Study on Vis/NIR spectra detecting system for watermelons and quality predicting in motion. *Spectroscopy and Spectral Analysis, 29*, 1536–1540.

Tian, X., Wang, Q., Huang, W., Fan, S., & Li, J. (2020). Online detection of apples with moldy core using the Vis/NIR full-transmittance spectra. *Postharvest Biology and Technology, 168*, 111269.

Torres, I., Sánchez, M. T., Haba, M. J. L., & Pérez-Marín, D. (2019). LOCAL regression applied to a citrus multispecies library to assess chemical quality parameters using near infrared spectroscopy. *Spectrochimica Acta Part A: Molecular and Biomolecular Spectroscopy, 217*, 206–214.

Walsh, K., Lu, R., & Nicolaï, B. (2021). Special issue: Recent advances in the use of visible and vibrational spectroscopy/imaging for measurement of postharvest quality. *Postharvest Biology and Technology, 171*, 111347. https://doi.org/10.1016/j.postharvbio.2020.111347

Walsh, K. B., Blasco, J., Zude-Sasse, M., & Sun, X. (2020). Visible-NIR 'point' spectroscopy in postharvest fruit and vegetable assessment: The science behind three decades of commercial use. *Postharvest Biology and Technology, 168*, 111246.

Walsh, K. B., McGlone, V. A., & Han, D. H. (2020). The uses of near infra-red spectroscopy in postharvest decision support: A review. *Postharvest Biology and Technology, 163*, 111139.

Wang, Z., Künnemeyer, R., McGlone, A., & Burdon, J. (2020). Potential of Vis-NIR spectroscopy for detection of chilling injury in kiwifruit. *Postharvest Biology and Technology, 164*, 111160.

Watanawan, C., Wasusri, T., Srilaong, V., Wongs-Aree, C., & Kanlayanarat, S. (2014). Near infrared spectroscopic evaluation of fruit maturity and quality of export Thai mango (Mangifera indica L. var. Namdokmai). *International Food Research Journal, 21*(3), 1073–1078.

Wedding, B. B., Wright, C., Grauf, S., Gadek, P., & White, R. D. (2019). The application of FT-NIRS for the detection of bruises and the prediction of rot susceptibility of 'Hass' avocado fruit. *Journal of the Science of Food and Agriculture, 99*, 1880–1887.

Williams, P., Antoniszyn, J., & Manley, M. (2019). *Near infrared technology: Getting the best out of light*. African Sun Media, Sun Press.

Xu, L., Li, J., & Zhang, D. (2018). Near-infrared light penetration depth analysis inside melon with thick peel by a novel strategy of slicing combining with least square fitting method. *Food Processing Engineering, 41*, e12886.

Xudong, S., Hailiang, Z., & Yande, L. (2009). Nondestructive assessment of quality of Nanfeng mandarin fruit by a portable near infrared spectroscopy. *International Journal of Agricultural and Biological Engineering, 2009*(2), 65–71.

Zeb, A., Waqar, S., Qureshi, W. S., Ghafoor, A., Malik, A., Imran, M., Iqbal, J., & Alanazi, E. (2021). Is this melon sweet? A quantitative classification for near-infrared spectroscopy. *Infrared Physics and Technology, 114*, 103645.

Near-Infrared Spectroscopy for Pomegranate Quality Measurement and Prediction

Umezuruike Linus Opara and Ebrahiema Arendse

Abstract Although the pomegranate has been cultivated and utilized in food, health, and other applications since time immemorial, it is only during the last couple of decades that commercial production and global trade in pomegranate fruit and processed coproducts have taken off. Heightened market demand for a consistent and year-round supply of fresh pomegranate fruit and value-added products has resulted in the need for fast and accurate nondestructive methods for the detection, measurement, and prediction of the external and internal quality attributes. Nondestructive measurement and prediction methods, such as near-infrared spectroscopy (NIRS), offer tremendous prospects for application in measuring and predicting the organoleptic and nutritional quality of pomegranate. Unlike other deciduous fruits such as apple and pears with edible skin/peel, the unique structure of pomegranate fruit is comprised of nonedible thick rind/peel/husk, and edible arils containing juice and seed which presents challenges in measurements to characterize fruit quality. This chapter examines the recent technological advances in NIRS for the measurement and prediction of external and internal quality attributes of pomegranate, including appearance features, total soluble solids, titratable acidity, and vitamin C. The application of NIRS, combined with chemometrics, to measure and predict external defects on whole fruit, and quality attributes of fruit peel, arils, and juice are discussed. Prospects of NIRS for on-site quality detection, measurement and control of pomegranates are highlighted.

Keywords *Punica granatum* · Infrared-spectroscopy · Nondestructive measurement · Fruit quality · Postharvest management

U. L. Opara (✉)
SARChI Postharvest Technology Research Laboratory, Africa Institute for Postharvest Technology, Faculty of AgriSciences, Stellenbosch University, Stellenbosch, South Africa

UNESCO International Centre for Biotechnology, Nsukka, Enugu State, Nigeria
e-mail: opara@sun.ac.za

E. Arendse
SARChI Postharvest Technology Research Laboratory, Africa Institute for Postharvest Technology, Faculty of AgriSciences, Stellenbosch University, Stellenbosch, South Africa

P. B. Pathare, M. S. Rahman (eds.), *Nondestructive Quality Assessment Techniques for Fresh Fruits and Vegetables*, https://doi.org/10.1007/978-981-19-5422-1_10

1 Introduction

Pomegranate (*Punica granatum* L.) is a fruit-bearing deciduous shrub that is native to Persia (Iran) and is widely cultivated in the Mediterranean region (Holland et al., 2009). Since time immemorial, the fruit has been utilized in many cultures as food, and as a medical therapy. However, it is only during the last few decades, that scientific research has revealed that regular consumption of pomegranate may provide antioxidant, anti-bacterial, anti-inflammatory, and anti-hypertension activities (Fuhrman & Aviram, 2006; Lansky & Newman, 2007; Viuda-Martos et al., 2010). These bioactive properties have been attributed to the higher levels of antioxidant polyphenol compounds when compared to other types of fruits such as apples, grapes, and blueberries (Seeram et al., 2008). Thus, the high global demand for pomegranate fruit has propelled commercial production, trade, research, and the development of various coproducts.

Unlike other deciduous fruits such as apples and pears with edible skin/peel, the structure of pomegranate fruit comprises of a nonedible thick leathery peel/husk with an interior consisting of edible portions known as the arils that are separated by membranous walls into different compartments (Aindongo et al., 2014). The arils consist of a woody portion also known as a kernel or seed (Al-Said et al., 2009). The unique structure of pomegranate fruit poses challenges in the measurement and characterization of external and internal fruit quality attributes. The heightened market demand for a consistent and year-round supply of fresh pomegranate fruit including the supply of microbially safe, nutritious, and traceable processed products has prompted the need for the implementation of more rapid, and cost-effective technologies for the assessment of different fruit quality attributes.

A recent trend has shown that modern agricultural practices are shifting the way quality attributes for fruit and vegetables are being assessed to faster, more accurate, and more objective nondestructive testing (Magwaza & Opara, 2015; Nicolaï et al., 2014; Walsh et al., 2021). Near-infrared spectroscopy (NIRS) is a nondestructive analytical tool used for qualitative and quantitative analysis of a wide range of biological materials (Magwaza & Opara, 2015). NIRS, the basic principle involves the interaction between light and molecular bonds (C–H, C–C, O–H, N–H) in biological materials that occurs within the region of 800–2500 nm of the electromagnetic spectrum (Williams et al., 2019). The obtained NIR spectra of a biological material would consist of broad bands arising from overlapping absorptions which result in overtones, and therefore advanced chemometric tools are used to obtain any relevant and useful information (Arendse et al., 2018c; Cozzolino et al., 2011). As an objective nondestructive tool for quality control and inspection, NIRS and associated chemometrics have several advantages. For instance, it is a rapid technique that is capable of assessing fruit at speeds as high as 10 per second on commercial sorting lines, it allows repeated testing of the same sample, requires minimal or no sample preparation and the technique is versatile, when coupled with an optical fiber probe it can operate inline or online, in addition to the available portable instruments (Arendse et al., 2018c; Williams et al., 2019).

For fruit and vegetables, NIRS is the most widely used nondestructive method for internal quality assessment (Brázio et al., 2016; Chandrasekaran et al., 2019; Walsh et al., 2021; Walsh et al., 2020a, 2020b). Some of the most successful application of NIRS in agricultural produce includes the prediction of composition in terms of constituents such as water, carbohydrates, fat, protein, cellulose, polysaccharides, and various other constituents which have an abundance of O–H, C–H, and N–H bonds (Williams et al., 2019). Furthermore, NIRS has been extensively used for authentication and detection of adulteration in processed horticultural products (Arendse et al., 2020; Danezis et al., 2016). Therefore, nondestructive measurement and prediction methods, such as NIRS, offer tremendous prospects for application in measuring and predicting the organoleptic and nutritional quality of pomegranate. This chapter evaluates the recent technological advancements and applications of near-infrared spectroscopy for the measurement and prediction of external and internal quality attributes of pomegranate fruit and its coproducts (arils, dried arils, juice).

2 Quality Attributes of Pomegranate Fruit: An Overview

Quality attributes of fruits or vegetables may be evaluated using instrumental analysis or through sensory evaluation using a trained panel or a combination of both techniques. Quality assessment of fruit and vegetables is a combination of appearance, taste, aroma, and textural properties (Kader, 2000). These characteristics are indices that govern the marketability of the fruit and sets benchmarks to ensure that industry standards are met (Kader, 2006). For pomegranates, several quality parameters have been used to evaluate their internal and external properties. External properties of the fruit include fruit weight or mass, fruit shape, and skin appearance (skin color, texture, free from defects). Internal properties are mainly associated with total soluble solids, titratable acidity, flavor (sugar/acid ratio) and astringency (Fawole & Opara, 2013b; Kader, 2000, 2006).

On a global scale, over 1000 pomegranate cultivars exist and have been identified (Levin, 1994); therefore, the determinant benchmark for the assessment of different quality attributes is crucial to the pomegranate industry. Growing demand for pomegranate fruit requires that the industry must agree and elucidate benchmarks for quality. This section provides a brief overview of various quality indices used to evaluate and grade pomegranate fruit.

2.1 *Physical and Textural Properties*

Consumers judge the quality of pomegranate fruit based on appearance (physical, visual, and textural) at the time of initial purchase. This may include fruit mass/weight, shape, color, gloss, and freedom of defects (Kader, 2000). For producers, packers, and processors, fruit mass/weight has been used to grade fruit into different tier classes based on fruit size (Fawole & Opara, 2013c; Kader, 2006). Textural properties of pomegranate fruit such as firmness, crispiness, juiciness, and toughness are not only important for their eating quality but also for their shipping ability. These textural properties also help provide information on the mechanical handling and storage of fruit both during the preharvest and postharvest stage of fruit production (Chen & Opara, 2013a, 2013b). The color properties of fruits and vegetables are important quality attributes as it influence consumer preference (Opara et al., 2009; Pathare et al., 2013). However, studies have reported that the outer skin color of the peel has no interrelationship with the inside color of the arils (Holland et al., 2009). The external color attributes may differ significantly as a result of cultivar differences and stage of fruit maturity. These colors may range from yellow, green, or pink overlain with pink to deep red or indigo to fully red (Holland et al., 2009). Other physical properties associated with the quality evaluation of pomegranate fruit include freedom of cracks and defects (Kader, 2006). Table 1 provides a summary of the various physical, textural, and color properties used for quality evaluation.

2.2 Chemical Properties

Pomegranate fruit is non-climacteric and does not ripen off the tree; therefore, harvesting fruit at the correct maturity stage is of utmost importance to ensure that fruit reaches the consumer in the best organoleptic condition. Several studies have reported that maturity indices of pomegranate fruit are affected by various factors such as differences in cultivars, geographical region, and maturity stage (Fawole & Opara, 2014; Shwartz et al., 2009). For pomegranates, to meet market standards maturity indices such as sweetness, acidity, and flavor (sugar to acid ratio) are often used for measurement of fruit quality (Cristosto et al., 2000; Holland et al., 2009; Kader, 2006). The flavor and aroma profile of pomegranate fruit have been ascribed to be a combination of sweetness, sourness, and astringency. For pomegranates, Kader (2006) reported that maximum acidity levels in sweet varieties may be 1% and for sweet-sour varieties range between 1.5% and 2%, whereas the minimum total soluble solids (TSS) vary from 15% to 17%. Fawole and Opara (2013a) established a maturity index for "Ruby" grown between 132 and 139 days of full bloom in South Africa, based on their finding, low TA levels (0.3%) and TSS values greater than 15% was highly desirable for market standards. Other internal quality attributes associated with pomegranate fruit include astringency which is correlated to total phenolic and total tannin content (Fawole & Opara, 2014). Table 1 provides a summary of the various chemical parameters and methods used to assess the maturity and quality of pomegranate fruit.

Table 1 Summary of methods used for evaluating maturity and quality attributes of pomegranate fruit (Fawole & Opara, 2013b)

Properties	Quality parameters	Method of determination	Subjective	Objective	Destructive	Nondestructive	References
Physical	Fruit mass	Measuring devices		x		x	Opara et al. (2009)
	Color	Light reflectance (colorimeter)		x		x	Shwartz et al. (2009) and Fawole and Opara (2014)
		Visual Munsell chart	x				Ben-Arie et al. (1984) and Shulman et al. (1984)
	Length, diameter,	Vernier caliper		x		x	Fawole and Opara (2014)
	Fruit volume	Water displacement method		x		x	Al-Maiman and Ahmad (2002) and Arendse et al. (2016b)
Textural	Firmness	Fruit texture analyzer		x	x	x	Al-Said et al. (2009)
	Crispness/ crunchiness	Sensory evaluation	x	x	x		Fawole and Opara (2013d)
Chemical	Total soluble solids	Digital refractometer		x	x	x	Fawole et al. (2012)
	Titratable acidity	Titration		x	x	x	Ben-Arie et al. (1984) and Fawole et al. (2012)
	Juice absorbance	Spectrophotometer		x	x	x	Al-Said et al. (2009) and Fawole and Opara (2013e)
	pH	pH meter		x	x	x	Al-Said et al. (2009) and Fawole et al. (2012)
	Total phenolic content	Chemical test (Folin C assay)		x	x	x	Shwartz et al. (2009) and Fawole et al. (2012)
	Total anthocyanin content	Chemical test (pH differential assay)		x	x	x	Fawole and Opara (2014) and Arendse et al. (2018d)
	Aroma and volatiles	Gas chromatography		x	x		Vázquez-Araújo et al. (2011)
		Sensory evaluation	x	x	x		Nanda et al. (2001) and Calín-Sánchez et al. (2011)

3 NIRS Prediction of Whole Pomegranate Fruit Quality

Pomegranate fruit has a thick leathery rind and a unique heterogeneous internal structure that is packed and separated into compartments. Thus, making it difficult to evaluate and characterize fruit quality. This section provides a general overview of NIRS and associated chemometrics applied to whole pomegranate fruit, including a discussion on the spectral characteristics and model interpretation.

3.1 NIR Characteristics

An example of a near-infrared spectrum for pomegranate fruit "Wonderful" is shown in Fig. 1. The pattern of the absorption curve in the NIR region for pomegranates is like that for other fruit such as kiwifruit (McGlone & Kawano, 1998), passion fruit (De Oliveira et al., 2014), and tomatoes (De Oliveira et al., 2014). The NIR spectra comprise of broad bands arising from overlapping absorptions, corresponding mainly to overtones and combinations of vibrational modes that involve C–H, O–H, and N–H chemical bonds of a molecule (Nicolaï et al., 2014; Williams et al., 2019). The spectral bands in the NIR region displayed contours having prominent absorbance bands in the following wavelength regions 950, 1200, 1450, 1780, and 1930 nm (Fig. 1). Fruits such as pomegranates comprise more than 80–90% water and their constituents are characterized by various interactions between hydrogen bonds of water, sugars, protein and other biochemical constituents (Williams et al., 2019). Therefore, in the NIR spectral region, molecular bonds especially O–H and C–H within molecules provide strong absorption bands, as the spectral profile is mainly dominated by water. For water molecules, the region of 950–1450 nm have

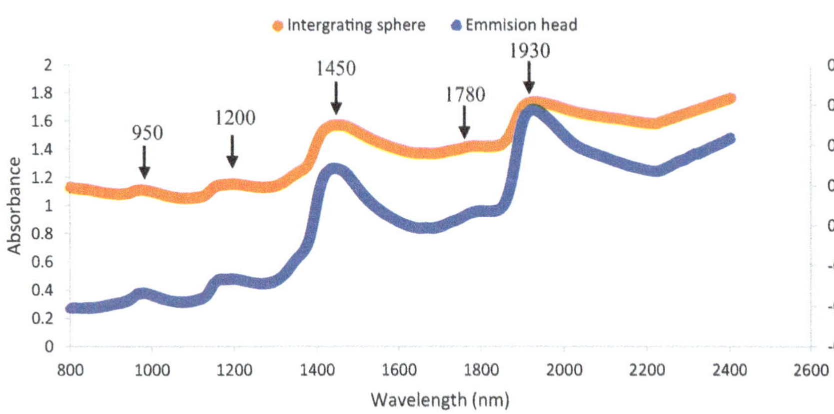

Fig. 1 Absorbance spectra of pomegranate fruit in the near-infrared region of 800–2400 nm from different spectral acquisition methods, an integrating sphere (orange solid line) and emission head (blue solid line) (Arendse et al., 2018d)

been reported to be related to the stretching of O–H bonds (Clément et al., 2008; Lestander & Geladi, 2005). For carbohydrates like sugars, the absorption bands can be prominent in the region of 1200–1930 nm which have been reported to correspond to the stretching of C–H bonds through second and first overtones, including the third overtone associated with the deformation of OH, CH, and CH_2 bonds. Several researchers have reported that sugar molecules display spectral bands in the region of 1100–1600 and 1700–2300 nm (Golic et al., 2003; Gómez et al., 2006; Louw & Theron, 2010).

3.2 Measurement of External Quality Attributes

Pomegranate fruit quality is primarily evaluated on its external appearance; these parameters are based on the measurement of fruit mass/weight, rind color, firmness, and freedom of external defects. In NIRS, the measurement of fruit mass has been reported to be indirectly related to the fruit water content (Louw & Theron, 2010). In pomegranates, Arendse et al. (2018d) applied NIRS, a contactless method coupled with an optic fiber emission head for the prediction of fruit weight. The partial least square regression (PLSr) algorithm used for model development of fruit weight gave an R^2 = 0.62 and RMSEP = 0.013 kg; however, the developed model was characterized as providing low prediction accuracy due to its low RPD values (1.62) (Table 2). Contrary to their results, Magwaza et al. (2013) reported that a contactless optic fiber emission head provided satisfactory results in predicting fruit mass (R^2 = 0.96, RMSEP = 10.45 g, and RPD = 3.53) of Valencia oranges. The authors concluded that the relatively high accuracy in predicting the fruit mass was based on the fruit curvature and size, both of these parameters are closely related to its mass.

Fruit firmness can be determined indirectly through the light scattering properties of the tissue (Nicolaï et al., 2007). In a study to predict fruit firmness in pomegranates, Arendse et al. (2018d) applied NIRS and chemometrics within the region of 1064–1333 and 1639–1835 nm. The authors observed good prediction statistics when comparing two different spectral acquisition methods, emission head (R^2 = 0.83, RMSEP = 7.45 N) and integrating sphere (R^2 = 0.82, RMSEP = 7.31 N). In a similar study using both reflectance and transmittance modes of visible to near-infrared spectroscopy (VIS-NIRS), Khodabakhshian et al. (2019) evaluated pomegranate fruit firmness and reported high prediction accuracy (R^2 = 0.94, RPD = 5.33) in reflectance mode (Table 2). These researchers recommended acquiring spectral data in reflectance mode and reported difficulty in predicting fruit firmness using transmittance. The authors observed that fruit size and the amount of light penetration may have influenced the prediction accuracy. Contrary to these results, several researchers have reported that transmission mode is appropriate for the assessment of various internal quality attributes of thick rind fruit such as oranges and mandarins (Cayuela, 2008; Miyamoto et al., 1998). For transmission mode, the NIR light intensity needs to be high to penetrate the fruit. Normally, increased exposure time and high light intensity levels may cause thermal damage to the rind of the fruit.

Table 2 Applications of NIR spectroscopy to predict the external fruit quality of pomegranate fruit

Parameter	Statistical analysis	Preprocess	Wavelength range (nm)	Predictors accuracy	Reference
Fruit weight (kg)	PCA, PLSr	MSC	1064–1333	$R^2 = 0.62$, RMSEP = 11.90, RPD = 1.63	Arendse et al. (2018d)
Firmness (N)	PCA, PLSr	SD	1064–1333, 1639–1835	$R^2 = 0.83$, RMSEP = 7.45, RPD = 2.43	Arendse et al. (2018d)
	PCA, PLSr	SNV	400–1100	$R^2 = 0.94$, RMSEP = 0.68, RPD = 5.33	Khodabakhshian et al. (2019)
Colour: a* (redness)	PCA, PLSr	SD	1064–1333, 1640–2174	$R^2 = 0.90$, RMSEP = 1.62, RPD = 3.34	Arendse et al. (2018d)
Chroma	PCA, PLSr	FD	1064–2355	$R^2 = 0.83$, RMSEP = 2.15, RPD = 2.43	Arendse et al. (2018d)
Hue	PCA, PLSr	None	1064–1333, 2174–2355	$R^2 = 0.83$, RMSEP = 1.67, RPD = 2.50	Arendse et al. (2018d)

R^2 coefficient of determination, *PLSr* partial least squares regression, *PCA* principal component analysis, *RMSEP* root mean square error of prediction, *SNV* vector normalization, *FD* first derivative, *SD* second derivative, *MSC* multiplicative scattering correction

For fruit and vegetables, color is the primary quality attribute that impacts consumer purchasing choices. The measurement of color properties can be used to indirectly measure various other quality attributes that have been shown to correlate well with various biochemical constituents and color pigments (Pathare et al., 2013). For color in pomegranates, Arendse et al. (2018d) developed PLSr models for predicting the fruit skin's color properties (a*, C*, h°) (Table 2). The authors were able to develop models using a wavelength range of 1064 and 2355 nm and reported a prediction accuracy for a* (R^2 = 0.90 and RPD = 3.34), C* (R^2 = 0.83 and RPD = 2.43), and h° (R^2 = 0.83 and RPD = 2.50). Literature suggests that the detection of fruit color is unlikely to occur outside the visible range (Clément et al., 2008; Magwaza et al., 2013; Ruiz et al., 2008). Therefore, the detection of color properties within the following regions (1064–1836, 2174–2355 nm) could be related to a secondary correlation with other biochemical compounds that are related to fruit maturity such as anthocyanin pigments.

3.3 Measurement of Internal Quality Attributes

Pomegranate quality parameters such as TSS, SSC, and TA have been associated with flavor and sensory properties. These quality parameters are associated with molecules that contain bonds such as C–H, O–H, C–O, and C–C therefore NIRS can detect and predict their concentration. For pomegranates, NIRS combined with chemometrics has been applied to measure various internal quality attributes (Table 3). Khodabakhshian et al. (2017) applied PLSr within the wavelength range of 400–1100 nm to predict TSS (R^2 = 0.92) and pH (R^2 = 0.85) in "Acco" pomegranates. Arendse et al. (2018d) evaluated the application of PLSr to measure the internal quality attributes (TSS, TA, and pH) of pomegranate (cv. Wonderful). The authors used the wavelength range of 1064–1333 and 1640–1836 nm and developed models for quantitative (RPD > 2.0) predictions. In a later study, Khodabakhshian et al. (2019) applied VIS-NIRS in predicting the TSS and pH values of "Ashraf" pomegranate fruit. The authors reported higher prediction statistics for TSS (R^2 = 0.96) and pH (R^2 = 0.86) when spectral data was collected in reflectance mode (Table 3). TSS:TA ratio is affected by the fact that flavor arises because of interactions among several constituents (Williams et al., 2019). Thus, the prediction of TSS:TA ratio is based on an indirect correlation between the spectral data obtained for each sample and the sugar to acid value ratio. Arendse et al. (2018d) develop accurate calibration models (R^2 = 0.86, RMSEP = 0.74 and RPD = 2.72) for the prediction of TSS:TA ratio.

For phytochemicals, Arendse et al. (2018d) used the wavelength range between 1064 and 2355 nm and obtained good, predicted statistics for total phenolic concentration (R^2 = 0.88, RMSEP = 0.11 g L^{-1}, and RPD = 2.91); however, low prediction accuracy was obtained for total anthocyanin concentration which was characterized by a high number of latent variables and a low RPD (1.64–1.59) values. NIR spectroscopy is a powerful nondestructive technique used for the detection of various compounds, the NIR spectrum provides information on the vibrational absorption of hydroxyl (O–H), amido (N–H), and C–H bonds. Therefore, NIRS can be used to measure ascorbic acid (vitamin C) concentration, since it contains four hydroxyl groups and several bands can be assigned to the ascorbic acid (Liu et al., 2006). For pomegranates, Arendse et al. (2018d) developed PLSr models for the prediction of vitamin C concentration (Table 3). The authors reported quantitative predictions (R^2 = 0.76, RPD = 2.06) for vitamin C concentration. For other fruits such as apple model development for vitamin C concentration (R^2 = 0.80, RPD = 3.40) have been successfully achieved within a similar wavelength range (Pissard et al., 2013).

It is evident from the reviewed literature that NIRS can be used to evaluate various internal and external quality characteristics of the pomegranate fruit. For improvement in calibration models for whole pomegranate fruit, future studies need to focus on applying different NIR settings and analytical frameworks. Studies also need to focus on the development of more robust models through the integration of different cultivars, seasonality, and fruit maturity. Future development of NIR systems needs to focus on the improvement of light intensity sources, photodetectors, optics, and faster integration times which could lead to improved model prediction for the internal quality attributes of the whole pomegranate fruit.

Table 3 Applications of NIR spectroscopy to predict the internal fruit quality of pomegranate fruit

Parameter	Statistical analysis	Preprocess	Wavelength range (nm)	Predictors accuracy	Reference
TSS (%)	PCA, PLSr	SNV	1064–1333, 1640–1836	$R^2 = 0.78$, RMSEP = 0.28, RPD = 2.17	Arendse et al. (2018d)
	PCA, PLSr	SNV	400–1100	$R^2 = 0.92$, RMSEP = 0.23, RPD = 6.38	Khodabakhshian et al. (2017)
	PCA, PLSr	SNV	400–1100	$R^2 = 0.94$, RMSEP = 0.21, RPD = 6.72	Khodabakhshian et al. (2019)
TA	PCA, PLSr	SD	1064–1333, 1640–1836	$R^2 = 0.76$, RMSEP = 0.13, RPD = 2.12	Arendse et al. (2018d)
	PCA, PLSr	SNV	400–1100	$R^2 = 0.95$, RMSEP = 0.26, RPD = 5.31	Khodabakhshian et al. (2017)
TSS:TA ratio	PCA, PLSr	SNV	1064–1333, 1640–2174	$R^2 = 0.86$, RMSEP = 0.74, RPD = 2.72	Arendse et al. (2018d)
pH	PCA, PLSr	SD	1064–2355	$R^2 = 0.84$, RMSEP = 0.06, RPD = 2.57	Arendse et al. (2018d)
	PCA, PLSr	SNV	400–1100	$R^2 = 0.85$, RMSEP = 0.064, RPD = 4.94	Khodabakhshian et al. (2017)
	PCA, PLSr	SNV	400–1100	$R^2 = 0.86$, RMSEP = 0.069, RPD = 4.43	Khodabakhshian et al. (2019)
Vitamin C (g/L)	PCA, PLSr	SD	1064–1333, 1640–1732	$R^2 = 0.76$, RMSEP = 0.09, RPD = 2.06	Arendse et al. (2018d)
TPC (g/L)	PCA, PLSr	FD	1064–1640, 1836–2355	$R^2 = 0.88$, RMSEP = 0.11, RPD = 2.91	Arendse et al. (2018d)
TAC (g/L)	PCA, PLSr	SNV	1064–1836, 2174–2355	$R^2 = 0.62$, RMSEP = 0.09, RPD = 1.64	Arendse et al. (2018d)

R^2 coefficient of determination, *PLSr* partial least squares regression, *PCA* principal component analysis, *RMSEP* root mean square error of prediction, *SNV* vector normalization, *FD* first derivative, *SD* second derivative, *TSS* total soluble solids, *TA* titratable acidity, *TPC* total phenolic content, *TAC* total anthocyanin content

4 NIRS Measurement and Prediction of Fresh and Dried Aril Quality

Consumption of fresh pomegranate arils is a convenient way to increase the intake of biologically active compounds (Ayhan & Eştürk, 2009; Caleb et al., 2012). However, fresh arils have a relatively short shelf-life of 5–8 days (Caleb et al., 2013). Therefore, to overcome this limitation of short shelf-life the pomegranate industry has promoted research and development of value-added pomegranate coproducts such as dried pomegranate arils. Limited research on the application of NIRS on fresh and dried pomegranate arils has been published (Arendse et al., 2017).

Typical NIR spectra of pomegranate arils are presented in Fig. 2 (Arendse et al., 2017). The absorption pattern is similar to fruits such as grape berries (Musingarabwi et al., 2016) and dried pomegranate arils (Okere, 2020). The contours show absorbance bands in the following regions of 950, 1200, 1400, 1789, and 1876 nm. A shift in the absorption bands could be a result of chemical differences within the sample, the interaction among constituents, or changes in the optical path in the spectrometer (Arendse et al., 2018c; Zude et al., 2008). The reported absorption bands within the region of 950 nm could be due to water and carbohydrate, as these molecules have been reported to strongly absorb at 958 nm (McGlone & Kawano, 1998; Williams & Norris, 1987). According to Williams et al. (2019), other water absorption bands exist at 1410, 1460, and 2486 nm, and in pomegranate arils these absorption bands occur at 1400 nm. Absorption bands at 1200, 1789 and 1876 nm have been reported to correspond to the stretching of second and first overtones of CH bonds found in carbohydrates as well as the third overtones of OH, CH, and CH_2 that are relatively strong absorbers.

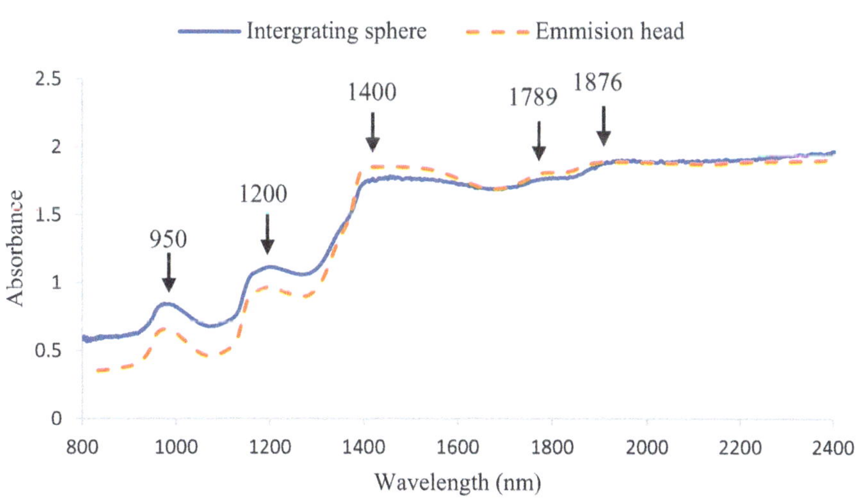

Fig. 2 Absorbance spectra of pomegranate arils were obtained in the near-infrared region of 800–2400 nm from different NIR spectral acquisition methods, an integrating sphere (blue solid line) and emission head (broken orange line) (Arendse et al., 2017)

Table 4 provides a summary of NIRS technology for evaluating fresh and dried pomegranate arils. For fresh pomegranate arils, Arendse et al. (2017) compared two different NIRS spectral acquisition methods, namely, an integrating sphere from a Multi-Purpose Analyser and an optic fiber emission head from a Matrix™-F spectrometer. The authors applied PLSr and several preprocessing methods and developed reasonably accurate calibration models were developed for TSS (R^2 = 87.55, RPD = 2.65), TA (R^2 = 85.59, RPD = 2.34), pH (R^2 = 85.18, RPD = 2.40), and TSS:TA ratio (R^2 = 83.43, RPD = 2.40). Other physicochemical properties that have been successfully evaluated using NIRS include color attributes (a*, chroma, hue) and aril firmness (Arendse et al., 2017). The authors reported that spectral acquisition methods played a major role in accurately predicting quality parameters using an optic fiber emission head from the Matrix™-F spectrometer which surpassed the prediction ability of the integrating sphere from the Multi-Purpose Analyser. The authors observed the emission head irradiated a larger surface of pomegranate arils compared to the integrating sphere, and consequently provided the best predictive ability.

For dried pomegranate arils, Okere (2020) compared different regression algorithms to evaluate various quality attributes using Fourier-transform near-infrared (FT-NIR) spectroscopy (Table 4). The authors applied both PLSr and support vector machine (SVM) within a spectral range of 800–2500 nm observed that SVM algorithm could predict titratable acidity (R^2 = 0.85, RMSEP = 0.04%) and several color components such as aril redness (R^2 = 0.72, RMSEP = 1.82) and color intensity (Chroma) (R^2 = 0.70, RMSEP = 1.99). The models developed using PLSr predicted pH value (R^2 = 0.86, RMSEP = 0.13) and TSS:TA ratio (R^2 = 0.74, RMSEP = 1.68). Their study has shown that both visual key visual (a*, C*) and sensory attributes (titratable acidity, TSS:TA ratio) of dried pomegranate arils can be predicted with NIRS.

The results from these studies suggest that NIRS and chemometrics can be used for the assessment of both fresh and dried pomegranate arils. Although both authors included several orchards from the same cultivar to develop robust models. However, for any given commodity, model development requires the assessment of robustness across different populations of fruit grown under differing conditions (Guthrie et al., 2005). Nicolaï et al. (2007) defined robustness as a model's prediction that should be insensitive to unknown changes. For commercial application, future research on fresh and dried pomegranate arils should try to improve model robustness by adding orchard variability, different stages of fruit maturity, and seasonality as some of the main factors which may affect the model performance.

5 NIRS Measurement and Prediction of Juice Quality

Fresh pomegranate juice has been reported to contain considerable amounts of sugars, organic acids, vitamins, mineral elements, and a diverse array of biochemical compounds (Arendse et al., 2015). Pomegranate juice is obtained by either whole

Table 4 Application of NIR spectroscopy to predict the quality parameters of fresh and dried pomegranate arils

Product	Parameter	Statistical analysis	Wavelength range (nm)	Predictors accuracy	Reference
Fresh arils	TSS (%)	PCA, PLSr	800–2400	$r = 0.96$, RMSEP = 0.0092°Brix	Khodabakhshian et al. (2015)
		PCA, PLSr	1064–1333, 1587–1724	$R^2 = 0.84$, RMSEP = 0.30, RPD = 2.65	Arendse et al. (2017)
Dried arils		PCA, SVM	800–2400	$R^2 = 0.22$, RPD = 0.83, Bias = −0.31	Okere (2020)
Fresh arils	TA (%)	PCA, PLSr	800–2400	$r = 0.92$, RMSEP = 0.19%	Khodabakhshian et al. (2015)
		PCA, PLSr	1064–1333, 1740–1835	$R^2 = 0.81$, RMSEP = 0.1, RPD = 2.34	Arendse et al. (2017)
Dried arils		PCA, SVM	800–2400	$R^2 = 0.85$, RPD = 2.50, Bias = −0.01	Okere (2020)
Fresh arils	pH	PCA, PLSr	800–2400	r = 0.92, RMSEP = 0.089	Khodabakhshian et al. (2015)
		PCA, PLSr	1064–2100	$R^2 = 0.80$, RMSEP = 0.10, RPD = 2.40	Arendse et al. (2017)
Dried arils		PCA, SVM, PLSr	800–2400	$R^2 = 0.86$, RPD = 2.38, Bias = −0.03	Okere (2020)
Fresh arils	TSS:TA ratio	PCA, PLSr	1064–2355	$R^2 = 0.77$, RMSEP = 1.03, RPD = 2.13	Arendse et al. (2017)
Dried arils		PCA, SVM	800–2400	$R^2 = 0.77$, RPD = 1.68, Bias = −0.35	Okere (2020)
	BrimA	PCA, PLSr	1064–1333, 1640–1835	$R^2 = 0.78$, RMSEP = 0.43, RPD = 2.22	Arendse et al. (2017)
Fresh arils	TPC	PCA, PLSr	1064–1333	$R^2 = 0.82$, RMSEP = 0.11, RPD = 2.47	Arendse et al. (2017)
	TAC	PCA, PLSr	1064–1333, 1640–1835	$R^2 = 0.68$, RMSEP = 0.13, RPD = 1.79	Arendse et al. (2017)
	Vitamin C	PCA, PLSr	1064–1333	$R^2 = 0.62$, RMSEP = 0.09, RPD = 1.63	Arendse et al. (2017)
Fresh arils	Firmness (N)	PCA, PLSr	1064–2355	$R^2 = 0.61$, RMSEP = 6.71, RPD = 1.62	Arendse et al. (2017)

(continued)

Table 4 (continued)

Product	Parameter	Statistical analysis	Wavelength range (nm)	Predictors accuracy	Reference
Dried arils		PCA, PLS	800–2400	$R^2 = 0.30$, RPD = 1.22, Bias = −0.99	Okere (2020)
Fresh arils	Colour (a*)	PCA, PLS	1064–1183, 1640–1732	$R^2 = 0.70$, RMSEP = 1.67, RPD = 1.86	Arendse et al. (2017)
Dried arils		PCA, SVM	800–2400	$R^2 = 0.72$, RPD = 1.71, Bias = −0.56	Okere (2020)

R^2 coefficient of determination, *PLSr* partial least squares regression, *SVM* support vector machine, *PCA* principal component analysis, *RMSEP* root mean square error of prediction, *SNV* vector normalization, *FD* first derivative, *SD* second derivative, *TSS* total soluble solids, *TA* titratable acidity, *TPC* total phenolic content, *TAC* total anthocyanin content

pomegranate fruit or carefully extracting the arils (Mphahlele et al., 2016). As an analytical tool for quality control, NIRS in combination with chemometric software has been extensively applied in the beverage industry. Research applications of NIRS for the evaluation of pomegranate juice quality are limited. Arendse et al. (2018a) evaluated NIRS capability to predict the quality attributes of pomegranate juice within a wavelength range of 12,500–4000 cm^{-1} (800–2400 nm) (Table 5). For model development, PLSr combined with preprocessing methods was applied to predict organoleptic properties of pomegranate juice such as TSS (R^2 = 0.92, RPD = 3.62), TA (R^2 = 0.86, RPD = 2.70), and the ratio of TSS:TA (R^2 = 0.76, RPD = 2.72). These models were developed in the region of 9400–5452 cm^{-1} (1063–1834 nm) and 4600–4250 cm^{-1} (2173–2352 nm) (Fig. 3). The prominent band observed at 6950 cm^{-1} (1438 nm) corresponds to O–H bonds in water molecules and the observed band at 5600 cm^{-1} (1785 nm) has been associated with the stretching of C–H bonds in sugars as reported by Golic et al. (2003).

Other parameters that have been predicted with NIRS include vitamin C concentration (R^2 = 0.70, RMSEP = 0.11, RPD = 1.85) which was within the wavelength region of 9400–4425 cm^{-1} (1063–2259 nm) (Table 5). This wavelength region is in agreement with previous studies to develop models for vitamin C concentration in oranges (Magwaza et al., 2013). The comparison of three different instruments (MPA-NIRS, Alpha-MIRS, and Winescan-MIRS) for predicting the TSS and TA were also investigated by Arendseet al. (2018a) who applied two regression algorithms, Bland and Altman and Passing-Bablok. The developed models for all three instruments showed reliability and repeatability in predicting TSS and TA content in pomegranate juice. The two studied algorithms showed no statistical differences, suggesting that all three instruments provided similar prediction statistics even though there was variation within the prediction statistics (RMSEP and RPD) between the three instruments.

Table 5 Applications of NIR spectroscopy to predict the quality of pomegranate juice

Parameter	Statistical analysis	Preprocess	Wavelength range (nm)	Predictors accuracy	Reference
TSS (%)	PCA, PLSr	MSC	9400–5450, 4600–4250	R^2 = 0.92, RMSEP = 0.31, RPD = 3.62	Arendse et al. (2018a)
TA	PCA, PLSr	FD + MSC	7500–5450	R^2 = 0.86, RMSEP = 0.11, RPD = 2.70	Arendse et al. (2018a)
TSS:TA ratio	PCA, PLSr	FD + SNV	9400–7500, 6098–5450	R^2 = 0.76, RMSEP = 0.74, RPD = 2.72	Arendse et al. (2018a)
pH	PCA, PLSr	MSC	9400–7500, 6098–5450	R^2 = 0.67, RMSEP = 0.17, RPD = 1.74	Arendse et al. (2018a)
Vitamin C (g/L)	PCA, PLSr	SD	9400–5450, 4600–4250	R^2 = 0.70, RMSEP = 0.11, RPD = 1.85	Arendse et al. (2018a)
TPC (g/L)	PCA, PLSr	SNV	8375–7855, 6309–5800	R^2 = 0.59, RMSEP = 0.18, RPD = 1.57	Arendse et al. (2018a)
TAC (g/L)	PCA, PLSr	SNV	9400–5777	R^2 = 0.66, RMSEP = 0.19, RPD = 1.73	Arendse et al. (2018a)
Colour (a*)	PCA, PLSr	FD + SNV	9400–4425	R^2 = 0.81, RMSEP = 3.78, RPD = 2.33	Arendse et al. (2018a)

R^2 coefficient of determination, *PLSr* partial least squares regression, *PCA* principal component analysis, *RMSEP* root mean square error of prediction, *SNV* vector normalization, *FD* first derivative, *SD* second derivative, *TSS* total soluble solids, *TA* titratable acidity, *TPC* total phenolic content, *TAC* total anthocyanin content

6 NIRS Measurement and Prediction of Physiological Rind Disorders and Diseases

The visual appearance of fruit and vegetables is the primary parameter to be subjectively and objectively used to evaluate external quality, including the presence of skin defects or pests. Currently, commercial grading systems for pomegranate fruit grade sound fruit with fruit containing slight defects thereby reducing the entire batch quality. Alternatively, sound fruit is graded and removed with fruit that has either slight defects or seriously damaged fruit due to insect damage or disease infestation causing huge economic losses. Pomegranate fruit is also susceptible to various postharvest physiological rind disorders and pests (Arendse et al., 2016a; Munhuweyi et al., 2016).

A review of the literature showed that limited NIRS studies have been performed to evaluate physiological rind disorders, diseases, and pests in pomegranate fruit. One pest, known as the false codling moth (*Thaumatotibia leucotreta*), causes

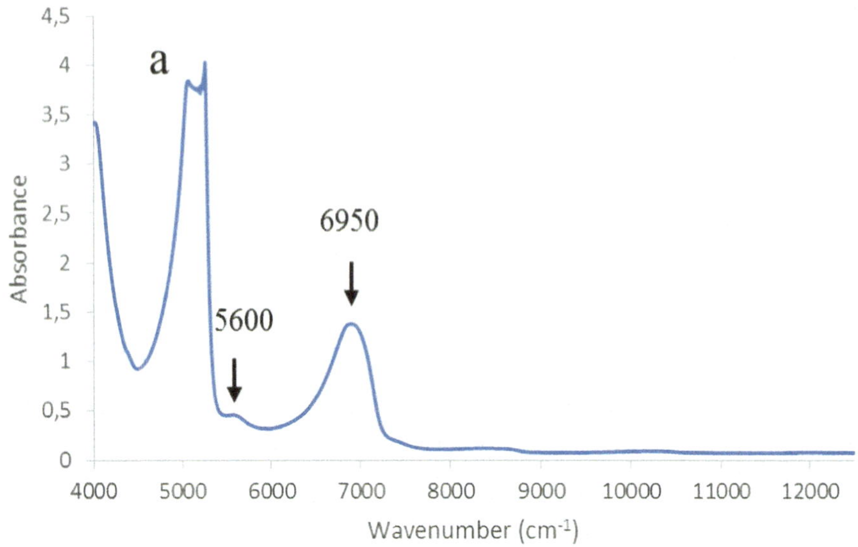

Fig. 3 Absorbance spectra of pomegranate juice were obtained in the near-infrared region of 12,000–4000 cm^{-1} (800–2400 nm) (Arendse et al., 2018a)

pomegranate infestation that may develop within the fruit without showing any external symptoms. Khodabakhshian et al. (2016) investigated the application of VIS-NIRS (400–1100 nm) to detect and discriminate between affected and healthy pomegranate fruit "Ashraf" at different fruit maturity stages. The authors applied partial least squares discriminant analysis (PLS-DA) and achieved a classification accuracy of 88%, 90%, and 86% for detecting carob moth infestation, using different harvesting times. This study has shown that VIS-NIRS combined with classification techniques can be used to predict infestation in pomegranate fruit.

Pomegranate fruit is highly susceptible to the development of rind disorders such as husk scald. Scalding is a rind disorder that manifests during prolonged cold storage, which usually corresponds with the commercial shipping period (Arendse et al., 2015; Magwaza et al., 2014a, 2014b). Arendse et al. (2018b) evaluated the development of husk scald during cold storage and then assessed the application of FT-NIRS using diffuse reflectance to predict the development of husk scald. The authors reported that storage duration significantly influenced the development of husk scald, with peel browning occurring as a result of the breakdown of tannin and phenolic compounds due to the enzymatic activity of polyphenol oxidase; (PPO) and peroxidase (POD). Orthogonal partial least squares discriminant analysis (OPLS-DA) was applied to qualitatively discriminate between healthy and scalded fruit. High classification accuracy was achieved between healthy (100%) and severe scald fruit (93%). The authors applied variable importance of projection (VIP) and observed that specific wavelength ranges were responsible for the separation of the different classes, which were 1350, 1450, 1830–1950, and 2150–2250 nm. These wavelength bands have been associated with compounds such as phenolics and anthocyanins. For pomegranates, various other nondestructive techniques have been successfully used to evaluate oily spot disease (Chavan et al., 2016), bacterial blight and wilt complex (Sannakki & Rajpurohit, 2015), and blackheart disease (Arendse et al., 2016a).

7 Prospects for NIRS in Pomegranate Quality

NIRS is the most advanced nondestructive technology that has been applied in the agricultural industry for quality assessment. In this chapter, the application of NIRS with chemometrics has been highlighted as a novel nondestructive tool to predict external and internal quality attributes of whole pomegranate fruit, and its coproducts (juice and arils). Considering that pomegranate fruit structures are heterogeneous, the biochemical composition may vary between the center and different internal parts of the fruit flesh, and thus the future application of NIRS instruments should consider the inherent variability of heterogeneous fruit. Therefore, to improve the models predictability future research should focus on the reliability of reference measurements with more emphasis should go into improving the robustness of multivariate calibration models by including seasonality, cultivar differences, and geographical locations.

NIR technology has been available commercially over the last few years; however, the limited adoption of NIR technology in the industry may be due to high equipment cost, setup, and technical knowledge requirements. These problems are slowly being eliminated through advancements and improvements in inexpensive and portable instrumentation. For pomegranates, future research prospects should also focus on evaluating other processed coproducts such as juice powder and seed oil during valorization to reduce losses and waste. Limited information on NIR technology exists for the assessment of authenticity and adulteration of processed pomegranate products. Given the considerable consumer interest and market potential of pomegranates, research needs to shift towards noninvasive evaluation and detection of fraud or adulteration within processed pomegranate products. This will provide novel opportunities to develop NIRS-based methods that could provide high potential for online or inline commercial applications.

Acknowledgment This work is based on the research supported by the National Research Foundation of South Africa (Grant Numbers: 64813) and UNESCO ICB, Nsukka. The opinions, findings, and conclusions or recommendations expressed are those of the author(s) alone, and the NRF accepts no liability whatsoever in this regard.

References

Aindongo, W. V., Caleb, O. J., Mahajan, P. V., Manley, M., & Opara, U. L. (2014). Effects of storage conditions on transpiration rate of pomegranate aril-sacs and arils. *South African Journal of Plant and Soil, 31*, 7–11.

Al-Maiman, S. A., & Ahmad, D. (2002). Changes in physical and chemical properties during pomegranate (*Punica granatum* L.) fruit maturation. *Food Chemistry, 76*, 437–441.

Al-Said, F. A., Opara, U. L., & Al-Yahyai, R. A. (2009). Physico-chemical and textural quality attributes of pomegranate cultivars (*Punica granatum* L.) grown in the Sultanate of Oman. *Journal of Food Engineering, 90*, 129–134.

Arendse, E., Fawole, O. A., & Opara, U. L. (2015). Effects of postharvest handling and storage on physiological attributes and quality of pomegranate fruit (*Punica granatum* L.): A review. *International Journal of Postharvest Technology and Innovation, 5*, 13–31.

Arendse, E., Fawole, O. A., Magwaza, L. S., & Opara, U. L. (2016a). Estimation of the density of pomegranate fruit and their fractions using X-ray computed tomography calibrated with polymeric materials. *Biosystems Engineering, 148*, 148–156.

Arendse, E., Fawole, O. A., Magwaza, L. S., & Opara, U. L. (2016b). Non-destructive characterization and volume estimation of pomegranate fruit external and internal morphological fractions using X-ray computed tomography. *Journal of Food Engineering, 186*, 42–49.

Arendse, E., Fawole, O. A., Magwaza, L. S., Nieuwoudt, H. H., & Opara, U. L. (2017). Development of calibration models for the evaluation of pomegranate aril quality by Fourier-transform near infrared spectroscopy combined with chemometrics. *Biosystems Engineering, 159*, 22–32.

Arendse, E., Fawole, O. A., Magwaza, L. S., & Opara, U. L. (2018a). Comparing the analytical performance of near and mid infrared spectrometers for evaluating pomegranate juice quality. *LWT - Food Science and Technology, 91*, 180–190.

Arendse, E., Fawole, O. A., Magwaza, L. S., & Opara, U. L. (2018b). Evaluation of biochemical markers associated with the development of husk scald and the use of diffuse reflectance NIR spectroscopy to predict husk scald in pomegranate fruit. *Scientia Horticulturae, 232*, 240–249.

Arendse, E., Fawole, O. A., Magwaza, L. S., & Opara, U. L. (2018c). Non-destructive prediction of internal and external quality attributes of fruit with thick rind: A review. *Journal of Food Engineering, 217*, 11–23.

Arendse, E., Fawole, O. A., Magwaza, L. S., Nieuwoudt, H., & Opara, U. L. (2018d). Fourier transform near infrared diffuse reflectance spectroscopy and two spectral acquisition modes for evaluation of external and internal quality of intact pomegranate fruit. *Postharvest Biology and Technology, 138*, 91–98.

Arendse, E., Nieuwoudt, H. H., Nturambirwe, J. F. I., Fawole, O. A., Magwaza, L. S., & Opara, U. L. (2020). Recent advancements on vibrational spectroscopic techniques for the detection of authenticity and adulteration in horticultural products with a specific focus on oils, juices, and powders. *Food and Bioprocess Technology, 14*, 1–22.

Ayhan, Z., & Eştürk, O. (2009). Overall quality and shelf life of minimally processed and modified atmosphere packaged "ready-to-eat" pomegranate arils. *Journal of Food Science, 74*, 399–405.

Ben-Arie, R., Segal, N., & Guelfat-Reich, S. (1984). The maturation and ripening of the 'Wonderful' pomegranate. *Journal of the American Society for Horticultural Science, 109*, 898–902.

Brázio, A., Cavaco, A. M., Antunes, M. D., & Guerra, R. (2016). Evaluation of the internal quality of pomegranates using non-invasive visible/near infrared transmittance spectroscopy. *Actas Portuguesas de Horticulturan, 28*, 261–268.

Caleb, O. J., Opara, U. L., Mahajan, P. W., Manley, M., Mokwena, L., & Tredoux, A. G. J. (2013). Effect of modified atmosphere packaging and storage temperature on volatile composition and postharvest life of minimally processed pomegranate arils (cvs. 'Acco' and 'Herskowitz'). *Postharvest Biology and Technology, 79*, 54–61.

Caleb, O. J., Opara, U. L., & Witthuhn, C. (2012). Modified atmosphere packaging of pomegranate fruit and arils: A review. *Food and Bioprocess Technology, 5*, 15–30.

Calín-Sánchez, A., Martínez, J. J., Vázquez-Araújo, L., Burló, F., Melgarejo, P., & Carbonell-Barrachina, A. A. (2011). Volatile composition and sensory quality of Spanish pomegranates (*Punica granatum* L.). *Journal of the Science of Food and Agriculture, 91*, 586–592.

Cayuela, J. A. (2008). VIS–NIR soluble solids prediction in intact oranges (*Citrus sinensis* L.) cv. Valencia late by reflectance. *Postharvest Biology and Technology, 47*, 75–80.

Chandrasekaran, I., Panigrahi, S. S., Ravikanth, L., & Singh, C. B. (2019). Potential of near-infrared (NIR) spectroscopy and hyperspectral imaging for quality and safety assessment of fruits: An overview. *Food Analytical Methods, 12*, 2438–2458.

Chavan, N. P., Pandey, R., Nawani, N., Nanda, R. K., Tandon, G. D., & Khetmalas, M. B. (2016). Biocontrol potential of actinomycetes against *Xanthomonas axonopodis pv. Punicae*, a causative agent for oily spot disease of pomegranate. *Biocontrol Science and Technology, 26*, 351–372.

Chen, L., & Opara, U. L. (2013a). Approaches to analysis and modeling texture in fresh and processed foods—A review. *Journal of Food Engineering, 119*, 497–507.

Chen, L., & Opara, U. L. (2013b). Texture measurement approaches in fresh and processed foods—A review. *Food Research International, 51*, 823–835.

Clément, A., Dorais, M., & Vernon, M. (2008). Non-destructive measurement of fresh tomato lycopene content and other physicochemical characteristics using visible-NIR spectroscopy. *Journal of Agricultural and Food Chemistry, 56*, 9813–9818.

Cozzolino, D., Cynkar, W. U., Shah, N., & Smith, P. (2011). Multivariate data analysis applied to spectroscopy: Potential application to juice and fruit quality. *Food Research International, 44*, 1888–1896.

Cristosto, C. H., Mitcham, E. J., & Kader, A. A. (2000). *Pomegranate: Recommendations for maintaining postharvest quality. Produce facts*. Postharvest Research and Information Centre, University of California.

Danezis, G. P., Tsagkaris, A. S., Camin, F., Brusic, V., & Georgiou, C. A. (2016). Food authentication: Techniques, trends & emerging approaches. *Trends in Analytical Chemistry, 85*, 123–132.

De Oliveira, G. A., Bureau, S., Renard, C. M. G. C., Pereira-Netto, A. B., & de Castilhos, F. (2014). Comparison of NIRS approach for prediction of internal quality traits in three fruit species. *Food Chemistry, 143*, 223–230.

Fawole, O. A., & Opara, U. L. (2013a). Changes in physical properties, chemical and elemental composition, and antioxidant capacity of pomegranate (cv. 'Ruby') fruit at five maturity stages. *Scientia Horticulturae, 150*, 37–46.

Fawole, O. A., & Opara, U. L. (2013b). Developmental changes in maturity indices of pomegranate fruit: A descriptive review. *Scientia Horticulturae, 159*, 152–161.

Fawole, O. A., & Opara, U. L. (2013c). Fruit growth dynamics, respiration rate and physico textural properties during pomegranate development and ripening. *Scientia Horticulturae, 157*, 90–98.

Fawole, O. A., & Opara, U. L. (2013d). Harvest discrimination of pomegranate fruit: Postharvest quality changes and relationships between instrumental and sensory attributes during shelf life. *Journal of Food Science, 78*, 1264–1272.

Fawole, O. A., & Opara, U. L. (2013e). Seasonal variation in chemical composition, aroma volatiles and antioxidant capacity of pomegranate during fruit development. *African Journal of Biotechnology, 12*, 4006–4019.

Fawole, O. A., & Opara, U. L. (2014). Physico-mechanical, phyto-chemical and free radical scavenging properties and volatile compounds in eight pomegranate cultivars and classification by principal component and cluster analyses. *British Food Journal, 116*, 544–567.

Fawole, O. A., Opara, U. L., & Theron, K. I. (2012). Chemical and phytochemical properties and antioxidant activities of three pomegranate cultivars grown in South Africa. *Food and Bioprocess Technology, 5*, 2934–2940.

Fuhrman, B., & Aviram, M. (2006). Protection against cardiovascular diseases. In N. P. Seeram et al. (Eds.), *Pomegranates: Ancient roots to modern medicine* (pp. 63–80). CRC Press.

Golic, M., Walsh, K. B., & Lawson, P. (2003). Short-wavelength near-infrared spectra of sucrose, glucose, and fructose with respect to sugar concentration and temperature. *Applied Spectroscopy, 57*, 139–145.

Gómez, A. H., He, Y., & Pereira, A. G. (2006). Non-destructive measurement of acidity, soluble solids and firmness of Satsuma mandarin using VIS–NIR spectroscopy techniques. *Journal of Food Engineering, 77*, 313–319.

Guthrie, J. A., Reid, D. J., & Walsh, K. B. (2005). Assessment of internal quality attributes of mandarin fruit. 2. NIR calibration model robustness. *Australian Journal of Agricultural Research, 56*, 417–426.

Holland, D., Hatib, K., & Bar-Ya'akov, I. (2009). Pomegranate: Botany, horticulture, breeding. *Horticultural Reviews, 35*, 127–191.

Kader, A. A. (2000). Quality of horticultural products. *Acta Horticulturae, 517*, 17–18.

Kader, A. A. (2006). Postharvest biology and technology of pomegranates. In N. P. Seeram et al. (Eds.), *Pomegranates: Ancient roots to modern medicine* (pp. 211–221). CRC Press.

Khodabakhshian, R., Emadi, B., Khojastehpour, M., & Golzarian, M. R. (2016). Carob moth, Ectomyelois ceratoniae, detection in pomegranate using visible/near infrared spectroscopy. *Computers and Electronics in Agriculture, 129*, 9–14.

Khodabakhshian, R., Emadi, B., Khojastehpour, M., & Golzarian, M. R. (2019). A comparative study of reflectance and transmittance modes of VIS/NIR spectroscopy used in determining internal quality attributes in pomegranate fruits. *Journal of Food Measurement and Characterization, 13*, 3130–3139.

Khodabakhshian, R., Emadi, B., Khojastehpour, M., Golzarian, M. R., & Sazgarnia, A. (2015). Quick quality evaluation of pomegranate arils using NIR spectroscopy. *Iranian Research Organisation for Food Science and Technology, 2*, 103–114. (Arabic).

Khodabakhshian, R., Emadi, B., Khojastehpour, M., Golzarian, M. R., & Sazgarnia, A. (2017). Non-destructive evaluation of maturity and quality parameters of pomegranate fruit by visible/near infrared spectroscopy. *International Journal of Food Properties, 20*, 41–52.

Lansky, E. P., & Newman, R. A. (2007). Punica granatum (pomegranate) and its potential for prevention and treatment of inflammation and cancer. *Journal of Ethnopharmacology, 109*, 177–206.

Lestander, T. A., & Geladi, P. (2005). NIR spectral information used to predict water content of pine seeds from multivariate calibration. *Canadian Journal of Forest Research, 35*, 1139–1148.

Levin, G. M. (1994). Pomegranate (*Punica granatum*) plant genetic resources in Turkmenistan. *Plant Genetic Resources Newsletter, 97*, 31–36.

Liu, H., Xiang, B., & Qu, L. (2006). Structure analysis of ascorbic acid using near infrared spectroscopy and generalized two-dimensional correlation spectroscopy. *Journal of Molecular Structure, 794*, 12–17.

Louw, E. D., & Theron, K. I. (2010). Robust prediction models for quality parameters in Japanese plums (*Prunus salicina* L.) using NIR spectroscopy. *Postharvest Biology and Technology, 58*, 176–184.

Magwaza, L. S., Landahl, S., Cronje, P. J. R., Nieuwoudt, H. H., Mouazen, A. M., Nicolaï, B. M., Terry, L. A., & Opara, U. L. (2014a). The use of VIS/NIRS and chemometric analysis to predict fruit defects and postharvest behaviour of 'Nules Clementine' mandarin fruit. *Food Chemistry, 163*, 267–274.

Magwaza, L. S., & Opara, U. L. (2015). Analytical methods for determination of sugars and sweetness of horticultural products—A review. *Scientia Horticulturae, 184*, 179–192.

Magwaza, L. S., Opara, U. L., Cronje, P. J. R., Landahl, S., Nieuwoudt, H. H., Mouazen, A. M., Nicolaï, B. M., & Terry, L. A. (2014b). Assessment of rind quality of 'Nules Clementine' mandarin fruit during postharvest storage: 2. Robust VIS/NIRS PLS models for prediction of physico-chemical attributes. *Scientia Horticulturae, 165*, 421–432.

Magwaza, L. S., Opara, U. L., Terry, L. A., Landahl, S., Cronje, P. J. R., Nieuwoudt, H. H., Hanssens, A., Saeys, W., & Nicolaï, B. M. (2013). Evaluation of Fourier transform-NIR spectroscopy for integrated external and internal quality assessment of 'Valencia' oranges. *Journal of Food Composition and Analysis, 31*, 144–154.

McGlone, V. A., & Kawano, S. (1998). Firmness, dry-matter and soluble solids assessment of postharvest kiwifruit by NIR spectroscopy. *Postharvest Biology and Technology, 13*, 131–141.

Miyamoto, K., Kawauchi, M., & Fukuda, T. (1998). Classification of high acid fruits by PLS using the near infrared transmittance spectra of intact Satsuma mandarins. *Journal of Near Infrared Spectroscopy, 6*, 267–271.

Mphahlele, R. R., Fawole, O. A., Mokwena, L. M., & Opara, U. L. (2016). Effect of extraction method on chemical, volatile composition and antioxidant properties of pomegranate juice. *South African Journal of Botany, 103*, 135–144.

Munhuweyi, K., Lennox, C. L., Meitz-Hopkins, J. C., Caleb, O. J., & Opara, U. L. (2016). Major diseases of pomegranate (*Punica granatum* L.), their causes and management—A review. *Scientia Horticulturae, 211*, 126–139.

Musingarabwi, D. M., Nieuwoudt, H. H., Young, P. R., Eyéghè-Bickong, H. A., & Vivier, M. A. (2016). A rapid qualitative and quantitative evaluation of grape berries at various stages of development using Fourier-transform infrared spectroscopy and multivariate data analysis. *Food Chemistry, 190*, 253–262.

Nanda, S., Rao, D. V. S., & Krishnamurthy, S. (2001). Effects of shrink film wrapping and storage temperature on the shelf life and quality of pomegranate fruits cv. Ganesh. *Postharvest Biology and Technology, 22*, 61–69.

Nicolaï, B. M., Beullens, K., Bobelyn, E., Peirs, A., Saeys, W., Theron, I. K., & Lammertyn, J. (2007). Non-destructive measurement of fruit and vegetable quality by means of NIR spectroscopy: A review. *Postharvest Biology and Technology, 46*, 99–118.

Nicolaï, B. M., Bulens, I., De Baerdemaker, J., De Ketelaere, B., Hertog, M. L. A. T. M., Verboven, P., & Hertog, M. L. A. T. (2014). Non-destructive evaluation: Detection of external and internal attributes frequently associated with quality and damage. In D. Florkowiski (Ed.), *Postharvest handling: A systems approach* (pp. 421–442). Academic Press, Elsevier.

Okere, E. E. (2020). *Non-invasive measurement of quality attributes of processed pomegranate products* [Master's thesis]. Stellenbosch University, South Africa.

Opara, L. U., Al-Ani, M. R., & Al-Shuaibi, Y. S. (2009). Physico-chemical properties, vitamin C content, and antimicrobial properties of pomegranate fruit (*Punica granatum* L.). *Food and Bioprocess Technology, 2*, 315–321.

Pathare, P. B., Opara, U. L., & Al-Said, F. A. J. (2013). Colour measurement and analysis in fresh and processed foods: A review. *Food and Bioprocess Technology, 6*, 36–60.

Pissard, A., Pierna, J. A. F., Baeten, V., Sinnaeve, G., Lognay, G., Mouteau, A., & Dupont, P. (2013). Non-destructive measurement of vitamin C, total polyphenol and sugar content in apples using near-infrared spectroscopy. *Journal of the Science of Food and Agriculture, 93*, 238–344.

Ruiz, D., Reich, M., Bureau, S., Rebard, C. M. G. C., & Audergon, J. M. (2008). Application of reflectance colorimeter measurements and infrared spectroscopy methods to rapid and nondestructive evaluation of carotenoids content in apricot (*Prunus armeniaca* L.). *Journal of Agricultural and Food Chemistry, 56*, 4916–4922.

Sannakki, S. S., & Rajpurohit, V. S. (2015). Classification of pomegranate diseases based on back propagation neural network. *International Journal of Advanced Research in Computer, 2*, 309–316.

Seeram, N. P., Aviram, M., Zhang, Y., Henning, S. M., Feng, L., Dreher, M., & Heber, D. (2008). Comparison of antioxidant potency of commonly consumed polyphenol-rich beverages in the United States. *Journal of Agricultural and Food Chemistry, 56*(4), 1415–1422.

Shulman, Y., Fainberstein, L., & Lavee, S. (1984). Pomegranate fruit development and maturation. *The Journal of Horticultural Science and Biotechnology, 59*, 265–274.

Shwartz, E., Glazer, I., Bar-Ya'akov, I., Matityahu, I., Bar-Ilan, I., Holland, D., & Amir, R. (2009). Changes in chemical constituents during the maturation and ripening of two commercially important pomegranate accessions. *Food Chemistry, 115*, 965–973.

Vázquez-Araújo, L., Chamber, E., Adhikari, K., & Carbonell-Barrachina, A. A. (2011). Physico-chemical and sensory properties of pomegranate juices with pomegranate albedo and capillary membranes homogenate. *LWT - Journal of Food Science and Technology, 44*, 2119–2125.

Viuda-Martos, M., Fernández-López, J., & Pérez-Álvarez, J. A. (2010). Pomegranate and its many functional components as related to human health: A review. *Comprehensive Reviews in Food Science and Food Safety, 9*, 635–665.

Walsh, K. B., Blasco, J., Zude-Sasse, M., & Sun, X. (2020a). Visible-NIR 'point' spectroscopy in postharvest fruit and vegetable assessment: The science behind three decades of commercial use. *Postharvest Biology and Technology, 168*, 111246. https://doi.org/10.1016/j.postharvbio.2020.111246

Walsh, K. B., Lu, R., & Nicolaï, B. (2021). Special issue: Recent advances in the use of visible and vibrational spectroscopy/imaging for measurement of postharvest quality. *Postharvest Biology and Technology, 171*, 111347. https://doi.org/10.1016/j.postharvbio.2020.111347

Walsh, K. B., McGlone, V. A., & Han, D. H. (2020b). The uses of near infra-red spectroscopy in postharvest decision support: A review. *Postharvest Biology and Technology, 163*, 111139. https://doi.org/10.1016/j.postharvbio.2020.111139

Williams, P., Antoniszyn, J., & Manley, M. (2019). *Near infrared technology: Getting the best out of light*. African Sun Media, Sun Press.

Williams, P. C., & Norris, K. H. (1987). Qualitative applications of near-infrared reflectance spectroscopy. In P. C. Williams & K. H. Norris (Eds.), *Near-infrared technology in the agricultural and food industries* (pp. 241–246). American Association of Cereal Chemistry.

Zude, M., Pflanz, M., Kaprielian, C., & Aivazian, B. (2008). NIRS as a tool for precision horticulture in citrus industry. *Biosystems Engineering, 99*, 455–459.

X-Ray Computed Tomography (CT) for the Internal Quality Evaluation of Fresh Produce

Mohsen Azadbakht

Abstract It is mentioned in the introduction of this chapter that Computed Tomography (CT) scan is a method to take two-dimensional X-ray images of different slices of a body in order to produce a visual representation of its internal partsor organs. The core idea of this method is that the data collected from multiple X-ray images of a body can be used to reconstruct its internal structure in a computer. Then the main components of the CT device are introduced and explained, which include X-ray source, detectors, data acquisition, data processing and CT numbers. Since then, different generations of CT scans have been introduced from the beginning to the present, including rotate/translate, pencil beam; rotate/translate, narrow fan beam; rotate/rotate, wide fan beam; rotate/stationary; stationary/stationary, electron beam CT; helical or spiral; multiple detector array. Finally, three applications of CT scan on pear fruit have been done by the author and his colleagues that have evaluated the pear bruising due to quasi-static and impact load with CT scan and have investigated its relationship with some physical characteristics of pear.

Keywords Computed tomography · Nondestructive · Impact load · Quasi-Static · Pear fruit

1 Introduction

Non-destructive quality evaluation of agricultural products has long been a popular subject of research and debate among researchers across the world. This discussion revolves around the question of how to assess the quality of a batch of harvested fruits or crops safely and efficiently without damaging the product. Destructive methods are time-consuming and expensive, and many instances require chemicals. Indeed, destructive quality evaluation tests are being gradually replaced with non-destructive alternatives, such as CT scans, ultrasound tests, optical tests, and

M. Azadbakht (✉)
Department of Bio-System Mechanical Engineering, Gorgan University of Agricultural Sciences and Natural Resources, Gorgan, Iran
e-mail: azadbakht@gau.ac.ir

P. B. Pathare, M. S. Rahman (eds.), *Nondestructive Quality Assessment Techniques for Fresh Fruits and Vegetables*, https://doi.org/10.1007/978-981-19-5422-1_11

MRI (Abbott, 1999). The CT scan method not only has the ability to determine the extent of internal damage but can also estimate the volume of various internal features. In addition to the two-dimensional radiography used in medicine and the linear scan radiography used in calibration machines, X-ray computed tomography is another powerful method in terms of the cross-sectional area of an object using a set of X-ray source and moving detection to collect data from a thin section of the sample.

2 Working Principles of CT Scanners

Computed Tomography (CT) scan is a method to take two-dimensional X-ray images of different slices of a body in order to produce a visual representation of its internal parts or organs. The core idea of this method is to collect data from multiple X-ray images of a body and then these can be used to reconstruct its internal structure on a computer (Jha, 2010).

Once the scanner is properly configured, it produces an X-ray beam (pulse) that passes through the body, losing some of its energy in the process. The part of the beam that reaches the other side is captured by a detector, which turns it into binary data and sends it to the computer for storage and processing. Once this measurement is over, the X-ray lamp rotates at a very small angle and sends another X-ray beam, which is again captured on the other side and stored in the computer memory. This process is repeated hundreds or even thousands of times (depending on the type of scanner), generating a database of X-ray beams produced and captured at different angles in the computer memory. Using this database, the computer estimates the amount of radiation absorbed by a certain volume of tissue, which is called a voxel and is equivalent to a few cubic millimeters of the body. Each CT cross-section is subdivided into an array of these small voxels. Each voxel is assigned a number depending on the amount of radiation it has absorbed. These numbers represent radiodensity on the grayscale (i.e., from white to black). The representation of each voxel on the monitor is called a pixel (i.e., pixels can be thought of as two-dimensional representations of three-dimensional voxels). An image with a higher number of pixels is said to have a higher resolution, this means that it has higher quality and offers more distinct details. The numbers assigned to tissues based on their absorbed radiation are called CT numbers or Hounsfield numbers. Since these numbers are on the grayscale, the tissues with lower numbers appear black and those with higher numbers (like bone) appear white on the CT image (Saunders & Ohlerth, 2011).

3 Main Components of CT Scanners

The main components of a CT scanner are X-ray tubes, X-ray detectors, and a data acquisition system. These components are placed inside a chamber called a gantry. The gantry can rotate 360° on a circular path, causing the X-ray source and detectors to rotate as well. As shown in Fig. 1, the X-ray source and the detectors are positioned on opposite sides facing each other and can be rotated 360°.

The X-ray detectors can measure the amount of radiation passing through the body at any point during this rotation, which allows the body to be scanned from all angles (Jha, 2010; Saunders & Ohlerth, 2011).

3.1 X-Ray Source

In modern CT scanners, X-ray is produced by the X-ray tube, which operates by accelerating electrons in a vacuum and making them collide with an obstacle to convert them into X-rays. When electrons hit the surface of the anode, about 99% of their energy is turned into heat and the remaining 1% is converted to X-rays. The produced rays spread out in the anode and some of them exit the tube through an outlet.

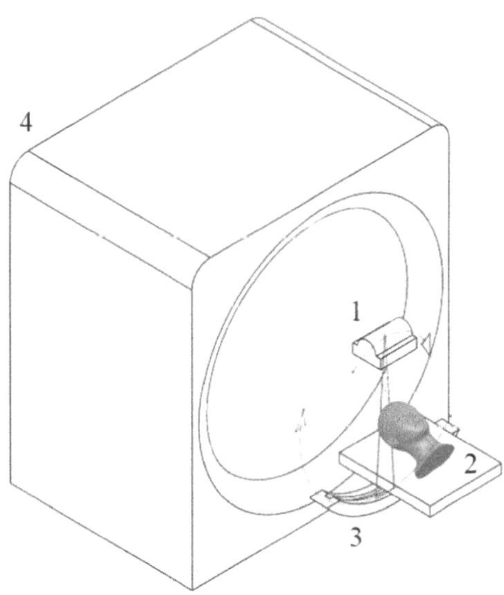

Fig. 1 Position of the X-ray tube and detectors in a CT scanner. (1) X-ray tube; (2) patient; (3) detectors; (4) gantry

3.2 Detectors

The X-rays produced by the source are directed to the body. A number of photons undergo physical phenomena, such as photoelectric absorption and Rayleigh scattering as they pass through the tissues. The photons are managed to reach the other side of the body and enter into the detectors, which measure the amount of X-rays absorbed and convert them into electrical signals, and then send them to the data acquisition system.

3.3 Data Acquisition

Data acquisition refers to the process of collecting and storing the measurements needed to produce an image. During the scan, the X-ray tube and detectors rotate around the body to collect X-ray absorption data from different angles. What is measured during this process is the ratio of radiation passing through the body to the produced radiation, or the radiation attenuation coefficient. These measurements are sent to a computer, where they are stored as raw data for the processing stage.

3.4 Data Processing and CT Numbers

The first stage of CT data processing is to convert the data into numbers that can be easily understood and processed by a computer. After scanning and data acquisition, the output is a series of linear attenuation coefficients for different voxels with extremely small (decimal) differences from each other. Since these numbers are not usable, they must be converted such that an image can be created on the grayscale. This conversion is done by the use of a standardized scale called the CT scale, which produces what is called CT numbers. To obtain the CT number, the computer calculates the linear attenuation coefficient of each pixel μ_p relative to the linear attenuation coefficient of water μ_w using Eq. 1.

$$\text{CT number} = \frac{K\left(\mu_p - \mu_w\right)}{\mu_w} \quad (1)$$

The coefficient K is called the contrast factor. In today's scanners, this value is considered to be 1000. CT numbers are expressed in terms of the Hounsfield Unit (HU). In the CT scale, the CT number of water is zero because its linear attenuation coefficient serves as the reference. The CT numbers of air and bone are -1000 and $+1000$, respectively. The computer takes the outputs of the detectors and converts them into CT numbers using the above formula, and then uses them to create an image on the grayscale, where tissues with higher CT numbers are whiter and those with lower CT numbers are blacker.

4 Generations of CT Scanners

Advanced imaging techniques are being used to diagnose in the field of medicine. However, CT scan has recently been used in other areas including the non-destructive quality evaluation of agricultural products. Since the development of the first CT scanners in the early 1970s, these machines have undergone many changes. As a result, CT scanners can be divided into seven generations, each with unique features in terms of the placement of the X-ray tube and detectors and how they move relative to each other (Bushberg, 2002).

4.1 First Generation: Rotate/Translate, Pencil Beam

In the early 1970s, Godfrey Hounsfield developed the first CT scanner with the help of EMI (Electric and Musical Industries). The first CT scanners, which were developed exclusively for head scans, were working with a rotation/translation system and a single X-ray beam called the pencil beam, which had parallel beam geometry. To produce such a thin beam, a pinhole collimator was placed in these scanners to ensure only one X-ray beam is emitted to the patient. These scanners had only two detectors installed on the opposite side of the X-ray source, which meant they could only measure the rays passing through the body for only two slices. Thus, to get a thorough image of an organ, tube and detectors had to be moved linearly to measure more slices and then rotated around the organ to take the image from different angles (Fig. 2).

Despite their good performance, these scanners had a major problem in the translation and rotation of tubes and detectors, since it could take a very long time to make the measurements and process them into an image (Bharath, 2008; Bushberg, 2002; RSNA, 2013).

4.2 Second Generation: Rotate/Translate, Narrow Fan Beam

The first major changes in CT scanners were made to reduce the time it takes to get a CT scan of the head. For this purpose, the scanner design was modified to use a narrow fan x-ray beam with an angle of about 10°. As a result, these scanners required a linear array of 30 detectors instead of only two detectors used in the previous generation. These modifications significantly reduced the time needed to perform a CT scan. Although their fan angle was small and they still required the

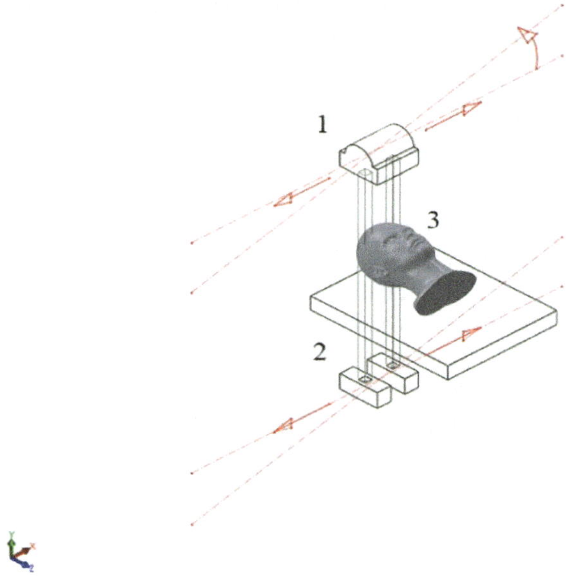

Fig. 2 Schematic diagram of the design of first-generation CT scanners. (1) X-ray tube; (2) detectors; (3) patient

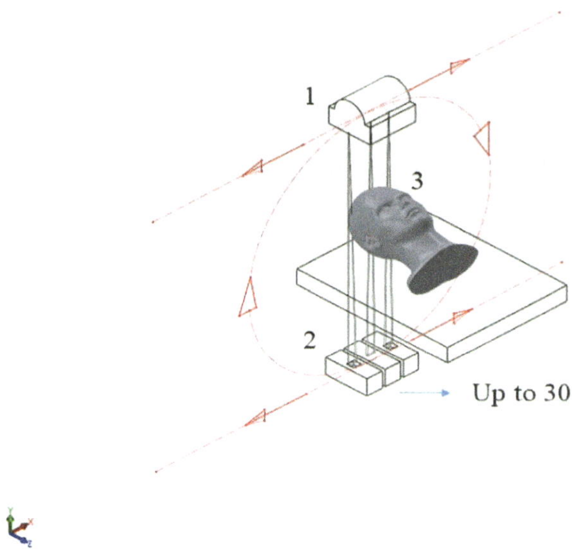

Fig. 3 Schematic diagram of the design of second-generation CT scanners. (1) X-ray tube; (2) detectors; (3) patient

translation of tubes and detectors, the second-generation scanners were 15 times faster in scanning than the first generation (Fig. 3).

A major problem of the second-generation CT scanners was the type of beam and higher number of detectors used (compared to the first generation). They were exposed to more scattered radiation, which could reduce the resolution of the obtained images (Bharath, 2008; Bushberg, 2002; RSNA, 2013).

4.3 Third Generation: Rotate/Rotate, Wide Fan Beam

The next development in this field was the elimination of the translational motion of the first and second-generation scanners, which was quite time-consuming so that imaging time can be reduced to less than 20 s. These faster scans were necessary for better imaging of the head and also to be able to scan organs in the abdomen with the patient holding his breath. This modification was done by using wide fan beams so that images of all target slices at each angle can be captured simultaneously thus allowing the tube and detectors to freely rotate without having to go through the translation stage. Since fan beams of third-generation CT scanners were 40°–60° wide, they needed a larger linear array of detectors (as was the case in the second generation). In this generation, the number of detectors varied between 400 and 1000. The greatest advantage of third-generation CT scanners was the significant reduction in scan time (Fig. 4).

However, these scanners also had two major drawbacks: (i) These scanners were very expensive as they needed 400–1000 detectors. (ii) Because of the large number of detectors and poor calibration between them, the images taken by these scanners tend to have a characteristic artifact called ring artifact (Bushberg, 2002; RSNA, 2013).

4.4 Fourth Generation: Rotate/Stationary

The fourth-generation CT scanners were designed specifically to address the ring artifact problem of the previous generation (Fig. 5).

This was done by removing the detectors from the rotating section and placing them in a stationary ring around the patient to resolve their calibration problem. However, this design required using an even higher number of detectors (about

Fig. 4 Schematic diagram of the design of third-generation CT scanners. (1) X-ray tube; (2) detectors; (3) patient

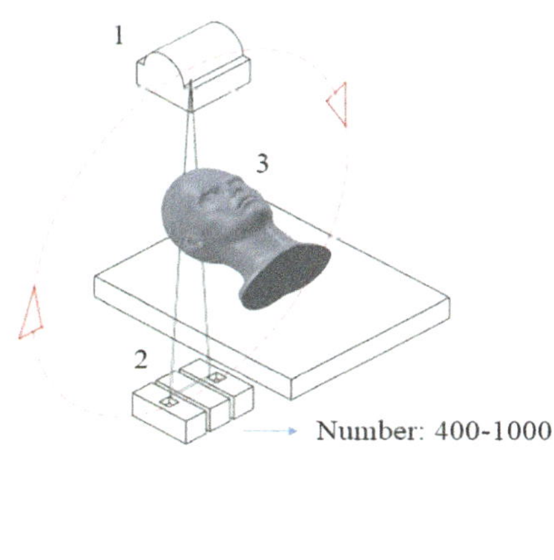

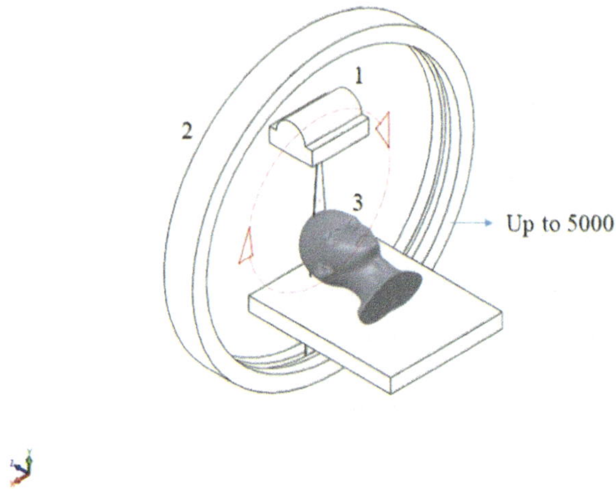

Fig. 5 Schematic diagram of the design of fourth-generation CT scanners. (1) X-ray tube; (2) detectors; (3) patient

5000). Given this modification in their design, fourth-generation CT scanners are called rotate-stationary (Bushberg, 2002; RSNA, 2013; Saunders & Ohlerth, 2011).

4.5 *Fifth Generation: Stationary/Stationary, Electron Beam CT*

Fifth-generation CT scanners, which are called cine-CT or electron beam CT scanners, were specifically designed to produce cross-sectional images of the heart and are still widely used for this purpose. These scanners were developed in response to the need for even faster scans to match the speed of the heartbeat based on the idea that the best way to reduce scan time is to keep all components stationary. Therefore, all imaging components of fifth-generation scanners are stationary. Instead of a rotating tube, these machines have a large X-ray ring in which the patient lies during the scan. These scanners emit an electron beam that hits the tungsten plate enclosing the patient, producing an X-ray that passes through the patient's chest and is received by detectors on the opposite side (Fig. 6).

The strength of these scanners was their high speed in taking heart scans. But since they were designed specifically for heart scans, they had a small target market, which made them very expensive and prevented them from becoming popular (RSNA, 2013).

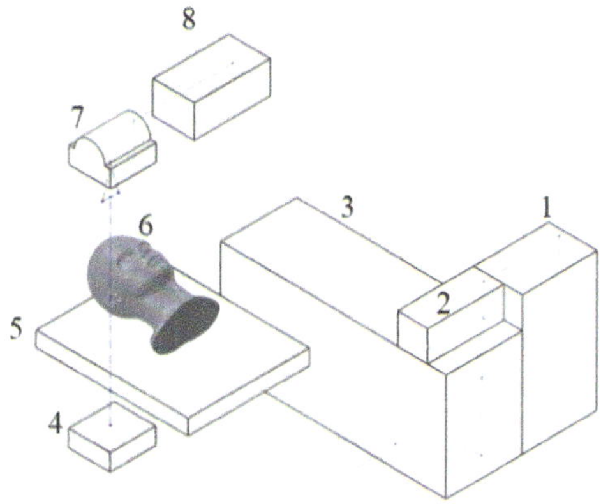

Fig. 6 Schematic diagram of the design of fifth-generation CT scanners. (1) Electron gun; (2, 3) detection coil; (4) target rings; (5) electron beam; (6) patient; (7) detectors; (8) data acquisitions system

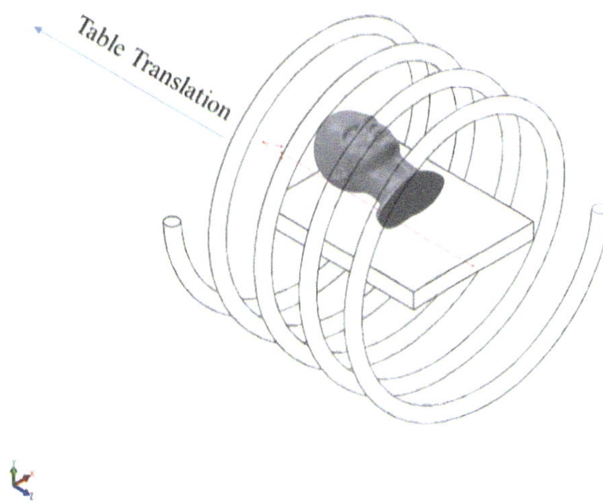

Fig. 7 Schematic diagram of the design of sixth-generation CT scanners

4.6 Sixth Generation: Helical or Spiral

In detectors of the previous generations, the gantry had to stop after each slice, which meant imaging was not a continuous process. Also, since X-ray tubes and detectors constantly use energy, they had to be connected to the power supply by wires, which inhibit their movement. This problem was solved with the introduction of slip ring technology to the field of medical imaging in the 1990s. With this technology, electricity can be supplied to rotating equipment without inhibiting their movement. Using this ring, the gantry could be rotated continuously around the patient covering all slices of the body, which led to faster scans. These changes led to the development of the sixth-generation CT scanners called helical or spiral CT scanners (Fig. 7).

The main problem of this generation of CT scanners is the acquired helically of data, it is not possible to collect full slices of data to produce planar sections (Bharath, 2008; Bushberg, 2002; Goldman, 2007; RSNA, 2013; Saunders & Ohlerth, 2011).

4.7 Seventh Generation: Multiple Detector Array

The latest generation of CT scanners make use of cone-shaped X-ray beams and multiple detector arrays. Unlike pencil beams and fan beams, conical beams do not pass through a narrow collimator, which means there will not be as much reduction in their initial intensity (Fig. 8).

Therefore, these beams interact better and more effectively with the detector array. To use the conical beam, the linear detector array of previous generations had to be modified into a flat panel detector or a multiple detector array. The combined use of conical beam and panel detectors in these scanners made it possible to scan a large number of slices in a very short time. However, these scanners need much more sophisticated and powerful processing components to collect and process the vast amount of data collected in very short time spans (Bushberg, 2002; RSNA, 2013).

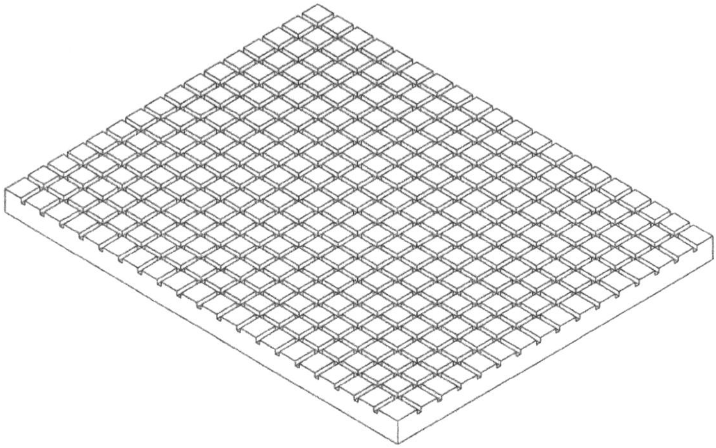

Fig. 8 Schematic diagram of linear detectors of previous generations and panel detectors of the seventh-generation CT scanners

4.7.1 Application of CT Scan in Non-destructive Quality Evaluation of Agricultural Products

Provided in the following is a review of previous studies on the use of CT scans in non-destructive quality evaluation of pear.

5 Assessment of Pear Bruising due to Quasi-Static Thin Edge Compressive Loading with CT Scan and Investigation of Its Relationship with Some Physical Characteristics of Pear

In this study, CT was used to investigate the relationship between the amount of bruising due to external force and some physical characteristics of pear as well as the storage period. For this purpose, 50 pears were examined with CT imaging and 27 of them that had zero bruising were selected for the experiment. Researchers then measured the dimensions of the selected pears (length, width, and thickness) and properties such as equivalent diameter, geometric mean diameter, sphericity, surface area, and aspect ratio. The selected pears were then subjected to quasi-static thin edge loading at three levels of 15, 20, and 25 N and 5, 10, and 15-day storage conditions to investigate the effect of applied loads on the pears. Then, after the loading, each pear was scanned. In the end, the results were used to calculate the amount of bruising in the pears (Azadbakht et al., 2019b). In this research the pears) Spadana variety (have been prepared from gardens around Gorgan, Golestan province, Iran. Samples were placed in separate boxes, and inside the boxes, pears were at a distance from each other. The life of the pear tree used in this research was 5 years old, the days after full bloom was about 120 days, and the pears were freshly harvested and the experiments were carried out as quickly as possible. Then Samples were taken to the laboratory of Gorgan University of Agricultural Sciences and Natural Resources. They were placed in an oven at 103 °C for 24 h and their moisture contents were measured (Azadbakht et al., 2016). The calculated moisture content of the pears was 77.92%. Environmental conditions for testing was conducted at a temperature of 18 °C and relative humidity of 72%. The purpose of this study was to determine the relationship between the physical characteristics of pear such as equivalent diameter, geometric mean diameter, sphericity, and surface area, and the amount of bruising that is caused by external loads during storage, in order to determine what physical characteristics and conditions play a bolder role in maintaining the quality of pears during storage and hence their sale value after storage (Azadbakht et al., 2019b).

5.1 Determination of Physical Characteristics

To obtain the average dimensions of the pears, the required geometric parameters including, the large diameter or length (L), the intermediate diameter or width (W), and the small diameter or thickness (T) were measured with a caliper with an accuracy of 0.05. These dimensions are illustrated in Fig. 9 (Azadbakht et al., 2019b).

The equivalent diameter (D_p), geometric mean diameter (D_g), sphericity ($\varnothing$), surface area (S), and aspect ratio (R) of the studied pears were calculated using the formulas given in Table 1.

Fig. 9 Length (L), Width (W), and Thickness (T) of pears. (Azadbakht et al., 2019b)

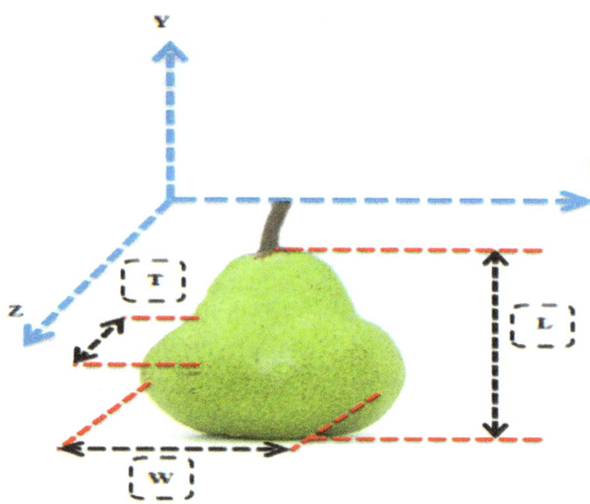

Table 1 Equations of equivalent diameter, geometric mean diameter, sphericity, surface area, roundness, and aspect ratio for pears (Azadbakht et al., 2019b)

Equation	Equation number	References
$D_p = \left(\frac{L(W+T)^2}{4}\right)^{\frac{1}{3}}$	(2)	Azadbakht et al. (2019b)
$D_g = (LWT)^{\frac{1}{3}}$	(3)	Azadbakht et al. (2019b)
$\varnothing = \frac{(LWT)^{\frac{1}{3}}}{L}$	(4)	Azadbakht et al. (2019b)
$S = \frac{\pi BL^2}{2L - B}$	(5)	Azadbakht et al. (2019b)
$B = (WT)^{0.5}$	(6)	Azadbakht et al. (2019b)
$R = \frac{W}{L} \times 100$	(7)	Azadbakht et al. (2019b)

5.2 *Quasi-Static Test*

In order to perform the wide and thin edge compression mechanical test, a pressure-deformation device (the Santam Indestrone-STM5-Made in Iran) with a load cell of 500 N was employed. The compression test, where the two circular plates were utilized, was performed at a speed of 5 mm/s with three forces of 70, 100, and 130 N and three repetitions. In this experiment, the pear was horizontally placed between the two plates and pressed, for the duration of the recorded measurement. Regarding the thin edge compression test, a double-jaw of plastic with a rectangular cross-section dimension of 0.3 × 1.5 cm was designed. The test was performed at a speed of 5 mm/s with three forces of 15, 20, and 25 N and three repetitions (Fig. 10).

By moving the movable jaw, the pressure operation was carried out until the force reached the desired level (Azadbakht et al., 2016).

5.3 *CT Imaging*

Pears were scanned with the Siemens Computed Tomography (CT) Scans of the SOMATOM Emotion 16-slice model, made in Germany. For imaging, the pears were taken to the test site and placed in a CT scan chamber (Fig. 11). Through the tube, X-rays were fired at the pears. Some of the radiated rays passed through the pear and were absorbed by the crystals inside the CT scan chamber. It was then

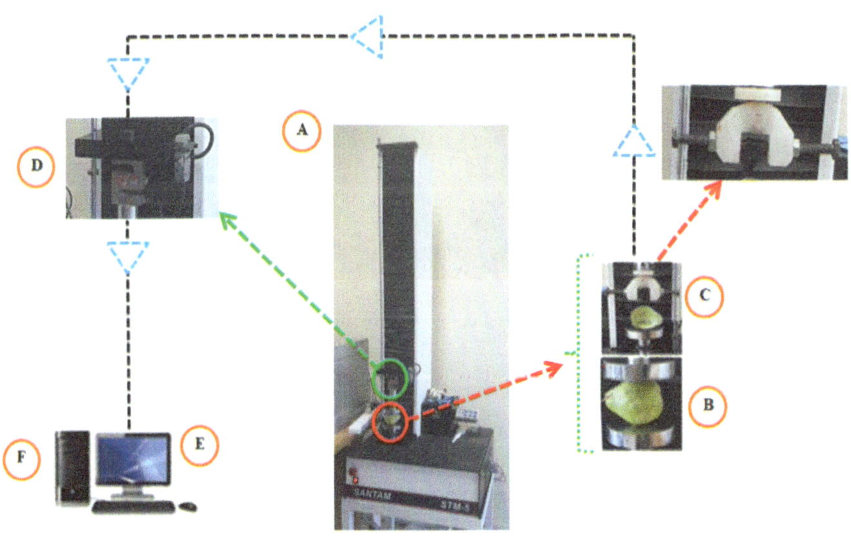

Fig. 10 Static quasi-load diagram of pear. (**a**) The force-deformation device (indestrone); (**b**) Jaw wide edges; (**c**) Jaw thin edges; (**d**) load cell; (**e**) computer; (**f**) information extract (Azadbakht et al., 2019b)

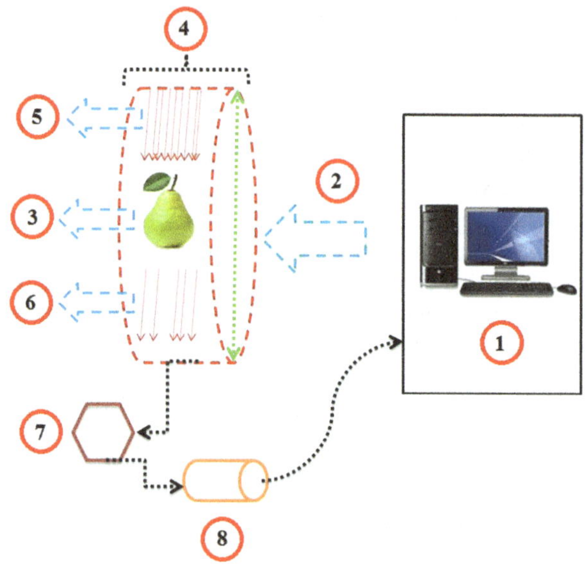

Fig. 11 CT imaging schematic. (1) Control room; (2) setting device; (3) pear sample; (4) X-rays tube; (5) X-ray input; (6) X-ray outlet from sample; (7) crystals CT scan; (8) converter (Azadbakht et al., 2019f)

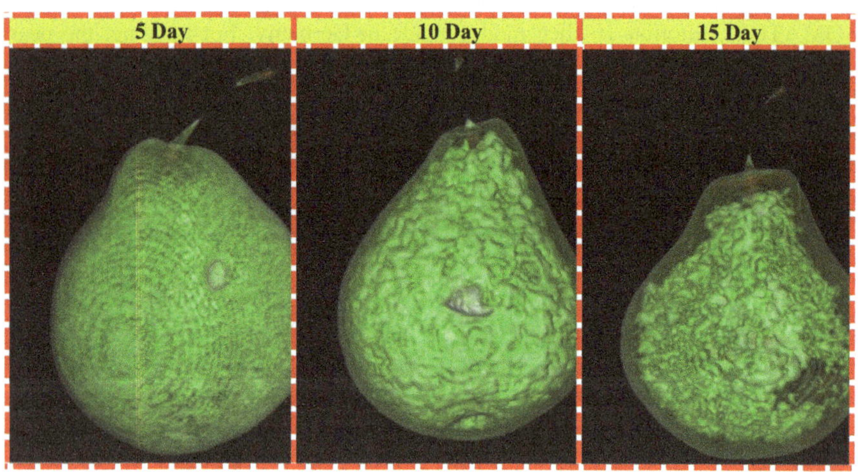

Fig. 12 Extraction images of fruit tissue, by CT in thin edge loading (Vahedi Torshizi & Azadbakht, 2020)

converted to image code using an optical converter and sent to a computer room for image reconstruction (Azadbakht et al., 2019f).

After quasi-static loading, the samples were stored for 5, 10, and 15 days and then imaged. Figure 12 shows the destruction of the texture of the studied pear during storage.

All experiments were performed in triplicates using the factorial experiment and completely randomized design methods and the results were analyzed in the statistical analysis software SAS. This software was used to run a correlation test for the relationship between bruising and geometric mean diameter, equivalent diameter, surface area, aspect ratio, and sphericity (Azadbakht et al., 2019b).

Table 2 shows the correlation between the geometric mean diameter, equivalent diameter, arithmetic mean diameter, surface area, aspect ratio, sphericity, and bruising of pears subjected to thin edge compression at three levels of 15, 20, and 25 N.

5.4 *Effect of Geometric Mean Diameter on Bruise Level*

The results of the analysis of variance of the geometric mean diameter of pears were found to have a significant effect at the 5% level on bruising in all storage periods. The same result was also obtained for the interaction effect of geometric mean diameter with loading Force.

For the pears subjected to the 15 N load, the highest geometric mean diameter, 66.27 mm, was observed in the 10-day storage sample group, and the lowest, 65.45 mm, was observed in the 15-day storage sample group (Fig. 13). As shown in Table 4, for these pears, geometric mean diameter had a significant positive correlation with equivalent diameter, arithmetic mean diameter, aspect ratio, and surface area.

For the pears subjected to the 20 N load, the highest geometric mean diameter, 69.75 mm, was observed in the 5-day storage sample group, and the lowest, 63.39 mm, was observed in the 10-day storage sample group (Fig. 13). For these pears, geometric mean diameter showed a significant positive correlation with equivalent diameter, arithmetic mean diameter, and surface area (Table 4).

For the pears subjected to the 25 N load, the highest geometric mean diameter, 68.42 mm, was observed in the 5-day storage sample group, and the lowest, 61.94 mm, was observed in the 15-day storage sample group (Fig. 13). For these pears, geometric mean diameter was found to have a significant positive correlation with equivalent diameter, arithmetic mean diameter, and surface area (Table 4) (Azadbakht et al., 2019b).

The correlation analysis results (Table 2) showed that the correlation between geometric mean diameter and bruise level of the pears subjected to thin edge compressive loads was not statistically significant. This correlation was negative for the 15 and 25 N load levels, meaning that the bruise level is inversely related to the geometric mean diameter, but was positive for the 25 N load, meaning that as this diameter increases, so does the bruise level. These results are plotted in Fig. 13 (Azadbakht et al., 2019b).

Table 2 Correlation between geometric mean diameter, equivalent diameter, arithmetic mean diameter, surface area, aspect ratio, sphericity, and bruise level of pears (Azadbakht et al., 2019b)

	Geometric diameter	Arithmetical diameter	Equivalent diameter	Sphericity	Aspect ratio	Surface area	Percentage of bruise
15 N							
Geometric diameter	1						
Arithmetical diameter	0.991**	1					
Equivalent diameter	0.99**	0.991**	1				
Sphericity	−0.486^{ns}	−0.851**	−0.469^{ns}	1			
Aspect ratio	−0.421	−0.531^{ns}	−0.423^{ns}	0.967**	1		
Surface area	0.999**	0.999**	0.999**	−0.465^{ns}	−0.419^{ns}	1	
Percentage of bruise	−0.03^{ns}	0.015^{ns}	−0.03^{ns}	−0.389^{ns}	−0.402^{ns}	−0.04^{ns}	1
20 N							
Geometric diameter	1						
Arithmetical diameter	0.998**	1					
Equivalent diameter	0.999**	0.998**	1				
Sphericity	0.024^{ns}	−0.023^{ns}	0.030^{ns}	1			
Aspect ratio	0.167^{ns}	0.122^{ns}	0.174^{ns}	0.931**	1		
Surface area	0.999**	0.997**	0.999**	0.036^{ns}	0.184^{ns}	1	
Percentage of bruise	0.172^{ns}	0.192^{ns}	−0.175^{ns}	0.434^{ns}	0.180^{ns}	−0.180^{ns}	1
25 N							
Geometric diameter	1						
Arithmetical diameter	0.999**	1					
Equivalent diameter	0.999**	0.999**	1				
Sphericity	−0.479^{ns}	−0.506^{ns}	−0.480^{ns}	1			

Aspect ratio	−0.213 ns	−0.221 ns	−0.214 ns	0.887**	1		
Surface area	0.999**	0.999**	0.999**	−0.478 ns	−0.211 ns	1	
Percentage of bruise	−0.750*	−0.749*	−0.750*	0.361 ns	−0.04 ns	−0.751*	1

*: Significant difference at the 1% probability level; **: Significant difference at the 5% probability level; ns: No significant difference

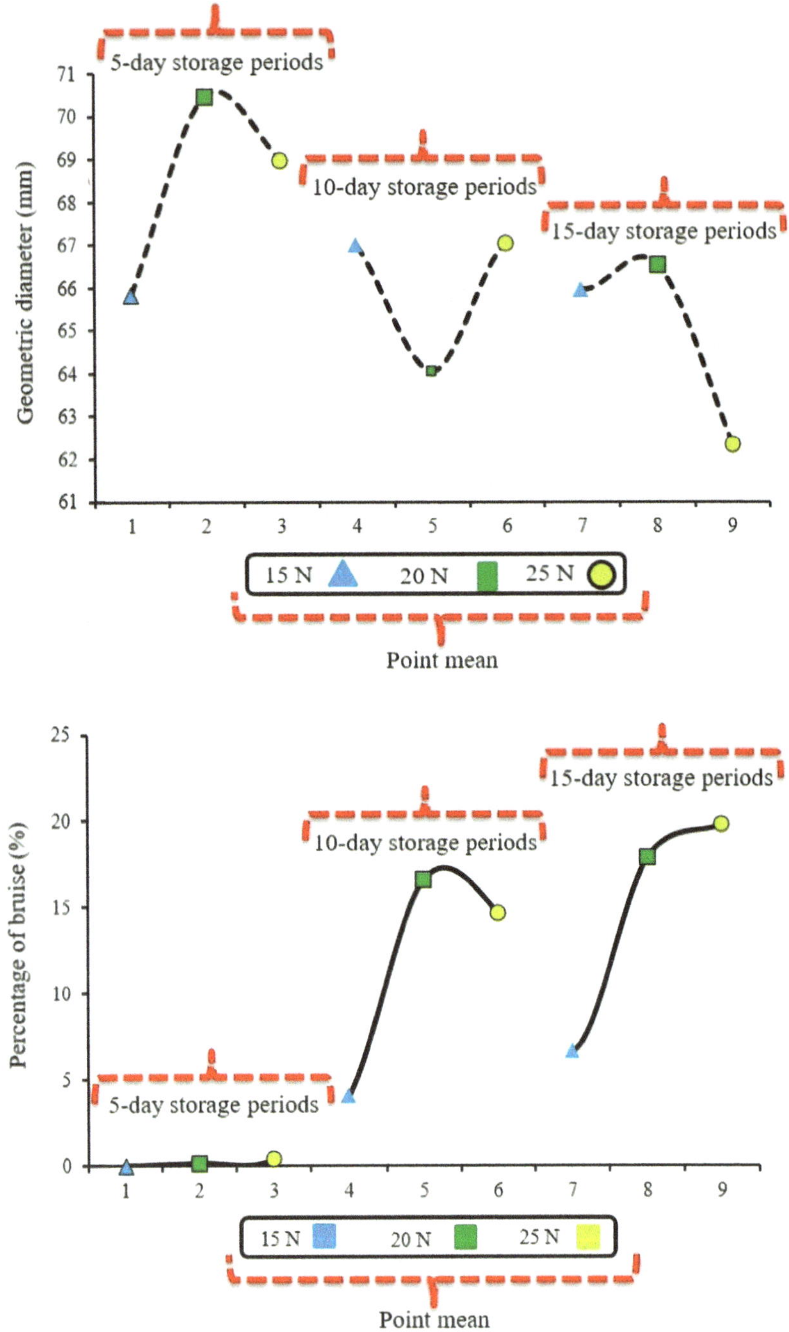

Fig. 13 Relationship between bruise level and geometric mean diameter of pears (Azadbakht et al., 2019b)

5.5 Effect of Equivalent Diameter on Bruise Level

The results of the analysis of variance of the equivalent diameter did not show a significant effect on the bruise level for any storage periods or loading level.

For the pears subjected to the 15 N load, the highest equivalent diameter was 40.33 mm, which was observed in the 10-day storage sample group, and the lowest was 39.74 mm, which was observed in the 15-day storage sample group (Fig. 14). As shown in Table 2, the equivalent diameter of these pears had a significant positive correlation with their geometric mean diameter, arithmetic mean diameter, aspect ratio, and surface area.

For the pears subjected to the 20 N load, the highest equivalent diameter was 42.31 mm, which was observed in the 5-day storage sample group, and the lowest was 38.49 mm, which was observed in the 10-day storage sample group (Fig. 14). For these pears, equivalent diameter showed a significant positive correlation with geometric mean diameter, arithmetic mean diameter, aspect ratio, and surface area (Table 2).

For the pears subjected to the 25 N load, the highest equivalent diameter was 41.51 mm, which was observed in the 5-day storage sample group, and the lowest was 32.56 mm, which was observed in the 15-day storage sample group (Fig. 14). The equivalent diameter of these pears showed a significant positive correlation with their geometric mean diameter, arithmetic mean diameter, aspect ratio, and surface area, and a negative correlation with their bruise level (Table 2) (Azadbakht et al., 2019b).

The correlation analysis results (Table 2) showed that the negative correlation between equivalent diameter and bruise level of the pears subjected to thin edge compressive load of 15 N was statistically insignificant. For the 25 load level, there was a statistically significant negative correlation between equivalent diameter and bruise level, meaning that bruise level is inversely related to equivalent diameter. For the 20 load level, there was a statistically insignificant negative correlation between these two parameters, meaning that bruising increases with the equivalent diameter. These results are presented in Fig. 14 (Azadbakht et al., 2019b).

5.6 Effect of Surface Area on Bruise Level

The results of the analysis of variance of the surface area of pears showed a significant effect at the 5% level on their bruising in all storage periods. The same result was also obtained for the interaction effect of this parameter with loading Force.

For the pears subjected to the 15 N load, the highest surface area, 13,832 mm^2, was observed in the 10-day storage sample group, and the lowest, 13,455 mm^2, was observed in the 15-day storage sample group (Fig. 15). However, as shown in Fig. 15, there was not much difference between the surface areas related to the

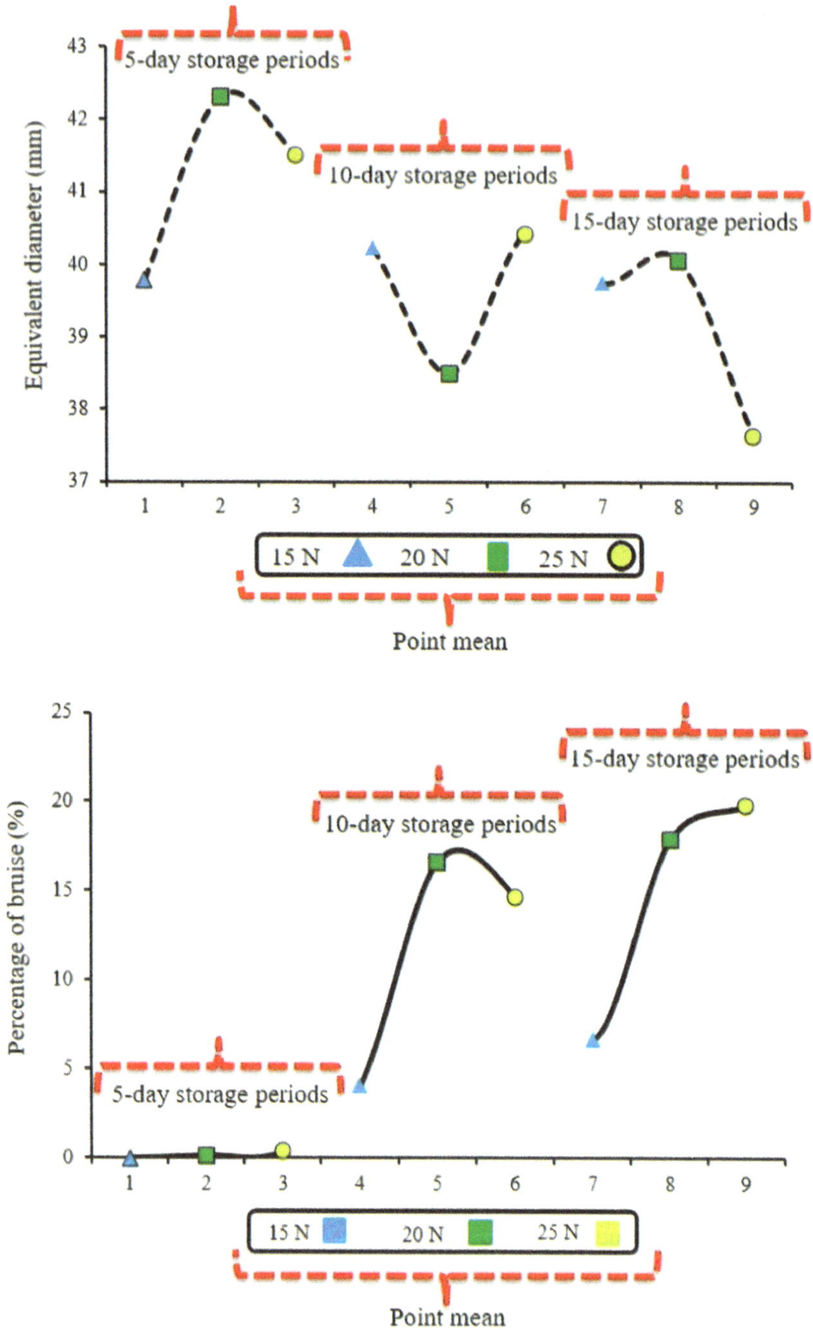

Fig. 14 Relationship between bruise level and equivalent diameter of pears (Azadbakht et al., 2019b)

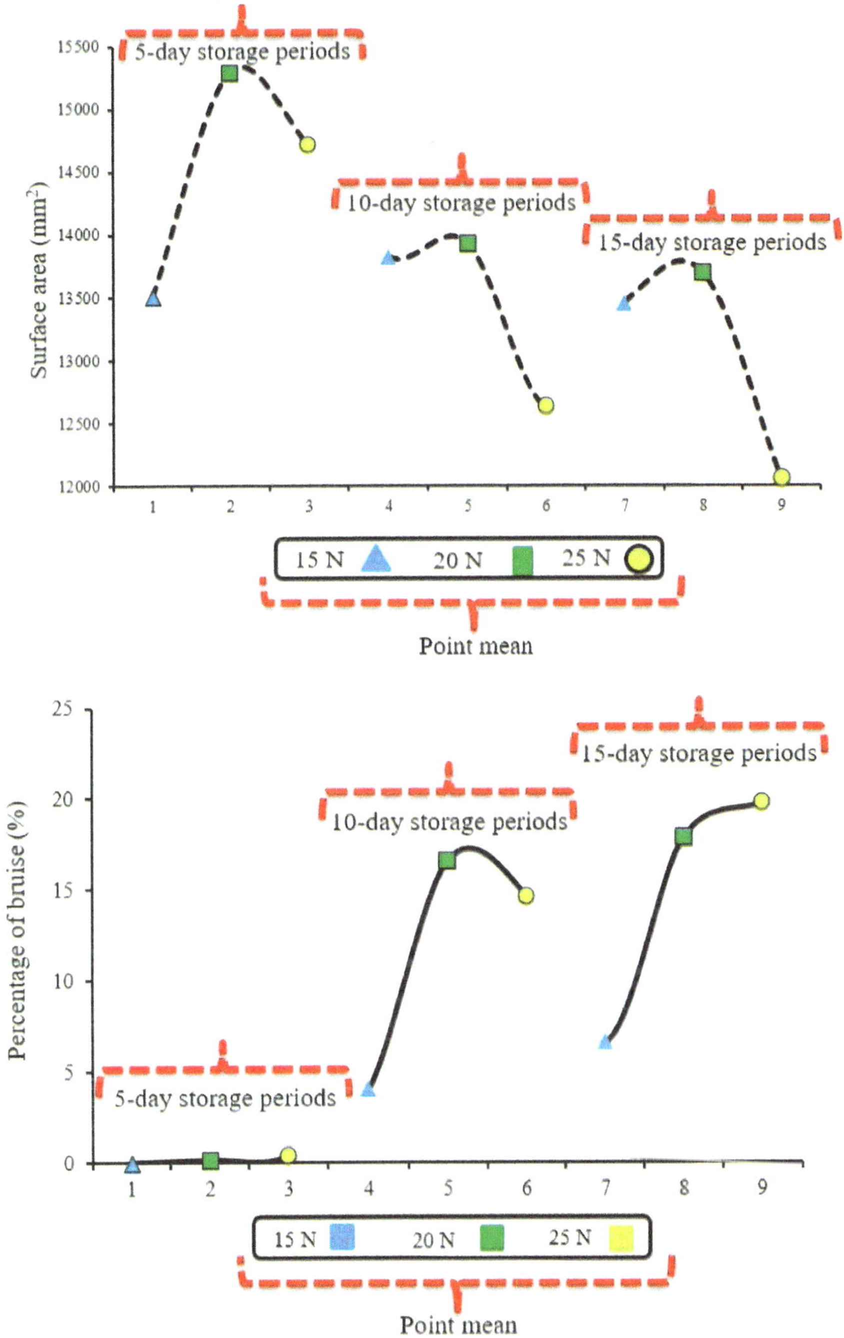

Fig. 15 Relationship between bruise level and surface area of pears (Azadbakht et al., 2019b)

three storage periods. The 10-day surface area was slightly higher than the 15-day surface area, which resulted in a small difference between their bruise levels. For these pears, the surface area was found to have a significant positive correlation with geometric mean diameter, equivalent diameter, arithmetic mean diameter, and aspect ratio.

For the pears subjected to the 20 N load, the highest surface area, 152,933.8 mm^2, was observed in the 5-day storage sample group, and the lowest, 13,699.3 mm^2, was observed in the 15-day storage sample group (Fig. 15). As shown in Fig. 15, the high surface area of pears in the 5-day storage sample group had no effect on their bruise level. Also, the surface area related to the 10-day period was higher than the one related to the 15-day period, causing a slight difference between their bruise levels. For these pears, surface area showed a significant positive correlation with geometric mean diameter, equivalent diameter, arithmetic mean diameter, and aspect ratio (Table 2).

For the pears subjected to the 25 N load, the highest surface area, 14,714 mm^2, was observed in the 5-day storage sample group, and the lowest, 12,051.6 mm^2, was observed in the 15-day storage sample group (Fig. 15). In Fig. 15, it can be seen that the high surface area related to the 5-day period had no effect on bruise level, and since the surface area related to the 10-day period was higher than the one related to the 15-day period, there was a slight difference between their bruise levels. In this case, it can be stated that the outcome has been more influenced by the storage period than by dimensions. The observed differences between surface areas can also be related to the internal texture of pears, which can affect how they respond to different factors. As shown in Table 2, for these pears, the surface area had a significant positive correlation with geometric mean diameter, equivalent diameter, and arithmetic mean diameter (Azadbakht et al., 2019b).

The correlation results (Table 2) showed a statistically insignificant negative correlation between surface area and bruise level of the pears subjected to thin edge compressive loads of 15 and 20 N. For the 25 load level, this negative correlation was statistically significant, meaning that bruise level is inversely related to the surface area. The results related to this parameter are illustrated in Fig. 15 (Azadbakht et al., 2019b).

5.7 *Effect of Aspect Ratio on Bruise Level*

The results of the analysis of variance of the aspect ratio did not show a significant effect on the bruise level for any storage period or loading level.

For the pears subjected to the 15 N load, the highest aspect ratio was 0.817 and was observed in the 5-day storage sample group, and the lowest aspect ratio was 0.747 and was observed in the 10-day storage sample group (Fig. 16). The aspect ratio of pears in the 5-day storage sample group showed no effect on their bruising, the reason for which is clearly visible in Fig. 16. Since the aspect ratio related to the 10-day period was higher than the one related to the 15-day period, there was a slight

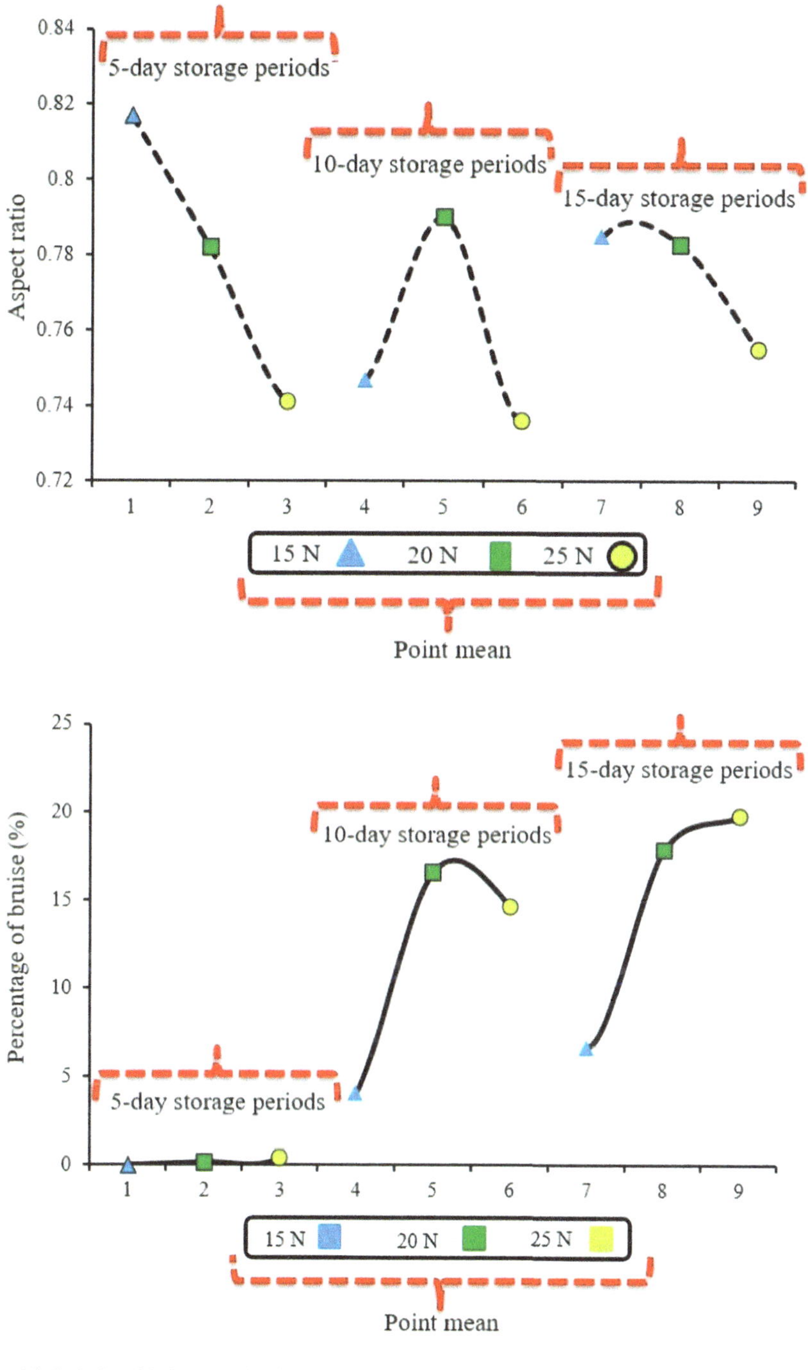

Fig. 16 Relationship between bruise level and aspect ratio of pears (Azadbakht et al., 2019b)

difference between their bruise levels. Also, as shown in Table 2, the aspect ratio of these pears was found to have a significant positive correlation with their sphericity.

For the pears subjected to the 20 N load, the highest aspect ratio was 0.790 and was observed in the 10-day storage sample group, and the lowest aspect ratio was 0.782 and was observed in the 5-day storage sample group (Fig. 16). As Fig. 16 shows, since the 10-day pears had a higher aspect ratio than the 15-day pears, they were also expected to have higher bruise levels, but this did not occur perhaps because thin edge loading only damages pear cells in a limited area whereas broad edge pressure affects a larger area. As shown in Table 2, the aspect ratio of these pears too had a significant positive correlation with their sphericity.

For the pears subjected to the 25 N load, the highest aspect ratio was 0.755 and was observed in the 15-day storage sample group, and the lowest aspect ratio was 0.741 and was observed in the 10-day storage sample group (Fig. 16). As shown in Fig. 16, the 15-day pears had a higher aspect ratio and higher bruise levels than the 10-day pears. For these pears, too, the aspect ratio showed a significant positive correlation with sphericity (Table 2) (Azadbakht et al., 2019b).

As shown in Table 2, the correlation results revealed an insignificant negative correlation between aspect ratio and bruise level of the pears subjected to thin edge compressive loads of 15 and 25 N, which means their bruise level is inversely related to their aspect ratio, but not in a statistically significant way. For the 20 load level, there was a statistically insignificant positive correlation between aspect ratio and bruise level. The results related to this parameter are presented in Fig. 16 (Azadbakht et al., 2019b).

5.8 *Effect of Sphericity on Bruise Level*

The results of the analysis of variance of sphericity did not show a significant effect on the bruise level for any storage period or loading level.

For the pears subjected to the 15 N load, the highest sphericity was 0.880, which was observed in the 5-day storage sample group, and the lowest was 0.828, which was observed in the 10-day storage sample group (Fig. 17). The 15-day pears had higher sphericity than the 10-day pears, resulting in higher bruise levels. As shown in Table 2, for these pears, sphericity showed a significant positive correlation with arithmetic mean diameter and aspect ratio.

For the pears subjected to the 20 N load, the highest sphericity was 0.841, which was observed in the 5-day storage sample group, and the lowest was 0.822, which was observed in the 10-day storage sample group (Fig. 17). As shown in Table 2, the sphericity of these pears showed a significant positive correlation with their aspect ratio.

For the pears subjected to the 25 N load, the highest sphericity was 0.857, which was observed in the 15-day storage sample group, and the lowest was 0.819, which was observed in the 5-day storage sample group (Fig. 17). The 15-day pears had

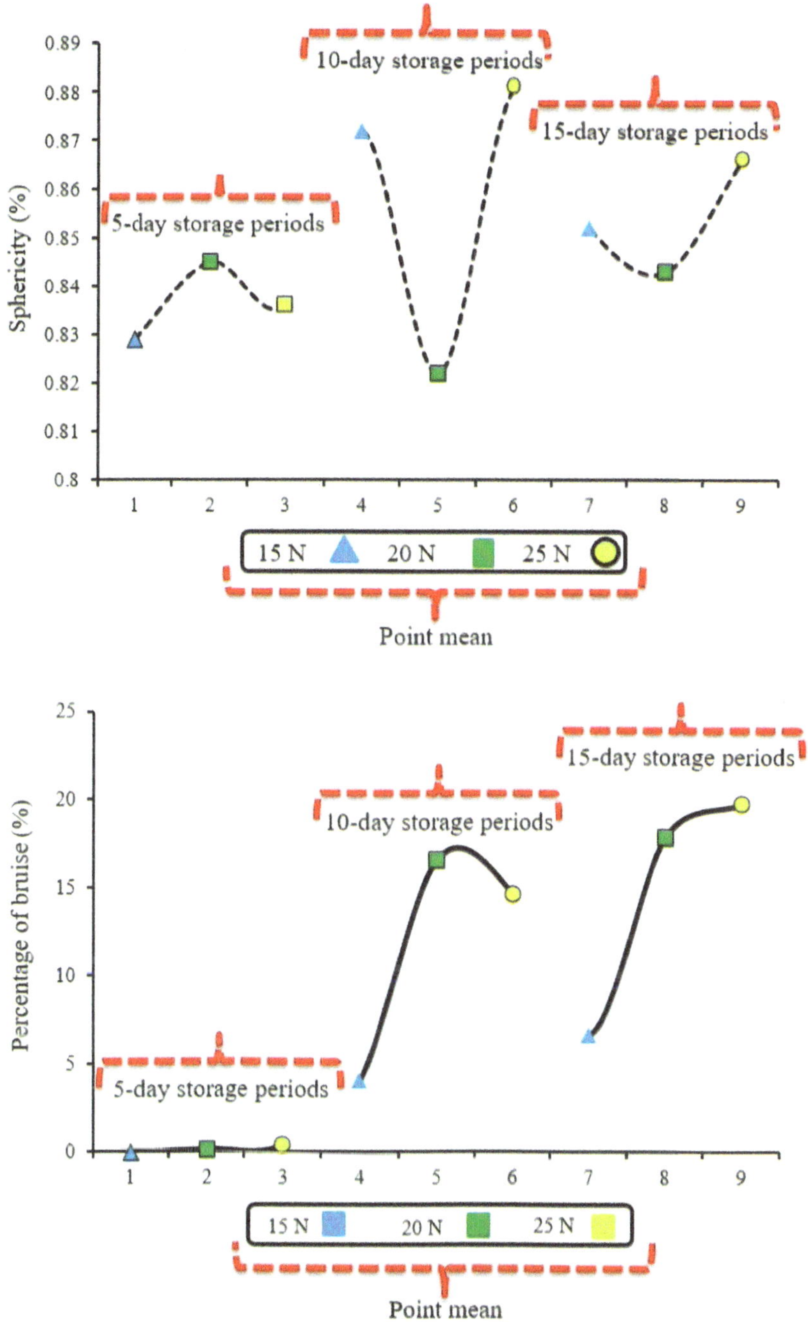

Fig. 17 Relationship between bruise level and sphericity of pears (Azadbakht et al., 2019b)

higher sphericity than the 10-day pears, leading to higher bruise levels. For these pears, too, sphericity showed a significant positive correlation with aspect ratio.

As the correlation results of Table 2 show, an insignificant negative correlation was detected between sphericity and bruise level of the pears subjected to thin edge compressive loads of 15 N. For the pears subjected to the 20 and 25 N loads, there was a statistically insignificant positive correlation between these parameters. These results are presented in Fig. 17 (Azadbakht et al., 2019b).

6 Assessment of Pear Bruising Due to Quasi-Static Broad Edge Compressive Loading with CT Scan and Investigation of Its Relationship with Some Physical Characteristics of Pear

This study was similar to the one described in Sect. 5, except that the researchers used broad jaws instead of thin ones in the compression apparatus (Fig. 10). Figure 18 shows the variations of pear texture during wide edge compression loading at different storage periods.

The results of the analysis of variance of physical characteristics of pear including geometric mean diameter, equivalent diameter, and surface area in this study showed no significant difference between the means of dependent variables. Table 3 shows the correlation between geometric mean diameter, equivalent diameter, surface area, and bruise level of pears subjected to broad edge compression at three loading levels of 70, 100, and 130 N (Azadbakht et al., 2019a).

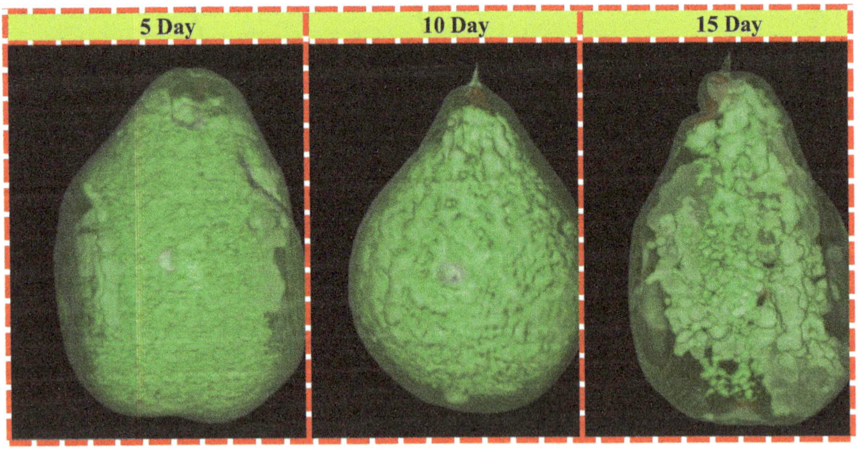

Fig. 18 Extraction images of fruit tissue, by CT in wide edge loading (Vahedi Torshizi & Azadbakht, 2020)

Table 3 Correlation between geometric mean diameter, equivalent diameter, and surface area of pear and its bruise level (Azadbakht et al., 2019a)

	Geometric Diameter	Diameter	Surface	Bruise
70 N				
Geometric diameter	1			
Diameter	0.999^{**}	1		
Surface	0.999^{**}	0.999^{**}	1	
Bruise	-0.108^{ns}	-0.090^{ns}	0.106^{ns}	1
100 N				
Geometric diameter	1			
Diameter	0.999^{**}	1		
Surface	0.999^{**}	0.997^{**}	1	
Bruise	0.018^{ns}	-0.013^{ns}	0.010^{ns}	1
130 N				
Geometric diameter	1			
Diameter	0.996^{**}	1		
Surface	0.999^{**}	0.996^{**}	1	
Bruise	0.162^{ns}	-0.129^{ns}	-0.161^{ns}	1

*: Significant difference at the 1% probability level; **: Significant difference at the 5% probability level; ns: No significant difference

6.1 *Effect of Geometric Mean Diameter on Bruise Level*

For the pears subjected to the 70 N load, as shown in Fig. 19, the highest geometric mean diameter, 67.78 mm, was observed in the 15-day storage sample group, and the lowest, 64.91 mm, was observed in the 10-day storage sample group. As shown in Table 3, the geometric mean diameter of these pears had a significant positive correlation with their equivalent diameter, arithmetic mean diameter, and surface area (Azadbakht et al., 2019a).

For the pears subjected to the 100 N load, the highest geometric mean diameter, 65.97 mm, was observed in the 15-day storage sample group, and the lowest, 64.91 mm, was observed in the 10-day storage sample group (Fig. 19). For these pears, geometric mean diameter showed a significant positive correlation with equivalent diameter, arithmetic mean diameter, and surface area (Table 3).

For the pears subjected to the 130 N load, the highest geometric mean diameter, 67.39 mm, was observed in the 5-day storage sample group, and the lowest, 61.24 mm, was observed in the 15-day storage sample group (Fig. 19). The geometric mean diameter of these pears was found to have a significant positive correlation with their equivalent diameter, arithmetic mean diameter, and surface area.

The correlation results of Table 3 showed that the correlation between geometric mean diameter and bruise level of the pears subjected to broad edge compressive loads was not statistically significant. The correlation between these parameters was negative for the 70 and 130 N load levels, indicating that bruise level is inversely

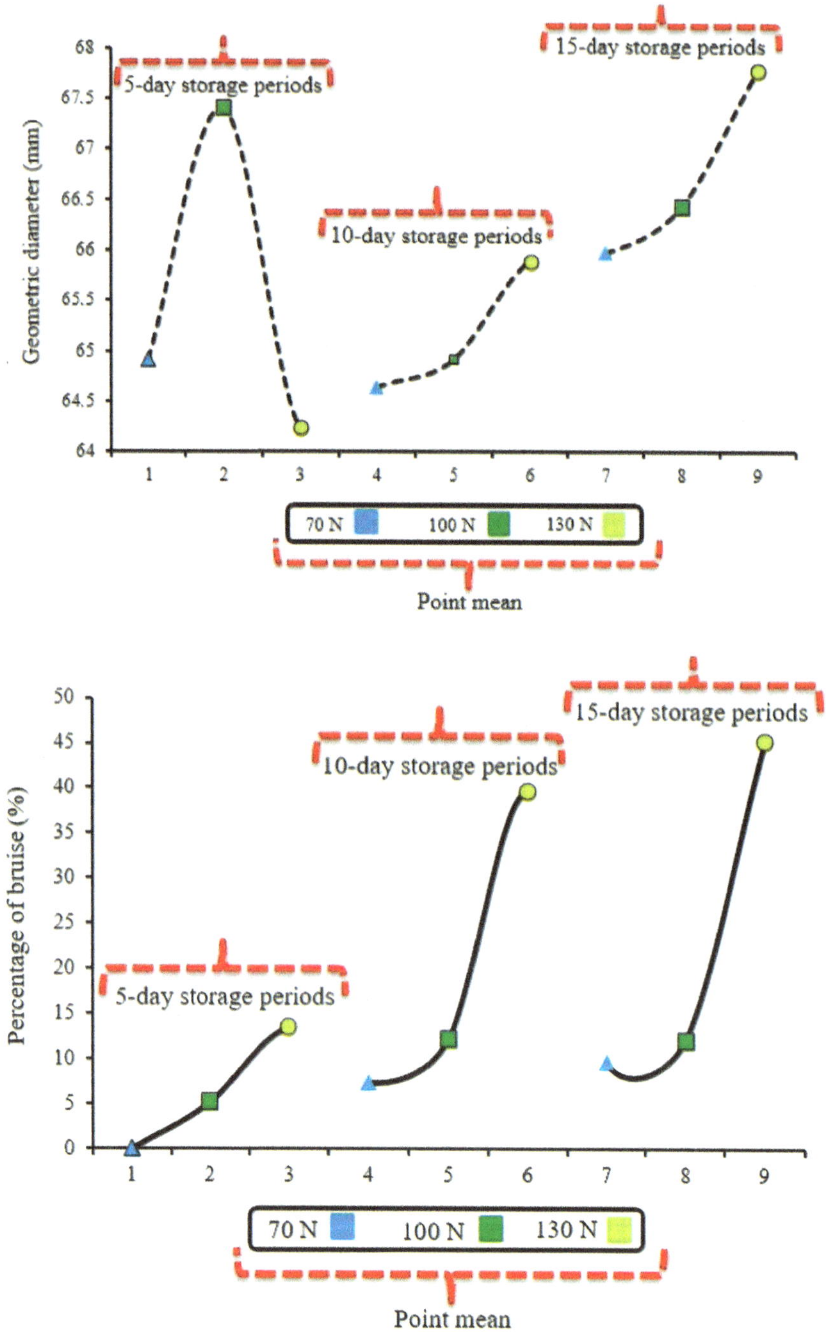

Fig. 19 Relationship between bruise level and geometric mean diameter of pears subjected to static loading (Azadbakht et al., 2019a)

related to geometric mean diameter. For the 100 N load however this correlation was positive, meaning that the bruise level increases with the increase in geometric mean diameter. This is due to the loss of moisture and therefore reduction of geometric mean diameter during storage and the existence of an inverse relationship between moisture content and bruise level. Thus, it can be concluded that bruise level decreases with decreasing geometric mean diameter (Kabas et al., 2006). These results were consistent with the finding of a study by Azadbakht et al. (2019) on pears, where bruising was found to be increasing with increasing storage time (Azadbakht et al., 2019), and also with the results of Kabas et al. (2006), which showed a reduction in the geometric mean diameter of pear during storage (Kabas et al., 2006).

6.2 *Effect of Surface Area on Bruise Level*

For the pears subjected to the 70 N load, the highest surface area was 14,473 mm^2, which was observed in the 15-day storage sample group, and the lowest surface area was 13,234 mm^2, which was observed in the 10-day storage sample group (Fig. 20). As shown in Fig. 20, the bruise level increased with the storage time, but there was no obvious relationship between surface area and bruise level for the 5-day storage period. In Fig. 20, it can also be seen that 15-day pears had higher surface area than the 10-day ones, and therefore showed higher bruise levels. The surface area of this group of pears showed a significant positive correlation with their geometric mean diameter, equivalent diameter, and arithmetic mean diameter (Azadbakht et al., 2019a).

For the pears subjected to the 100 N load, the highest surface area, 13,697 mm^2, was observed in the 10-day storage sample group, and the lowest, 13,147 mm^2, was observed in the 15-day storage sample group (Fig. 20). As shown in Fig. 25, among these pears, 10-day samples had higher surface area and thus higher bruise levels than the 15-day samples. This can be explained by the fact that the larger the contact area is, the greater will be the damage to the product (Li et al., 2017). For these pears, surface area showed a significant positive correlation with geometric mean diameter, equivalent diameter, and arithmetic mean diameter (Table 3).

For the pears subjected to the 130 N load, the highest surface area was 14,276 mm^2 and was observed in the 5-day storage sample group, and the lowest surface area was 12,960 mm^2, which was observed in the 15-day storage sample group (Fig. 20). In this case, Fig. 20 shows that the surface area of 10-day pears is only slightly higher than that of 15-day pears, resulting in little difference between their bruise levels. This difference can be attributed to storage time, because enzymatic activity in damaged fruit increases with storage time, leading to higher bruise levels. As shown in Table 3, the surface area of these pears had a significant positive correlation with their geometric mean diameter, equivalent diameter, and arithmetic mean diameter (Azadbakht et al., 2019a).

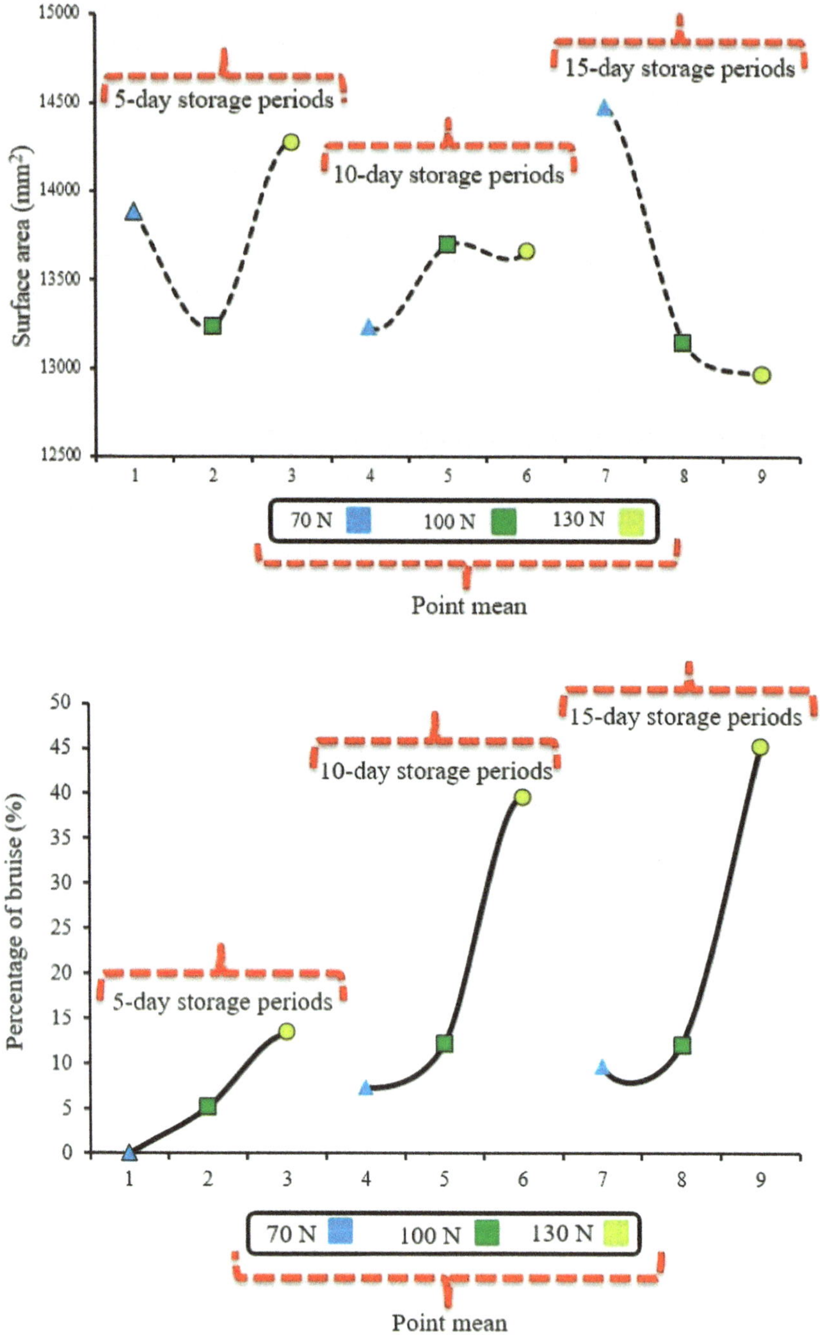

Fig. 20 Relationship between bruise level and surface area of pears subjected to static loading (Azadbakht et al., 2019a)

The correlation analysis results (Table 3) showed that the correlation between surface area and bruise level of the pears subjected to broad edge compressive loads were statistically insignificant. For the 70 and 130 N load levels, this correlation was negative, indicating that bruise level is inversely related to the surface area. For the 100 N load level however this correlation was positive, which means these parameters increase and decrease together. Considering the direct relationship between surface area and moisture content and the inverse relationship between moisture content and bruise level, it can be concluded that there is an inverse relationship between surface area and bruise level (Azadbakht et al., 2019a; Lownds et al., 1993).

6.3 Effect of Equivalent Diameter on Bruise Level

For the pears subjected to the 70 N load, the highest equivalent diameter was 40.91 mm, which was observed in the 5-day storage sample group, and the lowest was 39 mm, which was observed in the 15-day storage sample group (Fig. 21). As shown in Table 3, the equivalent diameter of these pears had a significant positive correlation with their geometric mean diameter, arithmetic mean diameter, and surface area.

For the pears subjected to the 100 N load, the highest equivalent diameter, 41.13 mm, was observed in the 15-day storage sample group, and the lowest, 39.42 mm, was observed in the 10-day storage sample group (Fig. 21). For these pears, too, equivalent diameter showed a significant positive correlation with geometric mean diameter, arithmetic mean diameter, and surface area (Table 3).

For the pears subjected to the 130 N load, the highest equivalent diameter was 40.30 mm, which was observed in the 10-day storage sample group, and the lowest was 39.24 mm, which was observed in the 15-day storage sample group (Fig. 21). As shown in Fig. 21, there was not much difference between the equivalent diameters related to the 10-day and 15-day storage periods. As Table 3 shows, the equivalent diameter of these pears showed a significant positive correlation with their geometric mean diameter, arithmetic mean diameter, and surface area (Azadbakht et al., 2019a).

The correlation results of Table 3 showed that the correlation between equivalent diameter and bruise level of the pears subjected to all broad edge compressive loads was statistically insignificant. For the 70 and 130 load levels, this correlation was negative, indicating an inverse relationship between the two factors. For the 100 load level, this correlation was positive, meaning that the bruise level increases with the equivalent diameter. This is consistent with the results of the study of Azadbakht et al. (2019b) on pears (Azadbakht et al. 2019b). As shown in Fig. 21, the bruise level was found to be increasing with the applied load.

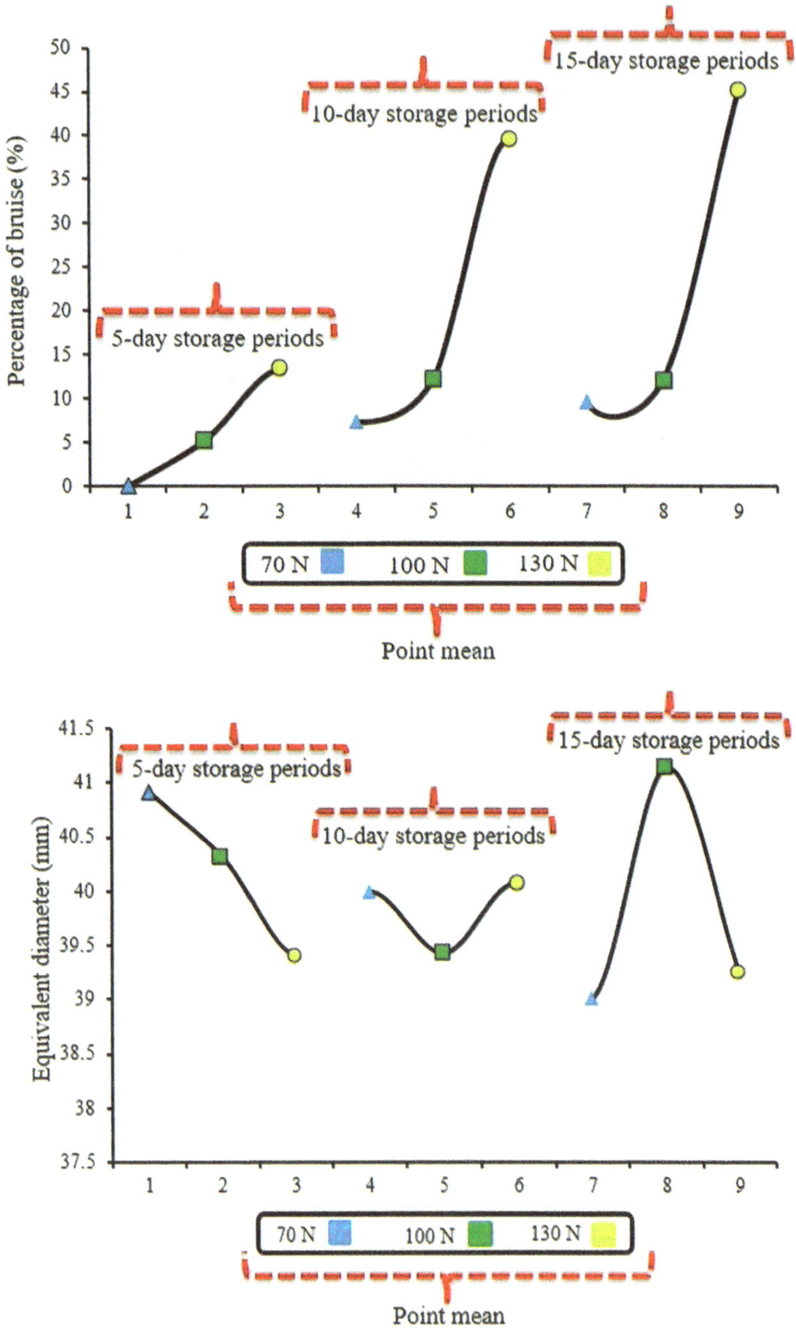

Fig. 21 Relationship between bruise level and equivalent diameter of pears subjected to static loading (Azadbakht et al., 2019a)

7 Assessment of Pear Bruising Due to Impact Load with CT Scan and Investigation of Its Relationship with Physical Characteristics of Pear

This study used CT scan as a non-destructive examination method to investigate the relationship between the physical characteristics of pear and the level of bruise they experience under different external loads and storage periods. For this purpose, researchers measured the dimensions (length, width, and thickness) of a group of pears and subjected them to impact loads and 5, 10, and 15-day storage conditions. After loading and storage, a CT scan was performed to determine the pear volume before and after loading, pear bruise volume, and also the ratio of bruise volume to the total volume of each pear (Azadbakht et al., 2019c). Figure 22 shows the variation of pears texture during storage where the effect of impact after storing for 15 days is shown.

The goal of this study was to examine the relationship of physical characteristics of pear such as equivalent diameter, geometric mean diameter, sphericity, surface area, and roundness with its bruising due to impact load, and storage with the help of a CT scan. In this study, it was stated that finding the relationship between the geometric dimensions of fruits and how they are bruised by impact loads can help determine the best storage time of fruits in stores and warehouses (Azadbakht et al., 2019c).

The methodology of this study was similar to the one described in Sect. 8, except that pears were subjected to impact loads instead of quasi-static loads.

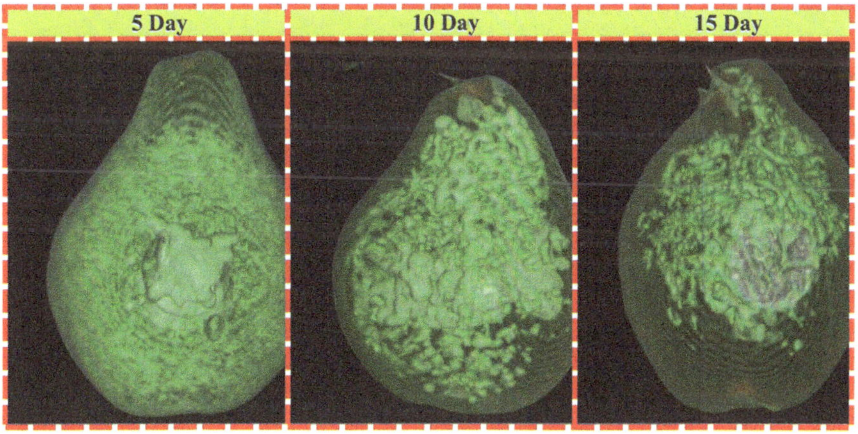

Fig. 22 Extraction images of fruit tissue, by CT in dynamic loading (Vahedi Torshizi & Azadbakht, 2020)

7.1 Impact Test

The pendulum and weights required for this test were fabricated in the workshop of the Biosystem-Mechanics department of Gorgan University (Fig. 23). The test was conducted by fixing the fruit in the tray, raising the arm of the device to the target angle (90°), and then releasing it in a controlled condition so that it would hit the pear. The pendulum device had a 200 g arm and three attachable weights of 100, 150, and 200 g for applying impact. In the calculations of this study, the effects of air resistance and friction were ignored (Azadbakht et al., 2019c).

7.2 Physical Characteristics

Table 4 shows the correlation between geometric mean diameter, equivalent diameter, arithmetic mean diameter, sphericity, aspect ratio, surface area, and bruising of pears subjected to the impact loads of 300, 350, and 400 g.

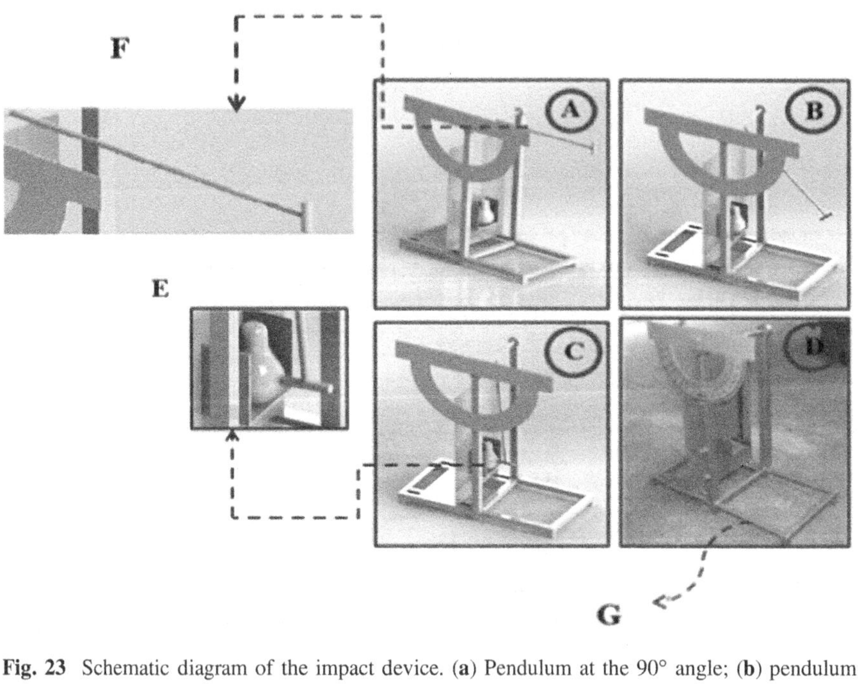

Fig. 23 Schematic diagram of the impact device. (**a**) Pendulum at the 90° angle; (**b**) pendulum motion along the path; (**c**) pendulum hitting the pear; (**d**) an overview of the device; (**e**) pear placement location; (**f**) the impact pendulum; (**g**) the body of the device (Azadbakht et al., 2019c)

Table 4 Correlation between geometric mean diameter, equivalent diameter, arithmetic mean diameter, sphericity, aspect ratio, surface area, and bruise level of pears (Azadbakht et al., 2019c)

	Equivalent diameter	Arithmetical diameter	Geometric diameter	Sphericity	Aspect ratio	Surface area	Percentage of bruise
300 g							
Equivalent diameter	1						
Arithmetical diameter	0.999^{**}	1					
Geometric diameter	0.999^{**}	0.998^{**}	1				
Sphericity	-0.516^{ns}	-0.563^{ns}	-0.513^{ns}	1			
Aspect ratio	-0.501^{ns}	-0.538^{ns}	-0.493^{ns}	0.922^{**}	1		
Surface area	0.999^{**}	0.998^{**}	0.999^{**}	-0.515^{ns}	−0.500	1	
Percentage of bruise	-0.638^{ns}	0.715^{*}	-0.629^{ns}	0.351^{ns}	0.348^{ns}	-0.621^{ns}	1
350 g							
Equivalent diameter	1						
Arithmetical diameter	0.995^{**}	1					
Geometric diameter	0.999^{**}	0.995^{**}	1				
Sphericity	0.00007^{ns}	0.097^{ns}	-0.001^{ns}	1			
Aspect ratio	0.115^{ns}	0.02^{ns}	0.112^{ns}	0.943^{**}	1		
Surface area	0.999^{**}	0.995^{**}	0.999^{**}	-0.005^{ns}	0.106^{ns}	1	1
Percentage of bruise	-0.555^{ns}	-0.590^{ns}	-0.553^{ns}	0.451^{ns}	0.162^{ns}	−0.550	
400 g							
Equivalent diameter	1					1	
Arithmetical diameter	0.995^{**}	1					
Geometric diameter	0.999^{**}	0.994^{**}	1				
Sphericity	0.108^{ns}	0.011^{ns}	0.118^{ns}	1			

(continued)

Table 4 (continued)

	Equivalent diameter	Arithmetical diameter	Geometric diameter	Sphericity	Aspect ratio	Surface area	Percentage of bruise
Aspect ratio	0.329^{ns}	0.239^{ns}	0.338^{ns}	0.948^{**}	1		
Surface area	0.999^{**}	0.993^{**}	0.999^{**}	0.12199^{ns}	0.339^{ns}	1	
Percentage of bruise	-0.589^{ns}	-0.605^{ns}	-0.583^{ns}	0.182^{ns}	0.155^{ns}	-0.587^{ns}	1

*: Significant difference at the 1% probability level; **: Significant difference at the 5% probability level; ns: No significant difference

7.3 Effect of Geometric Mean Diameter on Bruise Level

The results of the analysis of variance of the geometric mean diameter was found to have a significant effect at the 5% level on bruise level in all storage periods. However, the interaction effect of geometric mean diameter and impact load was not statistically significant.

For the pears subjected to the 300 g impact load, as shown in Fig. 24, the highest geometric mean diameter, 69.96 mm, was observed in the 5-day storage sample group, and the lowest, 68.57 mm, was observed in the 15-day storage sample group. As shown in Table 4, the geometric mean diameter of these pears had a significant positive correlation with their equivalent diameter, arithmetic mean diameter, and surface area.

For the pears subjected to the 350 g impact load, the highest geometric mean diameter, 71.09 mm, was observed in the 10-day storage sample group, and the lowest, 65.27 mm, was observed in the 15-day storage sample group (Fig. 24). For these pears, geometric mean diameter showed a significant positive correlation with arithmetic mean diameter and surface area (Table 4).

For the pears subjected to the 400 g impact load, the highest geometric mean diameter, 69.91 mm, was observed in the 5-day storage sample group, and the lowest, 66.11 mm, was observed in the 15-day storage sample group (Fig. 24). The geometric mean diameter of these pears was found to have a significant positive correlation with their equivalent diameter, arithmetic mean diameter, and surface area.

The correlation results presented in Table 4 showed that the correlation between geometric mean diameter and bruise level of the pears subjected to these three impact loads (300, 350, and 400 g) was not statistically significant. The correlation between geometric mean diameter and bruise level under these impact loads was negative, indicating they are inversely related, i.e., bruise level decreases as geometric mean diameter increases (Azadbakht et al., 2019c).

7.4 Effect of Equivalent Diameter on Bruise Level

The results of the analysis of variance of the equivalent diameter showed a significant effect on the bruise level for all storage periods. However, its interaction effect with the impact load was not found to be statistically significant.

For the pears subjected to the 300 g impact load, the highest equivalent diameter was 292.56 mm, which was observed in the 5-day storage sample group, and the lowest was 282.61 mm, which was observed in the 15-day storage sample group (Fig. 25). As shown in Table 4, the equivalent diameter of these pears had a significant positive correlation with their geometric mean diameter, arithmetic mean diameter, and surface area.

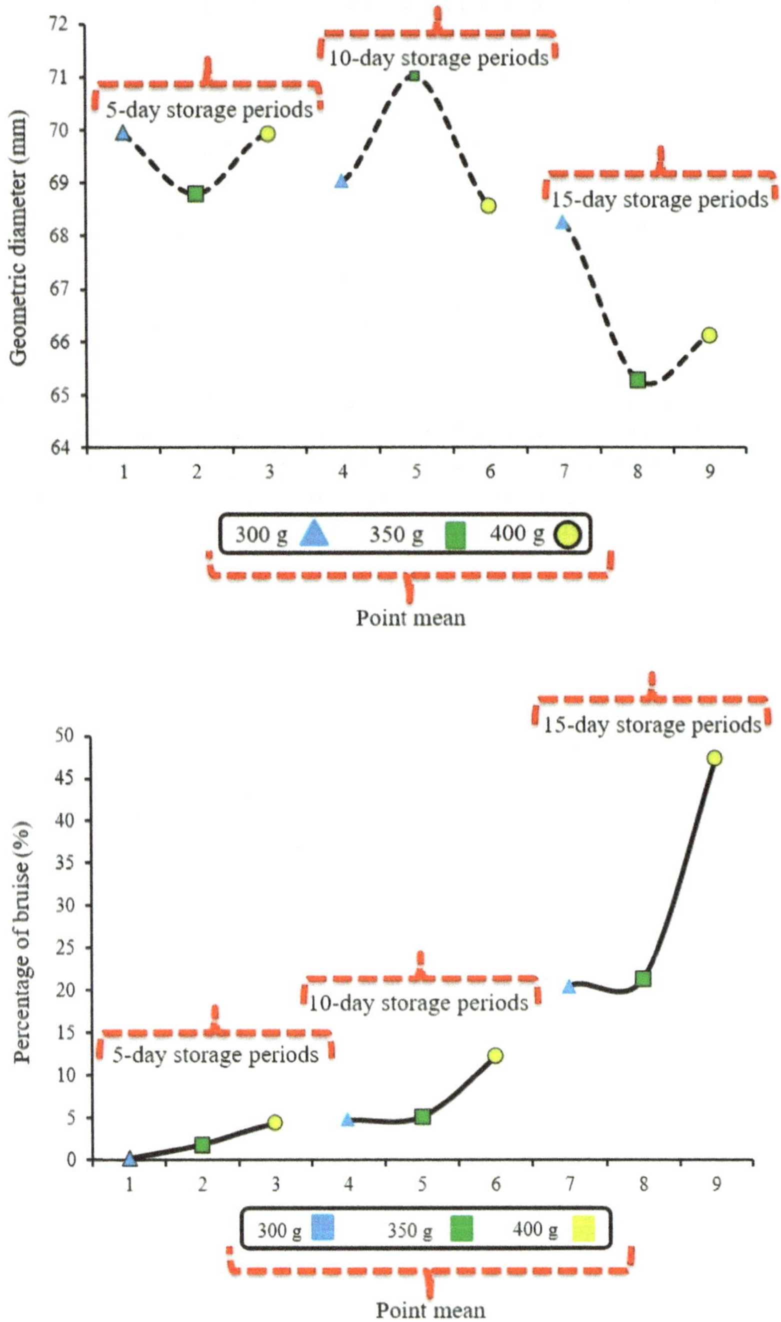

Fig. 24 Relationship between bruise level and geometric mean diameter of pears subjected to dynamic loading (Azadbakht et al., 2019c)

For the pears subjected to the 350 g impact load, the highest equivalent diameter was 299.34 mm, which was observed in the 10-day storage sample group, and the lowest was 263 mm, which was observed in the 15-day storage sample group (Fig. 25). For these pears, equivalent diameter showed a significant positive correlation with the geometric mean diameter and surface area (Table 4).

For the pears subjected to the 400 g impact load, the highest equivalent diameter, 292.37 mm, was observed in the 5-day storage sample group, and the lowest, 269.15 mm, was observed in the 15-day storage sample group (Fig. 25). The equivalent diameter of these pears showed a significant positive correlation with their geometric mean diameter, arithmetic mean diameter, and surface area (Table 4).

The correlation analysis results (Table 4) showed that the negative correlation between equivalent diameter and bruise level of the pears subjected to all three impact loads was statistically insignificant. The negative correlation between these parameters means that the bruise level is inversely related to the equivalent diameter, meaning that the former decreases as the latter increases (Azadbakht et al., 2019c).

7.5 *Effect of Surface Area on Bruise Level*

The results of the analysis of variance of the surface area of pears was found to have a significant effect at the 5% level on their bruise level in all storage periods. But its interaction effect with the impact load was not statistically significant (Azadbakht et al., 2019c).

For the pears subjected to the 300 g impact load, the highest surface area, 215,365 mm^2, was observed in the 5-day storage sample group, and the lowest, 14,669 mm^2, was observed in the 15-day storage sample group (Fig. 26). For these pears, surface area showed a significant positive correlation with geometric mean diameter, equivalent diameter, and arithmetic mean diameter (Table 4).

For the pears subjected to the 350 g impact load, the highest surface area, 15,844 mm^2, was observed in the 10-day storage sample group, and the lowest, 13,383 mm^2, was observed in the 15-day storage sample group (Fig. 26). The results of the analysis of variance of the surface area of these pears showed a significant positive correlation with their geometric mean diameter, equivalent diameter, and arithmetic mean diameter.

For the pears subjected to the 400 g impact load, the highest surface area, 153,531 mm^2, was observed in the 5-day storage sample group, and the lowest, 13,748 mm^2, was observed in the 15-day storage sample group (Fig. 26). For these pears, too, the surface area had a significant positive correlation with geometric mean diameter, equivalent diameter, and arithmetic mean diameter.

The correlation results presented in Table 4 showed a statistically insignificant negative correlation between surface area and bruise level of the pears subjected to all impact loads. This negative correlation indicates an inverse relationship between bruise level and surface area, meaning that bruise level due to impact loads decreases as the surface area increases (Azadbakht et al., 2019c).

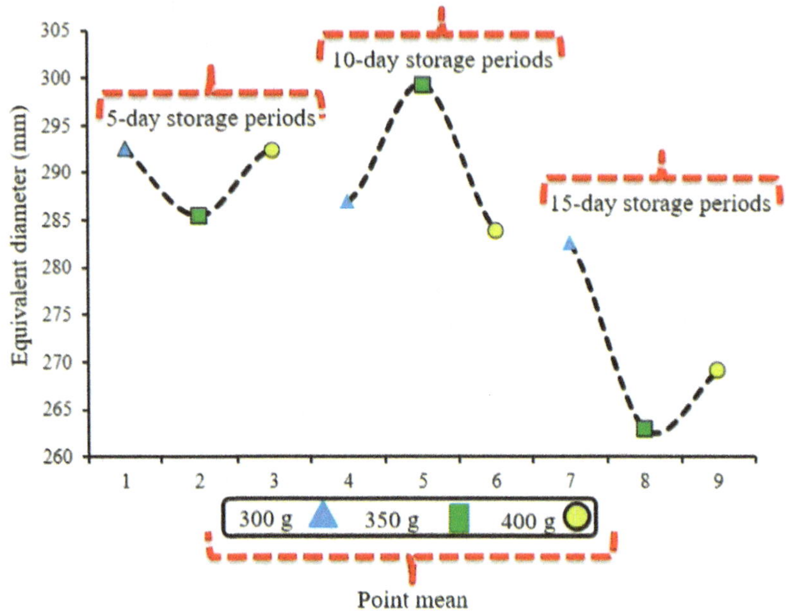

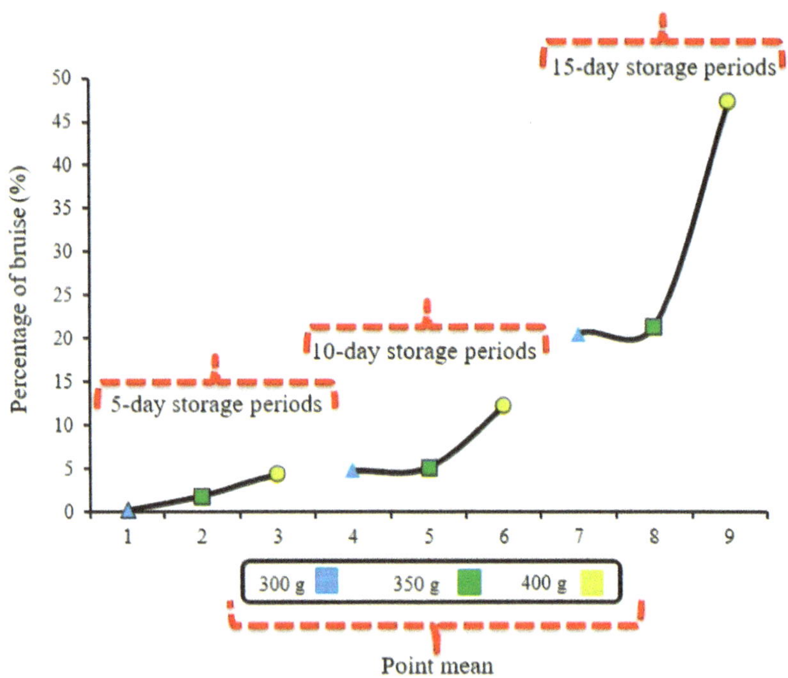

Fig. 25 Relationship between bruise level and equivalent diameter of pears subjected to dynamic loading (Azadbakht et al., 2019c)

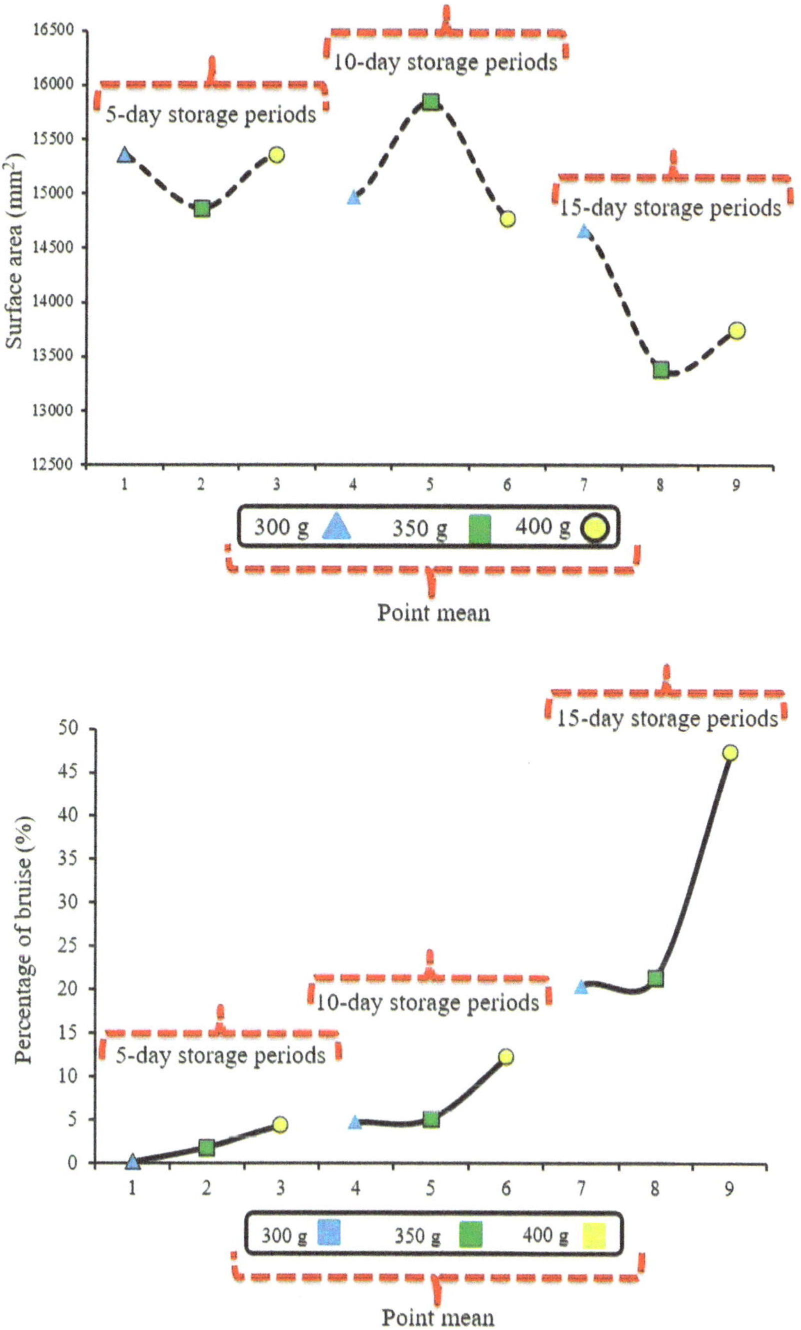

Fig. 26 Relationship between bruise level and surface area of pears subjected to dynamic loading (Azadbakht et al., 2019c)

7.6 Effect of Aspect Ratio on Bruise Level

The results of the analysis of variance of the aspect ratio did not show a significant effect on the bruise level for any storage period or loading level. The interaction effect of aspect ratio with these parameters was also insignificant.

For the pears subjected to the 300 g impact load, the highest aspect ratio was 0.831 and was observed in the 10-day storage sample group, and the lowest aspect ratio was 0.760, which was observed in the 5-day storage sample group (Fig. 27). As shown in Table 4, the aspect ratio of these pears showed a significant positive correlation with their sphericity.

For the pears subjected to the 350 g impact load, the highest aspect ratio, 0.798, was observed in the 5-day storage sample group, and the lowest, 0.748, was observed in the 10-day storage sample group (Fig. 27). For these pears, too, there was a significant positive correlation between aspect ratio and sphericity (Table 4).

For the pears subjected to the 400 g impact load, the highest aspect ratio was 0.843 and was observed in the 10-day storage sample group, and the lowest aspect ratio was 0.774 and was observed in the 5-day storage sample group (Fig. 27). The aspect ratio of these pears also showed a significant positive correlation with their sphericity (Table 4).

As shown in Table 4, the correlation results showed a statistically insignificant positive correlation between aspect ratio and bruise level of the pears subjected to all impact loads. However, as Fig. 27 shows, for the 300 and 400 g load levels, the highest bruise level was related to the 10-day period, but the percentage of bruises was higher for the 15-day period. This can be because the bruise level is more affected by the storage period (Azadbakht et al., 2019c).

7.7 Effect of Sphericity on Bruise Level

The results of the analysis of variance of the sphericity did neither have a significant effect on the bruise level for any storage period or loading level, nor had any significant interaction effect with these parameters (Azadbakht et al., 2019c).

For the pears subjected to the 300 g impact load, the highest sphericity was 0.872, which was observed in the 10-day storage sample group, and the lowest was 0.829 and was observed in the 5-day storage sample group (Fig. 28). As shown in Table 4, for these pears, sphericity only showed a significant positive correlation with aspect ratio.

For the pears subjected to the 350 g impact load, the highest sphericity, 0.845, was observed in the 5-day storage sample group, and the lowest, 0.822, was observed in the 10-day storage sample group (Fig. 28). The sphericity of these pears also showed a significant positive correlation with their aspect ratio (Table 4).

For the pears subjected to the 400 g impact load, the highest sphericity was 0.881 and was observed in the 10-day storage sample group and the lowest was 0.836 and

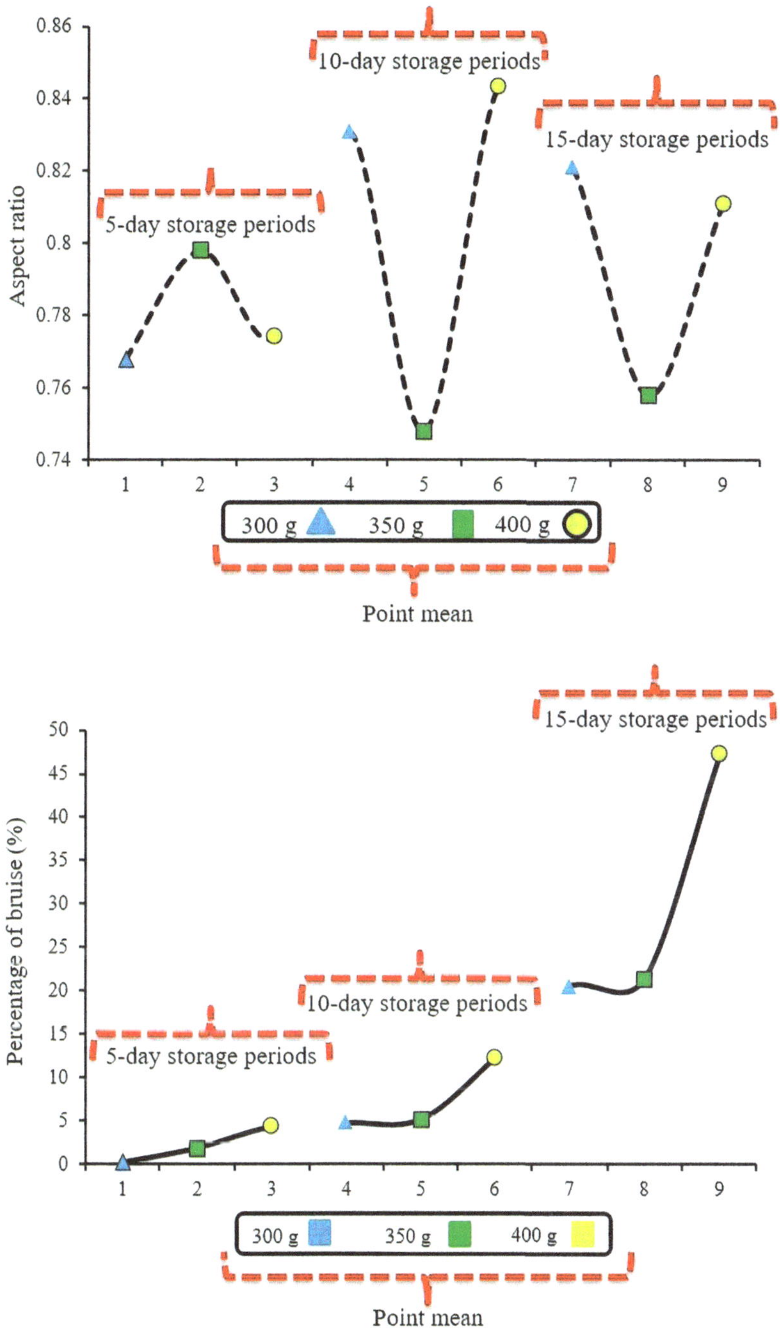

Fig. 27 Relationship between bruise level and aspect ratio of pears subjected to dynamic loading (Azadbakht et al., 2019c)

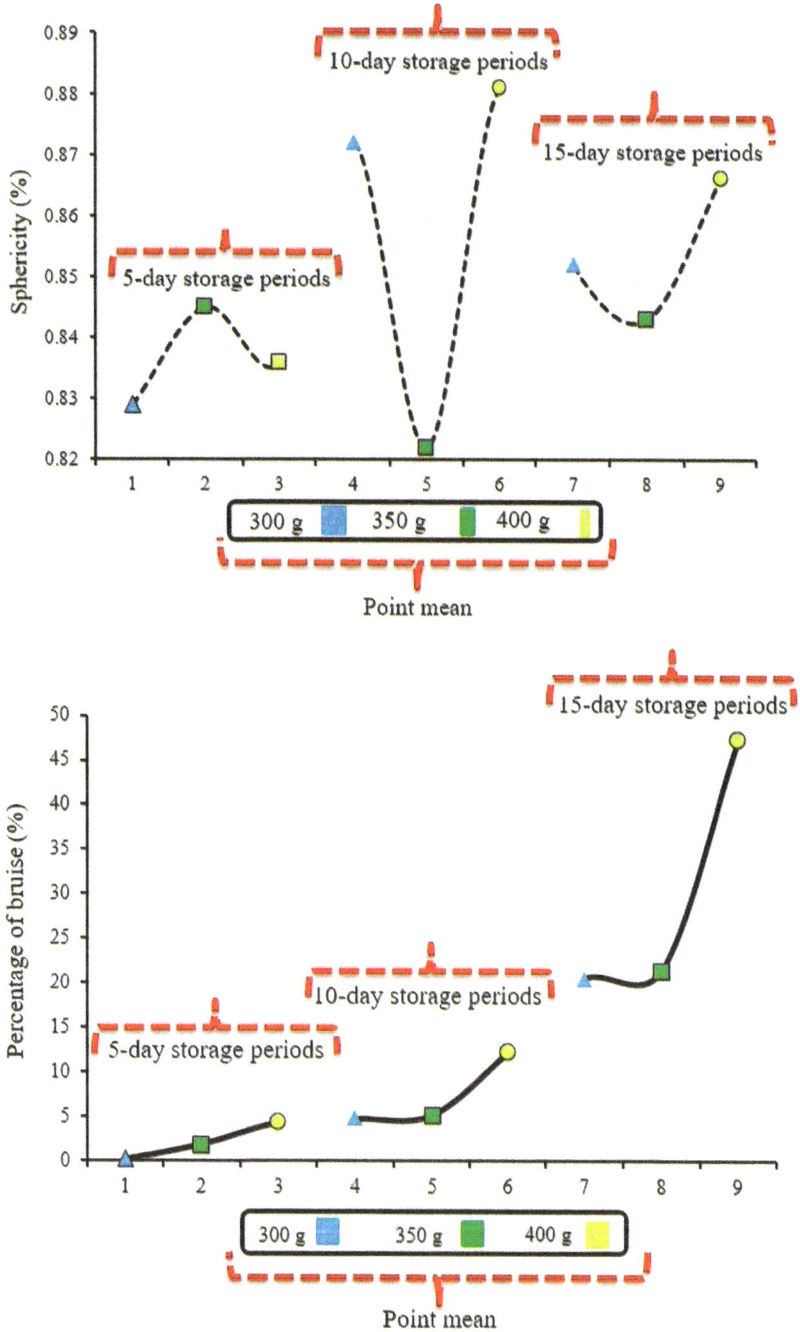

Fig. 28 Relationship between bruise level and sphericity of pears subjected to dynamic loading (Azadbakht et al., 2019c)

was observed in the 5-day storage sample group (Fig. 28). For these pears, too, sphericity only showed a significant positive correlation with the aspect ratio (Table 4).

The correlation results (Table 4) showed an insignificant positive correlation between sphericity and bruise level of the pears subjected to all impact loads. But as shown in Fig. 28, while the highest bruise level for the 300 and 400 g load levels was related to the 10-day period, for the 350 load level, it was related to the 15-day period. To explain this, it can be argued that the bruise level is more affected by the storage period (Azadbakht et al., 2019c).

The following research could be recommended for further reading: (i) Relationship of Pears' Dielectric Properties and Rates of Pears' Bruise (Azadbakht et al., 2020), (ii) The use of CT scan imaging technique to determine pear bruise level due to external loads (Azadbakht et al., 2019f), (iii) The relation of pear volume and its bruised volume by CT scan imaging (Azadbakht et al., 2019e), (iv) Effects of different loading forces and storage periods on the percentage of bruising and its relation with the qualitative properties of pear fruit (Azadbakht et al., 2019), (v) Determination of the relationship between volume and weight of pear fruit with bruise due to impact load loading using non-destructive CT scan (Azadbakht et al., 2019d), (vi) Study on the firmness and textural changes of pear fruit when loading different forces and stored at different periods using artificial neural network and CT scan (Vahedi Torshizi & Azadbakht, 2020), (vii) Investigating the Effects of Qualitative Properties on Pears Dielectric Coefficient (Mahmoodi & Azadbakht, 2021), and (viii) Application of artificial neural network and non-destructive CT scan test in estimating the amount of pear bruise due to external loads (Azadbakht and Vahedi Torshizi, 2019).

8 Conclusion

The need to be aware of the quality of agricultural products and food industry has led to increased research activities in the field of quality. With the advancement of technology, measurement has become less destructive and research has shifted to non-destructive and the use of advanced and expert systems. Numerous methods have been developed for the non-destructive evaluation of agricultural and food products, but only some of them have been technically and economically justified. CT scan imaging is one of the most successful methods for determining non-destructive internal quality. The main basis of this method is that the internal structure of an object can be reconstructed by multiple X-ray imaging. This method is very sensitive, fast, and repeatable and is successfully used in the normal course of quality control on the line of a large number of samples. There are many examples of the use of CT scans in determining the quality of agricultural products.

References

Abbott, J. A. (1999). Quality measurement of fruits and vegetables. *Postharvest Biology and Technology, 15*, 207–225.

Azadbakht, M., Mahmoodi, M. J., & Vahedi Torshizi, M. (2019). Effects of different loading forces and storage periods on the percentage of bruising and its relation with the qualitative properties of pear fruit. *International Journal of Horticultural Science and Technology, 6*(2), 177–188.

Azadbakht, M., Mahmoodi, M. J., Vahedi Torshizi, M., & Asghartabar Kashi, P. (2020). Relationship of pears' dielectric properties and rates of pears' bruise. *Agricultural Engineering International: The CIGR e-Journal, 22*(1), 169–179.

Azadbakht, M., Torshizi, M. V., Ghajarjazi, E., & Ziaratban, A. (2016). Determination of some constant parameters during cutting of canola stem determination of some constant parameters during cutting of canola stem. *Agricultural Engineering International: The CIGR e-Journal, 18*, 351–359.

Azadbakht, M., & Vahedi Torshizi, M. (2019). Application of artificial neural network and non-destructive CT scan test in estimating the amount of pear bruise due to external loads. *Journal of Research and Innovation in Food Science and Technology, 8*(2), 177–188.

Azadbakht, M., Vahedi Torshizi, M., & Mahmoodi, M. J. (2019a). Determination amount of pear bruises due to wide edge pressure via CT scan method and relation them with some physical properties pear. *Food Science and Technology, 16*, 153–163.

Azadbakht, M., Vahedi Torshizi, M., & Mahmoodi, M. J. (2019b). Determination of pear bruises due to a thin edge compression load by CT scan method. *Journal of Innovative Food Technologies, 8*(3), 305–321.

Azadbakht, M., Vahedi Torshizi, M., & Mahmoodi, M. J. (2019c). Determination of pear bruises due to impact loading via computed tomography scan method and relation them with physical properties pear. *Iranian Journal of Biosystem Engineering, 50*(2), 451–462.

Azadbakht, M., Vahedi Torshizi, M., & Mahmoodi, M. J. (2019d). Determination of the relationship between volume and weight of pear fruit with bruise due to impact load loading using non-destructive CT scan. *Iranian Food Science and Technology Research Journal, 15*(2), 341–353.

Azadbakht, M., Vahedi Torshizi, M., & Mahmoodi, M. J. (2019e). The relation of pear volume and it's bruised volume by CT scan imaging. *Journal of Food Measurement and Characterization, 7*(3), 1089–1099.

Azadbakht, M., Vahedi Torshizi, M., & Mahmoodi, M. J. (2019f). The use of CT scan imaging technique to determine pear bruise level due to external loads. *Food Science & Nutrition, 7*(1), 273–280.

Azadbakht, M., Vehedi Torshizi, M., Ghajarjazi, E., & Ziaratban, A. (2016). Application of artificial neural network (ANN) in predicting mechanical properties of canola stem under shear loading. *Agricultural Engineering International: The CIGR e-Journal, 18*(2), 413–425.

Bharath, A. A. (2008). Introductory medical imaging (synthesis lectures on biomedical engineering). *Morgan and Claypool Publishers, 3*(1), 186.

Bushberg, J. T. (2002). *The essential physics of medical imaging*. Lippincott Williams & Wilkins.

Goldman, L. W. (2007). Principles of CT and CT technology. *Journal of Nuclear Medicine Technology, 35*(3), 115–128.

Jha, S. N. (2010). *Nondestructive evaluation of food quality*. Springer-Verlag.

Kabas, O., Ozmerzi, A., & Akinci, I. (2006). Physical properties of cactus pear (*Opuntia ficus India L.*) grown wild in Turkey. *Journal of Food Engineering, 73*, 198–202.

Li, Z., Miao, F., & Andrews, J. (2017). Mechanical models of compression and impact on fresh fruits. *Comprehensive Reviews in Food Science and Food Safety, 16*, 1296–1312.

Lownds, N. K., Banaras, M., & Bosland, P. W. (1993). Relationships between postharvest water loss and physical properties of pepper fruit (Capsicum annuum L.). *HortScience, 28*, 1182–1184.

Mahmoodi, M. J., & Azadbakht, M. (2021). Investigating the effects of qualitative properties on pears dielectric coefficient. *Journal of Agricultural Machinery, 11*(1), 71–81.

RSNA. (2013). *Computed tomography (CT or CAT scan).* http://radiologyinfo.org/en/sitemap/modal-alias.cfm?modal=CT

Saunders, J., & Ohlerth, S. (2011). CT physics and instrumentation—Mechanical design. In T. Schwarz & J. Saunders (Eds.), *Veterinary computed tomography* (pp. 1–8). Wiley.

Vahedi Torshizi, M., & Azadbakht, M. (2020). Study on firmness and texture changes of pear fruit when loading different forces and stored at different periods using artificial neural network. *Iranian Food Science and Technology Research Journal., 5*(6), 113–132.

Non-destructive Testing (NDT): Development of a Custom Designed Ultrasonic System for Fruit Quality Evaluation

Fikret Yildiz, Selman Uluisik, Ahmet Turan Özdemir, and Hakan İmamoğlu

Abstract Optimum maturity at harvesting time is a very important parameter to ensure the final quality and shelf life of fruits. The fruits harvested too early or too late causes lack of aromatic compounds, different physiological disorders, and shorter shelf life. Therefore, optimizing harvesting time judgment can increase the yield and decrease the on-field losses. Fruit quality evaluation includes external (i.e., size and volume) and internal (i.e., texture, sugar content, and nutritional value) characteristics of fruits. In recent years, various non-destructive techniques have been generated to measure the quality parameters of different fruit samples at the time of harvesting or postharvest storage. Fast, cheap, and portable system without damage to fruits is required during the fruit quality determination. In this study, first progress of recent Non-Destructive Testing (NDT) and destructive technologies and their applications for fruit quality evaluation are analyzed and compared by considering advantages and disadvantages. Second, performance of a developed custom-designed ultrasonic system is evaluated for internal quality evaluation of fruits. Ultrasonic system includes a programmable bipolar remote pulser unit (Ultrasonar, US 100), a couple of probes for signal transmission and reception, an oscilloscope, and a computer. Performance of the developed ultrasonic system is presented and

F. Yildiz (✉)
Faculty of Engineering, Department of Electrical and Electronics Engineering, Hakkari University, Hakkari, Turkey
e-mail: fikretyildiz@hakkari.edu.tr

S. Uluisik
Burdur Food Agriculture and Livestock Vocational School, Burdur Mehmet Akif Ersoy University, Burdur, Turkey
e-mail: suluisik@mehmetakif.edu.tr

A. T. Özdemir
Department of Electrical and Electronics Engineering, Erciyes University, Kayseri, Turkey
e-mail: aturan@erciyes.edu.tr

H. İmamoğlu
Department of Radiology, Erciyes University Medical School, Erciyes University, Kayseri, Turkey
e-mail: himamoglu@erciyes.edu.tr

P. B. Pathare, M. S. Rahman (eds.), *Nondestructive Quality Assessment Techniques for Fresh Fruits and Vegetables*, https://doi.org/10.1007/978-981-19-5422-1_12

discussed in the light of experiments based on the attenuation measurement using TT (Through Transmission) mode. In this study, four different tomato varieties have been used. Moreover, external characteristics (i.e., size and volume) of fruits are also determined using a developed image vision system for the packaging process of fruits. At the final stage of the study, firmness of four different tomato varieties (i.e., Salkım, Beef, Village, and Cherry) was measured using a commercially available ultrasonic scanner (Siemens Acuson S3000). As a summary of this study, the performance of a complete non-destructive quality evaluation system including custom designed (a) ultrasonic testing and (b) automatic machine vision system is reported and discussed.

Keywords NDT · Ultrasonic testing · Automatic machine vision system · Fruit's firmness measurement

1 Introduction

Fruits and vegetables are essential natural products, which are important sources of vitamins, minerals, and antioxidants (Rizzolo et al., 2010). With the increasing awareness of people in the field of health, consumption of fruits and vegetables is increasing day by day. Therefore, their quality level is a critical factor both for producers and consumers as it determines market acceptance, storage, postharvest processing operations (such as transporting), and price of the product. Main quality parameters when purchasing fruit and vegetables are based on external aspects, such as size, color, main integrity, and firmness (Opara & Pathare, 2014). Then internal quality parameters related to sugar, acidity, and aroma are evaluated by consumers (Minas et al., 2020). Based on these parameters, the maturity stage is an important time point to offer fruits and vegetables in optimum internal and external quality conditions to consumers. Therefore, determination of the harvesting time of fruits, which depends on species, varieties, and environmental conditions, is a fundamental task. Although most of the fruits are harvested at the commercial ripening stage, some of those are harvested at immature or over-ripening stages (Arunkumar et al., 2021). Consequently, knowledge of the proper harvesting time, evaluating, sorting, and grading of fruit crops are of great importance, as it reduces the quality deterioration of fruits and provides sufficient time for postharvest conditions.

Fruit ripening is a part of a complex developmental process that is highly coordinated and includes changes in color (chlorophyll degradation and increase in carotenoids), texture (cell wall loose), taste (decrease in organic acids and increase in sugar), and flavor (increase in volatile compounds). Measuring these parameters, such as titratable acidity, soluble solids contents, color index, and firmness is essential to monitor the fruit ripening process (Lu et al., 2020). Fruit firmness and color are among the most important fruit quality traits in fruit species, it is an indirect measurement of ripeness. In their early years, fruit growers evaluated the quality of fruits on their own intelligence based on the change in color, size, shape, and weight. After that, evaluations of fruit quality parameters based on their external (i.e., texture, color, shape, and size) and internal (i.e., sugar composition, aroma, and nutritional value) have been made manually, automatically, or chemically (Opara and Pathare, 2014). For example, fruit firmness has still been estimated mostly with portable equipment, based on the penetration of a cylindrical head into the flesh fruit to measure the maximum applied force. However, it was realized that most instrumental techniques measuring these properties are destructive, labor-intensive, and tedious. Furthermore, the use of destructive methods is known to give errors in measurements applied by different people, which may reduce the accuracy and inapplicable to in-line grading and sorting (Magwaza & Tesfay, 2015). Therefore, these destructive methods are not convenient for industries, such as packaging industry, as it ruptures the fruit tissue which cannot take a measurement from the whole fruit. Also, increasing consumer demand for both better and healthier fruits has spurred considerable interest in the fresh product industry to develop fast, cost-effective noninvasive instrumentation for detection and monitoring of fruit quality.

In recent years, there is a growing interest in designing new instruments capable of non-destructive, noninvasive, and noncontact measurements for fruit quality determination. These techniques are becoming more favored and practical as it provides several advantages compared to traditional destructive techniques (Li et al., 2017). For example, some fleshy fruits, such as tomato becomes more colorful, softer, sweeter, and aromatic in a few days. It is unrealistic to take simultaneous measurements on physical and chemical attributes that can be quantified during ripening including size, firmness, color, and several internal components (Reid, 1992). Therefore, simple representative non-destructive measurements are thus required to evaluate the ripeness of a fruit. Moreover, non-destructive sensor provides to obtain repeated measurements on the same fruit which reduces waste and contributes to big data containing potentially relevant information for use in fruit quality prediction models (Osinenko et al., 2021). In this way, improvements in non-destructive methods could allow for fruits on the plant or sorting line to be analyzed quickly and repeatedly.

The measurement of fruit maturity by non-destructive methodologies back in the early 1970s with the development of light transmittance techniques (Povey & Wilkinson, 1980). Since then, a variety of non-destructive techniques have been offered including colorimetry (Baltazar et al., 2008), visible imaging (Choi et al., 1995), fluorescence imaging (Cerovic et al., 2009), and Magnetic resonance imaging (MRI) (Zhang & McCarthy, 2012). Apart from those, several researches have been conducted on the internal quality of fruits non-destructively using different methods/techniques, which are commonly focused on firmness measurements (Khosroshahi et al., 2017; Mireei et al., 2015). Some of them are summarized below: mechanical-based techniques for kiwifruit texture analysis using falling impact (FAI), forced impact (FOI), and acoustic impulse-response (AIR) (Khosroshahi et al., 2017), a non-destructive impact testing for date firmness determination (Mireei et al., 2015), visible/near infrared reflectance spectroscopy (Vis/NIRS) (Cavaco et al., 2009; He et al., 2005), and laser-induced acoustic waves for firmness/maturity evaluation of apple (Hitchman et al., 2016), evaluation of surface acoustic waves for watermelon firmness (Ikeda et al., 2015), and magnetic resonance imaging system for observation of different maturity level of tomato (Zhang & McCarthy, 2012).

Among these techniques, ultrasonic-based technique has become a more common modality due to cost-effectiveness, robustness, reliability, and fruit safety concern (Kim et al., 2006; Lee et al., 2013; Morrison & Abeyratne, 2014; Valente et al., 2013). In ultrasonic-based non-destructive measurement systems, ultrasonic velocity and attenuation of the ultrasonic signal are used to find correlation between elastic modulus and firmness fruits (Ikeda et al., 2015; Srivastava et al., 2014; Vasighi-Shojae et al., 2018). In addition to applications of non-destructive testing of fruits to determine interior attributes, external characteristics of fruits, such as volume, area of color should be considered to meet high-quality fruits of producers and consumers. Computer vision-based noncontact measurement systems have already been introduced to address superior quality assessment of the physical appearance of fruits. Size of fruits is generally referring to the volume of fruits and mass because they have a key role to be considered packaging process and transportation and should be optimized for producers to increase profit. These computer-based innovation techniques can provide fast and accurate grading, sorting, and packaging of fruits and vegetables. Several previously published works exist in the literature addressing the volume and mass estimation of fruits with different shapes based on computer vision techniques (Ahmad Saad et al., 2015; Concha-Meyer et al., 2018; Gokul et al., 2015; Iqbal et al., 2011; Ireri et al., 2019; Islamadina et al., 2018; Khojastehnazhand et al., 2010; Sabliov et al., 2002; Siswantoro & Asmawati, 2016; Venkatesh et al., 2015) and mentioned advantages over conventionally used water displacement, mathematical model, and manual measurements of fruits volume and mass.

A study by Nyalala et al. (Nyalala et al., 2019) was first developed regression model between mass and volume. This relation was described using different images of tomatoes, which were captured using a single camera and features were extracted from 2D, 3D, and combination of 2D and 3D depth image results and evaluated using five different regression models: Linear SVM, Quadratic SVM, Cubic SVM, RBF-SVM, Bayesian-ANN. RBF-SVM model showed better results among other with an accuracy of 0.9706 (only 2D features) and 0.9694 (all features) in mass and volume estimation, respectively. Using predicted volume and mass were then applied to the established mass-volume power function (Nyalala et al., 2019). Another study is referring to challenging points of volume and mass estimation for irregular shape fruits and vegetables using single camera and develop innovative split and merge technique imaging techniques. This method includes a few preprocessing stages to reduce the noise and extract the object from the background. The segmented object after these preprocessing steps has been split into two parts along the major axis and the boundary was extracted. The volume was obtained by integrating the polynomial of each part of the valid boundary. Total volume of fruit or vegetable was then calculated by summing the volumes from the parts. Mass is computed by multiplying volume with density (Jana et al., 2019). A different study used three cameras to capture the image of tomato. Color, shape, and size were accepted as the key features for quality determination. More specifically, color classifier is determined using the Hue value distributions of tomatoes with different maturities such as mature, semi-mature, and immature tomatoes. Next, the shape classifier is established using the first-order Fourier descriptor (1D-FD) to describe the radius sequence of tomato contour and the irregularity of tomato contour was calculated by established equations. The relation between transverse diameters of actual tomatoes and tomato images was found using a linear regression equation and comprehensive classifier including color, shape, and size showed that 90.7% mean grading accuracy of the proposed method was achieved (Srivastava et al., 2014).

Tomato (Solanum lycopersicum), is the second most widely produced, consumed, and traded agricultural product in the world both in raw and processed forms. With its good appearance and nutritional value, which is high demand in recent years, tomatoes have become one of the most popular agricultural products in the world (Abdeldym et al., 2020). Worldwide production of fresh market and processed tomatoes has steadily increased during the last 10 years and reached an annual production of about 180 million tons with a net value of over $ 190.4 billion (Food and Agriculture Organization of the United Nations, n.d.). The leading countries for production of tomatoes in the world are the People's Republic of China (PRC) (62.8 million tons), India (19 million tons), Turkey (12.8 million tons), and the USA (10.8 million tons), respectively (Food and Agriculture Organization of the United Nations, n.d.). Tomato is usually harvested at the earlier stage of their growth (green-ripe stage) for long market life as they can ripen during transportation and lose their quality parameters (Li et al., 2020). It is a fact that tomato does not ripen at the same pace, even though they are harvested at the same time. Tomatoes in different ripening periods in the same storage environment will trigger each other's ripening, and thus the fruits will be obtained with different ripening and softening levels at the end of storage. This situation is undesirable for the seller and the consumer. Therefore, it is better to detect the quality parameters of tomatoes during storage and to classify them due to increased attractiveness of fruits. There is no present system or technique to measure the ripeness of tomatoes. The ripening stage affects variety of fruit quality attributes. The firmness of fruits and vegetables, for example, is directly related to the ripening stage. Therefore, firmness is considered a key internal quality factor for sorting, grading, and packaging (Vasighi-Shojae et al., 2018). Compression or penetration tests, which are destructive methods, are the conventional method to determine the firmness of fruits (Peng & Lu, 2005).

In this chapter, a non-destructive system including ultrasonic and machine vision methods for tomato quality evaluation is presented. Firstly, the performance of a custom-made US100 ultrasonic system was tested for tomato quality evaluation in terms of attenuation and propagation velocity of ultrasonic waves. Secondly, the automatic machine vision system is designed for volume calculation. Finally, firmness measurement was performed using commercial Siemens Acuson S3000 US scanner (Siemens Medical Solutions, Mountain View, CA, USA) by measurement of the propagation velocity of ultrasonic waves at fixed frequencies ranging from 4 to 9 MHz on tomato and firmness was determined based on propagation velocity. The specific objectives of this study were: (i) to develop an efficient custom designed ultrasonic measurement system, (ii) to develop an automatic vision system for volume estimation and feature extraction, and (iii) develop a tomato classifier depending on the firmness and volume.

2 Fruits Samples and Quality Assessment

An example of tomato quality is presented in this chapter. Four different tomato varieties (Salkım, Village, Cherry, and Beef) at their commercial ripe stage were purchased from a local market as these fruits differentiate in firmness. The fruits were transported into the laboratory, cleaned with deionized water, and then visually inspected to ensure that they were free of damage, uniform in color, shape, and size based on the variety they belong to. All the non-destructive and destructive evaluations of tomato quality parameters were carried out at Erciyes University, Kayseri, Turkey. The destructive measurements were taken after non-destructive (ultrasound) and noncontact (computer vision system) measurements for the use of the same tomato samples to make a more reliable comparison between the two techniques. First, the weights of all fruits were taken by analysis balance. The skin color of the tomato fruits was measured by using a Color Meter (PCE-CSM 1) and recorded as Hunter's L*, a*, and b* values (at least 15 fruits from each tomato variety). Fruit mechanical properties were investigated by pushing a probe (6 mm) into the pericarp (PCE-PTR 200 Penetrometer) and results were expressed in Newton's (N). The assessment was performed at two opposite locations along the fruit equator after peeling around a 2 cm^2. The average of maximum forces was recorded to represent the fruit firmness value.

2.1 Experimental Setup and Data Collection

2.1.1 Ultrasonic System

Non-destructive measurements were completed using two different ultrasonic systems: commercially available Siemens Acuson S3000 US scanner (Siemens Medical Solutions, Mountain View, CA, USA) and custom made US100 ultrasonic system, which was developed by Ultrasonar Inc. Stiffness of tomato samples were measured in terms of elastic modulus (E, kPa) and velocity (v, m/s) parameters using the commercially available ultrasonic system in Pulse-Echo (PE) mode as shown in Fig. 1a. As an example, Figs. 2 and 3 show stiffness measurement results. A custom-made ultrasonic scanning system was used to measure the attenuation and velocity of transmitted signal in Through Transmission (TT) mode. Thus, velocity and attenuation measurements are important ultrasonic parameters to determine the physical characteristic of fruits in the literature (Kuo et al., 2008; Mizrach, 2004, 2008).

Developed ultrasonic system basically consists of a pulser (US 1000), which was developed by Ultrasonar Defense and Aeronautics Technologies Inc., a couple of

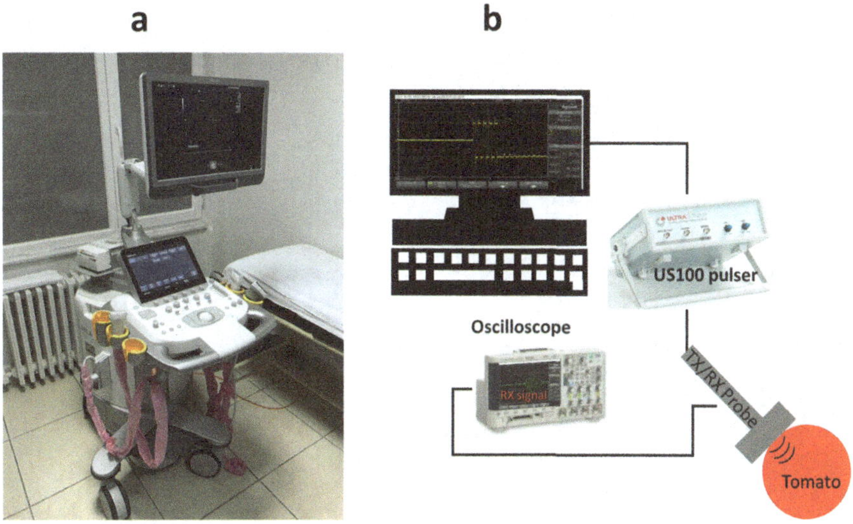

Fig. 1 (**a**) Commercially available Siemens Acuson S3000 US scanner (Siemens Medical Solutions, Mountain View, CA, USA) and (**b**) developed US100 pulser and measurement system

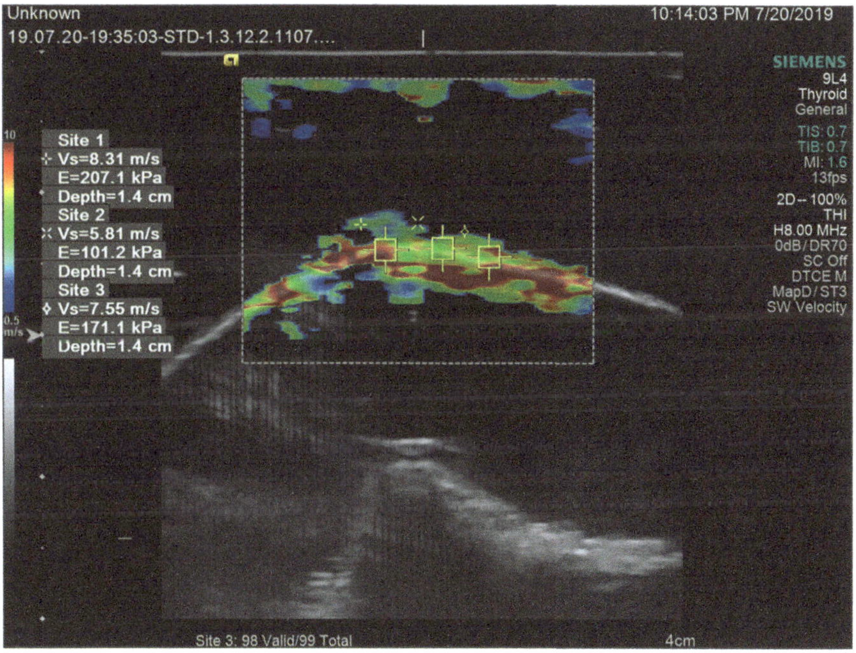

Fig. 2 Stiffness measurement results of a tomato sample using Siemens S3000

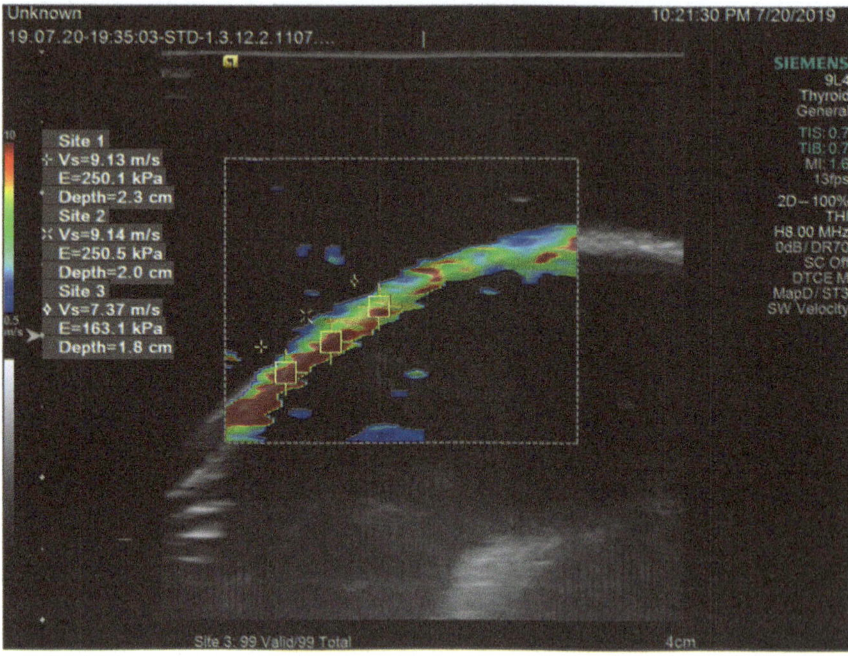

Fig. 3 Stiffness measurement results of a different tomato sample using Siemens S3000

transmitter/receiver probes, a PC for data processing and display the data. The developed experimental system is illustrated in Fig. 1. The US 1000 has outputs up to 400 Vpp with frequencies from 1 Hz to 10 MHz. The US 1000 pulser can be controlled via a PC over a USB port and give the users much flexibility during the measurements in terms of ultrasonic operation parameters such as frequency, Pulse Repetition Frequency (PRF), number of pulses in a burst, starting pulse (positive or negative) and internal or external triggering option. Pulser unit and experimental setup are shown in Fig. 1b. During the experimentation, fruits were placed between 55 kHz TX/RX transducers, TX probe was excited with the pulser unit, and RX probe was used to collect back-attenuated signals and connected to the oscilloscope in order to visualize data. Transducers were coupled to fruits with an ultrasonic gel in order to increase the amount of TX signal. More detail about developed ultrasonic system and experimentation can be found in the literature (Ozdemir, 2018; Uluisik et al., 2018; Yildiz et al., 2018, 2019).

2.1.2 Machine Vision System

Figure 4 shows the schematic of the machine vision system and Fig. 5 shows the different pictures of the fabricated image acquisition system (Uluisik et al., 2018). This system consists of three main parts: five high-resolution cameras, a LED light

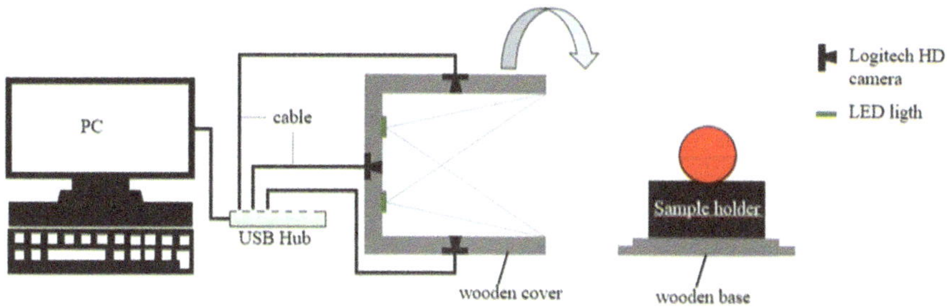

Fig. 4 Schematic of Machine Vision system

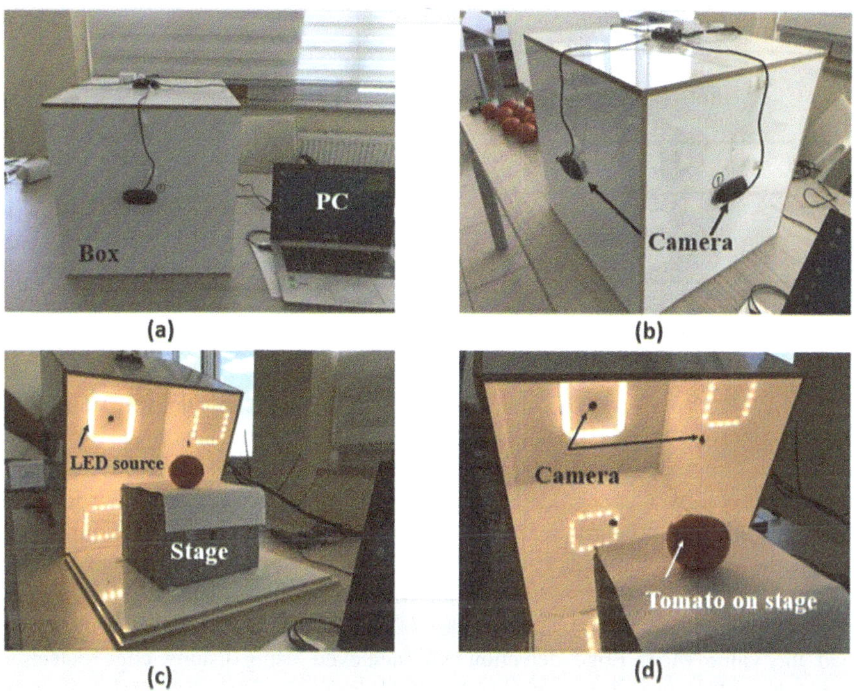

Fig. 5 Fabricated computer-based machine vision system. (**a**) General view, (**b**) side view, (**c**) interior view of image box, and (**d**) magnified view of (**c**)

source, and a cube-shaped white box. The wooden box was designed and fabricated with a size of 40 cm × 40 cm × 40 cm and was used to eliminate unwanted light sources from outside during the image acquisition from the fruits. Five images were taken from each tomato variety by using the five high-resolution cameras (Logitech), which were mounted on five sides of the box. A LED light source with a power of 50 W LED was placed at the top center of the box and used for the illumination of samples. A sample holder was used to place each fruit at the center of the camera's

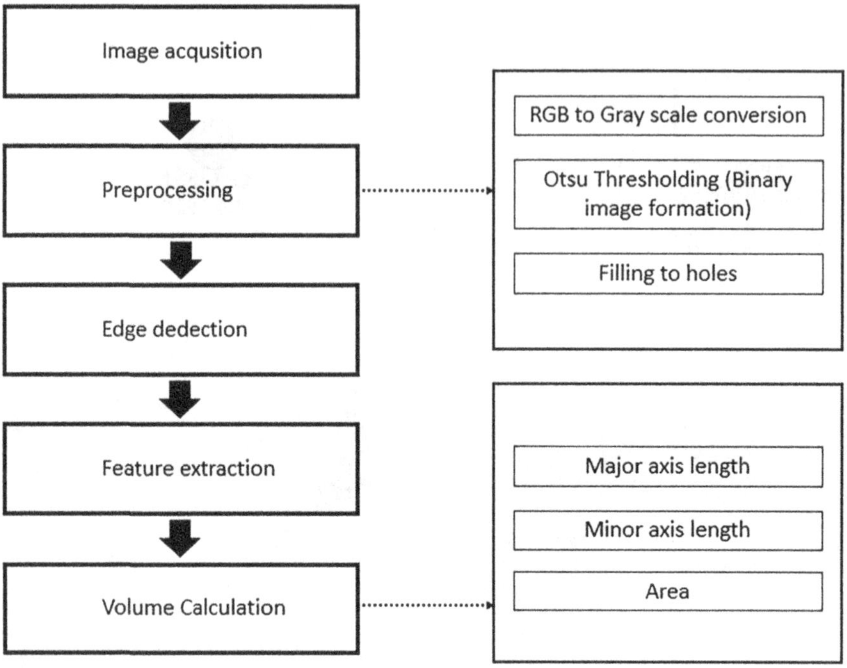

Fig. 6 Flowchart of image acquisition and processing

field of view. Captured RGB color images taken from each sample were transferred from cameras to PC via USB ports and saved in jpg format with a 1280 × 960 resolution. Preprocessing techniques were applied to the input (captured) images to remove noise as well as segment the object from the background. The RGB image is then transformed into a grayscale image. Using Otsu thresholding technique (Otsu, 1979), the segmented (binary) image is generated from the grayscale image. Tomato region is black, and the rest of the parts are white in the threshold image. Because the pixels showing the tomato had the value of 0 and the remaining pixels in the image had the value of 1. Edge detection was achieved using Canny edge detection methods and major, minor axis, and surface area of each captured image were extracted by completing the feature extraction process. It was assumed that tomato had an axisymmetric geometry and binary image includes small cylindrical image elements or pixels. The volume of the tomato was calculated by summing these image elements' volumes. Flowchart of image acquisition and processing steps are shown in Fig. 6. Figure 7 shows the captured image of different tomato varieties from different view using the developed vision system. More details about the volume estimation process of tomato was explained in our previous studies (Uluisik et al., 2018; Yildiz et al., 2019).

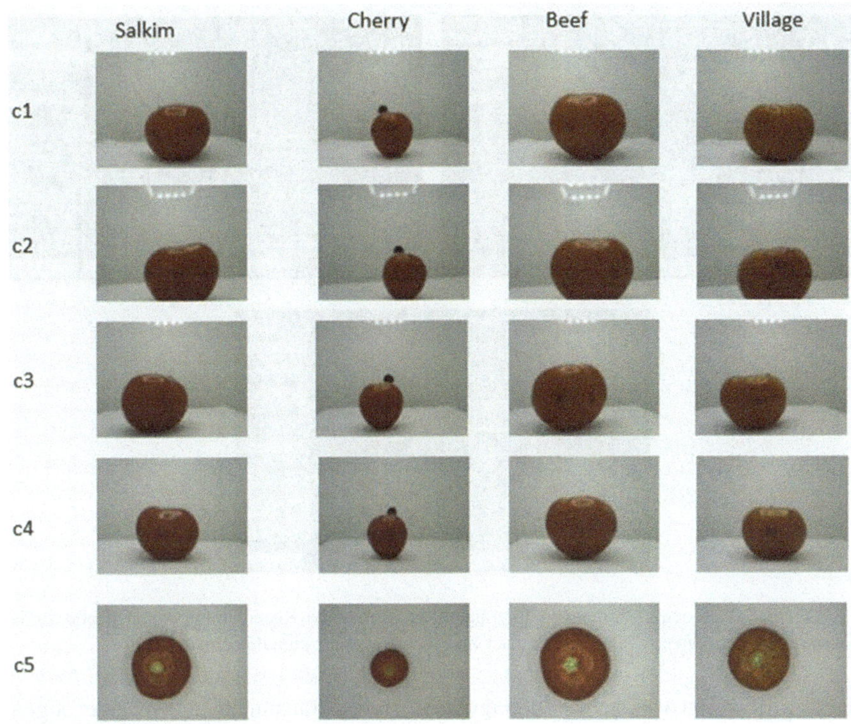

Fig. 7 Captured images of five different tomato samples from different view using the five cameras. c1, c2, c3, c4, and c5 present the number of camera mounted on the wooden box, respectively

2.1.3 Statistical Analysis

The paired t-test was used for testing whether the difference between the two measurements was significantly different. The important feature of this test is its ability to compare the measurements within each subject. These analyses were performed using the Excel Analysis Toolpack option (MS Corporation, Redmond, WA, USA).

2.2 *Ultrasonic Measurement (Attenuation and Velocity)*

Ultrasonic transducers were connected together and 180 Vpp Fig. 8a electrical pulses were applied to the transmitter. Transmitter ultrasonic vibrations actuated receiving transducer and 250 Vpp receiver responses measured by an oscilloscope as shown in Fig. 8b. This means of ultrasonic receiver is very sensitive and a good

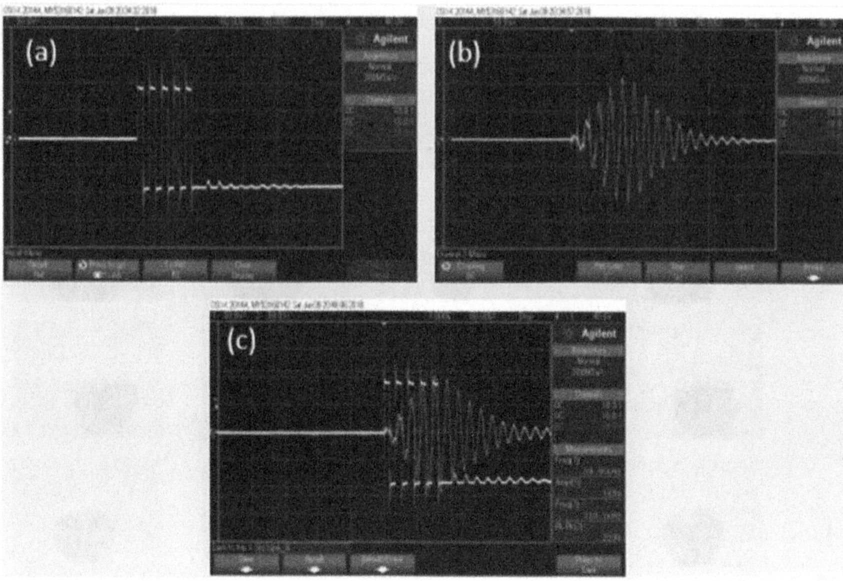

Fig. 8 (**a**) Pulser output connected to transmitter probe, (**b**) Receiver ultrasonic probe signal response, and (**c**) Directly connected pulser and receiver probe signal acquisitions

signal acquisition was achieved. Figure 8c shows transmitter and receiver signal acquisitions. This high sensitive ultrasonic probe couple cannot achieve to transmit and receive ultrasonic signal from a tomato. It was experimentally confirmed that ultrasonic pulses cannot pass through the fruit body as shown in Fig. 9. Similar results were obtained on the different fruits such as apricot and apple (Yildiz et al., 2019). More details about the results of ultrasonic measurement are explained below.

Fruits with and without shell have been used to measure attenuation and velocity of the transmitted ultrasonic signal. As expected, experimental results showed that the TX signal was strongly attenuated while the signal passes through the fruit. Before experimental measurements, reliability test was performed by contacting TX and RX probe with each other (there was no fruit sample between probes) and the received signal was observed on the oscilloscope. However, as it can be seen in Fig. 9, receiving probe did not collect the signal, even under the high amplitude TX signal (e.g., 120 V) as for tomatoes. Porous-uneven surface and pericarp of the fruit were assumed the main parameters preventing successful measurement of the TX signal due to high scattering of ultrasound on the surface of the fruits under measurements. Therefore, the measurements taken in the TT (Through Transmission) mode were carried out by peeling the fruits and applying ultrasound to the peel. Similar results were obtained and RX probe was not able to collect signals as fruits without skin were used for the experiment. However, it was realized that RX probe can collect the attenuated TX signal when only the skin of tomatoes or apricots were used for experimentation (Yildiz et al., 2019). Moreover, considering the high absorption of ultrasonic signal inside the fruits, TX probe with an aluminum horn

Fig. 9 Testing of developed US100 system for attenuation of TX signal in the TT (Through Transmission) mode

was designed and fabricated to focus on transmitting signal and achieve high-intensity TX signal (Yildiz et al., 2019). Same measurements were repeated after assembly with ultrasonic gel filling the horns mounted at the ends of the receiver/transmitter probes. However, we confirmed that using the ultrasonic horn has no significant impact on the improvement of the received signal.

In this study, it was concluded that TT (Through Transmission) mode ultrasonic testing is suitable for the very thin slice of fruit samples rather than the whole fruit. On the other hand, in the literature (Mizrach, 2000, 2008), it was shown that TT (Through Transmission) mode ultrasonic testing gives successful results for fruit quality evaluation. In summary, we confirmed that our developed ultrasonic system works properly when fruits were not sandwiched between probes. However, when the measurements were repeated on the fruits, the same result was not observed compared to previous studies in the literature (Mizrach, 2000, 2008). More studies are required to confirm the developed TT (Through Transmission) mode ultrasonic system reliability for the fruit quality measurements.

2.3 *Volume Estimation*

As a representative, different images of "Salkim" variety were captured from five different angles (Fig. 10a) and original images were converted to grayscale images

Fig. 10 Original image and the images generated in each step of processing. (**a**) Original image, (**b**) grayscale image, (**c**) binary image, and (**d**) find contours (edge)

(Fig. 10b), respectively. Binary images and contour detection were the next steps to find out the volume of tomatoes (Fig. 10c, d). Volumes of tomatoes were obtained by considering the tomato fruit as an axis-symmetric object. Results were compared to theoretically calculated volumes by assuming tomato as an ellipsoid. It was shown that the deviation between theoretical and developed system predictions is more than 10% (Uluisik et al., 2018). In other words, the developed image processing system in this study estimates the volume of tomatoes 10% bigger than the theoretical calculations. Shadows at the bottom of the tomatoes due to the position of the LED light source and the concave shape of the fruits might be the potential reasons causing the volume differences between theory and image processing method. Moreover, the application of different image processing techniques might eliminate shadows and minimize the volume differences between developed image processing systems and theoretical values. In this study, preliminary results of developed image vision system were obtained and discussed to determine the outer characteristics (size and volume) of the fruits (Uluisik et al., 2018). This system provides practical, cheap, and fast analysis of physical properties (surface area/volume) of fruits; however, improvements of the developed system are required in terms of accuracy and reliability concerns compared to destructive methods and it also lacks other quality characteristics determination such as firmness. Thus, improvement of prediction performance of image processing method needs to be studied in the future.

Table 1 The values of measured parameters, weight (g), maximum penetration force (N), and color parameters of peel (L*, a*, b*) in the examined varieties of tomatoes. Different letters indicate significant differences (Student's t-test; $p < 0.05$) (n = minimum 15)

			Color		
Varieties	Weight (g)	Firmness	L	a	b
Salkım	111.16 ± 4.66^{a}	10.17 ± 0.34^{a}	36.4 ± 0.12^{a}	19.9 ± 0.12^{a}	15.7 ± 0.23^{a}
Beef	167.00 ± 5.83^{c}	8.68 ± 0.45^{c}	38.7 ± 0.32^{c}	21.4 ± 0.31^{c}	19.8 ± 0.41^{b}
Köy	184.50 ± 4.89^{dc}	8.61 ± 0.28^{dc}	39.5 ± 0.59^{d}	21.0 ± 0.54^{c}	19.3 ± 0.63^{b}
Cherry	42.33 ± 0.81^{b}	6.67 ± 0.20^{b}	37.3 ± 0.24^{b}	26.7 ± 0.41^{b}	19.6 ± 0.24^{b}

2.4 *Stiffness and Velocity Measurement*

The destructive quality analysis focused on the assessment of fruit color and firmness. Additionally, we also weighted all fruits that were used for the analysis. As expected, the average weight of cherry tomatoes was significantly less (42.3 g) compared to the other three varieties (Table 1). There was no statistical difference between the weight of "Beef" and "Village" tomato varieties. Colorimeters express colors in numerical terms along the L*, a*, and b* axes (from white to black, green to red, and blue to yellow, respectively) within the CIELAB color sphere which are usually mathematically combined to calculate the color indexes. The values of the L*, the parameter determining the peel brightness of the tested samples ranged from 36.4 (Salkim) to 39.5 (Village). Although there is not a huge difference for L* value among cultivars, all values are significantly different from each other. The a* value is the most important color range in tomato ripening as increasing this value indicates the loss of green color. The values of the a* parameter, corresponding to the red color, ranged from 19.9 (Salkim) to 26.7 (Cherry) for peel. Based on this, the "Cherry" variety could be the most ripened variety. Obviously, these parameters change from variety to variety. In summary, it can be said for the color measurement of four varieties, the "Cherry" has the most intense red color, but the "Village" variety has light red color. The a* value, at the level of 22.72–34.01, and the b* value, at the level of 14.53–28.86 were determined by the color of ground tomatoes (Zalewska-Korona and Jablonska-Rys, 2012).

In another study, the tomato color parameters a* and b* were at the levels of 9.55–21.09 and 4.91–18.85, respectively (Ordóñez-Santos et al., 2008). The color parameters were taken from oval, oblong, and cherry tomatoes and a* 19.8–26.7, b* 26.3–29.5, and L* 42.0–45.5 values were obtained (Marsic et al., 2011). In a recent study using different colors of tomato, the values ranged for L* (from 32.88 for Olmeca to 51.68 for Beef-orange), a* (from 2.56 for Cherry-orange to 22.09 for Malinowy), and b*(from 12.65 for Cherry-minimalina to 48.8 for Beef-orange) were determined (Bojarska et al., 2020). Obviously, the color of the tomato fruit, its intensity and saturation, depends on the variety, environmental conditions, and the harvested time point.

Table 2 Measurement results of elastic modulus (E) and velocity (m/s)

Varieties	E (kPa)	v (m/s)
Salkim	225.69 ± 23.70	8.40 ± 1.59
Beef	183.98 ± 24.82	7.71 ± 0.77
Village	165.94 ± 32.09	7.17 ± 0.79
Cherry	141.80 ± 29.43	6.75 ± 0.73

Firmness is one of the most important indicators of sensory quality, and it was the main target to compare the firmness evaluations measured with penetrometer to non-destructive ultrasound method. The texture of the tomato varieties was determined by analyzing the maximum penetration force (N), at the given test deformation (2). The parameter maximum force (N) was from 6.67 N (Cherry) to 10.17 N (Salkim). There was no significant difference in firmness value between "Beef" and "Village" tomato varieties. Therefore, it can be said that the small-fruit tomato variety (Cherry) was characterized by lower values of maximum penetration force, by approximately 40% of "Salkim." Maximum penetration force (Fmax) was from 8.80 N (Cherrymini Malina) to 21.40 N (Olmeca) supporting our results that small-fruited tomatoes were characterized by lower values of maximum penetration force and firmness. According to Kowalczyk et al., 2011, the firmness of the fruit pericarp depends on the harvest date, the culture medium in which they are grown, and their variety. As reported by (Cantwell et al., 2009), the lower firmness of small fruits is associated with a higher content of acids and sugars. Therefore, the higher a* value of "Cherry" variety, which is the main indicator of ripeness, could be associated with less pericarp firmness, and vice versa for "Salkim" tomato.

With regards to ultrasonic measurements elastic modulus (E) and velocity (v) data varied from 141.8 (kPa) to 225.7 (kPa) and 6.75 (m/s) to 8.4 (m/s), respectively (Table 2). The mean values of the measured firmness of the tomato fruits (Table 1) are aligned from the firmest (Salkim) with an apparent $E = 225.7$ kPa and $v = 8.4$ m/s values to the softest tomato samples (cherry) with an apparent $E = 141.8$ kPa and $v = 6.75$ m/s values. It can be seen that the changes in E and v values were clearly correlated with the tomato firmness measured by the destructive method. In other words, E ($R = 0.96$) and v ($R = 0.98$) values are highly correlated with destructive firmness results. Our results, with regard to elastic modulus and ultrasonic velocity, are in agreement with those from Kim et al., 2006, who reported a reduction in ultrasonic velocity with a decrease in fruit firmness for apples stored at 19–23 °C for 25 days. In another study supporting us, E and destructive firmness values, Shomali persimmon cultivar (non-astringent) were higher than those obtained for Karaj persimmon cultivar (astringent). Overall, these results appear to explain why non-destructive ultrasound method in predicting the mechanical properties of the fruit. Nonetheless, there are still many uncertainties in the current research that need further investigation. This study was performed to predict the mechanical properties of four different tomato varieties with elastic modulus and ultrasonic velocity. The apparent elastic modulus, velocity, and firmness values varied based on tomato varieties. The ultrasonic velocity and elastic modulus decreased with a reduction in firmness values of fruits. As mentioned in previous studies, it might be possible to assess maturity and estimate shelf life of tomato fruits in different storage conditions.

3 Conclusion

In this study, ultrasonic and machine vision systems were developed for the quality evaluation of tomato fruits. Firstly, attenuation and velocity measurement were conducted on tomato samples and then volume estimation was studied by custom-made ultrasonic (US100) and machine vision systems, respectively. Finally, destructive and non-destructive predictions of changes in the firmness of four different tomato varieties were evaluated. The obtained results demonstrated that ultrasound can be applied in the determination of tomato firmness by considering Elastic modulus (E, kPa) and velocity (v, m/s) measurements. Further work must be undertaken in order to improve the use of this novel technique for tomato firmness and other quality parameters evaluation. First of all, considering the high degree of variability of tomatoes in terms of textural changes has to be analyzed, taking into account the initial firmness and its changes during ripening. Secondly, an engineering improvement might be required for noninvasive inspection not only for the purposes of increasing the signal-to-noise ratio for a more accurate analysis but also for addressing the characterization of tomato and/or other fruits using the air-coupled ultrasound technique and transducers used in this study. Thus, the ultrasonic analysis and machine vision system, which were developed in this study, need more improvement to confirm reliability and accuracy when compared to previous studies. Results of this study would contribute to the in-line implementation of this technology for noninvasive and real time tomato and/or other fruits sorting at an industrial scale.

Acknowledgments The authors would like to acknowledge the support of Ultrasonar Company (http://www.ultrasonar.com.tr) for the design and applications of the ultrasonic system used in this study. This research was funded by the Scientific Research Projects (BAP) unit of Hakkari University (grant no. FM2017BAP11). The authors declare that they have no conflict of interest.

References

Abdeldym, E. A., El-Mogy, M. M., Abdellateaf, H. R. L., & Atia, M. A. M. (2020). Genetic characterization, agro-morphological and physiological evaluation of grafted tomato under salinity stress conditions. *Agronomy, 10*(12), 1948. https://doi.org/10.3390/agronomy10121948

Ahmad Saad, F., Ibrahim, M. F., Shakaff, A. Y. M., Zakaria, A., & Abdullah, M. Z. (2015). Shape and weight grading of mangoes using visible imaging. *Computers and Electronics in Agriculture, 115*, 51. https://doi.org/10.1016/j.compag.2015.05.006

Arunkumar, M., Rajendran, A., Gunasri, S., Kowsalya, M., & Krithika, C. (2021). Non-destructive fruit maturity detection methodology—A review. *Materials Today: Proceedings*. https://doi.org/10.1016/j.matpr.2020.12.1094

Baltazar, A., Aranda, J. I., & Gonzalez-Aguilar, G. (2008). Bayesian classification of ripening stages of tomato fruit using acoustic impact and colorimeter sensor data. *Computers and Electronics in Agriculture, 60*(2), 113–121. https://doi.org/10.1016/j.compag.2007.07.005

Bojarska, J., Pilat, B., Majewska, K., Sobiechowska, D., & Narwojsz, A. (2020). Selected physical parameters and chemical compounds of different types of tomatoes. *Czech Journal of Food Sciences, 38*, 28–35. https://doi.org/10.17221/232/2019-CJFS

Cantwell, M., Nie, X., & Hong, G. (2009, April 8–12). *Impact of storage conditions on grape tomato quality* (pp. 1–8). In: Procceedings 6th ISHS Postharvest Symposium, Antalya, Turkey.

Cavaco, A., Pinto, P., Antunes, D., Marques da Silva, J., & Guerra, R. (2009). 'Rocha' pear firmness predicted by a Vis/NIR segmented model. *Postharvest Biology and Technology, 51*, 311. https://doi.org/10.1016/j.postharvbio.2008.08.013

Cerovic, Z. G., Goutouly, J., Hilbert, G., Destrac-Irvine, A., Martinon, V., & Moise, N. (2009). Mapping winegrape quality attributes using portable fluorescence-based sensors. *Fruit Science, 9*, 301–310.

Choi, K., Lee, G., Han, Y. J., & Bunn, J. M. (1995). Tomato maturity evaluation using color image analysis. *Transactions of the ASABE, 38*, 171.

Concha-Meyer, A., Eifert, J., Wang, H., & Sanglay, G. (2018). Volume estimation of strawberries, mushrooms, and tomatoes with a machine vision system. *International Journal of Food Properties, 21*(1), 1867. https://doi.org/10.1080/10942912.2018.1508156

Food and Agriculture Organization of the United Nations. (n.d.). http://www.fao.org/home/en/

Gokul, P., Raj, S., & Poornapushpakala, S. (2015). Estimation of volume and maturity of sweet lime fruit using image processing algorithm. In *2015 International Conference on Communications and Signal Processing (ICCSP)* (pp. 1227–1229). IEEE. https://doi.org/10.1109/ICCSP.2015.7322703.

He, Y., Zhang, Y., Pereira, A., Gomez, A., & Wang, J. (2005). Nondestructive determination of tomato fruit quality characteristics using vis/NIR spectroscopy technique. *International Journal of Information Technology, 11*, 97.

Hitchman, S., Wijk, K., & Davidson, Z. (2016). Monitoring attenuation and the elastic properties of an apple with laser ultrasound. *Postharvest Biology and Technology, 121*, 71. https://doi.org/10.1016/j.postharvbio.2016.07.006

Ikeda, T., Choi, P. K., Ishii, T., Arai, I., & Osawa, M. (2015). Firmness evaluation of watermelon flesh by using surface elastic waves. *Journal of Food Engineering, 160*, 28. https://doi.org/10.1016/j.jfoodeng.2015.03.020

Iqbal, S. M., Gopal, A., & Sarma, A. S. V. (2011). Volume estimation of apple fruits using image processing. In *2011 International Conference on Image Information Processing* (pp. 1–6). IEEE. https://doi.org/10.1109/ICIIP.2011.6108909.

Ireri, D., Belal, E., Okinda, C., Makange, N., & Ji, C. (2019). A computer vision system for defect discrimination and grading in tomatoes using machine learning and image processing. *Artificial Intelligence in Agriculture, 2*, 28. https://doi.org/10.1016/j.aiia.2019.06.001

Islamadina, R., Pramita, N., Arnia, F., & Munadi, K. (2018). Estimating fish weight based on visual captured. In *2018 International Conference on Information and Communications Technology (ICOIACT)* (pp. 366–372). IEEE. https://doi.org/10.1109/ICOIACT.2018.8350762.

Jana, S., Parekh, R., & Sarkar, B. (2019). A De novo approach for automatic volume and mass estimation of fruits and vegetables. *Optik, 200*, 163443. https://doi.org/10.1016/j.ijleo.2019.163443

Khojastehnazhand, M., Omid, M., & Tabatabaeefar, A. (2010). Determination of tangerine volume using image processing methods. *International Journal of Food Properties, 13*, 760. https://doi.org/10.1080/10942910902894062

Khosroshahi, B., Mireei, S., Sadeghi, M., & Hemmat, A. (2017). Multi-sensor data fusion in the nondestructive measurement of kiwifruit texture. *Measurement, 101*, 157–165. https://doi.org/10.1016/j.measurement.2017.01.024

Kim, K. B., Kim, M. S., Park, J. G., Lee, S., Kim, G. S., & Jung, H. M. (2006). *18th IMEKO World Congress 2006: Metrology for a Sustainable Development, 2*, 1224.

Kowalczyk, K., Gajc-Wolska, J., Radzanowska, J., & Marcinkowska, M. (2011). Assessment of chemical composition and sensory quality of tomato fruit depending on cultivar and growing conditions. *Acta Scientiarum Polonorum, Hortorum Cultus, 10*, 133–140.

Kuo, F. J., Sheng, C. T., & Ting, C. H. (2008). Evaluation of ultrasonic propagation to measure sugar content and viscosity of reconstituted orange juice. *Journal of Food Engineering, 86*, 84. https://doi.org/10.1016/j.jfoodeng.2007.09.016

Lee, S., Hasegawa, M., Kim, K. B., Park, J. G., & Cho, B. K. (2013). Evaluation of the firmness measurement of fruit by using a non-contact ultrasonic technique. *Journal of the Faculty of Agriculture, Kyushu University, 58*, 103. https://doi.org/10.5109/26168

Li, B., Lecourt, J., & Bishop, G. (2017). Advances in non-destructive early assessment of fruit ripeness towards defining optimal time of harvest and yield prediction—A review. *Plant, 7*, 3. https://doi.org/10.20944/preprints201712.0142.v1

Li, R., Sun, S., Wang, H., Wang, K., Yu, H., Zhou, Z., Xin, P., Chu, J., Zhao, T., Wang, H., Li, J., & Cui, X. (2020). FIS1 encodes a GA2-oxidase that regulates fruit firmness in tomato. *Nature Communications, 11*, 5844. https://doi.org/10.1038/s41467-020-19705-w

Lu, R., Van Beers, R., Saeys, W., Li, C., & Cen, H. (2020). Measurement of optical properties of fruits and vegetables: A review. *Postharvest Biology and Technology, 159*, 111003. https://doi.org/10.1016/j.postharvbio.2019.111003

Magwaza, L., & Tesfay, S. (2015). A review of destructive and non-destructive methods for determining avocado fruit maturity. *Food and Bioprocess Technology, 8*, 1995–2011. https://doi.org/10.1007/s11947-015-1568-y

Marsic, N. K., Gasperlin, L., Abram, V., Budic, M., & Vidrih, R. (2011). Quality parameters and total phenolic content in tomato fruits regarding cultivar and microclimatic conditions. *Turkish Journal of Agriculture and Forestry, 35*, 185. https://doi.org/10.3906/tar-0910-499

Minas, I., Cipollone, F., & Sterle, D. (2020). Accurate non-destructive prediction of peach fruit internal quality and physiological maturity with a single scan using near infrared spectroscopy. *Food Chemistry, 335*, 127626. https://doi.org/10.1016/j.foodchem.2020.127626

Mireei, S., Sadeghi, M., Heidari, A., & Hemmat, A. (2015). On-line firmness sensing of dates using a non-destructive impact testing device. *Biosystems Engineering, 129*, 288–297. https://doi.org/10.1016/j.biosystemseng.2014.10.012

Mizrach, A. (2000). Determination of avocado and mango fruit properties by ultrasonic technique. *Ultrasonics, 38*, 717. https://doi.org/10.1016/S0041-624X(99)00154-7

Mizrach, A. (2004). Assessing plum fruit quality attributes with an ultrasonic method. *Food Research International, 37*(6), 627. https://doi.org/10.1016/j.foodres.2003.12.015

Mizrach, A. (2008). Ultrasonic technology for quality evaluation of fresh fruit and vegetables in pre- and postharvest processes. *Postharvest Biology and Technology, 48*(3), 315. https://doi.org/10.1016/j.postharvbio.2007.10.018

Morrison, D., & Abeyratne, U. (2014). Ultrasonic technique for non-destructive quality evaluation of oranges. *Journal of Food Engineering, 141*, 107.

Nyalala, I., Okinda, C., Nyalala, L., Makange, N., Chao, Q., Chao, L., Yousaf, K., & Chen, K. (2019). Tomato volume and mass estimation using computer vision and machine learning algorithms: Cherry tomato model. *Journal of Food Engineering, 263*, 288. https://doi.org/10.1016/j.jfoodeng.2019.07.012

Opara, U., & Pathare, P. (2014). Bruise damage measurement and analysis of fresh horticultural produce—A review. *Postharvest Biology and Technology, 91*, 9–24. https://doi.org/10.1016/j.postharvbio.2013.12.009

Ordóñez-Santos, L. E., Arbones, E., Oderiz, M., Rodriguez, R., Gomez, J., & Lopez, M. (2008). Use of physical and chemical properties of commercial tomato (Lycopersicon esculentum Mill.) products for monitoring their quality. *Journal of AOAC International, 91*, 112. https://doi.org/10.1093/jaoac/91.1.112

Osinenko, P., Biegert, K., Mccormick, R., Ghort, T., Devadze, G., Streif, J., & Streif, S. (2021). Application of non-destructive sensors and big data analysis to predict physiological storage disorders and fruit firmness in 'Braeburn' apples. *Computers and Electronics in Agriculture, 183*, 106015. https://doi.org/10.1016/j.compag.2021.106015

Otsu, N. (1979). A threshold selection method from gray-level histograms. *IEEE Transactions on Systems, Man, and Cybernetics, 9*(1), 62. https://doi.org/10.1109/TSMC.1979.4310076

Ozdemir, A. (2018). A remote tone burst pulser design for automated ultrasonic scanning systems. *Journal of Aeronautics and Space Technologies, 11*, 87–93.

Peng, Y., & Lu, R. (2005). Modeling multispectral scattering profiles for prediction of apple fruit firmness. *Transactions of the American Society of Agricultural Engineers, 48*, 235. https://doi.org/10.13031/2013.17923

Povey, M. J. W., & Wilkinson, J. M. (1980). Application of ultrasonic pulse-echo techniques to egg albumen quality testing: A preliminary report. *British Poultry Science, 21*(6), 489–495. https://doi.org/10.1080/00071668008416701

Reid, M. S. (1992). Maturation and maturity indices. In *Postharvest technology of horticultural crops* (pp. 21–28). University of California, Division of Agriculture and Natural Resources. ISBN 09-318-76990.

Rizzolo, A., Vanoli, M., Spinelli, L., & Torricelli, A. (2010). Sensory characteristics, quality and optical properties measured by time-resolved reflectance spectroscopy in stored apples. *Postharvest Biology and Technology, 58*, 1. https://doi.org/10.1016/j.postharvbio.2010.05.003

Sabliov, C. M., Boldor, D., Keener, K. M., & Farkas, B. E. (2002). Image processing method to determine surface area and volume of axi-symmetric agricultural products. *International Journal of Food Properties, 5*(3), 641. https://doi.org/10.1081/JFP-120015498

Siswantoro, J., & Asmawati, E. (2016). A new framework for measuring volume of axisymmetric food products using computer vision system based on cubic spline interpolation. In *2016 2nd International Conference on Science in Information Technology (ICSITech)* (pp. 74–78). IEEE. https://doi.org/10.1109/ICSITech.2016.7852611

Srivastava, S., Vaddadi, S., & Sadistap, S. (2014). Non-contact ultrasonic based stiffness evaluation system for tomatoes during shelf-life storage. *Journal of Nutrition & Food Sciences, 4*, 3. https://doi.org/10.4172/2155-9600.1000273

Uluisik, S., Yildiz, F., & Ozdemir, A. (2018). Image processing based machine vision system for tomato volume estimation. In *2018 Electric Electronics, Computer Science, Biomedical Engineerings' Meeting (EBBT)* (pp. 1–4). IEEE. https://doi.org/10.1109/EBBT.2018.8391460.

Valente, M., Prades, A., & Laux, D. (2013). Potential use of physical measurements including ultrasound for a better mango fruit quality characterization. *Journal of Food Engineering, 116*, 57. https://doi.org/10.1016/j.jfoodeng.2012.11.022

Vasighi-Shojae, H., Gholami-Parashkouhi, M., Mohammadzamani, D., & Soheili, A. (2018). Ultrasonic based determination of apple quality as a nondestructive technology. *Sensing and Bio-Sensing Research, 21*, 22. https://doi.org/10.1016/j.sbsr.2018.09.002

Venkatesh, G. V., Iqbal, S. M., Gopal, A., & Ganesan, D. (2015). Estimation of volume and mass of axi-symmetric fruits using image processing technique. *International Journal of Food Properties, 18*(3), 608. https://doi.org/10.1080/10942912.2013.831444

Yildiz, F., Özdemir, A. T., & Uluısık, S. (2018). Custom design fruit quality evaluation system with non-destructive testing (NDT) techniques. In *2018 International Conference on Artificial Intelligence and Data Processing (IDAP)* (pp. 1–5). IEEE. https://doi.org/10.1109/IDAP.2018.8620769.

Yildiz, F., Özdemir, A. T., & Uluısık, S. (2019). Evaluation performance of ultrasonic testing on fruit quality determination. *Journal of Food Quality, 2019*, 6810865.

Zalewska-Korona, M., & Jablonska-Rys, E. (2012). Evaluation of processing usefulness of selected fruits of ground tomato cultivars. *Zywnosc Nauka Technologia Jakosc/Food Science Technology Quality, 19*, 77. https://doi.org/10.15193/zntj/2012/81/077-087

Zhang, L., & McCarthy, M. J. (2012). Measurement and evaluation of tomato maturity using magnetic resonance imaging. *Postharvest Biology and Technology, 67*, 37. https://doi.org/10.1016/j.postharvbio.2011.12.004

Acoustic Emission and Near-Infra Red Imaging Methods for Nondestructive Apple Quality Detection and Classification

Akinbode A. Adedeji, Nader Ekramirad, Alfadhl Y. Khaled, and Chadwick Parrish

Abstract The need to develop nondestructive techniques in the Agrifood industry is becoming more apparent because of the increasing difficulty of securing labor, the increased innovation around artificial intelligence (AI), sensor technology and machine learning, the efficiency of the system, less risk, reduced cost on the long term, and the reliability of the system among others. Apple is one of the most important and highly valued products and its most devastating pest is the codling moth that often defies the manual random method of sorting especially in organic apple farming. The industry is wary of organisms of quarantine concerns like CM being transferred across borders. The application of nondestructive detection and classification methods has the potential to address some of the inadequacies in conventional random manual methods of sorting for pests. This chapter reviewed two important nondestructive methods, acoustic sensing and hyperspectral imaging that are complementary, and present some of the challenges that need to be addressed.

Keywords Acoustic emission · Near-infrared imaging · Nondestructive testing · Apple

1 Introduction

Apple is one of the most valued and consumed fruits in the world. It is mostly consumed fresh and used for making all kinds of foods from snacks to main courses. In the USA, there are over 2500 apple varieties grown for both local and

A. A. Adedeji (✉) · N. Ekramirad · A. Y. Khaled
Department of Biosystems and Agricultural Engineering, University of Kentucky, Lexington, KY, USA
e-mail: akinbode.adedeji@uky.edu

C. Parrish
Department of Electrical and Computer Engineering, University of Kentucky, Lexington, KY, USA

P. B. Pathare, M. S. Rahman (eds.), *Nondestructive Quality Assessment Techniques for Fresh Fruits and Vegetables*, https://doi.org/10.1007/978-981-19-5422-1_13

international markets but only 10 are dominated among others. China is the world-leading producer of apples consistently in the last 10 years, followed by the USA. Similar statistics are reported for apple export globally. In the USA, 32 states are considered the major apple producers and about 69% of apples produced are from the North-Western corridor around Washington State. Apple is a very important fruit in the global horticultural market.

The problem of pests and invasive species continues to constitute a major threat to the plum industry, especially due to increased apple trading across international boundaries. There is high consumers demand for organic apples, which are more susceptible to pest infestation. Some of the prevalent ones are difficult to detect and often reduce the quality of apples when they infest them. The most devastating apple pest is the codling moth (CM). The current method of sorting apples for quality defects is mostly manual, a random sampling method that is inefficient, laborious, and often allows some of these organisms of quarantine concern to get into the supply chain. It is critical to maintain the supply of apples with the highest quality in terms of uniform geometry, color, variety, without blemishes or defects. Sorting for these quality attributes is often done manually except in the big apple processing plants where nondestructive technologies, such as computer vision, have been introduced to detect surface defects. While the apple industry has seen the introduction of nondestructive technologies, they come with some limitations. For example, computer vision is unable of detecting internal defects, nor can it detect CM activities in apples. In the last 5 years, our group at the University of Kentucky has developed two non-invasive methods to capture all quality attributes of apples, including the detection of internal defects caused by pest infestation. The approach is based on machine learning coupled with vibro-acoustic emission and hyperspectral imaging, two effective methods for nondestructive testing of apples. In this chapter, we have provided a succinct summary of the capability of these two methods in nondestructive evaluation of internal and external apple quality attributes and addressed limitations and future applications of the methods.

2 Background

Apple (*Malus domestica*) is one of the most consumed fruits in the world. It is reported to have originated from the Middle East, arrived in England around 1066 AD and English settlers brought it to North America in the eighteenth century (Geiling, 2014; Pickyourown, 2021). As they are the favorite fruit of the ancient Greeks and Romans, they remain a favorite of many today – they are the easiest fruit to pick and use, and the number one fruit consumed in the USA (U.S. Apple Association, 2021). It is estimated that global apple production between 2010 and 2019 ranged from 71.9 to 87.2 million metric tons (FAOSTAT, 2021). China and the USA are the largest producers of apples globally. China produces as much as eight times the quantity of apples in the USA, yet US apples are prime produce in China and other countries like Canada, Taiwan, Hong Kong, UK, New Zealand, and

Mexico (U.S. Apple Association, 2021). One in every three US apples is exported and it is worth about $1 billion in export value (Steward, 2009; U.S. Apple Association, 2021). The USA also imports apples in small quantities, especially during the late season or before fall harvest, mostly from the southern hemisphere (U.-S. Apple Association, 2021). Apples are the most popular fruit grown in the USA and they are grown in every state, with Washington State leading the pack with about 69% of the total national production. There are more than 7500 apple varieties grown worldwide, out of which 2500 varieties are grown in the USA. Among these, only 10 varieties constitute about 80% of the total annual production (Forsline et al., 2003; U.S. Apple Association, 2021). An estimated 33% of apples grown in the USA are processed into products like apple juice, applesauce, and frozen apples, and approximately 67% are consumed fresh. The yearly farm-gate (wholesale and retail) revenue of the US apple is close to $15 billion including downstream economic activities (U.S. Apple Association, 2021). Apples are indeed an important crop in the USA and other apple-producing countries.

Apples are mostly consumed fresh, and just like most other fruits, they are consumed for their nutritional composition and calorie source due to a relatively high amount of sugars in ripe apples. Apple is also an important ingredient in many food production processes. From apple pie to apple cider, apple juice, dehydrated apple, applesauce, canned apple, apple marmalade, and jam due to pectin content. Fresh apples compose of mostly water (about 83% by weight), followed by carbohydrates and trace amounts (<1%) of proteins, lipids, and ash (Lee & Mattick, 1989). Fresh apples come in different flavors depending on the variety – some are sweet and sour, and some are just sweet. The color, shape, and size of apples are a function of the cultivar. The commonest varieties grown commercially in the USA in the last few years include Gala, Red Delicious, Golden Delicious, Honeycrisp, Granny Smith, and Fuji (U.S. Apple Association, 2021).

Considering all these important uses of apples, pest attacks could reduce the quality of an apple by as much as 59% (Breth, 2009) and could be more costly if found in exported apples. The most devastating apple pest is the codling moth (*Cydia pomonella*, Lepidoptera: Tortricidae) (Dyck, 2010; Neven, 2012). Globalization of apple trading and the increased consumers' demand for organic apples are some of the factors driving concerns about CM and other pest infestation in apples. The conventional approach to sorting for infested apples is mostly manual and ineffective (Neven, 2012). The development of non-invasive technologies is critical to addressing this problem. Other quality characterizations, such as proximate content, surface defects, and physical characteristics like color and size are better done with non-invasive systems. There are a plethora of these technologies (e.g., acoustic emission and hyperspectral imaging) that have been developed in the last decade (Adedeji et al., 2020; Ekramirad et al., 2021b; Li et al., 2018; Rady et al., 2017) with capabilities to nondestructively classify apple based on their quality attributes.

3 Acoustic Techniques for Nondestructive Apple Testing

The acoustic technique is a major nondestructive quality evaluation approach used for fruits and vegetables postharvest processing. In the last three decades, there have been several studies on the application of acoustic method for quality assessment of agro-products. Moreover, there is a growing body of literature that investigates the further development of this method for various quality monitoring applications, especially for the detection and management of insect pest damage in fruits, vegetables, and grains (Mankin et al., 2021). The nondestructive acoustic techniques have some main advantages including low costs, rapidness, being environmentally friendly, and ease of automation (Adedeji et al., 2020). However, there are some challenges associated with this technique, such as effectively distinguishing the target signal from background noise, developing fast signal pre-processing and processing systems, and adapting the method for in-line applications at the industry level. With further development of inexpensive modern acoustic tools, development of improved machine learning, and deep learning methods this technique will continue to extend its application in the agro-industry. In the rest of this section, the discussion of the principles and processes of acoustic method was made, as well as its types and applications in quality assessments of fruit and vegetables with an emphasis on apple fruits.

3.1 Acoustic Principle Theory

Acoustic vibration waves can propagate through the product body providing some internal information about the specimen. The main acoustic parameters include natural/resonance frequency, propagation velocity, acoustic impedance, and attenuation coefficient among others. These acoustic parameters determine the acoustic vibration properties of a body. They vary for different products based on their mechanical characteristics (Sun et al., 2010; Zhang et al., 2018). Additionally, micro-level characteristics of biological tissues, such as cell size, turgor pressure, cell wall thickness, and cell wall strength affect their mechanical and structural properties (Harker et al., 1997). Therefore, it is expected that fruits show different vibrational and acoustic properties because of these changes during storage, ripening, or infestation.

To correlate the basic acoustic vibration parameters to the quality attributes of fruits, a theoretical analysis of elastic vibration is applied. For theoretical modeling purposes, an individual fruit is assumed to be elastic, which is represented by the mass-spring system (Taniwaki and Sakurai, 2010). Using this assumption, the natural frequency (f) of the fruit is modeled by Eq. (1):

$$f = \frac{1}{2\pi}\sqrt{\frac{4k}{m}} \tag{1}$$

where m is the mass, and k is the spring constant. Since k represent the elastic properties of the fruit, by converting the equation as a formula in term of k, it can be shown that the term f^2m (stiffness coefficient) is an index for elasticity and texture properties of the sample (Abbott et al., 1968; Liljedahl and Abbott, 1994). Many attempts have been made to measure the texture of fruits by determining the natural frequency term mentioned (Abbott, 1994; Abbott et al., 1992; Chen et al., 1992; Duprat et al., 1997; Fathizadeh et al., 2020; Huarng et al., 1993; Muramatsu et al., 1996; Schotte et al., 1999). Specifically, for apples, Abbott et al. (1968) reported that the texture of apples can be well measured using acoustic methods by the natural frequency term and the resonance behavior of fruits. However, later the index f^2m, which is a spring-mass system (dashpot) with one degree of freedom as a one-dimensional model, was modified to $f^2m^{2/3}$ by considering the sample to be a three-dimensional elastic sphere (Cooke, 1970; Cooke & Rand, 1973). Recently, Fathizadeh et al. (2020) modeled the structural properties of apples using an acoustic vibration system by combining the $f^2m^{2/3}$ term with sound response signals from fruit. Overall, the theoretical and experimental results show that the elastic quality characteristics of fruits can be effectively determined using their vibrational acoustic properties. Consequently, the acoustic vibration technique can be effectively used for measuring the characteristics of fruit, which are related to elasticity or internal quality (Zhang et al., 2018).

3.2 Acoustic Measurement Processes

Generally, there are two ways by which the acoustic vibration signals can be originated, namely passive and active modes. In the passive mode, the source of vibration or emissions is from an internal phenomenon such as structural failure or natural changes like ripening occurring to the tissue inside the sample. Other sources could be the displacement of foreign bodies like pest larva (Ekramirad et al., 2021b). This is defined as acoustic emission (AE) in which acoustic waves are generated and propagated in a solid body as a result of rapid irreversible internal damages (Muravin, 2009). In particular, AE is considered an acoustic phenomenon in the sonic and ultrasonic frequency range up to 1 MHz propagating in solids (Martini et al., 2017). Based on the definition, in the case of agro-products, the AE signals can originate from insects' feeding activities inside the specimen or biochemical (microbial and enzymatic) reactions or gas emissions in the infested samples. However, in the active mode, there is an external source of excitation or impulse to obtain vibrational responses from the sample. For the active mode, three main parts are considered the excitation device, signal acquisition, and signal processing systems (Zhang et al., 2018). Furthermore, the excitation device is divided into impact (impulse) and forced methods (Taniwaki & Sakurai, 2010). In the impact method, an impulse is applied to the sample in a short excitation duration using devices, such as a mechanical hammer, stick, pendulum, or electrical solenoid. The advantages of the impact method are its simplicity and low cost, and these make them suitable for fast online applications (Zhang et al., 2018). In the forced vibration method, an alternating force is applied to the sample during measurement using an electrodynamic vibrator (Abbaszadeh et al., 2013; Blahovec et al., 2007; Taniwaki et al., 2010; Terasaki et al., 2001; Zhang et al., 2018). While the forced vibration method has the advantage of better repeatability, this method demands more time compared to the impact technique and does not have the fast detection ability required for online processes (Zhang et al., 2018).

The acoustic techniques are also classified based on the signal acquisition mode and the type of the vibration measurement sensors as contact and non-contact sensors. In the contact mode, where there is direct contact with the surface of the specimen, either accelerometers or piezoelectric sensors are used. The contact sensors have the disadvantage and challenge of affecting the sample and its original vibration by inducing external force. On the other hand, a microphone can be used as a non-contact sensor for many applications (Chen et al., 1992; Mendoza et al., 2012; Valente et al., 2009; Zude et al., 2006). However, microphones have the downside of being influenced by ambient noise. A relatively recent non-contact vibrational system is the Laser Doppler Vibrometer (LDV) which consists of an emitting laser that beam toward the surface of the vibrating sample to acquire its vibrational parameters from the Doppler shift induced in the reflected laser beam (Johansmann et al., 2005). It was reported that the vibration parameters obtained using an LDV were more accurate than those using the accelerometer (Muramatsu et al., 1997). However, the LDV is costly and demands a strong and stable reflected laser beam makes it a costly technique. Some advantages of the LDV method include being unaffected by the environmental noise, no interference between sensor and sample, no need to attach the sensor to the sample, a non-touch sensor, and a wide frequency range and working distance (Liu et al., 2017; Zorović and Čokl, 2015). On the other hand, in most cases, the LDV system needs an excitation force making this method not suitable for fast in-line applications. Therefore, further studies are needed to expand the application of LDV method in a variety of agricultural processes.

4 Applications of Vibro-Acoustic Method in Fruit Quality Evaluation

4.1 Texture Evaluation

Much of the previous research on fruit quality assessment using the acoustic method has been conducted in the evaluation of texture parameters including firmness, stiffness, crispness, chewiness, juiciness, etc. (Chen & Opara, 2013; Lawless &

Heymann, 2013; Taniwaki & Sakurai, 2010; Zhang et al., 2018). Overall, the acoustic technique is the most commonly used nondestructive detection method for fruit texture evaluation (Lashgari et al., 2017; Zhang et al., 2014). This is because of the high correlation between the acoustic vibrational characteristics of fruits and their texture attributes, such as modulus of elasticity, mass, density, Poisson ratio, and shape (Song et al., 2006; Zhang et al., 2016, 2018;). In the case of the acoustic parameters, unlike the signal amplitude, which depends on the integrity of sample-to-vibrator and sample-to-detector contacts, and the sample size and shape, resonance frequencies (especially the second resonance frequency) primarily reflect the mechanical texture properties of the sample (Abbott et al., 1968, 1992; ; Finney Jr, 1970; Finney Jr et al., 1978). In order to further compensate for the effect of sample size in the acoustic evaluation of fruits, the combination of resonance frequency and mass can be used as the evaluation term, such as stiffness coefficient f^2m or $f^2m^{2/3}$ (Abbott et al., 1992; Zhang et al., 2018). The resonance frequencies of fruits are obtained by converting the time domain response signals to the frequency domain with Fast Fourier Transformation (FFT) to acquire the power spectrum and identify characteristic frequencies (Tesch et al., 1996).

For apples, Abbott et al. (1992) obtained a good correlation coefficient of greater than 0.8 between acoustic indices and the firmness values measured by the Magness–Taylor (MT) method. They concluded that while the inspectors' sensory scores showed a relatively higher correlation with measured firmness than with acoustic indices, acoustic measurements are nondestructive, non-site-specific, and representative of the entire fruit. In another study on apples, Tiplica et al. (2010) showed that the acoustic impulse response method could effectively classify 10 different apple varieties with a high classification rate of more than 90% using quadratic discriminant analysis. The 18 features they applied were extracted from the power band of the signal recorded with a microphone in the impact test. They discussed that not only did each variety of apple had its spectral signature related to the texture properties of their tissue but also the direction of the impact on fruit affected the characteristics of the acoustic signature of apples. They assumed that this "side effect" could be due to several reasons such as the heterogeneities of the internal structure of apples, density, and composition variation for the sun-exposed side compared to the shade side of the fruit.

4.2 Ripeness Prediction

Since the ripening and storage processes cause noticeable changes in the textural properties of fruits, the acoustic vibration method can be applied for ripeness detection and the peak of ripeness determination. Because there are no significant changes to the size, geometry, and density of fruits over storage time, the elastic modulus of the apple flesh is the only attribute affecting the changes in the textural properties leading to the fruit softening (Takeda et al., 1970). The softening and change in elastic modulus may be attributed to weakening intercellular bonds and cell walls, and/or loss of turgor pressure. Many authors have studied the effect of storage and ripening on the textural properties of agro-products. For example, Taniwaki and Sakurai (2010) used the vibro-acoustic vibration method to evaluate the ripeness of persimmons and melons. To do that, the predicted elasticity index of samples using the acoustic method over time was correlated to the rate of ripening. They reported that the optimum ripening time could be predicted by the acoustic elasticity index. The same elasticity index was applied by Fumuro et al. (2013) to predict the optimum harvest time of pitayas fruit. In another study, an acoustic sensor fusion method was used to classify three different types of peaches based on their softness with a classification accuracy of 87% (Vursavus et al., 2015).

Takeda et al. (1970) developed a specialized instrument for nondestructive ripeness evaluation by measuring acoustic vibrational signals when a product like an apple is tapped by an impact. The relationship between samples' natural frequency and apple maturity is investigated to draw a correlation with ripeness. Liljedahl and Abbott (1994) studied the relationship between the resonance frequency of apples with their softening over an accelerated ripening period using an acoustic measurement method composed of an electromagnetic vibrator and an accelerometer sensor. Their results showed a consistent monotonic decrease in frequency with time while the attenuated pattern in the amplitude was not uniform for all fruits. Symoneaux et al. (2004) studied the impact of harvest date at 3 ripeness stages on apple quality. They correlated the acoustic indices and sensory data as well as destructive measurements. They showed that the acoustic parameters such as stiffness factor were effective in classifying apples with different levels of ripeness as well as freshly harvested and stored ones. However, they concluded that their method was insufficient to identify the levels of ripeness in some apple cultivars. Lashgari et al. (2017) examined the potential of the acoustic impulse response method in nondestructive classification of apples with different degrees of freshness over storage time using an artificial neural network. Their results showed an F1 score of 84.9% and 84.7% for Golden Delicious and Red Delicious, respectively. They suggested decreasing the possible measurement errors and applying other acoustic features to improve their classification accuracy.

4.3 Internal Defect Detection in Apples

The initial hypothesis was that the acoustic vibration response of fruits with internal damage differs from the healthy samples due to textural changes, structural defects such as holes and cavities, and ongoing biochemical reactions as a result of the internal defect. Based on this many studies have been conducted and various devices were developed for the detection of internal damages in agricultural commodities using the acoustic vibrational response method (Armstrong et al., 1997; Diezma-Iglesias et al., 2004; Ekramirad et al., 2021b; Elbatawi, 2008; Li et al., 2018; Rao et al., 2004; Stone et al., 1996). Elbatawi (2008) investigated the potential of the acoustic impact test to detect and classify hollow heart disorder in potatoes using a linear discriminant analysis classifier based on three features extracted from the microphone signal. They reported that the peak frequency of hollow potatoes was significantly lower than that of healthy ones. They reported a high classification accuracy of 98% for distinguishing hollow-hearted potatoes.

Kadowaki et al. (2012) introduced a nondestructive acoustic method to detect the core rot in "Kosui," Japanese pear, with a classification accuracy of 96.9% by classifying fruits with f^2 less than 500 Hz as having core rot of over 5% of the entire batch. Muramatsu et al. (1999) used the LDV technique to detect internally damaged citrus fruit. Their results showed that the phase shift in the laser beam from the defective fruit was significantly lower than healthy ones at all the frequencies used in their study. Su et al. (2008) introduced a method for quick sorting of internally defective apples using the acoustic impulse response technique. The stiffness coefficient obtained based on the first response frequency and the mass of the fruit had a good correlation with the decayed area of the defected apples with a coefficient of determination, R^2 value of 0.998 for Red Fuji and 0.878 for Golden Delicious apples using the logistic regression method. In another attempt to detect internally defective apples, Li et al. (2018) applied a passive AE approach to detect and classify codling moth-infested apples using machine learning algorithms. The AE system used a high-frequency contact piezoelectric sensor with a sampling rate of 1 MHz. They used 11 time-frequency AE features as the independent input variables for linear discriminant analysis (LDA) and adaptive boosting classifiers. They achieved the test classification accuracies in the range of 83–100% for the signal window that ranged from 0.5 to 120 s. In a similar application with a different approach, Ekramirad et al. (2020) used a low-frequency vibro-acoustic method consisting of two low-cost contact-piezoelectric PZT sensors to detect and authenticate the signature of signals originating from different activities of codling moth larvae in apples. Following this study, in a complementary work with the same vibro-acoustic system as shown in Fig. 1, they developed pattern recognition and classification models to classify codling moth-infested apples using multi-domain vibro-acoustic features by machine learning using the larval behavior templates (Ekramirad et al., 2021a). As presented in Table 1, the results show an approach that gave accuracies up to 97.47% for 10 s signal duration. When a matched filter was applied, the classification result improved further to 100% accuracy for the 5 s signal duration. As shown in Table 1, the signal segmentation time decreases from 2 min to 5 s, generally the classification rates increases due to the better matching of the window size with the active moments of the larvae. Furthermore, Khaled et al. (n.d.) applied acoustic impulse response method at a frequency range between 35 and 100 kHz to detect codling moth infestation in apples. In this study, apples to be tested for infestation were placed between a large gripper that has the PZT sensor between them and the grip, secured firmly with actuating screws. Figure 2 shows the different data collection techniques for the passive and impulse methods used in Khaled et al. (n.d.) study. The results of classification of CM-infested and control samples among Gala apple cultivars were found to be 99% for a short time data collection range, 0.06–0.08 s.

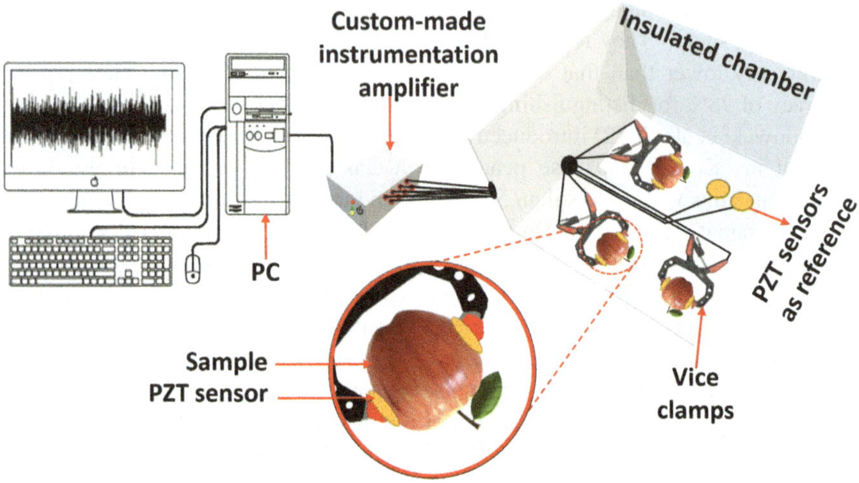

Fig. 1 Schematic of the vibro-acoustic system for detecting codling moth-infested apples (Ekramirad et al., 2021b)

5 Near-Infra Red Imaging

Agricultural techniques have developed significantly, and this development has led to an increase in the demand for the application of non-invasive techniques to inspect the quality of crops after harvest. One of the most applied techniques is spectroscopy, which is the study of the interaction of electromagnetic waves to know the absorption and emission of light by matter, such as visible (VIS), and infrared spectra. Normally, the results of spectroscopy data can be shown as a function of wavelength or frequency. Near-infrared (NIR) spectroscopy (780–2500 nm) is one of the most promising non-invasive techniques, which received wide acceptance in different areas because of its unique advantages as compared to other analytical techniques. It is an effective approach to revealing both the external and internal quality of agro-products. Based on the NIR wavelength range, hyperspectral imaging (HSI), which combines both spatial and spectral data, is considered the best technique to represent this range in the postharvest quality and safety inspection. This is so because of its inability to detect surface defects but also internal damages. Thus, this section will focus on the concept and the application of HSI in apple quality evaluation and prediction.

Table 1 Classification results for CM-infested and healthy apples using vibro-acoustic method (Ekramirad et al. 2021b)

Signal length	Classifier	Test Data accuracy (%)	Sensitivity (%)	Specificity (%)	Precision (%)
2 min	kNN	90.48	91.67	88.89	91.67
	Decision tree	71.43	70.83	72.22	77.27
	LDA	76.19	79.17	72.22	79.17
	SVM (quadratic)	73.81	70.83	77.78	80.95
	NB	52.38	58.33	44.44	58.33
	Ensemble (bagged tree)	80.95	87.5	72.22	80.77
1 min	kNN	79.75	91.3	63.64	77.78
	Decision tree	83.54	89.13	75.76	83.67
	LDA	65.82	76.09	51.52	68.63
	SVM (cubic)	74.68	80.43	66.67	77.08
	NB	73.42	84.78	57.58	73.58
	Ensemble (boosted trees)	87.34	93.48	78.79	86
30 s	kNN	85.62	93.94	72.13	84.55
	Decision tree	90	92.93	85.25	91.09
	LDA	73.75	79.8	63.93	78.22
	SVM (cubic)	84.38	83.84	85.25	90.22
	NB	75.62	70.71	83.61	87.5
	Ensemble (GentleBoost)	90	89.9	90.16	93.68
10 s	kNN	93.89	96.64	89.27	93.81
	Decision tree	90.11	93.96	83.62	90.61
	LDA	76.42	79.19	71.75	82.52
	SVM (Gaussian)	95.16	96.64	92.66	95.68
	NB	77.26	77.52	76.84	84.93
	Ensemble (bagged trees)	**97.47**	**96.98**	**98.31**	**98.97**
5 s	kNN	91.66	91.22	92.37	95.15
	Decision tree	91.12	90.19	92.66	95.27
	LDA	74.12	76.42	70.34	80.87
	SVM (Gaussian)	93.16	94.15	91.53	94.8
	NB	72.41	68.33	79.1	84.29
	Ensemble (GentleBoost)	95.94	96.39	95.2	97.05

CM codling month, *kNN* k-nearest neighbors, *LDA* linear discriminant analysis, *SVM* support vector machine, *NB* Naïve Bayes

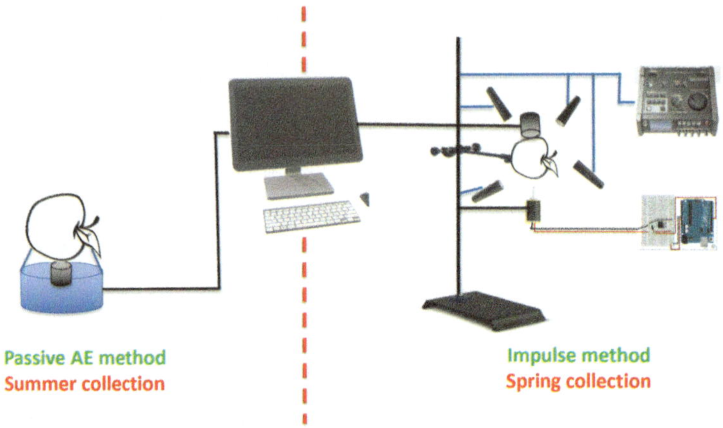

Fig. 2 The schematic diagram of the acoustic impulse system for detecting codling moth-infested apples (Khaled et al., n.d.)

6 Overall Concept of HSI

Hyperspectral imaging (HSI) technique has been gaining traction in recent years, with one major area of use being in nondestructive food quality evaluation (Saha & Annamalai, 2021). HSI combines the usage of spatial and spectral data simultaneously to evaluate the external and internal characteristics of a given sample (Shi et al., 2019). This allows for potentially better identification of more complicated features than can be done by the usage of exclusively spatial or spectral information in USA alone (Gowen et al., 2007). Due to advances in HSI equipment and machine learning, HSI has grown in popularity and accuracy within the past 5 years, becoming a powerful inspection tool for nondestructive quality assessment and safety evaluation of foods.

A typical HSI system is made of many pieces, such as an electromagnetic energy source (often a multispectral light), a spectrograph, a camera, a setup to hold the sample, and a computer that can process the acquired data, as shown in Fig. 3. A typical HSI system will emit multispectral light from its source, which will reflect off of the subject, transmit through the lens, and will get scanned by the spectrograph to ultimately get captured by the camera. This will get stored in a three-dimensional (3D) space with two spatial dimensions and one spectral dimension. This data is converted to a digital representation using the CCD camera. If the spectrograph and camera are well calibrated, data can be easily converted to radiometric qualities such as radiance, reflectance, and emissivity. There are three commonly applied methods for scanning objects to obtain spatial-spectral data: point-to-point scanning (also called whiskbroom imaging), line scanning (also called pushbroom imaging), and area scanning (also called staring imaging). The scanning methodology dictates the procedure in which the camera will sample spatially versus spectrally. In point-to-point scanning, the image is sampled at each pixel over the entire light spectrum

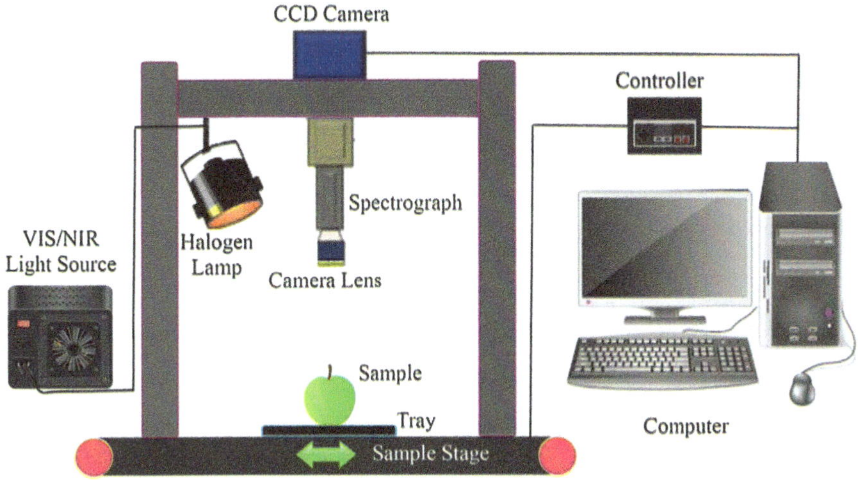

Fig. 3 Schematic diagram of the HSI system (Rady et al., 2017)

being captured. Line scanning will sample an entire line of pixels at the spectrum of interest, and area scanning will sample all pixels at one wavelength before moving to the next wavelength.

6.1 Analysis Process of HSI

HSI provides its information in (3D) hypercubes (x, y, λ), where x and y are spatial dimensions and λ is the spectral dimension. Hypercube data can be viewed in two ways: the entire image at a single wavelength sample, or each pixel over the entire spectrum of interest. Due to the large volume of data, hypercube analysis can be tricky. Any models built will often need to first apply dimensionality reduction techniques to greatly reduce the quantity of data that will be processed (Cheng et al., 2017). Some common ways of achieving this are interference correction, wavelength selection, and feature extraction (Da Silva et al., 2014). Spectral pretreatment techniques are used to reduce and correct interferences from sources such as baseline drift, scattering, and overlapping bands. Techniques to correct these issues include multiplicative scatter correction (MSC), smoothing, and baseline removal. Other selection techniques, such as PCA or genetic algorithms (GA), are often applied to select the most informative spectral features to further simplify the model. Finally, HSI analysis concludes with a performance evaluation through quantitative means, such as machine learning. This can be used to correlate, classify, predict, and/or validate the models, depending on the problem of interest. Figure 4 is a visualization of this process as a whole.

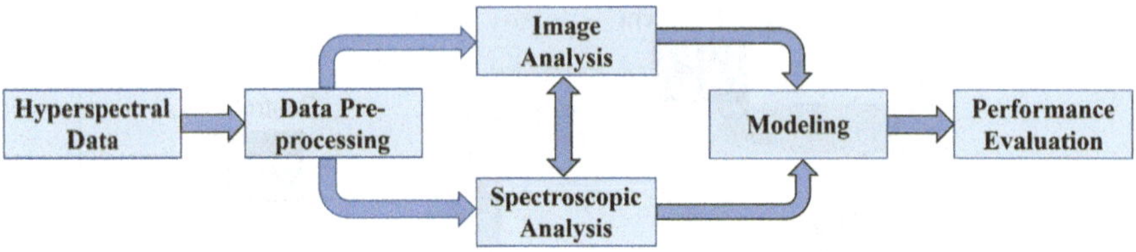

Fig. 4 Main steps flowchart for hyperspectral image analysis

6.2 *Application of HSI for Nondestructive Quality Assessment*

Applications of HSI mainly fall into three categories: external and internal physicochemical quality evaluation, and detection of diseases and pest infestation. An overview of these three areas is provided in this section, with some recent research findings done in each area.

6.2.1 External Quality Evaluation

Many external quality characteristics are commonly evaluated in applies, including physical appearances, shape, color, size, and the presence or absence of surface defects. These attributes affect the pricing of these products on the market. Normally, shape and size can be detected using computer vision, while HSI is mostly used to detect color and defects due to its high effectiveness for detecting subsurface features. The different modes of HSI (reflectance, transmittance, and interactance) are also helpful in detecting different types of defects, such as those on the surface, shortly below the surface, or deep in the tissue. Different transmission modes and wavelengths can be used for different applications with much success.

Apple bruising can directly influence storage stability, deterioration and reduce market value due to a flawed appearance suggesting a quality reduction to consumers. Accurate prediction or identification of bruised regions is critical for grading different tiers of apple quality. Slight bruises will commonly have little change in internal chemical components, and will also be difficult to detect. While detection of larger apple bruises is easy, this phenomenon makes detection of all bruises difficult. Image processing techniques are commonly used for bruise detection on apples with algorithms such as thresholding or feature vector extraction, but these have limitations imposed by complex or subtle bruises. Another large challenge is that it can be hard to truly evaluate the classification accuracy of the model because that will require grading each pixel as damaged or intact (Che et al., 2018). The detection of apple bruises mostly occurs in the range of short-near-infrared (SNIR), notably between 675 and 960 nm which represents the content of carotenoids and chlorophyll pigments in apples and the content of sugar and water (El Masry et al., 2008). This indicates that apple bruises cause an absorption of water and total sugar contents during the early stage of bruise development, which eventually leads to the consumption of chlorophyll and carotenoids (Wang et al., 2007). Thus, the application of HSI to predict and detect the bruise on apples can reduce the losses and preserve the high quality. Table 2 lists examples of many studies that have utilized HSI for the evaluation of different parameters in apple.

6.2.2 Internal Quality

The internal quality of an apple refers to its nutritional value, texture, and flavor. These quality features cannot be evaluated using visual inspection, they often require destructive physicochemical analysis such as testing total soluble content, SSC with a Brix refractometer, and firmness with the Magness–Taylor test. Generally, nutrients such as vitamins, fibers, and antioxidants are of interest with these methods (Wills & Golding, 2016). Firmness is the most commonly predicted textural attribute of horticultural products. Secondary are US sensory properties such as sourness, sweetness, and bitterness to form the flavor characteristic of apples. It is easy to see why evaluation of these internal quality features has been an important focus in nondestructive apple quality assessment. In Lu, 2007, Lu applied HSI to evaluate the SSC and the firmness of two types of apple varieties, namely golden delicious and delicious. The author used the artificial neural network (ANN) model to analyze the data and found that R^2 for SSC and firmness prediction was 0.79 and 0.76 for golden delicious and red delicious apples, respectively. They concluded that the poor predictions for apple SSC might be attributed to the irregular fruit shape, which could have negatively affected the scattering measurements. Light scattering, as done in HSI in Lu's study, can enhance structural features. This would lead to a stronger ability to predict firmness but does not immediately indicate an ability to predict SSC, perhaps because of the lower signal-to-noise ratio of the HSI method. A detailed description of the analysis of hyperspectral scatter images and characterization of scattering signature for fruit quality evaluation is available in Lu et al. (2020). Also, Mo et al. (2017) applied HSI from 400 to 1000 nm to evaluate SSC in Fuji apples. The authors used partial least square regression (PLSR) for the prediction and reported an R^2 of 94.7%.

6.2.3 Infestation Detection in Apples

The Codling moth (CM) is the most invasive and destructive species in apples. The manual method of detection is not effective and could be costly for both local and export apple, especially the latter where organisms of quarantine concerns are heavily controlled across international borders. Our group at the University of Kentucky developed (Ekramirad et al., 2017, 2021b; Rady et al., 2017) nondestructive techniques based on acoustic and HSI signaling to assess the quality of CM insect infestation) of different apple varieties (Fuji, Granny Smith, Goldrush, and Gala). In these studies, all the apples, including the infested and control samples,

Table 2 Studies on apple quality detection using the hyperspectral imaging technique

Apple variety	Quality attribute	Spectroscopy type	Wavelength range, nm	Image method	Experimental design	Statistical approach	Significant results	Wavelength selection, nm	Reference
Fuji	Bruise region	VIS and NIR	400–1000	Pixel-based	Data collected after 0, 12, and 18 h of the damage experiment	PCA, RF, SVM Decision tree, Stochastic gradient descent, RF and gradient tree boosting	99.9%	675, 960	Che et al. (2018)
Red Fuji	Bruising degree	VIS and NIR	400–1000	Pixel-based	Data collected immediately after applying 8 degrees f bruising (I, II, III, IV, V, VI, VII, VIII)	Successive projections algorithm (SPA), SVM, GS-SVM	95%	396, 500, 626, 631, 647, 719, 724, 734, 808, 882, 887, 898, 903, 908, 914, 967, 973, 978, 983, 989 [20]	Tan et al. (2018)
Golden delicious and pink lady	Early bruise detection	NIR	955–1700	Selected pixels	Data collected a day before bruising and after 0 h & 5 h, 1, 2, 3, 6, 7, 8 and 9 days of bruising	PLS-DA, PLS-DA	94.04%	–	Ferrari et al. (2015)
Fuji	Online detection of bruises	VIS and NIR	325–1100	Wavelength images and the PC images	Bruises created by dropping a steel ball from a 500 mm	PCA	96.2%	780, 850, 960 [3]	Huang et al. (2015)

					height and then data collected after 1, 12, and 24 h				
Fuji	Prediction of SSC	VIS and NIR	400–1000	–	Data collected from integrating spectra and textural features	Combined partial least square (CPLS)	93.27%	–	Fan et al. (2016)
GoldRush	Detection of codling moth infestation	VIS and NIR	400–900	ROI mean reflectance spectrum (MRS)	Samples were distributed into four different storage chambers set to 4, 10, 17, and 27 °C for monitoring temperature effect on codling moth	LDA, decision trees, kNN, PLSDA, feed-forward artificial neural networks (FFNN)	82%	434.0, 37.5, 538.3, 582.8914.5 [5]	Rady et al. (2017)
Fuji	Early detection of decay	VIS and NIR	400–1000	ROI	Four steps: (1) obtained infected apple with natural decay from fungi; (2) decayed apple tissues used as a culture medium to grow fungi spores; (3) fungal spore	PCA, improved watershed segmentation method	99%	400, 467, 575, 625, 684, 750, 813, 9,621,000 [9]	Li et al. (2019)

(continued)

Table 2 (continued)

Apple variety	Quality attribute	Spectroscopy type	Wavelength range, nm	Image method	Experimental design	Statistical approach	Significant results	Wavelength selection, nm	Reference
					solution was formed by dissolving water; (4) inoculated each apple using a 1 mL syringe				
					The infected samples were stored at 20 °C and 60% RH for 3–4 days to produce a decayed spot				
Jonagold	Early bruise detection	SWIR	1000–2500	Real-time pixel-based	Data collected before and 2 h after damage	PLS-DA	98%	–	Keresztes et al. (2016)
Fuji apples	Assessment of SSC	VIS and NIR	400–1000	ROI	First, 5 slices from each apple were marked with 29 sections established by coring with a hollow 12-mm diameter boring tool. The	PLSR	94.7%	–	Mo et al. (2017)

					cores were extracted post-imaging and SSC evaluation of each sample				
GoldRush apple	Prediction of firmness, TSS, and surface color parameters (L*, a*, b*)	VIS and NIR	400–1000	ROI	The apples divided equally into four groups and placed in storage conditions of 27 °C, 17 °C, 10 °C, and 4 °C at a relative humidity that ranged from 75% to 90%	PLS, LS-SVM, multivariate linear regression (MLR)	92%	–	Ekramirad et al. (2017)
Golden delicious apple	Prediction of moisture content and chromaticity of raw	VIS and NIR	400–1000	–	Apple slices were blanched as a pre-treatment with hot water at 50 °C and 70 °C, and in an acidic solution	PLS	99%	510, 540, 543, 664, 714, 817, 914, 966, 969, 977 [10]	Crichton et al. (2018)
Red Fuji apple	Quantitative evaluation of impact damage	NIR	90–1700	ROI	The impact from drop heights of H0.3, H0.6, H0.9, H1.2,	PLS	86%	967, 1001, 1100, 1154, 1190, 1407, 1443 [7]	Zhang et al. (2021)

(continued)

Table 2 (continued)

Apple variety	Quality attribute	Spectroscopy type	Wavelength range, nm	Image method	Experimental design	Statistical approach	Significant results	Wavelength selection, nm	Reference
					and H1.5 was set				
Red delicious apples	Prediction of firmness, TSS	VIS and NIR	400–1000	Pixel signatures	Mechanical impact method - manually	PCA	89.7%	–	Vetrekar et al. (2015)
Fuji apples	Identification accuracy of waxed	VIS and NIR	550–1710	Pixel-based	Half of the samples were scraped on their surface by a knife and then waxed removed gently without bruising the surface of the apples	LS-SVM, BPNN	100%	639.90, 806.85, 973.80, 1073.97, 1197.99, 1269.54, 1441.26 [7]	Wang et al. (2018)

The numbers in parenthesis are counts of the selected wavelength features

Fig. 5 Typical example of codling moth-infested GoldRush apple (Rady et al., 2017)

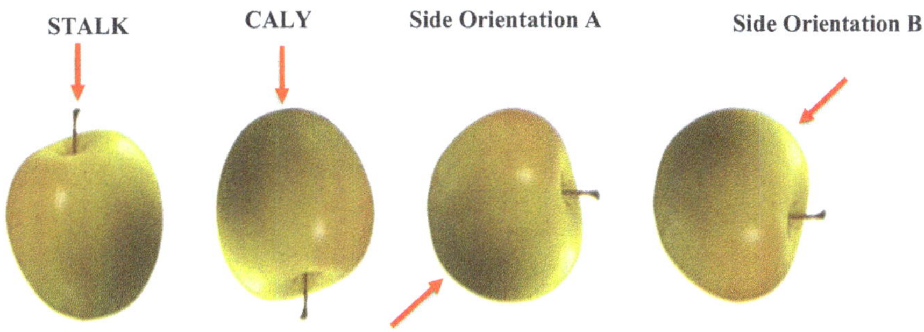

Fig. 6 Four apple orientations during HSI scanning (Rady et al., 2017)

were scanned using two HSI systems: (1) visible-near-infrared (VIS-NIR) HSI system with a wavelength range between 400 and 1000 nm, and (2) a NIR HSI system at a wavelength range from 900 to 1700 nm. To confirm the presence of CM inside the apples after scanning, each apple was cut open (Fig. 5) and they were labeled as one class and the non-infested apples as the other class. The apples were scanned from three different positions, namely stem, calyx, and the side as shown in Fig. 6 to determine the most effective point to scan an apple to obtain the best result. Figure 7 shows the basic steps for analyzing raw HSI data obtained from the apple sample scanned using the VIS-NIR imaging system. Different classifier models were applied to compare the test data classification accuracies. The combination orientations data set of apples produced the best performance with an overall accuracy of 81.04% using the PLS-DA classifier. Similarly, side and calyx orientations showed an overall classification accuracy of 80%; however, the lowest classification accuracy was for the stem orientation. These findings are slightly lower compared to the result of Rady et al. (2017) who achieved a maximum classification rate of 82% for the side orientation images. However, in the same study conducted by Ekramirad et al. (2021b), they obtained higher accuracies of 100% and 98.20% using the random forest (RF) classifier for training and testing sets, respectively. This

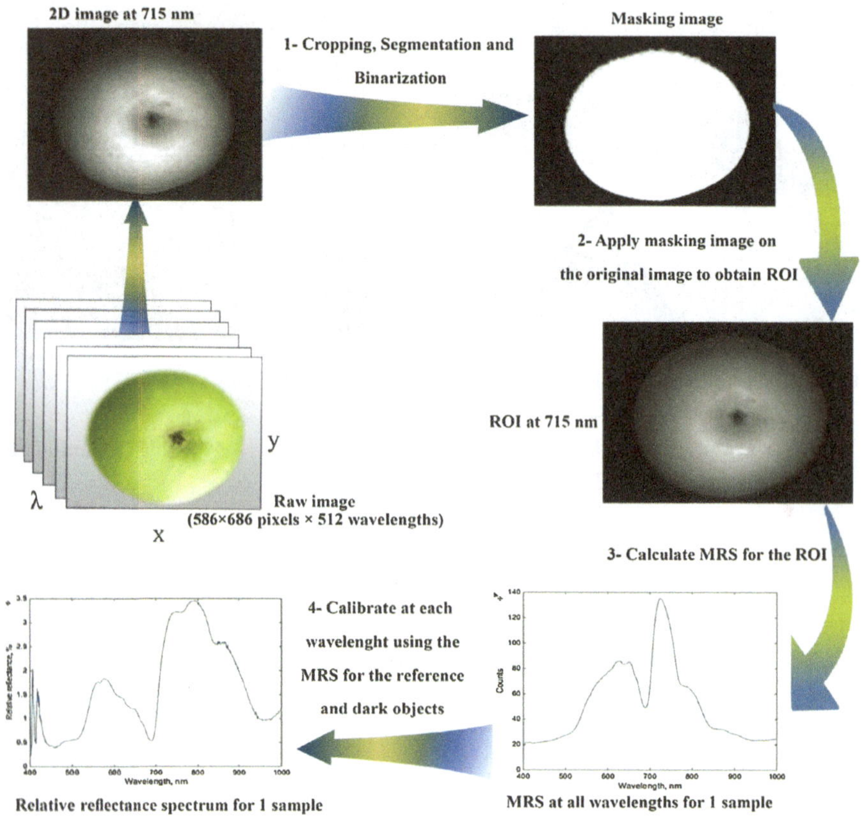

Fig. 7 Important steps for obtaining the raw spatial images and spectral data (hypercubes) of an apple by scanning with the VIS-NIR hyperspectral imaging system (Rady et al. 2017)

improvement in the classification rate in comparison to the mean spectra extraction method can be attributed to the spatial sub-segmentation (individual image pixel) of the infested area to build a better model.

6.2.4 Challenges and Future Directions

There are many avenues open for researchers to pursue further development of the nondestructive evaluation of apples. There are other technologies than the ones covered in this chapter, such as e-nose (Jia et al., 2019), which could hold great potential to provide additional information to improve the precision of measuring and predicting various qualities of apples. Furthermore, there are also challenges, which stem from the fruit themselves, which pose obstacles for researchers in the apple industry. One of the biggest examples is the heavy correlation between different desired quality characteristics. Previous sections discussed the quality evaluation of apples using two methods (vibro-acoustic and HSI), which is made very challenging if the quality factor being evaluated cannot be independently tracked and/or evaluated in the study. For example, a study is looking to track the water holding capacity of an apple over time, but the study cannot get data where the SSC stays the same over that period. Due to the inability to isolate water holding capacity, the model being built gets more complicated because it should also investigate how a changing SSC affects the data measurements being collected. Thus, the single output equation for water holding capacity then becomes a system of two equations, with the second being for SSC. With the ability to isolate one factor, the model would become more precise and much simpler.

Acoustic methods applications are still recent to the apple industry, where much of the work done in the last few years show how useful acoustic information can be in nondestructive apple quality evaluation. However, this method can be challenging to implement due to limitations in cost, noise tolerance, and activity. Depending on the frequency range and sensor technology, acoustic signal capture technology can have a very large range in prices (~\$100–~\$100,000). Cost can especially be an issue for higher-frequency systems, as they will need to take many more samples per second than lower-frequency systems. Higher sampling rates take more complex and expensive capture hardware. In the studies done in our group, we were able to obtain very high classification rates with the cheap low-frequency sensors, which could potentially address the issue of cost among other concerns. Another key issue in acoustic systems is their noise tolerance. Depending on the frequency range, it can be hard to isolate a subject from any background noises. This is especially true for very low frequencies, which will propagate large distances. The concept of activity with acoustic signals pertains to the percentage of time that the desired acoustic phenomena are generated. One issue with acoustic signal capture is that when the recording is occurring, the sounds may not be present. For example, insects will emit low-frequency vibrations when they move. If the insect does not happen to move during the capture interval, then the insect will not be captured and will be labeled falsely. Two ways we have addressed this are by using heat stimulation and active impulse acoustic generation (Khaled et al., n.d., 2021).

HSI methods have been explored more thoroughly than acoustic methods. The challenges in imaging technologies are well documented, both for quality assessment and infestation detection. Some of the major challenges in HSI apple quality assessment include low signal-to-noise ratio and heterogeneity of the sample (Saha and Annamalai, 2021). Many desirable bruises or defects can be subtle, which will need a clever algorithm to detect. Sample heterogeneity can cause the same issues as activity rate will with acoustic systems, as the defect may be missed if not enough of the sample is scanned. Insect detection presents a similar set of challenges, with one additional one: the age/size of the insect being detected. If the insect has only just infested an apple, then it may not be detectable using current HSI methods. One additional issue that will need to be addressed in the future for HSI systems is the robustness to non-ideal conditions. While many studies do consider outdoor applications, many do not. Further consideration will need to be given in all applications to how robust the models are, and how they will translate to uncontrolled conditions, especially in applications, which already have signal-to-noise ratio concerns.

Overall, the field of nondestructive analysis has improved greatly in the past few decades. However, the past 2 years of the COVID-19 pandemic have greatly emphasized that nondestructive analysis in the food industry is greatly important. Current destructive testing requires human labor, which will hamper the ability to process and ship apples if we experience another pandemic. Nondestructive methods are more conducive to having a sparse human presence, which would allow for plants to remain open without the risk of transmission in the plant. Many avenues are desirable for researchers to consider when developing robust nondestructive evaluation methods, including multi-modal detection approaches, better model development procedures, and robust testing protocols. As discussed previously, there are different advantages to both acoustic and hyperspectral detection methods. In combination (when fused), these two technologies should be able to detect a much wider range of phenomena in apples than either one alone. Some studies have been done in this domain, but there is still a lot of room for development and growth. One tool is the ability to better understand which machine learning and deep learning tools are the best fit for particular patterns. Quite often in the literature, several algorithms are applied without regard to the underlying patterns in the data; the best algorithm being best for a given dataset is often unexplained. More knowledge and training for biological scientists would greatly help the advancement of nondestructive techniques in this area. Robust testing protocols would also be very helpful when taking laboratory models into an outdoor/industrial setting. A full investigation of how much noise or variation a model can have and still meet the necessary accuracy standards required for usage will be very crucial to getting widespread usage and acceptance over traditional destructive testing. However, these are just a few suggestions in a field with wide possibilities. Within a few years, collaborative research could result in huge strides in nondestructive apple quality evaluation and their classifications.

References

Abbaszadeh, R., Rajabipour, A., Mahjoob, M., Delshad, M., & Ahmadi, H. (2013). Evaluation of watermelons texture using their vibration responses. *Biosystems Engineering, 115*(1), 102–105.

Abbott, J. A. (1994). Firmness measurement of freshly harvested 'Delicious' apples by sensory methods, sonic transmission, Magness-Taylor, and compression. *Journal of the American Society for Horticultural Science, 119*(3), 510–515.

Abbott, J. A., Affeldt, H. A., & Liljedahl, L. A. (1992). Firmness measurement of stored 'Delicious' apples by sensory methods, Magness-Taylor, and sonic transmission. *Journal of the American Society for Horticultural Science, 117*(4), 590–595.

Abbott, J. A., Childers, N. F., Bachman, G., Fitzgerald, J., & Matusik, F. (1968). *Acoustic vibration for detecting textural quality of apples* [Paper presentation]. Proceedings of the American Society for Horticultural Science.

Adedeji, A. A., Ekramirad, N., Rady, A., Hamidisepehr, A., Donohue, K. D., Villanueva, R. T., et al. (2020). Non-destructive Technologies for Detecting Insect Infestation in fruits and vegetables under postharvest conditions: A critical review. *Food, 9*(7), 927.

Armstrong, P., Stone, M., & Brusewitz, G. (1997). Nondestructive acoustic and compression measurements of watermelon for internal damage detection. *Applied Engineering in Agriculture, 13*(5), 641–645.

Blahovec, J., Kuroki, S., & Sakurai, N. (2007). Cooking kinetics of potato tubers determined by vibration techniques. *Food Research International, 40*(5), 576–584.

Breth, D. (2009). Reducing economic and environmental risk from codling moths and Oriental fruit moths. Cornell University College of Agriculture and Life Sciences (CALS) Newsletter for 2008–2009. http://impact.cals.cornell.edu/project/reducing-economic-and-environmentalriskcodling-moths-and-oriental-fruit-moths

Che, W., Sun, L., Zhang, Q., Tan, W., Ye, D., Zhang, D., & Liu, Y. (2018). Pixel based bruise region extraction of apple using Vis-NIR hyperspectral imaging. *Computers and Electronics in Agriculture, 146*, 12–21.

Chen, H., De Baerdemaeker, J., & Vervaeke, F. (1992). *Acoustic impulse response of apples for monitoring texture change after harvest* [Paper presentation]. Proceedings of the International Conference on Agricultural Engineering.

Chen, L., & Opara, U. L. (2013). Texture measurement approaches in fresh and processed foods—A review. *Food Research International, 51*(2), 823–835.

Chen, P., Sun, Z., & Huarng, L. (1992). Factors affecting acoustic responses of apples. *Transactions of ASAE, 35*(6), 1915–1920.

Cheng, J. H., Nicolai, B., & Sun, D. W. (2017). Hyperspectral imaging with multivariate analysis for technological parameters prediction and classification of muscle foods: A review. *Meat Science, 123*, 182–191.

Cooke, J. (1970). *A theoretical analysis of the resonance of intact apples* (ASAE Paper No. 70-345). American Society of Association Executives.

Cooke, J., & Rand, R. (1973). A mathematical study of resonance in intact fruits and vegetables using a 3-media elastic sphere model. *Journal of Agricultural Engineering Research, 18*(2), 141–157.

Crichton, S., Shrestha, L., Hurlbert, A., & Sturm, B. (2018). Use of hyperspectral imaging for the prediction of moisture content and chromaticity of raw and pretreated apple slices during convection drying. *Drying Technology, 36*(7), 804–816.

Da Silva, V. A., Talhavini, M., Peixoto, I. C., Zacca, J. J., Maldaner, A. O., & Braga, J. W. (2014). Non-destructive identification of different types and brands of blue pen inks in cursive handwriting by visible spectroscopy and PLS-DA for forensic analysis. *Microchemical Journal, 116*, 235–243.

Diezma-Iglesias, B., Ruiz-Altisent, M., & Barreiro, P. (2004). Detection of internal quality in seedless watermelon by acoustic impulse response. *Biosystems Engineering, 88*(2), 221–230.

Duprat, F., Grotte, M., Piétri, E., Loonis, D., & Studman, C. (1997). The acoustic impulse response method for measuring the overall firmness of fruit. *Journal of Agricultural Engineering Research, 66*(4), 251–259.

Dyck, V. A. (2010). *Rearing codling moth for the sterile insect technique*. FAO Plant Production and Protection Paper, pp. 6 of 199.

Ekramirad, N., Al Khaled, Y. A., Donohue, K., Villanueva, R., Parrish, C. A., & Adedeji, A. A. (2021a). NIR Hyperspectral Imaging with machine learning to detect and classify codling moth infestation in apples. A paper in the proceeding and presented (poster) during On-Demand Simulated Live Session at 2021 Annual International Virtual Meeting of American Society of Agricultural and Biological (ASABE) held online from July 12–15, 2021. Paper #: 2100066.

Ekramirad, N., Khaled, A. Y., Parrish, C. A., Donohue, K. D., Villanueva, R. T., & Adedeji, A. A. (2021b). Development of pattern recognition and classification models for the detection of vibro-acoustic emissions from codling moth infested apples. *Postharvest Biology and Technology, 181*, 111633. https://doi.org/10.1016/j.postharvbio.2021.111633

Ekramirad, N., Parrish, C. A., Villanueva, R. T., Donohue, K. D., & Adedeji, A. A. (2020). *Low frequency signal patterns for codling moth larvae activity in apples* [Paper presentation]. 2020 ASABE Annual International Virtual Meeting.

Ekramirad, N., Rady, A., Adedeji, A. A., & Alimardani, R. (2017). Application of hyperspectral imaging and acoustic emission techniques for apple quality prediction. *Transactions of the ASABE, 60*(4), 1391–1401.

Elbatawi, I. (2008). An acoustic impact method to detect hollow heart of potato tubers. *Biosystems Engineering, 100*(2), 206–213.

El Masry, G., Wang, N., Vigneault, C., Qiao, J., & ElSayed, A. (2008). Early detection of apple bruises on different background colors using hyperspectral imaging. *LWT - Food Science and Technology, 41*(2), 337–345.

Fan, S., Zhang, B., Li, J., Liu, C., Huang, W., & Tian, X. (2016). Prediction of soluble solids content of apple using the combination of spectra and textural features of hyperspectral reflectance imaging data. *Postharvest Biology and Technology, 121*, 51–61.

FAOSTAT. (2021). *Food and agricultural organization database*. Crops. http://www.fao.org/faostat/en/?#data/QC

Fathizadeh, Z., Aboonajmi, M., & Beygi, S. R. H. (2020). Nondestructive firmness prediction of apple fruit using acoustic vibration response. *Scientia Horticulturae, 262*, 109073.

Ferrari, C., Foca, G., Calvini, R., & Ulrici, A. (2015). Fast exploration and classification of large hyperspectral image datasets for early bruise detection on apples. *Chemometrics and Intelligent Laboratory Systems, 146*, 108–119.

Finney, E., Jr. (1970). Mechanical resonance within red delicious apples and its relation to fruit texture. *Transactions of ASAE, 13*(2), 177–0180. https://doi.org/10.13031/2013.38564

Finney, E., Jr., Abbott, J. A., Watada, A. E., & Massie, D. R. (1978). Nondestructive sonic resonance and the texture of apples. *Journal of the American Society for Horticultural Science, 103*, 158–162.

Forsline, P. L., Aldwinckle, H. S., Dickson, E. E., Luby, J. J., & Hokanson, S. C. (2003). Collection, maintenance, characterization, and utilization of wild apples of Central Asia. *Horticultural Reviews-Westport Then New York, 29*, 1–62.

Fumuro, M., Sakurai, N., & Utsunomiya, N. (2013). Improved accuracy in determining optimal harvest time for pitaya (*Hylocereus undatus*) using the elasticity index. *Journal of the Japanese Society for Horticultural Science, 82*(4), 354–361.

Geiling, N. (2014). *The real Johnny Appleseed brought apples—And booze—To the American frontier.* https://www.smithsonianmag.com/arts-culture/real-johnny-appleseed-brought-applesand-booze-american-frontier-180953263/

Gowen, A. A., O'Donnell, C. P., Cullen, P. J., Downey, G., & Frias, J. M. (2007). Hyperspectral imaging—An emerging process analytical tool for food quality and safety control. *Trends in Food Science & Technology, 18*(12), 590–598.

Harker, F. R., Redgwell, R. J., Hallett, I. C., Murray, S. H., & Carter, G. (1997). Texture of fresh fruit. *Horticultural Reviews, 20*, 121–224.

Huang, W., Li, J., Wang, Q., & Chen, L. (2015). Development of a multispectral imaging system for online detection of bruises on apples. *Journal of Food Engineering, 146*, 62–71.

Huarng, L., Chen, P., & Upadhyaya, S. (1993). Determination of acoustic vibration modes in apples. *Transactions of ASAE, 36*(5), 1423–1429.

Jia, W., Liang, G., Tian, H., Sun, J., & Wan, C. (2019). Electronic nose-based technique for rapid detection and recognition of moldy apples. *Sensors, 19*(7), 1526.

Johansmann, M., Siegmund, G., & Pineda, M. (2005). Targeting the limits of laser Doppler vibrometry. *Proceedings of the IDEMA, 2005*, 1–12.

Kadowaki, M., Nagashima, S., Akimoto, H., & Sakurai, N. (2012). Detection of core rot symptom of Japanese pear (*Pyrus pyrifolia* cv. Kosui) by a nondestructive resonant method. *Journal of the Japanese Society for Horticultural Science, 81*(4), 327–331.

Keresztes, J. C., Goodarzi, M., & Saeys, W. (2016). Real-time pixel based early apple bruise detection using short wave infrared hyperspectral imaging in combination with calibration and glare correction techniques. *Food Control, 66*, 215–226.

Khaled, Y. A., Ekramirad, N., Donohue, K., Doyle, L., Villanueva, R., Parrish, C. A., & Adedeji, A. A. (2021). Vibro-acoustic emission and heat stimulation effect on the detection of codling moth larvae in apples A paper in the proceeding and presented (poster) during On-Demand Simulated Live Session at 2021 Annual International Virtual Meeting of American Society of Agricultural and Biological (ASABE) held online from July 12–15, 2021. Paper #: 2100070. https://doi.org/10.13031/aim.202100070

Khaled, Y. A., Parrish, C. A., Eberhart, P. S., Doyle, L., Donohue, K. D., Villanueva, R., & Adedeji, A. A. (n.d.) Nondestructive detection of codling moth infestation in apples using acoustic impulse response signals. Biosystems Engineering. Under review for publication (Unpulished).

Lashgari, M., Maleki, A., & Amiriparian, J. (2017). Application of acoustic impulse response in discrimination of apple storage time using neural network. *International Food Research Journal, 24*(3), 1075–1080.

Lawless, H. T., & Heymann, H. (2013). *Sensory evaluation of food: Principles and practices*. Springer Science & Business Media.

Lee, C. Y., & Mattick, L. R. (1989). Composition and nutritive value of apple products. In *Processed apple products* (pp. 303–322). Springer. https://doi.org/10.1007/978-1-4684-8225-6_14.

Li, J., Luo, W., Wang, Z., & Fan, S. (2019). Early detection of decay on apples using hyperspectral reflectance imaging combining both principal component analysis and improved watershed segmentation method. *Postharvest Biology and Technology, 149*, 235–246.

Li, M., Ekramirad, N., Rady, A., & Adedeji, A. (2018). Application of acoustic emission and machine learning to detect codling moth infested apples. *Transactions of the ASABE, 61*(3), 1157–1164.

Liljedahl, L. A., & Abbott, J. A. (1994). Changes in sonic resonance of ‘Delicious’ and ‘Golden Delicious’ apples undergoing accelerated ripening. *Transactions of ASAE, 37*(3), 907–912.

Liu, H., Lee, S.-H., & Chahl, J. S. (2017). A review of recent sensing technologies to detect invertebrates on crops. *Precision Agriculture, 18*(4), 635–666.

Lu, R. (2007). Nondestructive measurement of firmness and soluble solids content for apple fruit using hyperspectral scattering images. *Sensing and Instrumentation for Food Quality and Safety, 1*(1), 19–27.

Lu, Y., Saeys, W., Kim, M., Peng, Y., & Lu, R. (2020). Hyperspectral imaging technology for quality and safety evaluation of horticultural products: A review and celebration of the past 20-year progress. *Postharvest Biology and Technology, 170*, 111318.

Mankin, R., Hagstrum, D., Guo, M., Eliopoulos, P., & Njoroge, A. (2021). Automated applications of acoustics for stored product insect detection, monitoring, and management. *Insects, 12*(3), 259.

Martini, A., Troncossi, M., & Rivola, A. (2017). Leak detection in water-filled small-diameter polyethylene pipes by means of acoustic emission measurements. *Applied Sciences, 7*(1), 2.

Mendoza, F., Lu, R., & Cen, H. (2012). Comparison and fusion of four nondestructive sensors for predicting apple fruit firmness and soluble solids content. *Postharvest Biology and Technology, 73*, 89–98.

Mo, C., Kim, M. S., Kim, G., Lim, J., Delwiche, S. R., Chao, K., et al. (2017). Spatial assessment of soluble solid contents on apple slices using hyperspectral imaging. *Biosystems Engineering, 159*, 10–21.

Muramatsu, N., Sakurai, N., Wada, N., Yamamoto, R., Takahara, T., Ogata, T., & Nevins, D. J. (1999). Evaluation of fruit tissue texture and internal disorders by laser Doppler detection. *Postharvest Biology and Technology, 15*(1), 83–88.

Muramatsu, N., Sakurai, N., Yamamoto, R., & Nevins, D. J. (1996). Nondestructive acoustic measurement of firmness for nectarines, apricots, plums, and tomatoes. *HortScience, 31*(7), 1199–1202.

Muramatsu, N., Tanaka, K., Asakura, T., Ishikawa-Takano, Y., Sakurai, N., Wada, N., & Nevins, D. J. (1997). Critical comparison of an accelerometer and a laser Doppler vibrometer for measuring fruit firmness. *HortTechnology, 7*(4), 434–438.

Muravin, B. (2009). Acoustic emission science and technology. *Journal of Building and Infrastructure Engineering of the Israeli Association of Engineers and Architects, 1*, 4–5.

Neven, L. G. (2012). Fate of codling moth (Lepidoptera: Tortricidae) in harvested apples held under short photoperiod. *Journal of Economic Entomology, 105*(2), 297–303.

Pickyourown. (2021). *U.S. Apple Crop facts*. https://www.pickyourown.org/USapplecrop.htm

Rady, A., Ekramirad, N., Adedeji, A. A., Li, M., & Alimardani, R. (2017). Hyperspectral imaging for detection of codling moth infestation in GoldRush apples. *Postharvest Biology and Technology, 129*, 37–44.

Rao, X., Ying, Y., Lv, F., & Jin, B. (2004). Development of a fruit quality inspecting system based on acoustic properties. *T. CSAM, 2*, 69–71.

Saha, D., & Annamalai, M. (2021). Machine learning techniques for analysis of hyperspectral images to determine quality of food products: A review. *Current Research in Food Science, 4*, 28–44.

Schotte, S., De Belie, N., & De Baerdemaeker, J. (1999). Acoustic impulse-response technique for evaluation and modelling of firmness of tomato fruit. *Postharvest Biology and Technology, 17*(2), 105–115.

Shi, C., Qian, J., Zhu, W., Liu, H., Han, S., & Yang, X. (2019). Nondestructive determination of freshness indicators for tilapia fillets stored at various temperatures by hyperspectral imaging coupled with RBF neural networks. *Food Chemistry, 275*, 497–503.

Song, H.-Z., Wang, J., & Li, Y.-H. (2006). Studies on vibration characteristics of a pear using finite element method. *Journal of Zhejiang University. Science. B, 7*(6), 491–496. https://doi.org/10.1631/jzus.2006.B0491

Steward, P. (2009). Chile is the most competitive: The U.S. apple industry ranks fifth in global competitiveness, an agricultural economist calculates. *Good Fruit Grower, 59*(1). http://www.goodfruit.com/issues.php?article=1813&issue=69.

Stone, M., Armstrong, P., Zhang, X., Brusewitz, G., & Chen, D. (1996). Watermelon maturity determination in the field using acoustic impulse impedance techniques. *Transactions of ASAE, 39*(6), 2325–2330.

Su, Z., Tu, K., Zou, X., Jing, W., & Pan, L. (2008). Detection of internal defect in apples by acoustic impulse technique. *Science and Technology of Food Industry, 29*(7), 235–238.

Sun, T., Huang, K., Xu, H., & Ying, Y. (2010). Research advances in nondestructive determination of internal quality in watermelon/melon: A review. *Journal of Food Engineering, 100*(4), 569–577.

Symoneaux, R., Royer, G., Madieta, E., & Jourjon, F. (2004). *Acoustic and sensory measurements of different ripeness of apples* [Paper presentaion]. The V International Postharvest Symposium 682.

Takeda, Y., Sawaji, M., & Yasukawa, J. (1970). The non-destructive measurement of ripeness of apples by the sonic characteristics. *Nippon Shokuhin Kogyo Gakkaishi, 17*(8), 358–360.

Tan, W., Sun, L., Yang, F., Che, W., Ye, D., Zhang, D., & Zou, B. (2018). Study on bruising degree classification of apples using hyperspectral imaging and GS-SVM. *Optik, 154*, 581–592.

Taniwaki, M., & Sakurai, N. (2010). Evaluation of the internal quality of agricultural products using acoustic vibration techniques. *Journal of the Japanese Society for Horticultural Science, 79*(2), 113–128.

Taniwaki, M., Tohro, M., & Sakurai, N. (2010). Measurement of ripening speed and determination of the optimum ripeness of melons by a nondestructive acoustic vibration method. *Postharvest Biology and Technology, 56*(1), 101–103.

Terasaki, S., Wada, N., Sakurai, N., Muramatsu, N., Yamamoto, R., & Nevins, D. J. (2001). Nondestructive measurement of kiwifruit ripeness using a laser Doppler vibrometer. *Transactions of ASAE, 44*(1), 81.

Tesch, R., Normand, M. D., & Peleg, M. (1996). Comparison of the acoustic and mechanical signatures of two cellular crunchy cereal foods at various water activity levels. *Journal of the Science of Food and Agriculture, 70*(3), 347–354. https://doi.org/10.1002/(SICI)1097-0010(199603)70:3<347::AID-JSFA510>3.0.CO;2-B

Tiplica, T., Vandewalle, P., Verron, S., Grémy-Gros, C., & Mehinagic, E. (2010). *Identification of apple varieties using acoustic measurements* [Paper presentation]. The Conférence Internationale en Métrologie (CAFMET'10).

U.S. Apple Association. (2021). *Industry at a glance and apple production, exports up for 2019 crop, says USApple*. https://usapple.org/industry-at-a-glance; https://usapple.org/news-resources/apple-production-exports-up-for-2019-crop-says-usapple

Valente, M., Leardi, R., Self, G., Luciano, G., & Pain, J. P. (2009). Multivariate calibration of mango firmness using Vis/NIR spectroscopy and acoustic impulse method. *Journal of Food Engineering, 94*(1), 7–13.

Vetrekar, N. T., Gad, R. S., Fernandes, I., Parab, J. S., Desai, A. R., Pawar, J. D., et al. (2015). Non-invasive hyperspectral imaging approach for fruit quality control application and classification: Case study of apple, chikoo, guava fruits. *Journal of Food Science and Technology, 52*(11), 6978–6989.

Vursavus, K. K., Yurtlu, Y. B., Diezma-Iglesias, B., Lleo-Garcia, L., & Ruiz-Altisent, M. (2015). Classification of the firmness of peaches by sensor fusion. *International Journal of Agricultural and Biological Engineering, 8*(6), 104–115.

Wang, H., Zhu, H., Zhao, Z., Zhao, Y., & Wang, J. (2018). The study on increasing the identification accuracy of waxed apples by hyperspectral imaging technology. *Multimedia Tools and Applications, 77*(20), 27505–27516.

Wang, Z., Sun, J., Liao, X., Chen, F., Zhao, G., Wu, J., & Hu, X. (2007). Mathematical modeling on hot air drying of thin layer apple pomace. *Food Research International, 40*(1), 39–46.

Wills, R. B., & Golding, J. (Eds.). (2016). *Advances in postharvest fruit and vegetable technology*. CRC press.

Zhang, P., Ji, H., Wang, H., Liu, Y., Zhang, X., & Ren, C. (2021). Quantitative evaluation of impact damage to apples using NIR hyperspectral imaging. *International Journal of Food Properties, 24*(1), 457–470.

Zhang, W., Cui, D., Liu, Z., & Ying, Y. (2016). Analysis of pear dynamic characteristics based on harmonic response. *Transactions of the ASABE, 59*(6), 1905–1913.

Zhang, W., Cui, D., & Ying, Y. (2014). Nondestructive measurement of pear texture by acoustic vibration method. *Postharvest Biology and Technology, 96*, 99–105.

Zhang, W., Lv, Z., & Xiong, S. (2018). Nondestructive quality evaluation of agro-products using acoustic vibration methods—A review. *Critical Reviews in Food Science and Nutrition, 58*(14), 2386–2397.

Zorović, M., & Čokl, A. (2015). Laser vibrometry as a diagnostic tool for detecting wood-boring beetle larvae. *Journal of Pest Science, 88*(1), 107–112.

Zude, M., Herold, B., Roger, J.-M., Bellon-Maurel, V., & Landahl, S. (2006). Non-destructive tests on the prediction of apple fruit flesh firmness and soluble solids content on tree and in shelf life. *Journal of Food Engineering, 77*(2), 254–260.

Chemometrics in Nondestructive Quality Evaluation

Md. Nahidul Islam

Abstract Because of the widespread use of nondestructive measurement techniques, such as spectroscopy, spectral imaging, which allow scientists to swiftly obtain a complete spectrum for a single sample. The datasets nowadays tend to have a smaller number of samples and a larger number of variables. In order to extract information from these high-dimensional and high-volume data, traditional univariate analysis is widely considered to be inadequate. Effective data preprocessing and applying the appropriate chemometric methods are required to gain insights and obtain essential information from these datasets. In this way, chemometrics has made substantial advancements and recognition in nondestructive quality assessment of foods. The purpose of this chapter is to introduce an understanding of the latest ideas, methodologies, techniques, and fundamental processes used during nondestructive analysis of fruits and vegetables, where chemometrics and/or multivariate analytical approaches were performed.

Keywords Principal component analysis (PCA) · Partial least squares regression (PLSR) · Partial least squares discriminant analysis (PLS-DA)

1 Introduction

Analytical techniques used in laboratories are frequently insufficient since they necessitate a large number of samples, a longer time to receive results, and highly technical personnel (Zou & Zhao, 2015). In an environment where speed is critical, engineering advances must require fewer samples or, at the very least, no one (nondestructive techniques): a) they must provide prompt, if not immediate, responses in order for the operator to make an informed decision on the next steps to regulate or release the product to the market; b) they must be simple to use in order to encourage their use across the manufacturing chain, where analytical laboratories

M. N. Islam (✉)
Department of Agro-Processing, Bangabandhu Sheikh Mujibur Rahman Agricultural University, Gazipur, Bangladesh
e-mail: nahidul.islam@bsmrau.edu.bd

P. B. Pathare, M. S. Rahman (eds.), *Nondestructive Quality Assessment Techniques for Fresh Fruits and Vegetables*, https://doi.org/10.1007/978-981-19-5422-1_14

are not always available. As a result, technology needs to be adjusted to a new strategy of production: the use of sensors and the necessarily associated information extraction system, which allows "measurement," to meet the demands of the agri-food stakeholders. Moreover, the manufacturers of technologies often provide devices that require calibration phases not always easy to perform but that are often the subject of actual researches. These are particularly complex when similar processes need to run repeatedly (Eriksson et al., 2013).

Hence, chemometrics techniques in nondestructive quality evaluation aim to produce an empirical or semi-empirical model from data that may be used to predict one or more chemical properties of a system from observations (Cocchi, 2017). Chemometrics employs mathematical and statistical methodologies to optimize experimental processes through the scientific design of experiments, to treat the experimental data and to extract as much relevant chemical information as possible from generated data (Guidetti et al., 2012). Chemical systems are often multivariate, which means that numerous information are obtained at the same time. As a result, the majority of chemometric procedures fall within the category of analytical methods known as multivariate statistical analysis, which contains many measurements on a number of individuals, objects, or data samples. Therefore, multiple measurements and analyses of the variable dependence are central to chemometrics (Marini, 2013). This chapter aims to present an overview and to provide a clear understanding of the chemometric methods with their advantages and disadvantages used in the nondestructive quality evaluation of fruits and vegetables.

2 Major Chemometric Tools in Food Analysis

Chemometric approaches are used to optimize the experimental process and extract relevant chemical information from massive quantities of data, identify hidden relationships, and provide a visual approach. There are several chemometric approaches: design of experiment (DoE), preprocessing, explorative analysis, classification, regression, validation, feature selection, multiway analysis, etc. These methods are utilized for nondestructive quality analysis of fruits and vegetables, as well as in other areas of food science and technology. The chosen method is determined by the challenge, type of experimental data, as well as by considering the pros and cons of that particular chemometric approach (Martens & Martens, 2001).

2.1 Design of Experiment

DoE technique ensures representativeness of the sample, allows for the evaluation of the primary sources of variability, and is the most effective way to optimize analytical measurement processes (Lawson, 2014). Experimental designs are

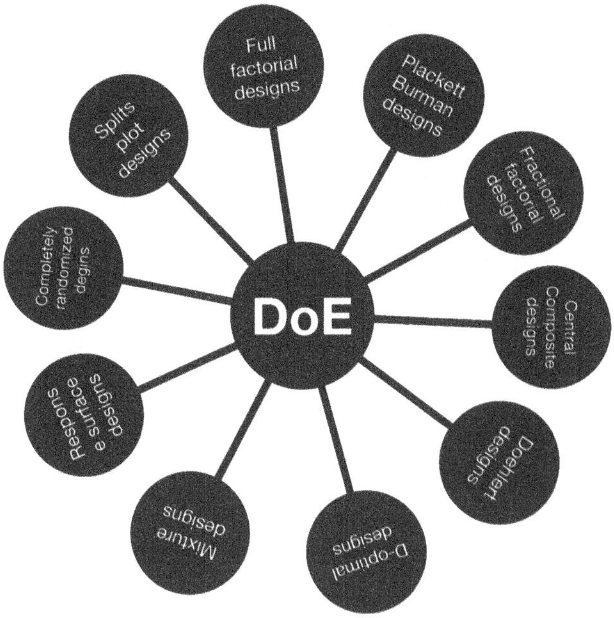

Fig. 1 Common experimental design methods in chemometrics

frequently neglected or undervalued; however, in order to illustrate the need for variable optimization as well as the development of adequate methods for carrying out the tests, a correct experimental design must be established in advance (Granato and de Araújo Calado, 2013; Leardi, 2009; Wold et al., 2004). It is essentially decided how to carry out scientific research. When developing new detection systems, the optimization protocol is especially important. A well-defined DoE not only allows scientists to investigate different factors and their interactions, but it also saves money (Granato and de Araújo Calado, 2013; Leardi, 2006, 2009; Wold et al., 2004). Figure 1 illustrates various methods of experimental designs used in chemometrics.

The use of a specific design depends on the specific problem statement. For example, some methods are used in optimization while some methods are used in screening experiments. The advantages and disadvantages of some common experimental designs are presented in Table 1.

2.2 *Preprocessing of Data*

After collecting data, data preprocessing is frequently the deciding factor between excellent and poor chemometric models (Rinnan, 2014). Preprocessing is used to reduce variation that is not connected to the topic of interest, allowing the variation

Table 1 Advantages and disadvantages of some common experimental designs

Experimental designs	Advantages	Disadvantages	References
Full factorial designs	Allows scientists to look at the effect of treatment	Difficult to ensure the elimination of in-group variations	(Ebrahimi-Najafabadi et al. (2014)
	Good for experiments with more than five factors	Need to report experiments in the standard order to avoid systematic error.	
		Bad for more than five factors	
Randomized block designs	Opportunity to block nuisance factors inside the block.	Need to eliminate the contribution of nuisance factors.	Huynh and Feldt (1976)
Plackett Burman designs	Good for more factors.	These designs have run numbers that are a multiple of four.	Vanaja and Shobha Rani (2007)
Fractional factorial design	Possible to give a first interpretation just looking at the highly informative data.	Require many runs.	Rakić et al. (2014)
	Use low-resolution designs for screening among main effects and use higher resolution designs when interaction effects and response surfaces need to be investigated.		
Response surface methodology	Quadratic models are almost always sufficient for industrial applications.	Response surface models may involve just main effects and interactions, or they may also have quadratic and possibly cubic terms to account for the curvature.	Bezerra et al. (2008)
Central composite design	Start with factorial or fractional factorial design (with center points) and add "star" points to estimate curvature.	The position of the star points is important.	Asghar et al. (2014)

of interest to stand out more and be more easily modeled (Islam et al., 2018b). There are some situations where enhancing spectral features is essential; for example, in a situation where intriguing spectral characteristics differ slightly from the global intensity, where small peaks are difficult to see in the presence of a large one and due to overlapped peaks. According to Roger et al. (2020), there are several types of systematic variations that are not related to topics of interest, for example, the shift of the baseline due to light scattering effects as a result of various particle sizes, offsets of baseline due to differences in instrumentation, and variations in the signal intensities due to size, shape, and volume of the sample. Figure 2 illustrates the available data preprocessing methods in chemometrics.

Among all the preprocessing methods, mean centering, standard normal variate (SNV) normalization, baseline correction, orthogonal signal correction (OSC),

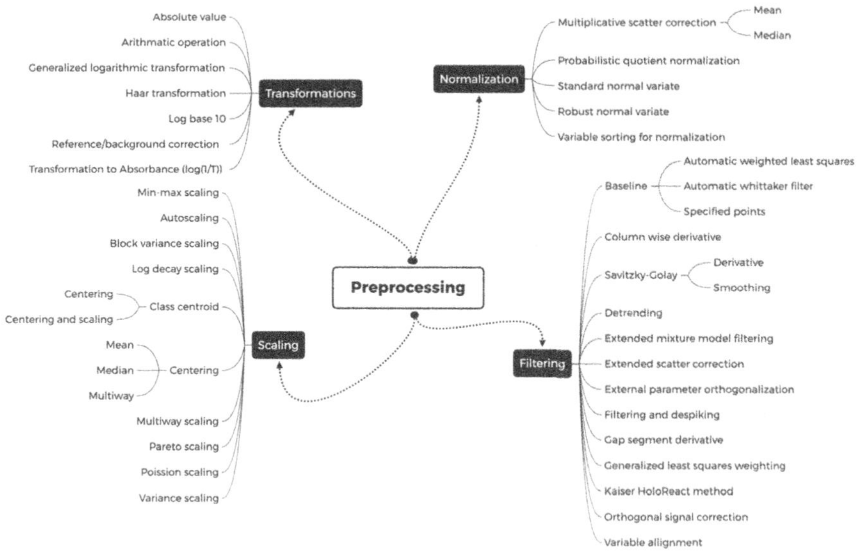

Fig. 2 Available preprocessing methods in chemometrics

Savitzky–Golay (SAV-GOL) smoothing and derivatization, and multiplicative signal correction (MSC) are the most common data preprocessing methods used in chemometrics (Vidal and Amigo 2012). The visualization of the data and the removal of severe bands that are driven by noise are the first steps in near-infrared (NIR) spectroscopic preprocessing. Then, to reduce any high-frequency noise, window-based smoothing techniques can be performed. The SAV-GOL algorithm is a widely used approach for reducing high-frequency noise. It includes fitting a polynomial of chosen order into a band of the defined size that is moved throughout the entire spectrum (Rinnan et al., 2009). In an ideal circumstance, the smoothed spectra should be ready for regression or classification modeling in the presence of just absorption features. The smoothing stage is frequently followed by scattering correction methods due to the prominence of scattering effects. The estimation of the second derivative of the spectra is the most frequent method since it can quickly remove first-order additive (baseline shift) effects and also show underlying peaks that would otherwise be invisible (Rinnan et al., 2009). Another widely used method is SNV, which includes subtracting each spectrum's mean spectral intensity from each intensity response and then dividing by its spectral-domain standard deviation (Barnes et al., 1989). SNV can be used to eliminate additive and multiplicative effects. In NIR modeling, both the second derivative and SNV are quite useful and usually increase model prediction performance. Another prominent method is MSC, which assumes the spectrum has a multiplicative, additive, and residual component (Isaksson & Næs, 1988). In order to describe these impacts, the Extended MSC (EMSC) model incorporates higher-order complex relations (Martens et al., 2003).

Table 2 Some common preprocessing methods with their pros and cons

Preprocessing methods	Advantages	Disadvantages	References
Mean centering	It may reduce model complexity by reducing factors. Provide better calibration model.	Risk of uncertainty. Poor prediction model outside of calibration space.	Enders and Tofighi (2007)
Autoscaling	All variables become equally important.	Inflation of the measurement errors.	van den Berg et al. (2006)
Variance scaling	A weak signal may provide information.	Noisy parts get weight.	Noda (2008)
		Difficult to interpret loading and beta coefficient.	
Pareto scaling	Isolated weak signal provides information.	Difficult to interpret loading and beta coefficient.	Kasprzak and Lewis (2001)
	Does not give much weights to noisy part as variance scaling.		
Logarithmic transformation	Transform multiplicative effect to additive effects.	Not applicable to data containing negative or zero values.	Bartlett and Kendall (1946)
SAV-GOL	Derivation may enhance less apparent spectral features.	A wide window may remove important information, while a narrow window keeps lots of noises.	Press and Teukolsky (1990)
MSC	Good at NIR reflectance measurement of fruits and vegetables.	Not good at predicting physical properties.	Isaksson and Næs (1988)
		Depends on the reference spectral data.	
SNV	Effective in correcting systematic effects, and NIR scattering effects.	Contains negative value in the processed data.	Barnes et al. (1989)
Detrending	Can deal with data collected over time where baseline/background may have drift.	Processed spectra have a negative value.	Tanabe et al. (2002)

MSC multiplicative scatter correction, *SNV* standard normal variate, *SAV-GOL* Savitzky–Golay

In many cases, the scattering effect is also important to perfectly describe the quality of fresh fruits and vegetables. Therefore, the removal of the scattering effect in those cases may lead to the wrong chemometric model (Mishra et al., 2021). Robust normal variate (RNV) (Guo et al., 1999), probabilistic quotient normalization (PQN) (Dieterle et al., 2006), and variable sorting for normalization (VSN) (Rabatel et al., 2020) have all been offered as improvements and alternatives to SNV. To summarize, there are numerous chemometric preprocessing approaches for removing/reducing scattering effects from spectral data. Table 2 summarizes the advantages and disadvantages of the common data preprocessing methods used in chemometrics.

3 Principal Component Analysis (PCA)

Principal component analysis (PCA) is one of the most important and powerful methods in chemometrics (Bro & Smilde, 2014). PCA is a bilinear reduction approach that may condense enormous amounts of data into a few parameters known as principal components (PCs) or latent variables, which reflect the levels, differences, and similarities among the samples and variables that make up the modeled data. A linear transformation is used to accomplish this goal, with the constraints of conserving data variance and imposing orthogonality on the latent variables (Smilde et al., 2005).

3.1 PCA Data Analysis

PCA can be used to visualize the X data matrix in the multivariate space, cluster identification and detection of outliers, reducing the dimensionality of the data and removing the noise. The starting point for PCA is a matrix of data with N rows (observations) and M columns (variables), here denoted by X. Technically, PCA seeks lines, planes, and hyperplanes in K-dimensional space that best approximate the data in terms of least squares. It is obvious that a line or plane that is the least squares approximation of a set of data points minimizes the variance of the coordinates on the line or plane (Wold et al., 1987).

The first PC is the line in K-dimensional space that best approximates the data in terms of least squares. The line intersects the mean point. Hence, each observation can be projected onto this line to obtain a coordinate value along the PC line. This new coordinate value is referred to as a score. A second PC is a line in K-dimensional variable space that is orthogonal to the first PC. This line likewise crosses through the average point and enhances the X-data approximation as much as feasible. If X is a data matrix with N rows and M columns, and with each variable being a column and each sample a row, PCA decomposes X as the sum of r t_i and p_i, and where r is the rank of the matrix X (Eq. 1).

$$X = t_1 pT_1 + t_2 pT_2 + \ldots\ldots + t_m pT_m + \ldots\ldots + t_r pT_r \quad (1)$$

$$r \le min\ \{M, N\}$$

$$X = t_1 pT_1 + t_2 pT_2 + \ldots\ldots + t_m pT_m + E \quad (2)$$

The amount of variance captured by t_i, p_i pairs are ordered. The scores are vectors that include information about how the samples relate to one another. The vectors are called loadings and they provide information on how the variables interact. In general, after m components, the PCA model is usually truncated, and the small variance factors are consolidated into a residual matrix E (Eq. 2).

The basic premise is that the investigated systems are "indirectly observable," meaning that the relevant phenomena that cause data variation/patterns are concealed and not directly measurable/observable. This is where the phrase "latent variables" comes from. Latent variables (PCs) can be expressed as scatter plots in the Euclidean plane once they have been discovered. A loading plot can be discussed in conjunction with the associated score plot, which is generated for the same pair of PCs, or it can be directly shown in the same figure, which is called a biplot. It becomes easier to explain the groups or patterns observed in the PC space in terms of the original variables in this way. Although the biplot format for spectral data is difficult to visualize, specific spectral regions that are responsible for the separation of process phases can be highlighted.

3.2 Outlier Detection

Q residuals are the sum of squared residuals for each sample. In other words, Q is a measure of the distance of a sample from the PCA model. Therefore, a higher Q value means a lower model fit. Hotelling's T^2 is the sum of normalized squared scores (Hotelling, 1947). T^2 is a measure of the variation in each sample within the PCA model (Fig. 3). Figure 4 presents Q residuals versus Hotelling's T^2 plot, which is very useful to determine the outlier sample.

The region of extreme samples (bottom right) exhibits unusual behavior since they adhere to the variable correlation structure recorded by the PCA model while achieving high scores in the scores space. Because they pull the PC axes toward them, these samples with high Hotelling's T^2 values are said to have strong leverage. The region far from model samples (top left): these samples, with high Q residuals values, appear to be "well behaving" when projected onto model space because they

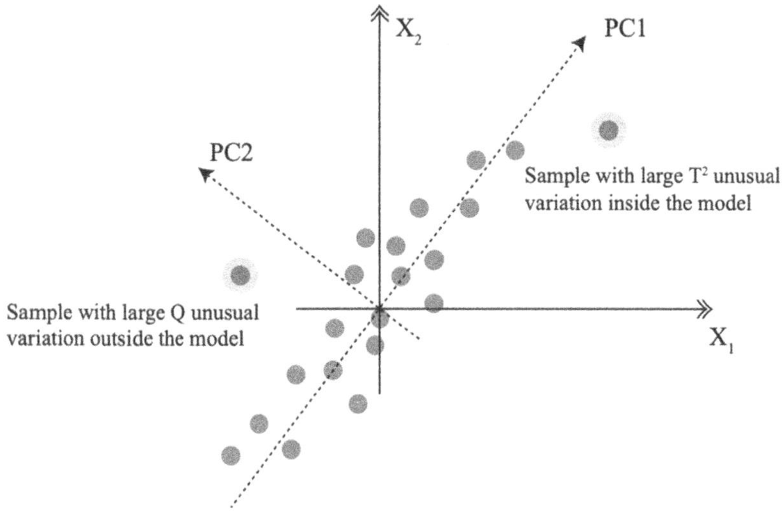

Fig. 3 Graphical representation of the principal components space for a two-component model

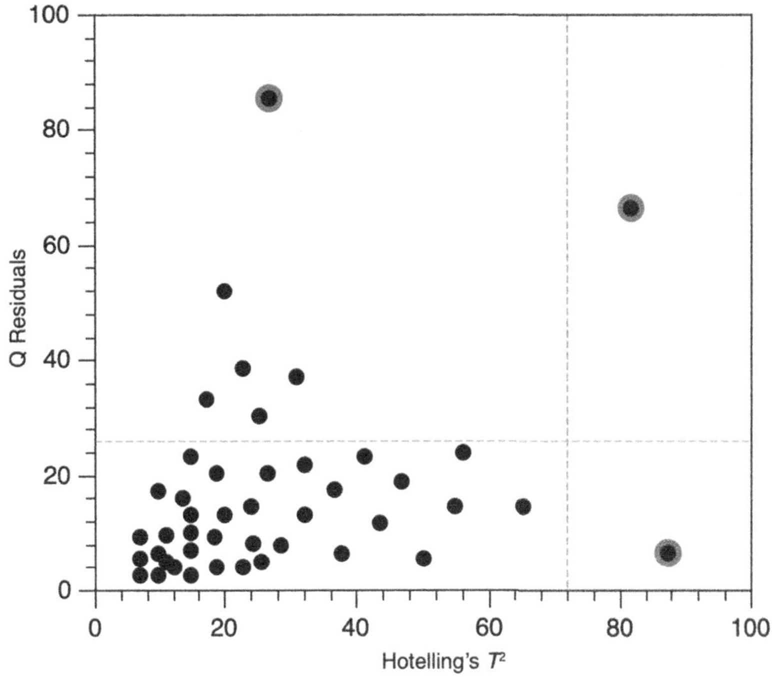

Fig. 4 Q residuals versus Hotelling's T^2 plot

share some characteristics with the modeled category, but they are not well modeled because part of their variation is not accounted for by the model. The anomalous, extreme, and non-modeled samples that have both T^2 and Q high values belong in the outliers region (top right) (Westerhuis et al., 2000).

4 Partial Least Squares Regression

Partial least squares regression (PLSR) is a regression extension of PCA, which is used to connect the information in two blocks of variables, *X* and *Y*, to each other (Wold et al., 2001). PLSR is a method of relating two data matrices, *X* and *Y*, to each other by a linear multivariate model. PLSR stands for projections to latent structures by means of partial least squares. It derives its usefulness from its ability to analyze data with many noisy, collinear, and even incomplete variables in both *X* and *Y*. For parameters related to the observations (samples, compounds, objects, items), the precision of a PLSR model improves with the increasing number of relevant *X*-variables. This corresponds to the intuition of most chemists, technicians, and engineers that many variables provide more information about the observations than just a few variables do (Martens & Naes, 1991).

PLSR can be seen as a particular regression technique for modeling the association between X and Y, but it can be seen as a philosophy of how to deal with complicated and approximate relationships (Geladi & Kowalski, 1986). Because PLSR considers not just the correlation between two variables but also the amount of variation in each, the criterion for defining the PLS latent variables is formulated using covariance, which is a good metric of interrelation, component-based criterion because converting it to a global loss function is quite challenging. As a result, PLSR is a sequential algorithm: PLS latent variables are computed in such a way that the first PLS component is the dependent variables' direction of maximum covariance. The second PLS component, for example, is orthogonal to the first and has the highest residual covariance, and so on (Wold et al., 1983).

Outlier samples that are far from the center within the space given by the PLS model can be detected using plots of leverage or Hotelling's T^2. The critical limit for Hotelling's T^2 statistics is based on an F-test (Hotelling, 1992), while the critical limit for Leverage is based on ad hoc knowledge (Martens & Naes, 1991). A predicted versus measured plot should, in a good PLS model, display a straight-line relationship between predicted and measured values, ideally with a slope of one and a correlation close to one. A residual plot may be plotted against the value of the y-variable to check that the residuals are not depending on the value of Y. Outliers of various types, such as samples with significant residuals and influential samples, are commonly detected using F residuals versus Hotelling's T^2 plot. Outliers are samples with high residual variance or those that lie at the top of the plot. Influential samples are those that have high leverage, that is, those that lie to the right of the plot (Rousseeuw & Leroy, 1987). This indicates that they are attracting the model in order for it to better describe them. Influential samples are not always risky if the variables follow the same pattern as the more "average" samples. A sample with significant residual variance and leverage is referred to as a "potential outlier," and in the presence of these outliers, the model focuses on the differences between that sample and the outlier rather than defining more general traits common to all samples.

5 Classification

Datasets are frequently made up of samples from various groups or "classes." Groups may differ for a variety of reasons, including variations in sample preparation, chemical constituent types such as aromatic, aliphatic, etc., or process conditions. A number of approaches for classifying samples based on measured responses have been developed, as shown in Fig. 5. Cluster analysis and unsupervised pattern recognition are methods for attempting to find groups or classes without the use of prior knowledge regarding class memberships. On the other hand, classification or supervised pattern recognition are terms used to describe methods that leverage known class memberships (Ballabio & Consonni, 2013).

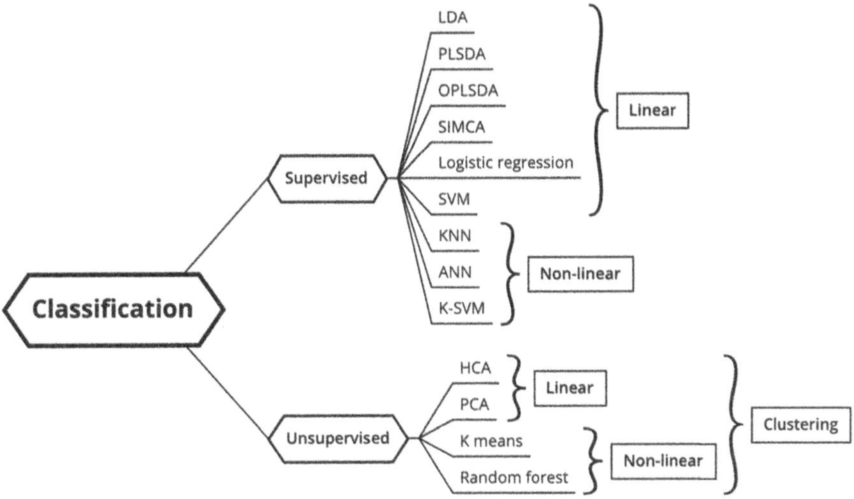

Fig. 5 Overview of the classification techniques in chemometrics

Most cluster analysis approaches are based on the concept that samples that are close together in the measurement space are comparable and so likely to belong to the same class. However, there are other ways to define the distance between samples. The most popular is the simple Euclidean distance. A Mahalanobis distance accounts for the fact that variance in some directions is substantially greater in some datasets than in others. As a result, distance in some directions is more relevant than the distance in others (De Maesschalck et al., 2000).

Soft Independent Modeling of Class Analogy (SIMCA) makes use of the model features and incorporates information about the calibration data types. A SIMCA model is made up of a set of PCA models, one for each class in the dataset (Wold, 1976). The number of major components in each class can vary. The number is determined by the data in the class. Each PCA sub-model includes all of the standard components of a PCA model, such as the mean vector, scaling information, preprocessing such as smoothing and derivatizing, and so on. The oldest and most studied supervised pattern recognition approach is linear discriminant analysis (LDA) (Fisher, 1936). It is a linear approach in the sense that the decision boundaries dividing the classes of variables in their multidimensional space are linear surfaces (hyperplanes). The purpose of LDA is to identify the ideal linear surface in a multidimensional space that corresponds to the best two-dimensional straight line. Partial least squares discriminant analysis (PLS-DA) is quite similar to LDA, another common discriminating approach. Indeed, Barker and Rayens (2003) demonstrated that PLS-DA is simply the inverse-least squares method to LDA, producing essentially the same result but with the noise reduction and variable selection benefits of PLS. PLS is used in PLS-DA to create a model that predicts the class number for each sample (Næs et al., 2002). Table 3 summarizes the advantages and

Table 3 Advantages and disadvantages of some common chemometric methods

Chemometric tools	Advantages	Disadvantages
PCA	PCA shows the similarities and differences between the samples as well as the relationships between the variables quickly and easily.	It does not allow for the classification of samples and the assignment of a class to each one.
PLSR	The ability to handle more descriptor variables than compounds, nonorthogonal descriptors, and multiple biological outcomes with more predictability and a considerably reduced probability of chance correlation.	Increased risk of missing "real" correlations and sensitivity to the relative scaling of the descriptor variables.
PLS-DA	The categorization model may be created quickly and easily, and the results are usually extremely good.	Classification errors might occur if the distinction between the areas of the various classes is not clear enough.
SIMCA	It can train a binary classification model exclusively with the target class since it defines an acceptance region that contains all of the target class's objects/samples.	There may be overlapping between acceptance areas that contain samples from distinct classes in models trained with two or more classes. As a result, some samples may be assigned to one or more classes.
OPLS-DA	To improve classification, increase the difference in means as well as the difference in within-class variance between the two classes.	It is impossible to see different treatment effects among the participants in the community.
KNN	Method of application that is simple to use.	Because the dominant class affects the classification, if there are more samples of one class than the other (skewed distribution of classes), the samples may be incorrectly classified.
SVM	When the demarcation between the regions of the different classes of samples is not sufficiently obvious, this can be used to get around the technical challenge.	Alternative kernel functions must be utilized for non-linear SVM models. As a result, the model's development is complex, and a lot of informatics resources are required.
Random forest	Because the variance is reduced, this is an excellent choice for unstable models or class imbalance issues. Overfitting is avoided as much as possible.	Because the classification is not displayed as a graphical tree, understanding the results is difficult.

PCA principal component analysis, *PLSR* partial least squares regression, *PLS-DA* partial least squares discriminant analysis, *SIMCA* soft independent modeling of class analogy, *OPLS-DA* orthogonal partial least squares discriminant analysis, *KNN* K-nearest neighbors, *SVM* support vector machines

disadvantages of the most common chemometric methods used in nondestructive quality evaluation.

6 Model Validation

The most conservative validation method is to run the model on a sufficiently large representative independent test set. Several methodologies can be used to quantify sources of variation that are in principle unknown for future objects in order to make a model more robust to changes in the sample matrix, raw materials, chemical reagents, and so on (Westad & Marini, 2015). Though the goal is to have enough items to set aside a decent amount as a test set, this is not always practicable due to factors such as sample costs or reference testing. Cross-validation is the best alternative to using an independent test set for validation (Westad & Kermit, 2003).

Cross-validation is a practical and reliable way to test the significance of a PLS model. This procedure has become standard in chemometric analysis and is incorporated in one form or another in most commercial software. With CV, the basic idea is to keep a portion of the data out of the model development, develop a number of parallel models from the reduced data, predict the omitted data by different models, and finally compare the predicted values with the actual ones. The square differences between predicted and observed values are summed to form the predictive residual sum of squares, which is a measure of the predictive power of the tested model (Stone, 1974). Various ways of cross-validation is available, for example, full cross-validation, segmented cross-validation, systematic segmented cross-validation, and validating across categorical information about the objects (Kos et al., 2003).

7 Model Performances

The number of latent variables in a PLSR model is determined by minimizing the root mean square error of cross-validation (RMSECV). Given the data and number of latent variables, overfitting is a possibility, but the purely data-driven strategy is the best option. The root mean square error of prediction (RMSEP) is a direct estimate of the model's prediction error in PLSR modeling. The RMSEP can be calculated using Eq. (3). Alternatively, the PLSR model's accuracy and precision are represented by the bias and standard error of performance (SEP), respectively. Equations (4) and (5) can be used to compute the SEP and bias, respectively, where, and are the predicted and measured values of the ith observation in the test set and n is the size of the validation set (Amigo, 2021).

The accuracy, precision, and linearity of the models can be used to assess their performance. The root mean square error of calibration (RMSEC), RMSECV, RMSEP, and bias can all be used to express the model's correctness. The SEP can be used to examine the PLSR model's precision, and R^2 can be used to assess linearity using a linear fit of predicted versus measured values. Low RMSEC, RMSECV, RMSEP, and SEP values, as well as a high R^2 value, indicate a good model (Islam et al., 2018a).

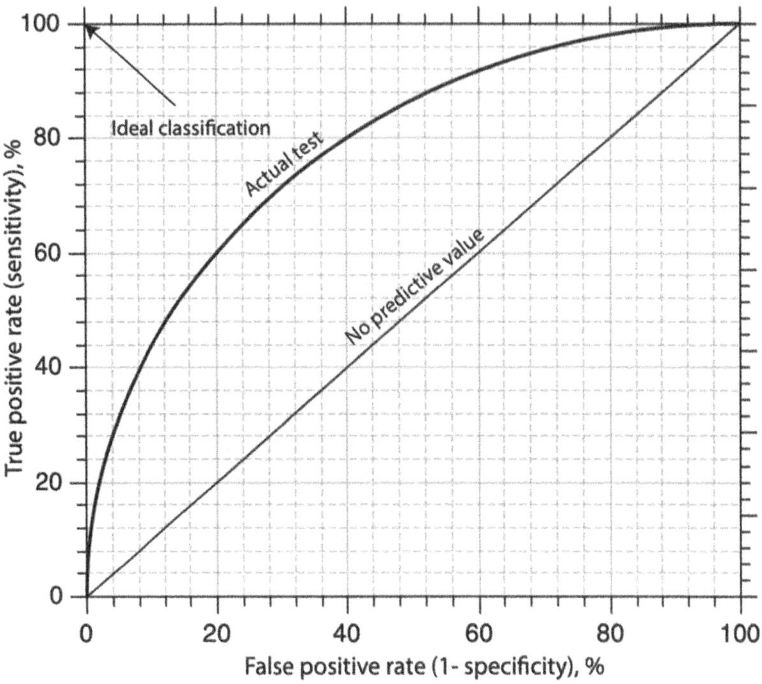

Fig. 6 Receiver operating characteristic (ROC) curve

$$RMSEP = \sqrt{\frac{\sum \left(\hat{y}_{cal} - y_{val}\right)^2}{n}} \quad (3)$$

$$SEP = \sqrt{\frac{\sum_{i=1}^{n} \left(\hat{y}_i - y_i - bias\right)^2}{n-1}} \quad (4)$$

$$bisa = \sum_{i=1}^{n} \frac{\left(\hat{y}_i - y\right)}{n} \quad (5)$$

The receiver operating characteristic (ROC) curve demonstrates the trade-off between sensitivity (or TPR) and specificity (1 − FPR). Classifiers that produce curves closer to the top-left corner perform better. A random classifier is expected to give points along the diagonal as a baseline (FPR = TPR). The test becomes less accurate when the curve approaches the ROC space's 45-degree diagonal. The class distribution has no bearing on the ROC. This makes it ideal for testing classifiers that anticipate infrequent events like rotten products. Using accuracy *(TP + TN)/(TP + TN + FN + FP)* to evaluate performance, on the other hand, would favor classifiers that always predict a negative outcome for uncommon events (Fig. 6).

The confusion matrix and ROC curves can also be used to measure a classifier's error rate. To understand the confusion matrix, consider a classification problem where there are two classes, *X* is the negative class, and *Y* is the positive class. And four possible outcomes: sample from class *X* is assigned to class *X*, the sample from class *Y* is assigned to class *Y*, the sample from class *X* is assigned to class *Y* (false positive), and sample from class *Y* is assigned to class *X* (false negative).

To keep track of these various outcomes, a confusion matrix (or contingency table) is utilized. The confusion matrix's columns correspond to the samples' actual classes, while the rows correspond to the assigned classes. The main diagonal of the matrix shows the number of correctly categorized samples in each class, while the off-diagonal elements show the number of wrongly classified samples. The off-diagonal members of the matrix are zero if the data is perfectly classified. The accuracy, sensitivity (also known as precision, recall, hit rate, or true-positive rate), false-positive rate (also known as false alarm rate), and specificity of the classification can all be determined using the confusion matrix.

(*Confusion matrix*)	*True Group X*	*True Group Y*
Classified as Group X	m	n
Classified as Group Y	o	p

Equation (6) gives the accuracy of the classification, where m is the number of samples from class ***X*** that are assigned to class ***X*** by the classifier, p is the number of samples from class ***Y*** that are assigned to class ***Y*** by the classifier, n is the number of samples from class ***Y*** assigned to class ***X*** by the classifier, and o is the number of samples from class ***X*** assigned to class ***Y*** by the classifier. The sensitivity of the classification is given by Eq. (7), and the false-positive rate is given by Eq. (8). The specificity of the classification is given by Eq. (9) (Islam et al., 2018b).

$$Accuracy = \frac{m+p}{m+n+o+p} \tag{6}$$

$$Sensitivity = \frac{p}{n+p} \tag{7}$$

$$False\ positive = \frac{o}{o+m} \tag{8}$$

$$Specificity = \frac{m}{m+o} \tag{9}$$

8 Variable Selection Methods

A model that uses the entire spectral range may be at threat of overfitting, resulting in decreased predictive performance. Furthermore, a spectrum contains a significant quantity of data, most of which is unnecessary. Given the large redundancy in spectroscopic data, variable selection can typically improve chemometric models (Fig. 7).

The advantages of variable selection have been concluded in the following three aspects: (a) improve the prediction accuracy of the model because of the elimination of uninformative variables that must lead to less precision as proved theoretically; (b) selecting wavelengths probably responsible for the property of interest makes the model more interpretative; and (c) enhance the computational efficiency for modeling with a small number of variables. The advantages and disadvantages of the most common variable selection methods are presented in Table 4.

9 Multiway Analysis

In some cases, data structures are more complex than typical. Most multivariate methods are designed to work with matrices, which can be thought of as data tables. If, on the other hand, the measurements for each sample are stored in a matrix, the structure of the data is then more effectively stored in a data "box." Such data is referred to as multiway data. If each sample produces a matrix of size $M \times N$ and there are L samples, then an $L \times M \times N$ three-way array is produced. There are several methods available to deal with these kinds of three-way data (Bro, 1998).

The Generalized Rank Annihilation Method (GRAM) is a simple method with various applications; many second-order analytical procedures, such as GC-MS, are bilinear, meaning that the data may be described as the outer product of concentration profiles and pure component spectra. The main issue with GRAM is that the concentration profiles in many systems alter due to drift in the analytical apparatus

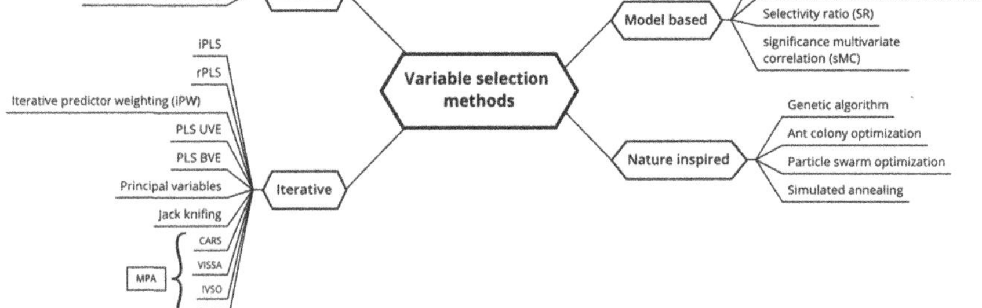

Fig. 7 Common variable selection methods in chemometrics

Table 4 Advantages and disadvantages of available variable selection methods

	Advantages	Disadvantages	References
Ad hoc	Models of causation.	Expert knowledge is required, and there are no indirect measurements (firmness, digestibility, obesity).	Mehmood et al. (2012)
Back/forward selection	Straightforward and simple.	Defining the limit is challenging.	Marcucci (1997)
*i*PLS	Simple procedure. Slow (particularly backward *i*PLS).	Autocorrelated data is required (i.e., spectra, chromatograms, process variables)	Nørgaard et al. (2000)
rPLS	Quick. Converges to a small number of variables, which is very useful for interpretation.		Rinnan et al. (2014)
VIP	Select essential variables in the PLSR model quickly.	Combines information that is associated with the parameter of interest with information that is orthogonal to it.	Wold et al. (1998)
SR	Select important variables for predicting Y quickly.	A bit too sensitive to small changes in Y pred.	Kvalheim (2010)
sMC	The sMC approach highlights variables that have little bias (in terms of parameter estimation) and are statistically significant in the model. The F-values of the variables are used to rank them.	For noisy data, it is dependable.	Tran et al. (2014)
Jack-knifing	Fast-paced. It focuses on the ambiguity in the X/Y relationship.	Data must be resampled (normally cross-validation, but Monte Carlo and bootstrap also works).	Martens and Martens (2000)
CARS	Takes into account the interactions between factors.	It is possible that critical variables will be removed forcibly. The calibration model's stability may be harmed by very collinear variables.	Li et al. (2009)
VISSA	VISSA theoretically optimizes the variable space at each phase. The variable space's statistical data is highlighted. It decreases the variable space smoothly, lowering the danger of omitting informative variables and ignoring variable combination effects. It results are indifferent to parameters like sample number		Deng et al. (2014)

(continued)

Table 4 (continued)

	Advantages	Disadvantages	References
	and model ratio, and the software stops immediately without further requirements.		
IVSO	Eliminate uninformative variables gradually, considering their interactions.		Wang et al. (2015)
BOSS	Reduces the chance of missing vital variables. Compensates for collinearity's influence.	Ignores high correlation between variables. Pick fewer options.	Al-Kaf et al. (2020)
PLS-UVE	Quickly (fast compared to backward *i*PLS, which does the same). Removes irrelevant variables (useful for large datasets).	It is not a variable selection method but rather a variable elimination method.	Centner et al. (1996)
Genetic algorithm	It looks into variable combinations that none of the "traditional" approaches looks into.	There are numerous parameters. Prone to overfitting. Best to combine with intervals or dimension reduction.	Leardi (2000)
Ant Colony optimization	Investigates variable combinations that none of the "traditional" approaches has looked into.	Prevent overfitting by combining intervals or reducing dimensions.	Shamsipur et al. (2006)
Particle swarm optimization	Investigates variable combinations which do not seek "standard" techniques	Numerous parameters. Prone to overfitting, preferable to be combined with intervals or reduced dimensions.	Lin et al. (2008)
Simulated annealing	Looks into variable combinations that none of the "normal" methods look into.	Lots of variables. Risk of overfitting, hence intervals or dimension reduction should be used in conjunction.	Meiri and Zahavi (2006)

iPLS interval partial least squares, *rPLS* recursive weighted partial least squares, *VIP* variable importance projection, *SR* selectivity ratio, *sMC* significance multivariate correlation, *CARS* competitive adaptive reweighted sampling, *VISSA* variable iterative space shrinkage approach, *IVSO* iteratively variable subset optimization, *BOSS* bootstrapping soft shrinkage, *PLS-UVE* partial least squares uninformative variable elimination

(changes in the GC column in a GC-MS, for example). These changes have the potential to rapidly damage GRAM solutions. GRAM's early implementations resulted in nonsensical fictitious solutions.

The most remarkable difference between Parallel Factor Analysis (PARAFAC) and PCA is that PARAFAC is unique in terms of scaling and permutation. Scaling ambiguity means that a column of A can be scaled by any value α as long as the corresponding column of B or C is scaled inversely i.e., by $1/\alpha$. Component one, component two, and vice versa can be called as results of permutation ambiguity. Aside from these minor uncertainties, the PARAFAC model is special in that it has only one solution. When compared to the non-uniqueness of a bilinear model, this uniqueness is the direct cause of much of PARAFAC's popularity (Bro, 1997). If the measured data fit a PARAFAC model, the model's underlying parameters can be calculated without rotational ambiguity.

The Tucker3 model, also known as the three-way PCA model, is among the most fundamental three-way models used in chemometrics (Tucker, 1966). The number three in the name Tucker3 refers to the fact that all three modes have been reduced. If the Tucker model is applied to a four-way dataset and all modes are decreased, the model will be called a Tucker4 model.

The so-called PARAFAC2 model, designed by Harshman (1972), is a more exotic yet extremely useful model. However, a workable method was not developed until 1999 (Kiers et al., 1999). A dataset may be ideally trilinear but not correspond to the PARAFAC model in some instances. It could be due to sampling issues or physical artifacts. Another issue arises when the array's slabs do not have the same row (or column) dimension. It turns out that the PARAFAC2 model can be used to solve both the problem of axis shifts and the problem of shifting axis diameters in some circumstances. (Amigo et al., 2008). One of the essential features of the PARAFAC2 model is that, like PARAFAC, it is unique in some situations. The PARAFAC2 model conditions for uniqueness have received far less attention than the PARAFAC model.

10 Tools for Chemometric Analysis

To perform chemometric analyses, several MATLAB toolboxes, R packages and software's are available. Different tools offer different functionalities. Common chemometric tools with their functionalities are listed in Table 5.

11 Conclusion

The growth in instrumentation is causing a data overload, and as a result, a large portion of the data is "wasted," meaning that no usable information is collected from it. The issue occurs with data compression as well as extraction. In general, laboratory and process measurements contain a lot of correlated or redundant data. This data must be gathered in such a way that keeps the relevant information while making it easier to show than each variable individually. Furthermore, crucial information is frequently found not in any particular element but in how the parameters vary in relation to each other; that is, the manner in which they co-vary. The information must be taken from the data in this scenario. Furthermore, in the presence of a lot of noise, it is always preferable to use some type of data processing. Therefore, a proper chemometric tool is essential for data cleaning,

Table 5 Available tools for Chemometrics

Software package	Available methods	Website
MATLAB environment		
PLS_toolbox	PCA, PLS, PLS-DA, SIMCA, SVM, HCA, PARAFAC, MLR, CLS, ANN, MCR DoE, preprocessing	https://eigenvector.com/software/pls-toolbox/
iToolbox	Interval PLS (*i*PLS), backward interval PLS (*bi*PLS), moving window PLS (*mw*PLS), synergy interval PLS (*si*PLS), and interval PCA (*i*PCA)	http://www.models.life.ku.dk/itoolbox
GAPLS toolbox	Genetic algorithm PLS	http://www.models.kvl.dk/GAPLS
SPA toolbox	Successive projections algorithm	http://www.ele.ita.br/~kawakami/spa/
LS-SVM lab	Kernel PCA, kernel CCA, and kernel PLS	https://www.esat.kuleuven.be/sista/lssvmlab/
Hypertools	Multispectral and hyperspectral image analysis	https://www.hypertools.org
A good number of essential source code/toolbox for chemometric analysis are freely available at http://www.models.life.ku.dk/algorithms		
R environment		
mdatools	Preprocessing, exploring data	https://mdatools.com
Chemometrics	PCA, PLSR	https://rdrr.io/cran/chemometrics/
ChemoSpec	Chemometrics for spectroscopy	https://cran.r-project.org/web/packages/ChemoSpec/index.html
prospectr	Preprocessing of data	https://cran.r-project.org/web/packages/prospectr/index.html
The Unscrambler	Mostly cover all chemometrics methods	https://www.aspentech.com/en/products/msc/aspen-unscrambler
SIMCA	Multivariate tools, data visualizations, and process intelligence	https://www.sartorius.com/en/products/process-analytical-technology/data-analytics-software/mvda-software/simca
Latentix	Preprocessing	https://www.latentix.com
Pirouette	Multivariate calibration and prediction	https://infometrix.com/pirouette/
PerClass	Spectral image analysis	https://www.perclass.com
OriginLab	PCA, DoE, logistic regression	https://www.originlab.com
DesignExpert	Experimental design	https://www.statease.com/software/design-expert/

preprocessing, and extracting the most relevant chemical information from the experimental data.

Acknowledgement The author acknowledges the support from the research management wing, Bangabandhu Sheikh Mujibur Rahman Agricultural University, Gazipur-1706, Bangladesh. The author declares no conflict of interest.

References

Al-Kaf, H. A. G., Alduais, N. A. M., Saad, A. M. H. Y., Chia, K. S., Mohsen, A. M., Alhussian, H., Mahdi, A. A. M. H., & Salam, W. S. I. W. (2020). A bootstrapping soft shrinkage approach and interval random variables selection hybrid model for variable selection in near-infrared spectroscopy. *IEEE Access, 8*, 168036–168052. https://doi.org/10.1109/ACCESS.2020.3023681

Amigo, J. M. (2021). Data mining, machine learning, deep learning, chemometrics definitions, common points and trends (spoiler alert: VALIDATE your models!). *Brazilian Journal of Analytical Chemistry, 8*, 22–38. https://doi.org/10.30744/brjac.2179-3425.ar-38-2021

Amigo, J. M., Skov, T., Bro, R., Coello, J., & Maspoch, S. (2008). Solving GC-MS problems with PARAFAC2. *TrAC Trends in Analytical Chemistry, 27*, 714–725. https://doi.org/10.1016/j.trac.2008.05.011

Asghar, A., Abdul Raman, A. A., & Daud, W. M. A. W. (2014). A comparison of central composite design and Taguchi method for optimizing Fenton process. *The Scientific World Journal, 2014*, 869120. https://doi.org/10.1155/2014/869120

Ballabio, D., & Consonni, V. (2013). Classification tools in chemistry. Part 1: Linear models. PLS-DA. *Analytical Methods, 5*, 3790–3798. https://doi.org/10.1039/c3ay40582f

Barker, M., & Rayens, W. (2003). Partial least squares for discrimination. *Journal of Chemometrics, 17*, 166–173. https://doi.org/10.1002/cem.785

Barnes, R., Dhanoa, M. S., & Lister, S. J. (1989). Standard normal variate transformation and de-trending of near-infrared diffuse reflectance spectra. *Applied Spectroscopy, 43*, 772–777. https://doi.org/10.1366/0003702894202201

Bartlett, M. S., & Kendall, D. (1946). The statistical analysis of variance-heterogeneity and the logarithmic transformation. *Supplement to the Journal of the Royal Statistical Society, 8*, 128–138. https://doi.org/10.2307/2983618

Bezerra, M. A., Santelli, R. E., Oliveira, E. P., Villar, L. S., & Escaleira, L. A. (2008). Response surface methodology (RSM) as a tool for optimization in analytical chemistry. *Talanta, 76*, 965–977. https://doi.org/10.1016/j.talanta.2008.05.019

Bro, R. (1997). PARAFAC. Tutorial and applications. *Chemometrics and Intelligent Laboratory Systems, 38*, 149–171. https://doi.org/10.1016/s0169-7439(97)00032-4

Bro, R. 1998. *Multi-way analysis in the food industry-models, algorithms, and applications* (Doctoral dissertation). Royal Veterinary and Agricultural University.

Bro, R., & Smilde, A. K. (2014). Principal component analysis. *Analytical Methods, 6*, 2812–2831. https://doi.org/10.1039/c3ay41907j

Centner, V., Massart, D.-L., de Noord, O. E., de Jong, S., Vandeginste, B. M., & Sterna, C. (1996). Elimination of uninformative variables for multivariate calibration. *Analytical Chemistry, 68*, 3851–3858. https://doi.org/10.1021/ac960321m

Cocchi, M. (2017). Chemometrics for food quality control and authentication. In *Encyclopedia of analytical chemistry* (pp. 1–29). Wiley. https://doi.org/10.1002/9780470027318.a9579.

De Maesschalck, R., Jouan-Rimbaud, D., & Massart, D. L. (2000). The Mahalanobis distance. *Chemometrics and Intelligent Laboratory Systems, 50*, 1–18. https://doi.org/10.1016/s0169-7439(99)00047-7

Deng, B.-C., Yun, Y. H., Liang, Y. Z., & Yi, L.-Z. (2014). A novel variable selection approach that iteratively optimizes variable space using weighted binary matrix sampling. *The Analyst, 139*, 4836. https://doi.org/10.1039/c4an00730a

Dieterle, F., Ross, A., Schlotterbeck, G., & Senn, H. (2006). Probabilistic quotient normalization as robust method to account for dilution of complex biological mixtures. Application in 1H NMR Metabonomics. *Analytical Chemistry, 78*, 4281–4290. https://doi.org/10.1021/ac051632c

Ebrahimi-Najafabadi, H., Leardi, R., & Jalali-Heravi, M. (2014). Experimental design in analytical chemistry—Part I: Theory. *Journal of AOAC International, 97*, 3–11. https://doi.org/10.5740/jaoacint.sgeebrahimi1

Enders, C. K., & Tofighi, D. (2007). Centering predictor variables in cross-sectional multilevel models: A new look at an old issue. *Psychological Methods, 12*, 121. https://doi.org/10.1037/1082-989x.12.2.121

Eriksson, L., Byrne, T., Johansson, E., Trygg, J., & Vikström, C. (2013). *Multi-and megavariate data analysis basic principles and applications*. Umetrics Academy. https://doi.org/10.1002/cem.713

Fisher, R. A. (1936). The use of multiple measurements in taxonomic problems. *Annals of Eugenics, 7*, 179–188. https://doi.org/10.1111/j.1469-1809.1936.tb02137.x

Geladi, P., & Kowalski, B. R. (1986). Partial least-squares regression: A tutorial. *Analytica Chimica Acta, 185*, 1–17. https://doi.org/10.1016/0003-2670(86)80028-9

Granato, D., & de Araújo Calado, V. M. (2013). The use and importance of design of experiments (DOE) in process modelling in food science and technology. In *Mathematical and statistical methods in food science and technology* (pp. 1–18). Wiley. https://doi.org/10.1002/9781118434635.ch01.

Guidetti, R., Beghi, R., & Giovenz, V. (2012). Chemometrics in food technology. In *Chemometrics in Practical Applications*. InTech. https://doi.org/10.5772/34148.

Guo, Q., Wu, W., & Massart, D. L. (1999). The robust normal variate transform for pattern recognition with near-infrared data. *Analytica Chimica Acta, 382*, 87–103. https://doi.org/10.1016/S0003-2670(98)00737-5

Harshman, R. (1972). PARAFAC2: Extensions of a procedure for "explanatory" factor-analysis and multidimensional scaling. *The Journal of the Acoustical Society of America, 51*, 111–111. https://doi.org/10.1121/1.1981298

Hotelling, H. (1947). Multivariate quality control-illustrated by the air testing of sample bombsights. In C. Eisenhart, M. Hastay, & W. Wallis (Eds.), *Techniques of statistical analysis* (pp. 111–184). McGraw-Hill.

Hotelling, H. (1992). The generalization of Student's ratio. In S. Kotz & N. L. Johnson (Eds.), *Breakthroughs in statistics* (pp. 54–65). Springer. https://doi.org/10.1007/978-1-4612-0919-5_4

Huynh, H., & Feldt, L. S. (1976). Estimation of the box correction for degrees of freedom from sample data in randomized block and split-plot designs. *Journal of Educational Statistics, 1*, 69–82. https://doi.org/10.3102/10769986001001069

Isaksson, T., & Næs, T. (1988). The effect of multiplicative scatter correction (MSC) and linearity improvement in NIR spectroscopy. *Applied Spectroscopy, 42*, 1273–1284. https://doi.org/10.1366/0003702884429869

Islam, M. N., Nielsen, G., Stærke, S., Kjær, A., Jørgensen, B., & Edelenbos, M. (2018a). Noninvasive determination of firmness and dry matter content of stored onion bulbs using shortwave infrared imaging with whole spectra and selected wavelengths. *Applied Spectroscopy, 72*, 1467–1478. https://doi.org/10.1177/0003702818792282

Islam, M. N., Nielsen, G., Stærke, S., Kjær, A., Jørgensen, B., & Edelenbos, M. (2018b). Novel non-destructive quality assessment techniques of onion bulbs: A comparative study. *Journal of Food Science and Technology, 55*, 3314–3324. https://doi.org/10.1007/s13197-018-3268-x

Kasprzak, E. M., & Lewis, K. E. (2001). Pareto analysis in multiobjective optimization using the collinearity theorem and scaling method. *Structural and Multidisciplinary Optimization, 22*, 208–218. https://doi.org/10.1007/s001580100138

Kiers, H. A., Ten Berge, J. M., & Bro, R. (1999). PARAFAC2—Part I. A direct fitting algorithm for the PARAFAC2 model. *Journal of Chemometrics: A Journal of the Chemometrics Society, 13*, 275–294. https://doi.org/10.1002/(sici)1099-128x(199905/08)13:3/4<275::aid-cem543>3.0.co;2-b

Kos, G., Lohninger, H., & Krska, R. (2003). Validation of chemometric models for the determination of deoxynivalenol on maize by mid-infrared spectroscopy. *Mycotoxin Research, 19*, 149–153. https://doi.org/10.1007/bf02942955

Kvalheim, O. M. (2010). Interpretation of partial least squares regression models by means of target projection and selectivity ratio plots. *Journal of Chemometrics, 24*, 496–504. https://doi.org/10.1002/cem.1289

Lawson, J. (2014). *Design and analysis of experiments with R*. Taylor & Francis. https://doi.org/10.1201/b17883

Leardi, R. (2000). Application of genetic algorithm–PLS for feature selection in spectral data sets. *Journal of Chemometrics, 14*, 643–655. https://doi.org/10.1002/1099-128X(200009/12)14:5/6<643::AID-CEM621>3.0.CO;2-E

Leardi, R. (2006). *D-optimal designs Encyclopedia of analytical chemistry: Applications, theory and instrumentation* (pp. 1–11). Wiley. https://doi.org/10.1002/9780470027318.a9646

Leardi, R. (2009). Experimental design in chemistry: A tutorial. *Analytica Chimica Acta, 652*, 161–172. https://doi.org/10.1016/j.aca.2009.06.015

Li, H., Liang, Y., Xu, Q., & Cao, D. (2009). Key wavelengths screening using competitive adaptive reweighted sampling method for multivariate calibration. *Analytica Chimica Acta, 648*, 77–84. https://doi.org/10.1016/j.aca.2009.06.046

Lin, S.-W., Ying, K.-C., Chen, S.-C., & Lee, Z.-J. (2008). Particle swarm optimization for parameter determination and feature selection of support vector machines. *Expert Systems with Applications, 35*, 1817–1824. https://doi.org/10.1016/j.eswa.2007.08.088

Marcucci, M. (1997). *Applied multivariate techniques*. Taylor & Francis. https://doi.org/10.2307/1270777

Marini, F. (2013). *Chemometrics in food chemistry*. Newnes. https://doi.org/10.1016/c2011-0-08492-2

Martens, H., & Martens, M. (2000). Modified Jack-knife estimation of parameter uncertainty in bilinear modelling by partial least squares regression (PLSR). *Food Quality and Preference, 11*, 5–16. https://doi.org/10.1016/s0950-3293(99)00039-7

Martens, H., & Martens, M. (2001). *Multivariate analysis of quality: An introduction*. Wiley. https://doi.org/10.1088/0957-0233/12/10/708

Martens, H., & Naes, T. (1991). *Multivariate calibration*. Wiley. https://doi.org/10.2307/2532682

Martens, H., Nielsen, J. P., & Engelsen, S. B. (2003). Light scattering and light absorbance separated by extended multiplicative signal correction. Application to near-infrared transmission analysis of powder mixtures. *Analytical Chemistry, 75*, 394–404. https://doi.org/10.1021/ac020194w

Mehmood, T., Liland, K. H., Snipen, L., & Sæbø, S. (2012). A review of variable selection methods in partial least squares regression. *Chemometrics and Intelligent Laboratory Systems, 118*, 62–69. https://doi.org/10.1016/j.chemolab.2012.07.010

Meiri, R., & Zahavi, J. (2006). Using simulated annealing to optimize the feature selection problem in marketing applications. *European Journal of Operational Research, 171*, 842–858. https://doi.org/10.1016/j.ejor.2004.09.010

Mishra, P., Rutledge, D. N., Roger, J.-M., Wali, K., & Khan, H. A. (2021). Chemometric pre-processing can negatively affect the performance of near-infrared spectroscopy models for fruit quality prediction. *Talanta, 229*, 122303. https://doi.org/10.1016/j.talanta.2021.122303

Næs, T., Isaksson, T., Fearn, T., & Davies, T. (2002). *A user-friendly guide to multivariate calibration and classification*. NIR. https://doi.org/10.1255/978-1-906715-25-0

Noda, I. (2008). Scaling techniques to enhance two-dimensional correlation spectra. *Journal of Molecular Structure, 883*, 216–227. https://doi.org/10.1016/j.molstruc.2007.12.026

Nørgaard, L., Saudland, A., Wagner, J., Nielsen, J. P., Munck, L., & Engelsen, S. B. (2000). Interval partial least-squares regression (iPLS): A comparative chemometric study with an example from near-infrared spectroscopy. *Applied Spectroscopy, 54*, 413–419. https://doi.org/10.1366/0003702001949500

Press, W. H., & Teukolsky, S. A. (1990). Savitzky-Golay smoothing filters. *Computers in Physics, 4*, 669–672. https://doi.org/10.1063/1.4822961

Rabatel, G., Marini, F., Walczak, B., & Roger, J.-M. (2020). VSN: Variable sorting for normalization. *Journal of Chemometrics, 34*, e3164. https://doi.org/10.1002/cem.3164

Rakić, T., Kasagić-Vujanović, I., Jovanović, M., Jančić-Stojanović, B., & Ivanović, D. (2014). Comparison of full factorial design, central composite design, and Box-Behnken design in chromatographic method development for the determination of fluconazole and its impurities. *Analytical Letters, 47*, 1334–1347. https://doi.org/10.1080/00032719.2013.867503

Rinnan, Å. (2014). Pre-processing in vibrational spectroscopy—When, why and how. *Analytical Methods, 6*, 7124–7129. https://doi.org/10.1039/c3ay42270d

Rinnan, Å., Andersson, M., Ridder, C., & Engelsen, S. B. (2014). Recursive weighted partial least squares (rPLS): An efficient variable selection method using PLS. *Journal of Chemometrics, 28*, 439–447. https://doi.org/10.1002/cem.2582

Rinnan, Å., Nørgaard, L., Berg, F. V. D., Thygesen, J., Bro, R., & Engelsen, S. B. (2009). Data pre-processing. In D.-W. Sun (Ed.), *Infrared spectroscopy for food quality analysis and control* (pp. 29–50). Academic Press. https://doi.org/10.1016/B978-0-12-374136-3.00002-X

Roger, J.-M., Boulet, J.-C., Zeaiter, M., & Rutledge, D. N. (2020). Pre-processing methods. In S. Brown, R. Tauler, & B. Walczak (Eds.), *Comprehensive chemometrics* (2nd ed., pp. 1–75). Elsevier. https://doi.org/10.1016/b978-0-12-409547-2.14878-4

Rousseeuw, P. J., & Leroy, A. M. (1987). *Robust regression and outlier detection*. Wiley. https://doi.org/10.1002/0471725382

Shamsipur, M., Zare-Shahabadi, V., Hemmateenejad, B., & Akhond, M. (2006). Ant colony optimisation: A powerful tool for wavelength selection. *Journal of Chemometrics: A Journal of the Chemometrics Society, 20*, 146–157. https://doi.org/10.1002/cem.1002

Smilde, A., Bro, R., & Geladi, P. (2005). Two-way component and regression models. In *Multi-way analysis with applications in the chemical sciences* (pp. 35–45). Wiley. https://doi.org/10.1002/0470012110.ch3.

Stone, M. (1974). Cross-validatory choice and assessment of statistical predictions. *Journal of the Royal Statistical Society: Series B: Methodological, 36*, 111–133. https://doi.org/10.1111/j.2517-6161.1974.tb00994.x

Tanabe, J., Miller, D., Tregellas, J., Freedman, R., & Meyer, F. G. (2002). Comparison of detrending methods for optimal fMRI preprocessing. *NeuroImage, 15*, 902–907. https://doi.org/10.1006/nimg.2002.1053

Tran, T. N., Afanador, N. L., Buydens, L. M. C., & Blanchet, L. (2014). Interpretation of variable importance in partial least squares with significance multivariate correlation (sMC). *Chemometrics and Intelligent Laboratory Systems, 138*, 153–160. https://doi.org/10.1016/j.chemolab.2014.08.005

Tucker, L. R. (1966). Some mathematical notes on three-mode factor analysis. *Psychometrika, 31*, 279–311. https://doi.org/10.1007/BF02289464

van den Berg, R. A., Hoefsloot, H. C., Westerhuis, J. A., Smilde, A. K., & van der Werf, M. J. (2006). Centering, scaling, and transformations: Improving the biological information content of metabolomics data. *BMC Genomics, 7*, 1–15. https://doi.org/10.1186/1471-2164-7-142

Vanaja, K., & Shobha Rani, R. (2007). Design of experiments: Concept and applications of Plackett Burman design. *Clinical Research and Regulatory Affairs, 24*, 1–23. https://doi.org/10.1080/10601330701220520

Vidal, M., & Amigo, J. M. (2012). Pre-processing of hyperspectral images. Essential steps before image analysis. *Chemometrics and Intelligent Laboratory Systems, 117*, 138–148. https://doi.org/10.1016/j.chemolab.2012.05.009

Wang, W., Yun, Y., Deng, B., Fan, W., & Liang, Y. (2015). Iteratively variable subset optimization for multivariate calibration. *RSC Advances, 5*, 95771–95780. https://doi.org/10.1039/c5ra08455e

Westad, F., & Kermit, M. (2003). Cross validation and uncertainty estimates in independent component analysis. *Analytica Chimica Acta, 490*, 341–354. https://doi.org/10.1016/s0003-2670(03)00090-4

Westad, F., & Marini, F. (2015). Validation of chemometric models—A tutorial. *Analytica Chimica Acta, 893*, 14–24. https://doi.org/10.1016/j.aca.2015.06.056

Westerhuis, J. A., Gurden, S. P., & Smilde, A. K. (2000). Generalized contribution plots in multivariate statistical process monitoring. *Chemometrics and Intelligent Laboratory Systems, 51*, 95–114. https://doi.org/10.1016/s0169-7439(00)00062-9

Wold, S. (1976). Pattern recognition by means of disjoint principal components models. *Pattern Recognition, 8*, 127–139. https://doi.org/10.1016/0031-3203(76)90014-5

Wold, S., Esbensen, K., & Geladi, P. (1987). Principal component analysis. *Chemometrics and Intelligent Laboratory Systems, 2*, 37–52. https://doi.org/10.1016/0169-7439(87)80084-9

Wold, S., Josefson, M., Gottfries, J., & Linusson, A. (2004). The utility of multivariate design in PLS modeling. *Journal of Chemometrics, 18*, 156–165. https://doi.org/10.1002/cem.861

Wold, S., Martens, H., & Wold, H. (1983). The multivariate calibration problem in chemistry solved by the PLS method. In B. Kågström & A. Ruhe (Eds.), *Matrix pencils* (pp. 286–293). Springer. https://doi.org/10.1007/bfb0062108

Wold, S., Sjöström, M., & Eriksson, L. (1998). Partial least squares projections to latent structures (PLS) in chemistry. In P. von Ragué Schleyer (Ed.), *Encyclopedia of computational chemistry* (pp. 2006–2021). Wiley. https://doi.org/10.1002/0470845015.cpa012

Wold, S., Sjöström, M., & Eriksson, L. (2001). PLS-regression: A basic tool of chemometrics. *Chemometrics and Intelligent Laboratory Systems, 58*, 109–130. https://doi.org/10.1016/s0169-7439(01)00155-1

Zou, X., & Zhao, J. (2015). NIR spectroscopy detection. In Z. Xiaobo & J. Zhao (Eds.), *Nondestructive measurement in food and agro-products* (pp. 57–126). Springer. https://doi.org/10.1007/978-94-017-9676-7_3

Electronic Nose for Fresh Produce Quality

Adinath Kate, Shikha Tiwari, and Debabandya Mohapatra

Abstract The electronic nose (E-nose) is an instrument usually used for the detection of simple or complex volatile organic compounds (VOCs) with the help of an array of sensors. Different types of sensors ranging from metal oxide sensors to polymers, conducting to nonconducting polymers, electrochemical to fluorescence-based sensors are being employed in many gas sensing applications. Some of the sensors are adapted to cater to the dynamics that are pertinent in the fresh agricultural produce. Fresh produce, still-living, undergoes various metabolic activities and is also susceptible to various synthetic and biochemical contaminations, as well as microbial infections. All these biological activities lead to the generation of various volatile metabolites and can be sensed through appropriate gas sensors. In this chapter, we have discussed various e-noses available commercially and their suitability for the quality monitoring of fresh produce. In addition, classification techniques used for developing algorithms, which is a major component of e-noses, are discussed.

1 Introduction

In recent decades the electronic nose (E-nose) is strongly drawing the attention of researchers due to its diverse application and advantages. E-nose can sense complex aroma compounds and volatiles efficiently with the help of sensor arrays, specific pattern-recognition systems, and sample handling parts based on electronic principles. The identification and amalgamation of particular sensors, with defined pattern-recognition, make the e-noses highly selective, sensitive, and reproducible making them compatible with various onsite applications in food, health, cosmetics, water, environmental, medicinal, agriculture, and biomedical industries.

The major application of E-nose is in situ detection of volatiles, which represents the physicochemical state of stored material under controlled conditions, especially perishable horticulture commodities, like fruits and vegetables. E-nose also detects

A. Kate · S. Tiwari · D. Mohapatra (✉)
ICAR-Central Institute of Agricultural Engineering, Bhopal, Madhya Pradesh, India

P. B. Pathare, M. S. Rahman (eds.), *Nondestructive Quality Assessment Techniques for Fresh Fruits and Vegetables*, https://doi.org/10.1007/978-981-19-5422-1_15

volatiles in a quantifiable form, which further plays a key role in decision-making for postharvest handling. The presence of high water content and degradable chemical constituents makes fruits and vegetables more susceptible to deterioration in various ways. Even a little change in the physiological condition of fresh material can instigate a change in the aroma, taste, and other quality attributes, which declines its salability. Fruits and vegetables can emit a wide range of volatiles, these may or may not be associated with aroma, and reflect their freshness stage or initiation of unwanted decay. E-nose is a potential solution for the nondestructive and continuous assessment of fruits and vegetables by sensing a complex mixture of volatiles and reflecting its in situ quality. The output digital pattern act as Electronic Aroma Signature Pattern (EASP) and generates a print of volatile compounds to identify the type, variety, physicochemical features, and condition of fresh produce quality. The general aroma signature print of fruits and vegetables are esters, lactones, terpenoids, phenolic compounds, fatty acids, and amino acid derivatives. The climatic conditions on the field during development, harvesting methods, postharvest handling, transport type, packaging, and storage are key factors that rule the class of volatile compounds released by a particular commodity. There is a direct link between volatile compound emission and internal biochemical pathways going on inside the selected category of fruit and vegetable. This chapter comprises the relations of volatile gaseous mixture emitted by fruits and vegetables, their relation with various biochemical pathways, and the E-nose technology for their characterization and possible quantification.

Before the 1990s, different olfactory factors were sensed by human senses. This method was convenient; however, it was imprecise, nonreproducible, and nonobjective, owing to the fact that human senses are highly affected by various parameters ranging from physical to psychological, environmental to cultural, influencing the test results (Pearce et al., 2003). In 1988, Gardner and Bartlett invented the term "electronic-nose," which was an electronic instrument with an array of sensors to measure chemical stimuli (Wilson & Baietto, 2009). A typical e-nose consists of three basic parts: (1) a sample handing or holding system, (2) a chemical gas sensor array, and (3) a pattern-recognition system (Amalia, 2010). Eventually, the e-nose became one of the most potent instruments to be used in the food industry. The analysis of the olfactory factors is an important part of quality control. The instrument-based quality analysis has been quite common in the food industry. This instrumental quality analysis was carried out both in a destructive as well as nondestructive manner using the techniques, like chromatography, image analysis, different types of microscopy, and texture analysis (Butz et al., 2005). The chromatographic and microscopic-based analyses have high sensitivity and accuracy, but often require pretreatment of the sample prior to the testing. Some of these pretreatments require complex extraction processes before testing, which were often time processes thus limiting their application for onsite rapid detection of deterioration of food quality.

Because of the nondestructive principle and volatile odor-based detection, electronic sensing system becomes popular in food industries and gradually extended to other sectors like chemical, textile industries (Thomas et al., 2012), clinical diagnostic (Scarlata et al., 2015; Schnabel et al., 2015), fermentation technology (Jiang et al., 2015), environmental pollution monitoring (Capelli et al., 2014a), agricultural and forestry (Wilson, 2013), with the aim of real-time, efficient detection. The wide area of application of electronic sensing systems truly opened the door for the development and application of electronic noses in food quality determinations likewise diagnosis of fresh and infected objects, classification of crops, flavor analysis, measurement of freshness, and any possible spoilage detection.

2 Electronic Nose

The electronic nose is defined as "an electronic instrument, which has an array of electronic sensors with partial specificity and specific system for pattern-recognition, which is capable of recognizing simple or complex odors comes out from source" (Gardner & Bartlett, 1994). In most cases, a single e-nose was developed and used for a specific purpose only. Compared with other analytical instruments used in food processing and allied sectors, E-nose looks more promising as it is more effective, easy to implement, and cost-effective. Its volatile compound-based discrimination and classification capability is sometimes superior to GC-MS analysis (Cevoli et al., 2011). The instrument of the electronic nose is fully capable of identification and detection of complex volatile smells because of the configuration of the array of gas sensors, which is composed of wide coordination of selective sensors, at the same time, the instrument can also be capable to work for detection of similar compounds belongs to different biological or chemical sources (Pearce et al., 2003). When E-nose is used for discriminating or detecting the problem of quality attributes of fresh produce, the volatile samples in the headspace of the sample holder are injected and fed to the sensing system, further the signal generated by the sensing system is sent to the appropriate signal handling system and processed into suitable forms. The conditioned signal develops an appropriate fingerprint as the output of the electronic nose. The fingerprint of VOC detection obtained from the electronic nose is needed to be further analyzed for interpretation of the information (Choi et al. 2016). The classification and discrimination techniques like Principal component analysis (PCA), Partial least square discrimination analysis (PLSDA), and Sparse partial least square discrimination analysis (S-PLSDA) are some of the techniques for further analysis of e-nose fingerprints detected by sensors. The applicability and superiority of any technique over others is the basis for the analysis of e-nose fingerprint data.

2.1 Sensing System in Electronic Nose

The sensing system diploid in the electronic nose can be a single sensor device or an array of several selective sensors or a combination of both. During the measurement when the detection sensor comes in contact with test volatiles, it chemically reacts with films of sensors. These chemical reactions cause the disturbance of electrons that is detected and quantified by the electrical signal (Arshak et al., 2004). Presently, different types of commercial gas sensors are available, like metal oxide semiconductors (MOS), organic conducting polymers, metal oxide semiconductor field effect transistors (FET), and piezoelectric crystal sensors. The selection criteria for these sensors are based on specific applications (Sanaeifar et al., 2016). Among all types, the metal oxide sensor is the most commonly used versatile sensor for a variety of applications (Zohora et al., 2013). Besides, sometimes sensors like conductive sensors, optical sensors (Zhao et al., 2016), chemoresistor sensors (Alizadeh, 2010), Surface Acoustic Wave (SAW) sensors (Verma & Yadava, 2015), Quartz Crystal Microbalance (QCM) (Ali et al., 2018; Papadopoulou et al., 2013) sensors are also used as combinations arrays for VOC-based quality determination problems. The output of the e-nose is not only influenced by the type of sensor but also influenced by the number of sensors present in the array.

Steps for Development of an E-Nose:

1. *Selection of sensors:*
   (a) The first step in the selection of sensors is to identify the key volatile compounds, through GC-MS. The collection of volatiles is either carried out through solid-phase microextraction (SPME) fiber or collecting it in a bottle with help of a syringe. The volatile compounds are then injected into the GC-MS system. The key volatile compounds are then identified using suitable multivariate analysis tools.
   (b) The sensors are exposed to fixed quantities of those identified chemical compounds, one by one finding out the responses in each case. The sensor responses are then analyzed to find out the sensitivity and selectivity.
2. *Development of the embedded software:*
   (a) Embedded software development for sensor data reading.
   (b) Selection of classifier algorithms.
   (c) Sensor data analysis.
   (d) Display of results.
3. *Physical model development:*
   (a) A physical system is developed for sniffing the volatiles, through suction and later purging of the metabolites, once sensing is completed.
   (b) The amplified conditioned responses are then processed through a data acquisition device. The data is processed through pattern-recognition algorithms.

Figure 1 shows the typical electronic sensing system with various functional components.

The key to the development of e-noses is sensor selection, based on its sensitivity, reproducibility, and stability. As most agricultural activities involve high humidity and dust, the sensors should be able to operate in those conditions and be consistent

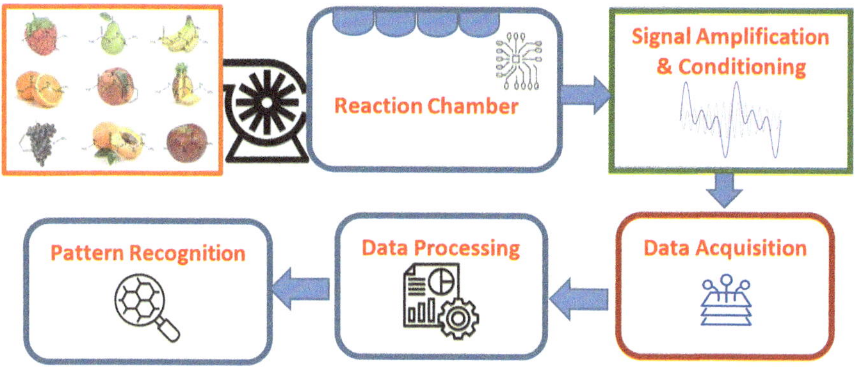

Fig. 1 Schematic diagram of a typical e-nose system

in giving their output. The aging of sensors is another hurdle; therefore, periodic calibration of these electronic devices is a requirement. Table 1 summarizes various types of sensors used in e-nose, sensitivity, and their typical response time.

2.2 Application of E-Nose in Fresh Produce

Even though there is a wide range of VOC/gas sensors available in the market, still many of the gas sensors are not been fully explored in quality monitoring, especially in the agricultural field. The sensors need to be stable and should be able to work in high humidity and moderate temperature level. Besides, it should have temperature and pH stability and quick response time, and be less expensive. Most of the commercially available e-noses for quality monitoring in the agricultural field employed metal oxide sensors. Table 2 delineates some of the commercialized e-noses for quality monitoring in the agricultural field.

The electronic nose application mainly is for quality determination and expanding rapidly in various areas of analytics, presently it has been reflected in different applications in the food industry, pharmaceutical industry, security, health care military fields, environmental testing, etc. In the area of fresh food analytics, it can be used for process monitoring, in situ monitoring for raw material during different stages of harvesting, processing, storage, logistics, and other processes. Fresh produce like fruits and vegetables emit a wide range of volatile organic compounds (VOCs) during the growth phase as a part of natural metabolism, and/or as secondary metabolism to combat the various abiotic and biotic stresses. Postharvest, the still-living fresh produce may still undergo senescence, ripening, sprouting, and maybe undergo microbial infection. These products are stored and transported under different temperatures and relative humidity conditions, which may follow different biochemical pathways, stimulating the release of distinct VOCs. Many of the infecting microorganisms emit secondary metabolites, which are substrate-specific and can vary with the change in the commodity. These VOCs might be the signature

Table 1 Type of E-nose, their mechanism, and components

E-nose	Detection mechanism	Sensitivity of sensor	Sensitive component	Time of response (seconds)	Advantages	Disadvantages	References
Calorimetric or catalytic bead (CB)	Change in absorbance	10–100 ppm	Catalytic bead (pellistor)	2–10	Inexpensive to manufacture, good response time, low cost, versatile range of gas detection	Susceptible to poisoning from silicates, oversaturation from high concentrations of gas, requires periodic calibration	Sherrit et al. (2003); Casey et al. (2003); Kirchner et al. (2012)
Carbon black composite (CBC)	Change in electrical resistance	<~20 ppm	Graphene nanotubes	2–4	Good absorptive capacity	Manufacturing difficulties, expensive, low precision	Schiffman and Pearce (2003); Kim et al. (2005)
Conducting polymers (CP)	Change in electrical resistance	0.1–100 ppm	Aromatic polymers in doped form	10–90	Inexpensive, ambient	Limited life, sensitive to T, RH	Vaschetto et al. (1999); Luo and Chakraborty (2009); Lee et al. (2012); Zohora et al. (2013)
Electrochemical (EC)	Change in voltage or current	0.1–100 ppm	Electrolytes		Stable, reliable, fast, require very little power, less expensive, not affected by other gases	Limited temperature range, life, affected by low humidity and high temperatures	Grieshaber et al. (2008); Xie et al. (2010)
Field effect transistor (MOSFET)	Change in electrical field	10–100 ppm	Catalytic bead (pellistor)	2–10	Thermostable, high efficiency, longer life, low noise	Expensive, lower switching time, performance is affected at higher frequencies	Sherrit et al. (2003); Kirchner et al. (2012)
Fluorescence sensors	Change in the emission of fluorescence	–	Fluorescence-sensitive detector	–	High sensitivity, high specificity, immune to light scattering	Susceptible to autofluorescence, short life span, sensitive to O_2 and pH changes, limitation associated with photostability, and loss of recognition capability	Wilson and Baietto (2009); Dung et al. (2018)
Graphene oxide	Change in conductivity		Ethylene gas		Ultra-high sensitivity, a wide detection range Simple equipment construction Signal testing	High cost, hydrophobicity, difficult workability, and small production	Mallick et al. (2018)

Infrared sensors	Changes in the absorption of infrared radiation	–	Infrared sensitive detector	–	Inexpensive, unaffected by environmental changes	Sensitive to T, RH, magnetic field	Capelli et al. (2014b)
Metal oxide semiconductor	Change in electrical resistance	5–500 ppm	Inorganic metal oxide	5–35	Low cost, moderately sensitive, rapid, long sensor life	High T operation and power consumption RH sensitive	Albert et al. (2000); Feng et al. (2012)
Micro-electromechanical systems (MEMS)	Change in voltage or current	–	Electrolytes	–	High sensitivity, accuracy, low cost, low power consumption	High fabrication cost, brittle material	Simon et al. (1996)
Optical fiber live cell (OF-LC)	Changes in optical waves or light modulation	<~10 ppb	Photodiode light sensitive	<10	Very high sensitivity, long life, can identify various compounds, unaffected by environmental changes	Complex, high cost, not very portable	Walt et al. (1998); Blumentritt et al. (2008); Dey and Goswami (2011)
Piezoelectric bulk acoustic wave (BAW)	Change in mass (frequency variation)	1 pg	Inorganic or organic films	~10	Greater power handling capability, selectivity, lower loss	Expensive, performance degrades at higher frequencies, more complex to fabricate	Lai and Huang (2006); Krishnamoorthy and Iliadis (2008)
Quartz crystal microbalance (QCM)	Change in mass (frequency variation)	1 ng	Inorganic or organic films	1 ng	Highly sensitive, precise, diverse range	Complex, poor-signal to noise ratio, sensitive to RH and T	Smith and Shirazi (2005); Pejcic et al. (2009, 2011); Zhu et al. (2013)
Surface acoustic wave (SAW)	Change in mass (frequency variation)	1 pg	Inorganic or organic film layers	~10	High sensitivities and short response times, cost-effective fabrication, structures flexibility, and reproducible quality	Instability for long time, irreversibility, poor selectivity, sensitive to T and RH	Lai and Huang (2006); Krishnamoorthy and Iliadis (2008)

Table 2 Various commercial e-noses for quality monitoring in the agricultural field

Agricultural applications	Smart sensor system	E-nose	Sensor types	References
Residual pesticides	Chemoresistance sensors	MOSES II AROMASCAN	8 MOS 32 CP	Baby et al. (2000); Wilson et al. (2004); Wilson (2012)
Fungal toxins of fruits	Microbial metabolites-based sensors	EOS 835 PEN 2	6 MOS 10 MOS	Falasconi et al. (2005); Solis-solis et al. (2007); Cheli et al. (2009); Campagnoli et al. (2009); Musatov et al. (2010)
Postharvest water loss in fruits and vegetables	O_2 sensors, CO_2 sensors	EXPERIMENTAL	8 QMB	Schaller et al. (1998); Santonico et al. (2010)
Fruit odor	Aromas sensors	E-nose	8 MOS	Tang et al. (2010)
Classification of fruits	Aromas sensors	MOSES II		Adak and Yumusak (2016)
Fruit ripening	pH chemical sensors, chemoresistance sensors	FOX 4000	18 MOS	Santos et al. (2010)
Fungal disease detection	Microbial metabolites, gas detection, and optical based sensors	AROMA SCAN	32 CP	Wilson and Lester (1997, 1998); Wilson et al. (2004); Baietto et al. (2010); Alam et al. (2021)
Bacterial disease detection	Microbial metabolites, gas detection, and optical-based sensors	CYRANOSE 320 EXPERIMENTAL	32 CBC GNP 14	Li et al. (2010a); Peng et al. (2010); Alam et al. (2021)

of a particular state of fresh produce; therefore, can be classified as biomarkers in the development of e-sensing systems (Tiwari et al., 2020). The VOCs associated with different biotic (Fig. 2) and abiotic stresses (Fig. 3) during the storage of fresh commodities differ widely.

As the identification of appropriate maturity conditions for harvesting and the investigation of the shelf life are the keys to processing and handling any fresh produce, it can be well assessed by using e-nose. There are different types of e-nose for the identification of fruit and vegetable maturity (Chen et al., 2018b), determination of the degree of freshness, storage conditions, and prediction of shelf life. Likewise, E-nose can discriminate against all kinds of alcoholic beverages, dairy products, different species, authenticity, adulteration, and so on (Peris & Escuder-Gilabert, 2016). Accordingly, the applications of e-nose for fresh produce quality assessment involve the classification of different growth stages of fruits (Qiu et al., 2015), varietal difference, fruit maturity (Chen et al., 2018c), freshness evaluation, spoilage conditions, and ripening stages commonly associated with fruits and vegetables (Table 3).

Fig. 2 Representation of volatiles associated with different biotic stresses in horticultural commodities during storage through subway plot

2.3 *Classification*

Among various sensory attributes, smell is one of the important characteristics considered during the food quality evaluation. Similarly, the particular smell is the function of specific VOCs. It can easily distinguish fresh food materials based on their odor characteristics, types, origin, processing modes, etc. As the electronic nose consists of an array of different sensors it can be one of the powerful tools for fast and accurate food classification.

In the area of fresh produce, the classification may be carried out based on the stages where physiological changes take place, i.e., at different stages of maturity, ripening stages, type and stage of microbial contaminations, and other physiological disorders. The VOC data contains multidimensional features, which needed to be

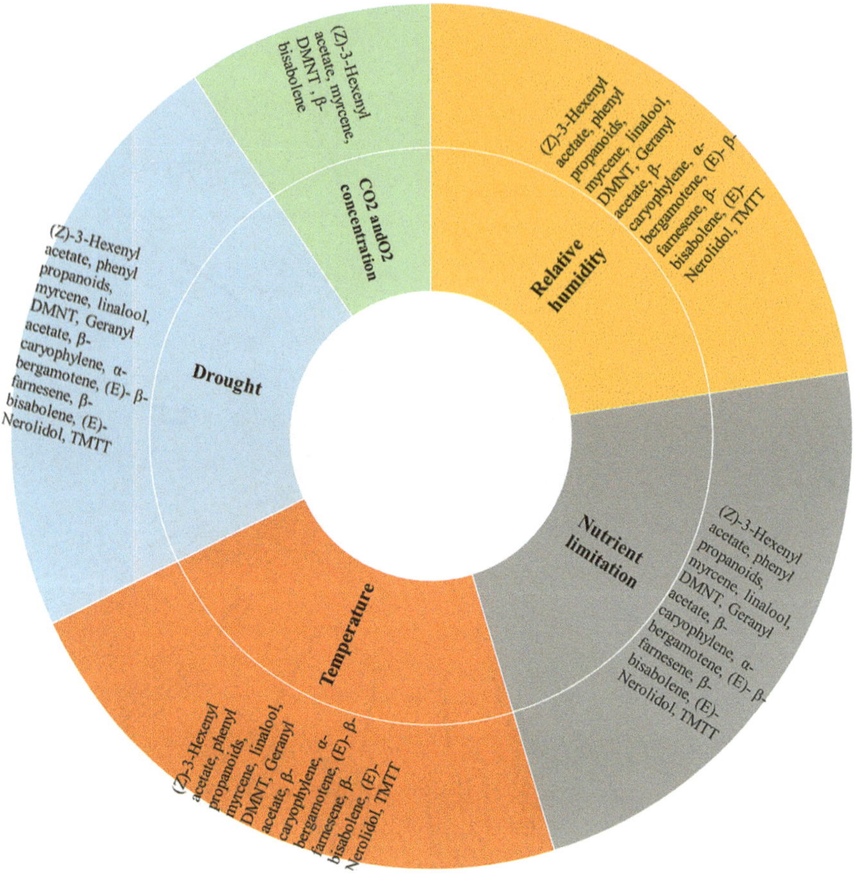

Fig. 3 Representation of volatiles associated with different abiotic stresses in horticultural commodities during storage through subway plot

extracted and their dimensionality needs to be reduced for ease of analysis and interpretation. The use of the appropriate statistical technique for sensor data analysis is very much essential for appropriate and meaningful distinction. Principal component analysis (PCA) with cluster analysis is the most applicable technique for the classification into desired groups. PCA classifies VOCs data based on their similar characteristics, it is often used as the base analytical tool for reducing dimensionality. Sometimes PCA fails to distinguish without overlapping of class groups; therefore, other techniques like partial least square differential analysis (PLSDA) and sparse partial least square differential analysis (sPLSDA) are used for distinct separation of groups. The technique PLSDA and sPLSDA classifies the VOCs data based on their variations. Classification of four fruits, i.e., melon, strawberry, lemon, and cherry was carried out using Artificial Neural Network (ANN) training with an artificial bee colony algorithm (ABC) (Adak & Yumusak,

Table 3 E-noses for fresh fruit and vegetable produce quality detection

Application area	Commodities	References
Bacterial infection	Chilli and papaya	Chang et al. (2014)
Rotting	Onion and potato	Ghosh et al. (2022)
Fruit classification	Mandarins and sweet orange	Qiu et al. (2015)
	Fresh mature guava, orange, banana fruits	Kanade and Shaligram (2014)
	Strawberry, lemon, cherry, and melon	Adak and Yumusak (2016)
Fruit maturity/ripening Storage quality	Four stages (overripe, fully ripe, half-ripe, unripe) of banana fruit	Chen et al. (2018c)
	Peaches	Voss et al. (2020)
	"Florida radiance" and "strawberry festival" and at five developmental stages in descending order (overripe, full-ripe, three-quarter ripe, half red, white)	Du et al. (2010)
	Tomato (unripe, half-ripe, full-ripe, and overripe)	Gómez et al. (2006)
	Tomato in storage	Gómez et al. (2008)
	Kiwi fruit	Du et al. (2019)
	Mango	Lihuan et al. (2017)
	Satsuma mandarin stored at different temperatures	Qiu and Wang (2015a, 2015b)
	Cherry tomatoes stored at different temperatures	Hong and Wang (2014)
	Banana quality in storage	Sanaeifar et al. (2016)
	Decay in peach fruit during cold storage	Wei et al. (2018)
	Apple shelf life	Saevels et al. (2004)
	Shelf life of lemon, apple, banana, and pear	Karamchandani et al. (2021)
Freshness	Fresh and stale peach	Guohua et al. (2012)
	Fresh and spoiled apple, pitaya, tribute citrus	Wen et al. (2022)
Fungal disease	Strawberry fruit infected with botrytis sp. (BC), Penicillium sp. (PE), Rhizopus sp. (RH) during storage	Zhu et al. (2013)
Growth stages (internal quality)	Two types of mandarins	Qiu et al. (2015)
Variety distinction	Red and white-fleshed peaches during ripening	Xin et al. (2018)
	Ten varieties of Chinese jujube	Chen et al. (2018a)
	Climacteric and non-climacteric melons	Chaparro-Torres et al. (2016)
	Six varieties of Chinese chives	Hanif et al. (2022)
	Three varieties of durians	Xiao et al. (2022)

2016). Convoluted neural network (CNN), support vector machine (SVM), and machine learning (ML) are some of the analytical tools used for classifying VOC data. Some of the widely used data analysis tools employed in the e-nose system with their features and limitations are given in Table 4.

Table 4 Advantages and limitations classifiers used in e-sensing systems

Analytical tools	Features	Limitations	References
PCA	• Most widely used basic feature extraction technique • Can analyze unsupervised data • Can eliminate multicollinearity between features • Recognizes the similarities of highly variance data to reduce the dimensionality	• Scalability • Fails to extract key features for achieving recognition, and when independent variables are correlated • Loss of data due to feature extraction and reduction of dimensionality • Extractable features may not be easy to interpret • Does not address nonlinearity among variables	Lever et al. (2017)
PLS	A supervised method where each group information is to be provided for data analysis • Takes care of skewness in data distribution • Ability to robustly handle more descriptor variables than compounds • Non-orthogonal descriptors and multiple biological results • Better predictive accuracy • Lower risk of chance correlation	• Overlooking the real correlations • Needs a large sample size • Need for normalized data	Cramer (1993)
ANN	• The whole data is stored on the network • It can address incomplete data or missing data • It has good fault tolerance • Slowing down of network is gradual	• Cannot perform data with complex dimensionality and noise • Dependency on hardware • Unknown duration of the network • Does not show problem points in the network	Mijwel (2018)
CNN	• Data-driven • Supervised learning • Higher accuracy than ANN • No or little domain knowledge is required for system construction • Simple algorithm	• Cannot work with a small database • Black box phenomena limit interpretation of results • No theoretical foundation • Supervised database to be built up	Li (2018)
SVM	• Performs better than PCA for large data. • High accuracy. • Has good tolerance for irrelevant attributes.	• Cannot use all training data points. • Fails when target features are overlapping or when target features for each data point exceed the number of training data points.	Tan and Xu (2020)

3 Conclusion

With the advancement of sensor technology, many sectors are now employing man-less equipment for quality monitoring. These electronic systems can work in adverse situations, and alleviate human drudgery to a great extent. Besides, it can address the issues of human error and fatigue involved in the quality monitoring of fresh produce, which is susceptible to various metabolic activities as well as external stimulations. E-noses, which detect the volatile pattern, must be stable at different temperature, RH, and pH conditions and should have a long life to be employed successfully in the quality monitoring of fresh produce, during preharvest and postharvest stages.

References

Adak, M. F., & Yumusak, N. (2016). Classification of E-nose aroma data of four fruit types by ABC-based neural network. *Sensors, 16*(3), 304.

Alam, A. U., Rathi, P., Beshai, H., Sarabha, G. K., & Deen, M. J. (2021). Fruit quality monitoring with smart packaging. *Sensors, 21*(4), 1509.

Albert, K. J., Lewis, N. S., Schauer, C. L., Sotzing, G. A., Stitzel, S. E., Vaid, T. P., & Walt, D. R. (2000). Cross-reactive chemical sensor arrays. *Chemical Reviews, 100*(7), 2595–2626.

Ali, S. B., Ghatak, B., Debabhuti, N., Sharma, P., Ghosh, A., Tudu, B., et al. (2018). Detection of β-caryophyllene in mango using a quartz crystal microbalance sensor. *Sensors and Actuators B: Chemical, 255*, 3064–3073.

Alizadeh, T. (2010). Chemiresistor sensors array optimization by using the method of coupled statistical techniques and its application as an electronic nose for some organic vapors recognition. *Sensors and Actuators B: Chemical, 143*(2), 740–749.

Amalia, B. (2010). Metal oxide sensors for electronic noses and their application to food analysis. *Sensors, 10*(4), 3882–3910.

Arshak, K., Moore, E., Lyons, G. M., Harris, J., & Clifford, S. (2004). A review of gas sensors employed in electronic nose applications. *Sensor Review, 24*(2), 181–198. https://doi.org/10.1108/02602280410525977

Baby, R. E., Cabezas, M., & De Reca, E. W. (2000). Electronic nose: A useful tool for monitoring environmental contamination. *Sensors and Actuators B: Chemical, 69*(3), 214–218.

Baietto, M., Wilson, A. D., Bassi, D., & Ferrini, F. (2010). Evaluation of three electronic noses for detecting incipient wood decay. *Sensors, 10*(2), 1062–1092.

Blumentritt, M., Melhorn, K., Flachsbarth, J., Kroener, M., Kowalsky, W., & Johannes, H. H. (2008). A novel fabrication method of fiber-optical planar transmission sensors for monitoring pH in concrete structures. *Sensors and Actuators B: Chemical, 131*(2), 504–508.

Butz, P., Hofmann, C., & Tauscher, B. (2005). Recent developments in noninvasive techniques for fresh fruit and vegetable internal quality analysis. *Journal of Food Science, 70*(9), R131–R141.

Campagnoli, A., Cheli, F., Savoini, G., Crotti, A., Pastori, A. G. M., & Dell'Orto, V. (2009). Application of an electronic nose to detection of aflatoxins in corn. *Veterinary Research Communications, 33*(1), 273–275.

Capelli, L., Dentoni, L., Sironi, S., & Del, R. R. (2014a). The need for electronic noses for environmental odour exposure assessment. *Water Science and Technology, 69*, 135–141.

Capelli, L., Sironi, S., & Del Rosso, R. (2014b). Electronic noses for environmental monitoring applications. *Sensors, 14*(11), 19979–20007.

Casey, V., Cleary, J., D'arcy, G., & McMonagle, J. B. (2003). Calorimetric combustible gas sensor based on a planar thermopile array: Fabrication, characterisation, and gas response. *Sensors and Actuators B: Chemical, 96*(1-2), 114–123.

Cevoli, C., Cerretani, L., Gori, A., Caboni, M. F., Toschi, T. G., & Fabbri, A. (2011). Classification of pecorino cheeses using electronic nose combined with artificial neural network and comparison with GC–MS analysis of volatile compounds. *Food Chemistry, 129*(3), 1315–1319.

Chang, K. P., Zakaria, A., Nasir, A. A., Yusuf, N., Thriumani, R., Shakaff, A. Y. M., & Adom, A. H. (2014). Analysis and feasibility study of plant disease using e-nose. In *2014 IEEE International Conference on Control System, Computing and Engineering* (ICCSCE 2014) (pp. 58–63). IEEE.

Chaparro-Torres, L. A., Bueso, M. C., & Fernández-Trujillo, J. P. (2016). Aroma volatiles obtained at harvest by HS-SPME/GC-MS and INDEX/MS-E-nose fingerprint discriminate climacteric behaviour in melon fruit. *Journal of the Science of Food and Agriculture, 96*(7), 2352–2365.

Cheli, F., Campagnoli, A., Pinotti, L., Savoini, G., & Dell'Orto, V. (2009). Electronic nose for determination of aflatoxins in maize. *BASE, 13*, 39–43.

Chen, L. Y., Wu, C. C., Chou, T. I., Chiu, S. W., et al. (2018a). Development of a dual MOS electronic nose/camera system for improving fruit ripeness classification. *Sensors, 18*(10), 3256.

Chen, Q., Song, J., Bi, J., Meng, X., & Wu, X. (2018b). Characterization of volatile profile from ten different varieties of Chinese jujubes by HS-SPME/GC–MS coupled with E-nose. *Food Research International, 105*, 605–615.

Chen, L. Y., Wong, D. M., Fang, C. Y., Chiu, C. I., Chou, T. I., Wu, C. C., … & Tang, K. T. (2018c, April). Development of an electronic-nose system for fruit maturity and quality monitoring. In *2018 IEEE International Conference on Applied System Invention* (ICASI) (pp. 1129–1130). IEEE.

Choi, S. I., Eom, T., & Jeong, G. M. (2016). Gas classification using combined features based on a discriminant analysis for an electronic nose. *Journal of Sensors*. https://doi.org/10.1155/2016/9634387. Article ID 9634387.

Cramer, R. D. (1993). Partial least squares (PLS): Its strengths and limitations. *Perspectives in Drug Discovery and Design, 1*, 269–278. https://doi.org/10.1007/BF02174528

Dey, D., & Goswami, T. (2011). Optical biosensors: A revolution towards quantum nanoscale electronics device fabrication. *BioMed Research International, 2011*, 348218.

Du, X., Bai, J., Plotto, A., Baldwin, E., Whitaker, V., & Rouseff, R. (2010). Electronic nose for detecting strawberry fruit maturity. *Proceedings of the Florida State Horticultural Society, 123*, 259–263.

Du, D., Wang, J., Wang, B., Zhu, L., & Hong, X. (2019). Ripeness prediction of postharvest kiwifruit using a MOS e-nose combined with chemometrics. *Sensors, 19*(2), 419. https://doi.org/10.3390/s19020419

Dung, T., Oh, Y., Choi, S. J., Kim, I. D., Oh, M. K., & Kim, M. (2018). Applications and advances in bioelectronic noses for odour sensing. *Sensors, 18*(1), 103.

Falasconi, M., Gobbi, E., Pardo, M., Della Torre, M., Bresciani, A., & Sberveglieri, G. (2005). Detection of toxigenic strains of fusarium verticillioides in corn by electronic olfactory system. *Sensors and Actuators B: Chemical, 108*(1–2), 250–257.

Feng, W., Hettiarachchi, R., Sato, S., Kakushima, K., Niwa, M., Iwai, H., et al. (2012). Advantages of silicon nanowire metal–oxide–semiconductor field-effect transistors over planar ones in noise properties. *Japanese Journal of Applied Physics, 51*(4S), 04DC06.

Gardner, J. W., & Bartlett, P. N. (1994). A brief history of electronic noses. *Sensors and Actuators B: Chemical, 18*(1–3), 210–211.

Ghosh, A., Ghosh, T. K., Das, S., Ray, H., Mohapatra, D., Modhera, B., Ghosh, D., Parua, S., Pal, S., Tiwari, S., Kate, A., Tripathy, M. K., Bhattacharyya, N., & Bandyopadhyay, R. (2022). Development of electronic nose for early spoilage detection of potato and onion during post-harvest storage. *Journal of Materials NanoScience, 9*(2), 101–114. https://www.pubs.thesciencein.org/journal/index.php/jmns/article/view/281

Gómez, A. H., Hu, G., Wang, J., & Pereira, A. G. (2006). Evaluation of tomato maturity by electronic nose. *Computers and Electronics in Agriculture, 54*(1), 44–52.

Gómez, A. H., Wang, J., Hu, G., & Pereira, A. G. (2008). Monitoring storage shelf life of tomato using electronic nose technique. *Journal of Food Engineering, 85*(4), 625–631.

Grieshaber, D., MacKenzie, R., Vörös, J., & Reimhult, E. (2008). Electrochemical biosensors-sensor principles and architectures. *Sensors, 8*(3), 1400–1458.

Guohua, H., Yuling, W., Dandan, Y., Wenwen, D., Linshan, Z., & Lvye, W. (2012). Study of peach freshness predictive method based on electronic nose. *Food Control, 28*(1), 25–32.

Hanif, M., Xie, B., Wei, S., Li, J., Gao, C., Wang, R., et al. (2022). Characterization of the volatile profile from six different varieties of Chinese chives by HS-SPME/GC–MS coupled with E. NOSE. *Journal of King Saud University-Science, 34*(4), 101971.

Hong, X., & Wang, J. (2014). Application of e-nose and e-tongue to measure the freshness of cherry tomatoes squeezed for juice consumption. *Analytical Methods, 6*(9), 3133–3138.

Jiang, H., Zhang, H., Chen, Q., Mei, C., & Liu, G. (2015). Recent advances in electronic nose techniques for monitoring of fermentation process. *World Journal of Microbiology and Biotechnology, 31*, 1845–1852.

Kanade, A., & Shaligram, A. D. (2014). Development of an E-nose using metal oxide semiconductor sensors for the classification of climacteric fruits. *International Journal of Scientific & Engineering Research, 5*(2), 467–472.

Karamchandani, S., Sekhani, B., Nair, K., & Shah, K. (2021, September). E-nose for shelf-life prediction of climacteric fruits. In *2021 IEEE 4th international conference on computing, power and communication technologies (GUCON)* (pp. 1–4). IEEE.

Kim, Y. S., Ha, S. C., Yang, Y., Kim, Y. J., Cho, S. M., Yang, H., & Kim, Y. T. (2005). Portable electronic nose system based on the carbon black–polymer composite sensor array. *Sensors and Actuators B: Chemical, 108*(1-2), 285–291.

Kirchner, P., Oberländer, J., Friedrich, P., Berger, J., Rysstad, G., Keusgen, M., & Schöning, M. J. (2012). Realisation of a calorimetric gas sensor on polyimide foil for applications in aseptic food industry. *Sensors and Actuators B: Chemical, 170*, 60–66.

Krishnamoorthy, S., & Iliadis, A. A. (2008). Properties of high sensitivity ZnO surface acoustic wave sensors on SiO2/(100) Si substrates. *Solid-State Electronics, 52*(11), 1710–1716.

Lai, F. D., & Huang, H. M. (2006). Fabrication of high frequency and low-cost surface-acoustic wave filters using near field phase shift photolithography. *Microelectronic Engineering, 83*(4-9), 1407–1409.

Lee, J., Kim, W. D., & Lim, H. (2012). Facile fabrication of conducting polymer nanowire based field effect transistor with controlled shape and position. *Microelectronic Engineering, 98*, 382–385.

Lever, J., Krzywinski, M., & Altman, N. (2017). Points of significance: Principal component analysis. *Nature Methods, 14*(7), 641–643.

Li, H. (2018). Deep learning for natural language processing: Advantages and challenges. *National Science Review, 5*(1), 24–26.

Li, C., Krewer, G. W., Ji, P., Scherm, H., & Kays, S. J. (2010a). Gas sensor array for blueberry fruit disease detection and classification. *Postharvest Biology and Technology, 55*(3), 144–149.

Li, S., Simonian, A., & Chin, B. A. (2010b). Sensors for agriculture and the food industry. *The Electrochemical Society Interface, 19*(4), 41–46.

Lihuan, S., Liu, W., Xiaohong, Z., Guohua, H., & Zhidong, Z. (2017). Fabrication of electronic nose system and exploration on its applications in mango fruit (M. indica cv. Datainong) quality rapid determination. *Journal of Food Measurement and Characterization, 11*(4), 1969–1977.

Luo, C., & Chakraborty, A. (2009). Effects of dimensions on the sensitivity of a conducting polymer microwire sensor. *Microelectronics Journal, 40*(6), 912–920.

Mallick, M., Hossain, S. M., & Das, J. (2018). Graphene oxide based fruit ripeness sensing e-nose. *Materials Today: Proceedings, 5*(3), 9866–9870.

Mijwel, M. M. (2018). *Artificial neural networks advantages and disadvantages*. Retrieved from LinkedIn https://www.linkedin.com/pulse/artificial-neuralnetwork

Musatov, V. Y., Sysoev, V. V., Sommer, M., & Kiselev, I. (2010). Assessment of meat freshness with metal oxide sensor microarray electronic nose: A practical approach. *Sensors and Actuators B: Chemical, 144*(1), 99–103.

Papadopoulou, O. S., Panagou, E. Z., Mohareb, F. R., & Nychas, G. J. E. (2013). Sensory and microbiological quality assessment of beef fillets using a portable electronic nose in tandem with support vector machine analysis. *Food Research International, 50*(1), 241–249.

Pearce, T. C., Schiffman, S. S., Nagle, H. T., & Gardner, J. W. (2003). *Handbook of machine olfaction: Electronic nose technology*. Wiley.

Pejcic, B., Barton, C., Crooke, E., Eadington, P., Jee, E., & Ross, A. (2009). Hydrocarbon sensing. Part 1: Some important aspects about sensitivity of a polymer-coated quartz crystal microbalance in the aqueous phase. *Sensors and Actuators B: Chemical, 135*(2), 436–443.

Pejcic, B., Crooke, E., Doherty, C. M., Hill, A. J., Myers, M., Qi, X., & Ross, A. (2011). The impact of water and hydrocarbon concentration on the sensitivity of a polymer-based quartz crystal microbalance sensor for organic compounds. *Analytica Chimica Acta, 703*(1), 70–79.

Peris, M., & Escuder-Gilabert, L. (2016). Electronic noses and tongues to assess food authenticity and adulteration. *Trends in Food Science & Technology, 58*, 40–54.

Peng, G., Hakim, M., Broza, Y. Y., Billan, S., Abdah-Bortnyak, R., Kuten, A., et al. (2010). Detection of lung, breast, colorectal, and prostate cancers from exhaled breath using a single array of nanosensors. *British Journal of Cancer, 103*(4), 542.

Qiu, S., Wang, J., & Gao, L. (2015a). Qualification and quantisation of processed strawberry juice based on electronic nose and tongue. *LWT- Food Science and Technology, 60*(1), 115–123.

Qiu, S., & Wang, J. (2015b). Effects of storage temperature and time on internal quality of Satsuma mandarin (Citrus unshiu marc.) by means of E-nose and E-tongue based on two-way MANOVA analysis and random forest. *Innovative Food Science & Emerging Technologies, 31*, 139–150.

Qiu, S., Wang, J., Tang, C., & Du, D. (2015). Comparison of ELM, RF, and SVM on E-nose and E-tongue to trace the quality status of mandarin (Citrus unshiu Marc.). *Journal of Food Engineering, 166*, 193–203.

Saevels, S., Lammertyn, J., Berna, A. Z., Veraverbeke, E. A., Di Natale, C., & Nicolaï, B. M. (2004). An electronic nose and a mass spectrometry-based electronic nose for assessing apple quality during shelf life. *Postharvest Biology and Technology, 31*(1), 9–19.

Sanaeifar, A., Mohtasebi, S. S., Ghasemi-Varnamkhasti, M., & Ahmadi, H. (2016). Application of MOS based electronic nose for the prediction of banana quality properties. *Measurement, 82*, 105–114.

Santonico, M., Bellincontro, A., De Santis, D., Di Natale, C., & Mencarelli, F. (2010). Electronic nose to study postharvest dehydration of wine grapes. *Food Chemistry, 121*(3), 789–796.

Santos, J. P., Lozano, J., Aleixandre, M., Arroyo, T., Cabellos, J. M., Gil, M., & del Carmen Horrillo, M. (2010). Threshold detection of aromatic compounds in wine with an electronic nose and a human sensory panel. *Talanta, 80*(5), 1899–1906.

Scarlata, S., Pennazza, G., Santonico, M., Pedone, C., & Antonelli, I. R. (2015). Exhaled breath analysis by electronic nose in respiratory diseases. *Expert Review of Molecular Diagnostics, 15*, 933–956.

Schaller, E., Bosset, J. O., & Escher, F. (1998). 'Electronic noses' and their application to food. *LWT-Food Science and Technology, 31*(4), 305–316.

Schiffman, S. S., & Pearce, T. C. (2003). Introduction to olfaction: Perception, anatomy, physiology, and molecular biology. In *Handbook of machine olfaction: Electronic nose technology* (pp. 1–32.

Schnabel, R. M., Boumans, M. L., Smolinska, A., Stobberingh, E. E., Kaufmann, R., Roekaerts, P. M., & Bergmans, D. C. (2015). Electronic nose analysis of exhaled breath to diagnose ventilator-associated pneumonia. *Respiratory Medicine, 109*, 1454–1459.

Sherrit, S., Bao, X., Bar-Cohen, Y., & Chang, Z. (2003, July).BAW and SAW sensors for in situ analysis. In *Smart structures and materials 2003: Smart sensor technology and measurement systems* (Vol. 5050, pp. 81–91). International Society for Optics and Photonics.

Simon, J. E., Hetzroni, A., Bordelon, B., Miles, G. E., & Charles, D. J. (1996). Electronic sensing of aromatic volatiles for quality sorting of blueberries. *Journal of Food Science, 61*(5), 967–970.

Smith, A. L., & Shirazi, H. M. (2005). Principles of quartz crystal microbalance/heat conduction calorimetry: Measurement of the sorption enthalpy of hydrogen in palladium. *Thermochimica Acta, 432*(2), 202–211.

Solis-Solis, H. M., Calderon-Santoyo, M., Gutierrez-Martinez, P., Schorr-Galindo, S., & Ragazzo-Sanchez, J. A. (2007). Discrimination of eight varieties of apricot (Prunus armeniaca) by electronic nose, LLE and SPME using GC–MS and multivariate analysis. *Sensors and Actuators B: Chemical, 125*(2), 415–421.

Tan, J., & Xu, J. (2020). Applications of electronic nose (e-nose) and electronic tongue (e-tongue) in food quality-related properties determination: A review. *Artificial Intelligence in Agriculture, 4*, 104–115.

Tang, K. T., Chiu, S. W., Pan, C. H., Hsieh, H. Y., Liang, Y. S., & Liu, S. C. (2010). Development of a portable electronic nose system for the detection and classification of fruity odors. *Sensors, 10*(10), 9179–9193.

Thomas, K., Christoph, Z., Niko, M., & Gerhard, T. (2012). An electronic nose on flexible substrates integrated into a smart textile. *Sensors and Actuators B: Chemical, 174*, 81–86.

Tiwari, S., Kate, A., Mohapatra, D., Tripathi, M. K., Ray, H., Akuli, A., et al. (2020). Volatile organic compounds (VOCs): Biomarkers for quality management of horticultural commodities during storage through e-sensing. *Trends in Food Science & Technology, 106*, 417–433.

Vaschetto, M. E., Monkman, A. P., & Springborg, M. (1999). First-principles studies of some conducting polymers: PPP, PPy, PPV, PPyV, and PANI. *Journal of Molecular Structure: THEOCHEM, 468*(3), 181–191.

Verma, P., & Yadava, R. D. S. (2015). Polymer selection for SAW sensor array based electronic noses by fuzzy c-means clustering of partition coefficients: Model studies on detection of freshness and spoilage of milk and fish. *Sensors and Actuators B: Chemical, 209*, 751–769.

Voss, H. G. J., Ayub, R. A., & Stevan, S. L. (2020). E-nose prototype to monitoring the growth and maturation of peaches in the orchard. *IEEE Sensors Journal, 20*(20), 11741–11750.

Walt, D. R., Dickinson, T., White, J., Kauer, J., Johnson, S., Engelhardt, H., & Jurs, P. (1998). Optical sensor arrays for odor recognition. *Biosensors & Bioelectronics, 13*(6), 697–699.

Wei, X., Zhang, Y., Wu, D., Wei, Z., & Chen, K. (2018). Rapid and non-destructive detection of decay in peach fruit at the cold environment using a self-developed handheld electronic-nose system. *Food Analytical Methods, 11*(11), 2990–3004.

Wen, J., Zhao, Y., Rong, Q., Yang, Z., Yin, J., & Peng, Z. (2022). Rapid odor recognition based on reliefF algorithm using electronic nose and its application in fruit identification and classification. *Journal of Food Measurement and Characterization, 16*(3), 2422–2433.

Wilson, A. D. (2012). Review of electronic-nose technologies and algorithms to detect hazardous chemicals in the environment. *Procedia Technology, 1*, 453–463.

Wilson, A. D. (2013). Diverse applications of electronic-nose Technologies in Agriculture and Forestry. *Sensors, 13*, 2295–2348.

Wilson, A., & Baietto, M. (2009). Applications and advances in electronic-nose technologies. *Sensors, 9*(7), 5099–5148.

Wilson, A. D., & Lester, D. G. (1997). Use of an electronic-nose device for profiling headspace volatile metabolites to rapidly identify phytopathogenic microbes. *Phytopathology, 87*, S116.

Wilson, A. D., & Lester, D. G. (1998). Application of aromascan analysis to detect and diagnose oak wilt in live oaks. *Phytopathology, 88*(suppl), S97.

Wilson, A. D., Lester, D. G., & Oberle, C. S. (2004). Development of conductive polymer analysis for the rapid detection and identification of phytopathogenic microbes. *Phytopathology, 94*(5), 419–431.

Xiao, Z., Niu, M., & Niu, Y. (2022). Comparative study on volatile compounds and taste components of different durian cultivars based on GC-MS, UHPLC, HPAEC-PAD, E-tongue and E-nose. *Molecules, 27*(4), 1264. https://doi.org/10.3390/molecules27041264

Xie, G., Sun, P., Yan, X., Du, X., & Jiang, Y. (2010). Fabrication of methane gas sensor by layer-by-layer self-assembly of polyaniline/PdO ultra thin films on quartz crystal microbalance. *Sensors and Actuators B: Chemical, 145*(1), 373–377.

Xin, R., Liu, X., Wei, C., Yang, C., Liu, H., Cao, X., et al. (2018). E-nose and GC-MS reveal a difference in the volatile profiles of white-and red-fleshed peach fruit. *Sensors, 18*(3), 765. https://doi.org/10.3390/s18030765

Zhu, N., Mao, S., Pan, L., Yuan, L., & Tu, K. (2013). Early detection of fungal disease infection in strawberry fruits by e-nose during postharvest storage. *Transactions of the Chinese Society of Agricultural Engineering, 29*(5), 266–273.

Zhao, Y., Li, X. G., Zhou, X., & Zhang, Y. N. (2016). Review on the graphene based optical fiber chemical and biological sensors. *Sensors and Actuators B: Chemical, 231*, 324–340.

Zohora, S. E., Khan, A. M., & Hundewale, N. (2013). Chemical sensors employed in electronic noses: A review. In *Advances in computing & information technology* (pp. 177-184). Springer.

Use of Electronic Tongue to Determine Quality and Safety of Fresh Produce

Zahir Al-Attabi, Nasser Al-Habsi, and Mohammad Shafiur Rahman

Abstract Safety and quality of fresh produce is considered a high concern to the consumers and food industry. Traditional methods, like gas chromatography (GC) and high-performance liquid chromatography (HPLC), are expensive, time-consuming, and require skills to operate. Similarly, the sensory evaluation method has limitations to apply in the routing quality control evaluation due to its required time for evaluation and its complicated process. Electronic tongue (e-tongue) is an emerging technology and it is considered as low cost, rapid and accurate, reproducible, and sensitive method. Its applications covered different areas in food science, including safety, adulteration, quality control, quality assurance, monitoring process, and spoilage. This chapter presents the applications of e-tongue in the context of different selected fresh food produce.

Keywords E-tongue · Sensory · Fresh produce · Quality · Adulteration · Safety

1 Introduction

Fresh fruits and vegetables, milk, meat, and seafood are considered perishable. Some of them are consumed without any processing, while others must be processed to secure safety and quality with a minimal effect on the nutritional value. In addition, meat and seafood must be cooked before their consumption. Sensory attributes (i.e., appearance, texture, color, smell, and taste) are major quality parameters that may be affected by processing. These are used as a primary quality indicator, and consumers assess the acceptability based on these attributes. Therefore, maintaining a desired level of the sensory attribute is very important. Furthermore, post-harvest storage conditions before and after treatment can also affect the shelf life of fresh produce. Human senses are used mainly to assess these attributes. However, health and other physiological status of humans can affect their responses. In addition, spoiled foods

Z. Al-Attabi (✉) · N. Al-Habsi · M. S. Rahman
Department of Food Science and Nutrition, College of Agricultural and Marine Sciences, Sultan Qaboos University, Al-Khodh, Oman
e-mail: zaherr@squ.edu.om

P. B. Pathare, M. S. Rahman (eds.), *Nondestructive Quality Assessment Techniques for Fresh Fruits and Vegetables*, https://doi.org/10.1007/978-981-19-5422-1_16

and adulterated or contaminated foods may raise health risks. Furthermore, human sensory has some drawbacks, like fatigue, saturation, stability, and reproducibility. The individual and group responses can change between the assessment sessions. Even though adequate training can minimize these errors but it is time-consuming, complex, and costly. Therefore, human sensory method is not practically suitable for routine quality control. In addition, safety evaluation of toxic foods cannot be performed by sensory method. Technologies, like electronic nose (e-nose), electronic tongue (e-tongue), and computer vision are being developed and implemented. The applications of e-nose and e-tongue simultaneously can give a better insight of the taste and flavor. In addition, signals from these sensors are correlated with human sensory perceptions. Therefore, these technologies can help in monitoring and predicting the sensory properties as well as other quality attributes of foods (Bleibaum et al., 2002). This chapter presents an overview of the application of e-tongue in assessing the safety and quality of fresh produce.

2 Electronic Tongue System

The human tongue is used to assess the taste of sweetness, bitterness, sourness, saltiness, and umami by taste bud receptors. In this process, components in foods are first dissolved in saliva or water, or oil, and interact with the taste buds' receptors. Then signals are transferred to the specific regions in the brain for their analysis and interpretation. Considering the same process of the perceptions by the human tongue, e-tongue system was developed to mimic this process. This system consists of a chemical sensor array, a reaction vessel, measuring devices, transducers, data acquisition devices, and data processing and pattern recognition algorithms (Fig. 1) (Tan & Xu, 2020). The sample must be in a liquid form and placed in the reaction vessel where then the chemical sensors (i.e., electrodes) react with the analytes. Voltage difference between working electrodes and reference electrode is detected

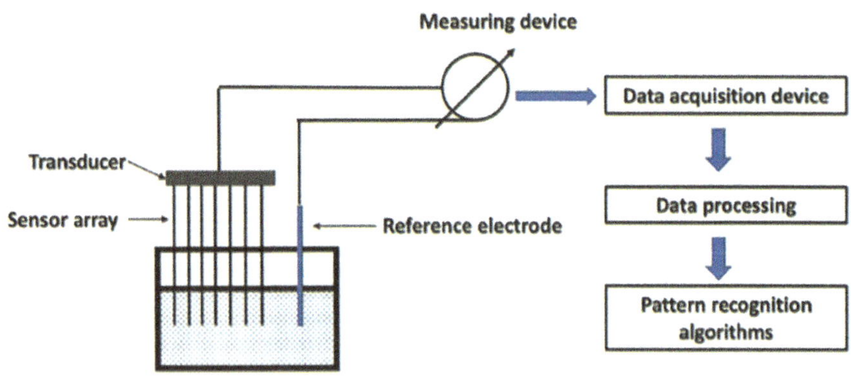

Fig. 1 Electronic tongue system (Tan & Xu, 2020)

by transducer. Electrical signals are generated and recorded by data acquisition device. These signals are then processed by different pattern recognition algorithms based in a computer.

It is successfully used as a qualitative and quantitative tool in taste evaluation, quality control, and process monitoring (Beullens et al., 2006, 2008; Tan & Xu, 2020). The advantages of applying e-tongue include their low cost, rapid and accurate prediction, reproducibility, and sensitivity; however, its pre-treatment steps are required for some samples, and the short life of the sensors is considered the main disadvantage (Tan & Xu, 2020; Wadehra & Patil, 2016). Its selectivity and specificity increased with the recent progress in the development of biosensors (Tan & Xu, 2020). E-tongue can be equipped with different sensors (e.g., potentiometric, voltammetric, and bioelectric sensors) and this uses different pattern recognition algorithms, such as principal component analysis (PCA), linear discriminant analysis (LDA), partial least square discriminant analysis (PLS-DA), canonical discriminant analysis (CDA), and artificial neural network (ANN) (Kalit et al., 2014; Tan & Xu, 2020).

3 Applications of e-Tongue

3.1 Tomato

The application of e-tongue to classify four tomato cultivars (*Lycopersicon esculentum* Mill.) (i.e., Aranca, Climaks, Clotilde, and DRW 73–29) was based on sugar and acid profile detection (Beullens et al., 2006). It was successfully separated anranca cultivar by indicating its difference in sugar and acid profile as compared to others (Fig. 2). Furthermore, other possible compounds could be attributed to this difference as e-tongue could be less sensitive to sugar as compared to minerals and acids. In addition, DRW 73–29 was also classified separately from climaks and clotilde cultivars; however, these two cultivars were overlapped. The e-tongue signals showed a 90% correlation with high-performance liquid chromatography (HPLC) data when canonical correlation analysis was applied. Besides, e-tongue was able to predict a few compounds, such as fumaric acid and malic acid.

Further work by Beullens et al. (2008) showed a successful classification of six tomato cultivars (*Lycopersicon esculentum* Mill.) based on sugar (i.e., glucose, fructose), acid (i.e., citric acid, malic acid, and glutamic acid), and mineral contents (i.e., Na, K) (Fig. 3). The clear separation of Amoroso cultivar was related to its high sugar content compared to other cultivars. The difference in sugars and minerals content caused a clear separation between Amoroso, Macarena, and Tricia cultivars. The electronic tongue results were correlated well with the data of trained sensory panel, which indicated a possible determination of tomato taste by e-tongue.

The freshness of squeezed *youbei* cherry tomato juice was tracked using e-tongue and e-nose (Hong & Wang, 2014b). The post-harvest storage conditions for cherry tomatoes were 25 ± 1 °C and 80 ± 5% relative humidity for 8 days. The juice was squeezed and analyzed every 2 days. Qualitatively, e-tongue was successfully

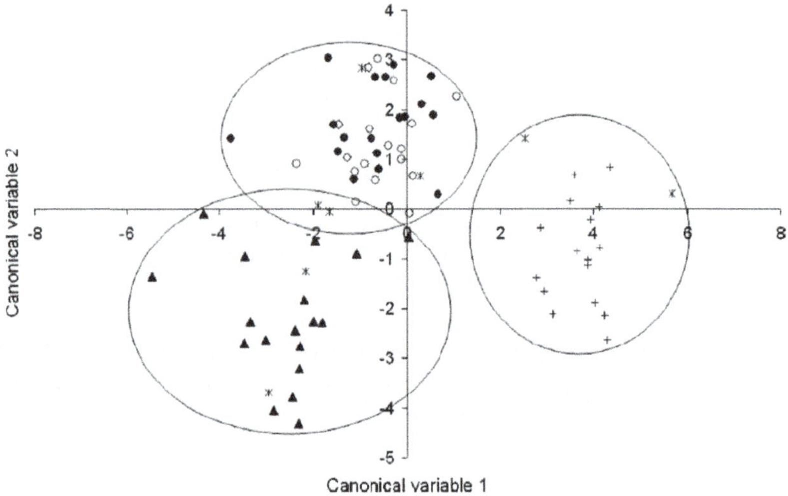

Fig. 2 Canonical discriminant analysis of ET measurements for four tomato cultivars. (*) Unknown, (+) Aranca, (●) Climaks, (○) Clotilde, (▲) DRW 73-29 (Beullens et al., 2006)

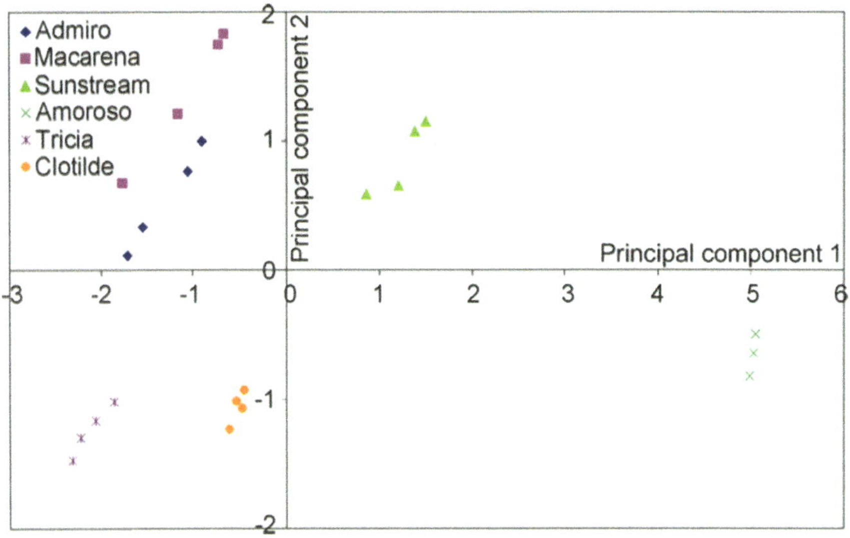

Fig. 3 PCA score plot of six tomato cultivars using ET (Beullens et al., 2008)

classified and discriminated juice samples (Fig. 4) using CDA and PCA. However, CDA provided better discrimination. Furthermore, learning vector quantization (LVQ) network and library support vector machine (lib-SVM) gave high classification accuracy with e-tongue data. Quantitatively, tracking and prediction

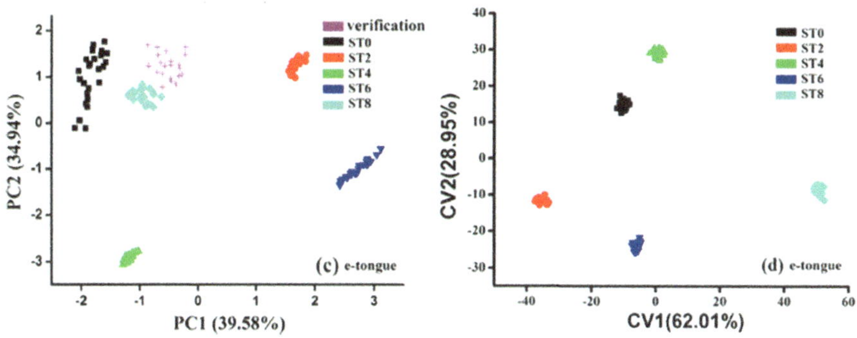

Fig. 4 PCA and CDA plots of juices squeezed from cherry tomatoes stored at different times (Hong & Wang, 2014b)

of physico-chemical qualities (i.e., pH, soluble solids content, vitamin C, and firmness) using principle components regression (PCR) gave good regression models with e-tongue. This study showed that the e-tongue gave better results than e-nose either qualitatively or quantitatively. However, applying both e-tongue and e-nose simultaneously was a better approach than individual one, which was also supported by Hong and Wang (2014a). The possibility of detecting adulteration of cherry tomato juice by adding different levels of overripe tomato juice (0–30%) was reported. E-tongue was able to discriminate the adulterated samples from each other by CDA (Hong & Wang, 2014a).

Recent work by Xu et al. (2018) showed the ability of e-tongue to distinguish tomato taste of different cultivars at different maturity stages. Moreover, authors tested the effect of post-harvest treatments, blanching (100 °C, 60 s) and refrigeration (5 °C), on their taste. The LDA showed a clear classification of the three used commercial tomato cultivars and differentiate between the six harvested maturities. The e-tongue detected some taste profiles in refrigeration and blanching, especially for the refrigeration treatment. The authors reported that sugar and acid contents were less affected by chilling in the ripening stage of tomatoes. Chilling injury occurred more in low mature tomatoes leading to several effects. In contrast, heating caused little change to both contents. Some of the e-tongue sensors correlated to titratable acidity (TA) and total soluble solids (TSS); however, TSS showed a higher correlation.

3.2 *Fruits*

Fruits are normally consumed as fresh or as freshly squeezed juice (100%). Therefore, regulatory and food control needs to ensure or authenticate the freshness of the commercially sold juices (Hong & Wang, 2014b). E-tongue was successfully discriminated between juices (i.e., apple, orange), made from concentrate by different

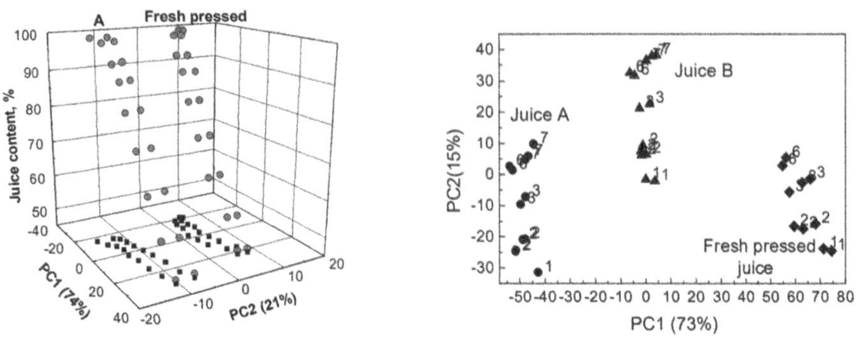

Fig. 5 PCA plot of adultered orange juice (A prepared from concentrate) with water (left) and monitoring orange juice spoilage over 7 days of storage (right) (Rudnitskaya et al., 2001)

manufacturers, and distinguished them from fresh pressed natural juice using PCA (Rudnitskaya et al., 2001). It was also found that e-tongue was sensitive to adulterated juice samples (i.e., the addition of sugar syrup or water) as well as monitoring aging and spoilage (Fig. 5). The upper points presented on the left plot of Fig. 5 indicated undiluted juice and the lower points represent the diluted ones.

Another study conducted by Kantor et al. (2008) showed the effect of post-harvest treatment with 1-methylcyclopropene (1-MCP) (i.e., block the ethylene action of ripening) and controlled atmosphere storage on taste and ripening level of three apricot varieties. In the first experiment, the control and treated apricot samples were stored for 4 days at 20 °C and 45% relative humidity (RH). This condition simulated the commercial distribution conditions and investigated whether the treatment could reduce the ripening. The results showed a significant ripening difference during the 4 days of storage between the controlled and treated ones. The treated samples showed higher sugar content, which was related to the delay of carbohydrates degradation. The sensory evaluation indicated higher sour flavor in all treated samples with only a significant difference in Pannonia variety. The second experiment investigated the effect of long storage (28 days) on control and treated fruits stored at 1 °C and 95% RH under controlled atmosphere conditions (i.e., four champers with different O_2, CO_2, and N_2 content). The CDA correctly classified the treated and controlled samples at 83.9% after 2 weeks and at 92.9% after 4 weeks. It was interesting to notice that the distance in CDA plot increased between the treated samples and the controlled samples. This was dependent on CO_2 content in each chamber, more CO_2 caused the shorter the distance. Generally, the classification of apricot varieties and determination of correlations between e-nose, chemical properties (pH, °Bx), and sensory analysis was successful. Moreover, this study indicated the sensitivity of e-tongue was higher as compared to sensory analysis data (i.e., sweet, sour, and overall impression).

In the case of apple juice quality as assessed by consumer, trained panels and combination of the electronic sensors (i.e., e-tongue and e-nose) showed a high correlation (Bleibaum et al., 2002). The combined data of e-tongue and e-nose was

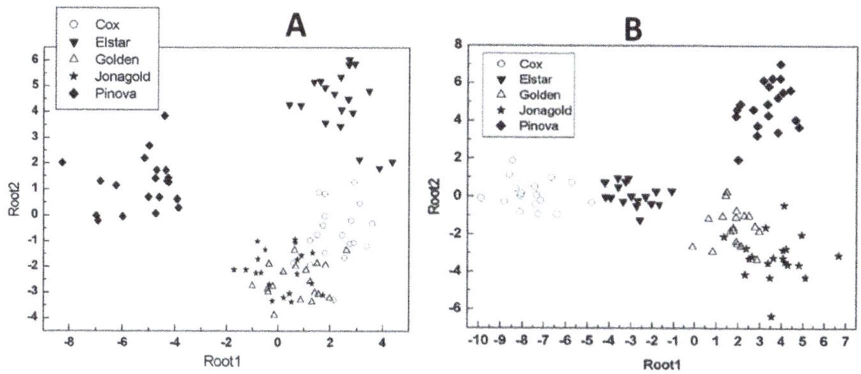

Fig. 6 Canonical roots plot for (**a**) ET & (**b**) ATR-FTIR data (Rudnitskaya et al., 2006)

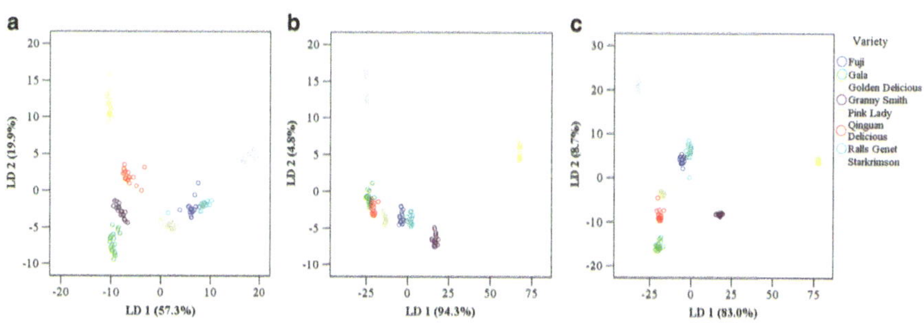

Fig. 7 LDA discrimination according to apple variety. (**a**) EN, (**b**) ET, (**c**) combined approach (Wu et al., 2018)

able to predict the apple juice sensory characteristics as per defined by the quality attributes assessed by consumers. This indicated the possibility of using the sensors technology in quality control and quality assurance. Rudnitskaya et al. (2006) were able to classify the five apple varieties including Cox, Elstar, Golden, Delicious, Jonagold, and Pinova using e-tongue, which was based on organic acids contents (Fig. 6). The overlap between Golden and Jonagold was due to their similarity as the composition of Jonagold, which is a hybrid cultivar obtained from crossing of Jonathan with Golden verities. This study was compared with the e-tongue and Fourier Transform Infrared Spectroscopy (ATR-FTIR). Better discrimination was found using ATR-FTIR as it was sensitive to both organic acids and sugars.

Recently, Wu et al. (2018) traced the apples according to their variety and geographical origin using e-tongue, e-nose, and combined approach (Fig. 7). It was clearly shown that the combined approach gave better separation based on variety. The same conclusion was found with geographical discrimination using LDA. Even though individual data set of e-tongue and e-nose showed high classification ability but the combined approach showed 100% classification ability. Furthermore, satisfactory discriminant was achieved by SVM and PLS-DA models.

3.3 Meat

The assessment of meat origin and meat quality is essential. Awareness of consumers demands meat products with high nutritional value with enriched minerals, vitamins, and essential polyunsaturated fatty acids, and low fat and cholesterol (Paulos et al., 2015). However, meat from dairy breeds is generally considered to be of low quality. The preference of consumer to know the type of meat they are eating is related to many factors, such as cost, preference, religion, dietary requirements, or simply pleasure. The meat quality of different types of animal and processed meat has been measured and compared in numerous studies. The quality attribute comprises sensory attributes (i.e., texture, color, and flavor), biological characteristics of the carcass [i.e., pH, water activity, dry matter content (DMC), water-holding capacity (WHC), and juiciness]. These quality markers are currently being monitored and determined through different series of methods, such as spectrophotometric techniques (color), pH differential methods (pH), texture analyzers (for tenderness, juiciness, drip loss), Soxhlet (crude fat), Kjeldahl analysis (crude protein), and the filter-paper press method (WHC). The adulteration of processed meat with other inexpensive types of meat or other ingredient are performed for the purpose of incrassating profit and it is common nowadays (Cozzolino & Murray, 2004; Kamruzzaman et al., 2013). Therefore, the need for one single device to detect adulteration and evaluate meat quality is of necessity.

E-tongue has been used to detect this type of fraud and it has proved its reliability. Moreover, the combination of e-tongue and e-nose has also been used to detect the quality of meat and showed better accuracy. Tian et al. (2019) studied the efficacy of data fusion methods for e-nose and e-tongue in the authentication of mutton in consonance to its fraction of pork and to relate the fusion data set with pork fraction in minced mutton. The data from both equipment were normalized using stepwise linear discrimination analysis (step-LDA) and PCA. The results of discrimination were then evaluated and compared using CDA and Bayesian discriminant analysis (BAD). The capability of classification of both instrument was superior or equal to that obtained when two instrument were used separately. E-tongue (classification error 0–2.5%) system obtained higher accuracy than e-nose (classification error 0.83–10.83%). E-tongue was also used to discriminate different species of cattle meat namely Angus, domestic buffalo, Hungarian Grey, Hungarian Spotted cattle, and Holstein beef meat samples from the chuck steak part of the animals (Surányi et al., 2021). The results of e-tongue were correlated with conventional methods of analysis. It was shown that the e-tongue could also be used to predict some of the physico-chemical characteristics with acceptable accuracies. Moreover, the classification models (PCA, LDA) used were able to classify most of the cattle with 100% accuracy. E-tongue was sensitive enough to identify and distinguish from the beef breed. Some studies have recommended that expansions in automated genotyping approaches may ultimately favor the use of dialysis tags. These techniques can permit scanning of hundreds of loci, but this is time-consuming and costly as compared to e-tongue measurements (Blott et al., 1999). E-tongue fitted with

TS-5000Z taste sensing system was used to evaluate meat quality based on flavor recognition (Zhang et al., 2015). In this study, 18 beef cattle (six Wagyu breed cattle, six Angus breed cattle, and six Simmental breed cattle) were used to correlate their chemical compositions and e-tongue responses. The results showed different chemical compositions and flavors in different breeds of cattle. Good and rapid classification was reported for the classification of different breeds of cattle using PCA. Other types of meat, such as poultry and red meat, were also analyzed using e-tongue after applying the standard extraction technique (Zaukuu et al., 2021). It is a high challenge to standardize meat extraction method to detect the adulteration level in meat. Zaukuu et al. (2021) studied the optimal dilution level of meat extract for e-tongue evaluations and also developed a standard method for extraction. They used a mixture of pork and beef as one mixture and used turkey and chicken as another mixture. The optimum dilution factor of 1% liquid meat extract was used to discriminate 1%, 3%, 5%, 10%, and 20% w/w chicken in turkey and pork in beef. A recognition of 78.13% and validation of 64.73 were achieved using LDA. They also found that the PLS model could predict both sets of meat mixture with higher accuracies. It reveals that the best extraction method using cooking can be applied for meat quality evaluation using e-tongue. Both e-tongue and e-nose were also used to study the effect of different wood chips on the volatile compound and sensory characteristics of smoked bacon (Du et al., 2021). This study revealed that the type of wood chips influenced the volatile compound. The combination of two instruments resulted in an effective discrimination in the flavor of bacon smoked with different types of wood chips. The capability of e-tongue to discriminate and identify adulteration is very sensitive and it can be more accurate when this is combined with e-nose. These former studies reveal that e-tongue and e-nose could replace the human sensory when they are combined together.

3.4 Fish

Fish is a highly perishable food commodity. Its freshness is highly important to be monitored after harvesting as it affects both consumers and industry. Fish spoilage is a high concern due to its potential for food poisoning. Fish freshness is normally assessed by sensory evaluation. However, fast and reliable technologies, like e-tongue and e-nose, have been tested and reported in the literature. Gil et al. (2008) studied the filleted fish freshness using sea bream (*Sparus Auratus*) for 14 days at 4 °C. The PCA gave clear discrimination throughout the 14 days of storage. In addition, fish freshness indicators, such as texture, pH, color, microbial analysis, total volatile basic nitrogen (TVB-N), and biogenic amines, were studied. PLS is used to develop a correlation between the e-tongue data and these quality indicators. A very high correlation (i.e., higher than 0.95) was reported with total biogenic amines, pH, microbial analysis, and TVB-N. In contrast, texture and color had a correlation coefficient lower than 0.9. The TVB-N and total viable count (TVC) indicated spoilage after day 7. Similarly, a high prediction correlation was achieved using support vector regression (SVR) for TVB-N (0.95) and TVC (0.90) (Han et al., 2014b). Additionally, fish freshness was better classified using a support vector machine (SVM) than Fisher LDA. Combining both e-tongue and e-nose resulted in a higher TVC predicted correlation than individual instrument (Han et al., 2014a).

Apetrei et al. (2013) stored four fish (i.e., Pontic shad) at 4 °C for 10 days and every day 1 g of muscle sample was analyzed. PCA and PLS-DA successfully classified the fish freshness in which the fish was described as fresh up to day 4 and thereafter on days 5 and 6 the off odors were observed and fish collected on days 7 till 10 was classified as spoiled. Furthermore, PCA clearly discriminates five model solutions of different biogenic amines including ammonia, histamine, dimethylamine, trimethylamine, and cadaverine.

Successful involvement of e-tongue in evaluating the taste of fish muscle was reported. Mabuchi et al. (2019) used e-tongue to evaluate the taste of different fish species. Different taste attributes including sourness, acidic bitterness, umami, and saltiness were assessed. Their intensities were also measured by e-tongue and they were different between fish muscles from different fish. E-tongue data and GC-MS results were correlated. Some of these taste attributes were highly correlated with some compounds, such as phosphoric acid, lactic acid, and creatinine. Furthermore, the safety of fish meat in relation to heavy metal is vital. Han et al. (2019a) were able to detect three heavy metals residue including Pb, Cd, and Hg using optical e-tongue system, which was based on colorimetric sensors array. Inductively coupled plasma/mass spectrometry (ICP-MS) was used as a reference method. Extreme learning machine (ELM) gave a better correlation coefficient in the prediction of heavy metals as compared with partial least square regression (PLS).

3.5 Milk

Several applications of e-tongue in milk and dairy products were reported, including milk classification (e.g., based on brand, processing, and fat content), adulteration, spoilage, effect of storage conditions, safety, and quality control. The e-tongue was able to classify milk samples from different producers (Hruškar et al., 2009) as well as different processed milk (homogenized/pasteurized, UHT milk, UHT milk with low fat content) (Paixão & Bertotti, 2009). Safety is a major concern for milk; therefore, the e-tongue showed the ability to detect antibiotic residue in milk (Wei & Wang, 2011). Bovine milk sample spiked with six types of antibiotic at four different concentration levels. Discriminant function analysis (DFA) gave better separation than PCA for all antibiotics. Furthermore, PCR, PLSR, and LS-SVM showed a good prediction of the antibiotics concentrations. Monitoring the quality of milk by e-tongue was achieved (Wei et al., 2013). Unsealed pasteurized milk samples stored at 25 °C were analyzed at 0, 12, 24, 36, 48, 60, and 72 h. The data was clearly classified by PCA and cluster analysis (CA). In addition, PLSR and LS-SVM predicted the total bacterial count and viscosity with a high correlation (over 0.94).

Milk taste is one of the primary sensory attributes used to assess milk quality. The effect of storage at 27 °C up to 12 h (0, 2, 4, 6, 8, 10, and 12 h) on raw bovine and goat milk was investigated by Tazi et al. (2017). The e-tongue data with pH sensors correctly classified bovine milk (100%) and goat milk (98.6%) when LDA was used. Therefore, e-tongue is a powerful and rapid tool to assess milk freshness. Furthermore, e-tongue was used to study the effect of storage on the taste of pasteurized milk. The taste was unchanged within 3 days of storage at 0, 4, or 10 °C, while its sweetness significantly dropped when stored at 15 or 25 °C due to milk spoilage (Ruixue et al., 2019). The effect of light on milk flavor using e-tongue and e-nose was also studied. Exposing high hydrostatic pressure to cow milk and ultraviolet light C showed insignificant differences in flavor between the control and treated sample. Furthermore, simultaneous application of e-nose and e-tongue gave an acceptable classification of milk (Hu et al., 2017).

Milk adulteration is one of the areas in which e-tongue showed a very successful application. The addition of water is one of the major adulterations in dairy. E-tongue was able to identify adulterated milk with water (Paixão & Bertotti, 2009). Goat milk is expensive with low production compared to cow milk; therefore, milk producers tend to mix both and sell milk as goat milk. Dias et al. (2009) successfully identified the adulterated goat milk with cow milk samples using e-tongue (Fig. 8). Recently, Han et al. (2019b) mixed cow milk with goat milk with up to 50% adulteration. Therefore, adulterated goat milk samples represented as 100%, 90%, 80%, 70%, 60%, and 50%. These authors used e-tongue equipped with eight sensors and data analyzed by kernel principal component analysis (KPCA). The adulterated goat milk samples clearly and accurately distinguished between them (Fig. 9). Furthermore, high accuracy in the quantitative prediction model for goat adulteration purity was achieved using an improved artificial fish swarm-extreme learning machine (IAFSA-ELM). The addition of formaldehyde, urea, and melamine in milk as preservative and protein sources was reported. E-tongue gave a good discrimination at concentrations above 10.0, 4.16, and 0.95 mmol/L for formaldehyde, urea, and melamine, respectively (Bueno et al., 2014).

Skim milk flavor is different from whole milk due to fat separation. Therefore, Pan et al. (2019) studied the possibility of improving the skim flavor by optimizing the pre-heating conditions before defatting milk using sensory evaluation and ET. The PCA classified pre-heating conditions of 50 °C for 30 min with raw milk indicating the similarity of taste. The sensory evaluation of the milks subjected to the selected condition was the most preferred sample.

Classification of milk based on fat content using ET was successfully reported. In a recent work by Pérez-González et al. (2021), e-tongue with five sensors only was developed and used to evaluate the quality of 13 milk samples with different fat content (skimmed, semi-skimmed, and whole) and nutritional content (classic, calcium-enriched, lactose-free, folic acid-enriched, and enriched in sterols of vegetal origin). In addition, it is used to predict the physico-chemical properties of milk including acidity, density, proteins, lactose, and fat. The PCA classified different milk samples based on their nutritional content regardless of fat content (skimmed, semi-skimmed, and whole) (Fig. 10). The prediction of the physico-chemical

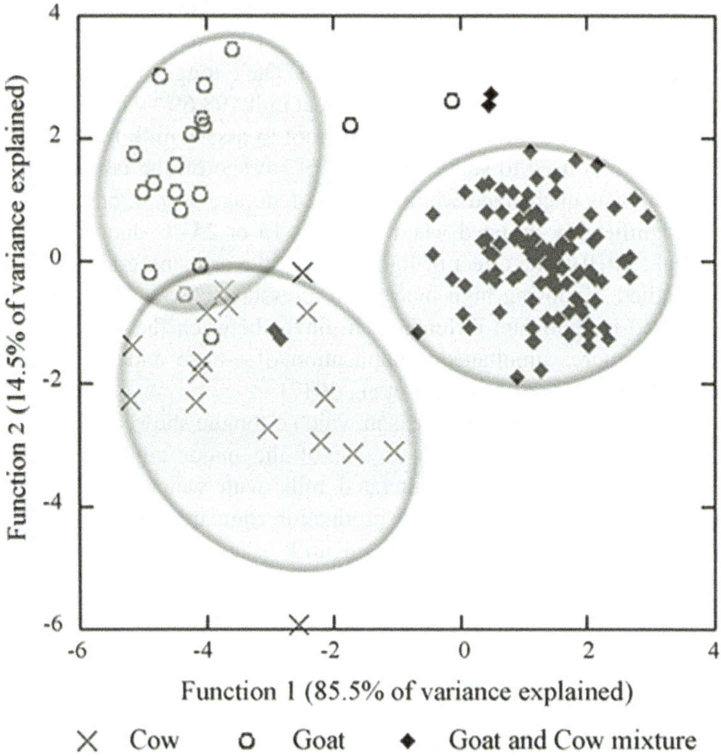

Fig. 8 Discriminant analysis of adulterd skim raw goat milk samples (Dias et al., 2009)

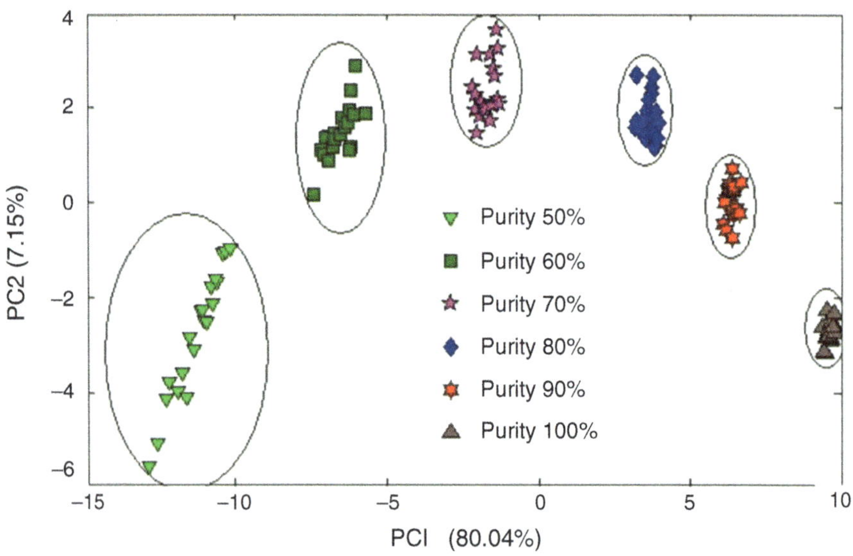

Fig. 9 Kernel principal component analysis of adultterd goat milk samples (Han et al., 2019a)

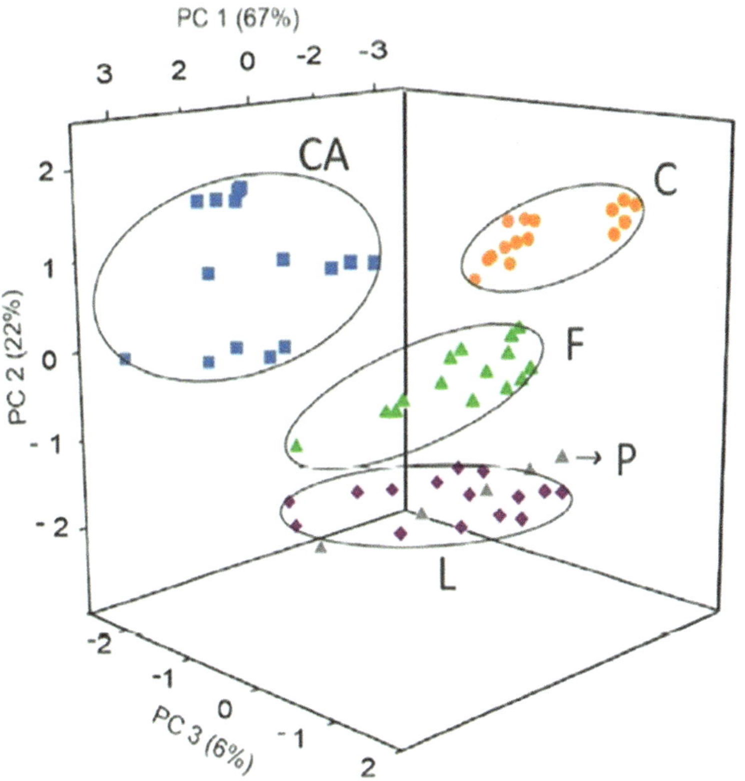

Fig. 10 PCA score plot of 13 milk samples with different fat and nutritional content. C classic milk, CA calcium-enriched milk, L lactose-free milk, F folic acid-enriched milk, P vegetal sterol-enriched milk

properties of milk was done by using support vector machine regression (SVMR). Correlation coefficient of both calibration and prediction models are reported respectively for acidity (0.93, 0.91), protein content (0.94, 0.91), lactose content (0.98, 0.97), and density (0.87, 0.81), and fat (0.64, 0.36). In the case of fat, results indicated a poor correlation between the data provided by e-tongue and the physical-chemical methods. Therefore, e-tongue was unable to predict fat content. The authors explained the formation of a double layer of fat around the sensors, which created membrane potential, and therefore, the access fat could contribute to the signal. Furthermore, e-tongue was able to identify the unsealed spoiled milk.

4 Conclusion

Research on e-tongue has proven its ability as a qualitative and quantitative tool for quality analysis of foods. A high prediction correlation was reported with traditional methods with good discrimination. E-tongue successfully classified fruits based on variety and geographical origin as well as juice freshness. Furthermore, adulteration of juices, milk, and meat was effectively detected by e-tongue or a combination of e-tongue and e-nose. Assessment of fish freshness and monitoring its spoilage was successfully achieved by e-tongue. Meat quality, origin, and flavor based on the breed can be traced and discriminated by e-tongue Furthermore, the effects of storage conditions and post-harvest treatment were clearly demonstrated. Future applications of e-tongue in the food industry are promising due to its advantages over the traditional methods. E-tongue can be equipped with different types of sensors and advanced pattern recognition systems can transform it into a very powerful analytical tool. A portable e-tongue may enhance its field applications.

Acknowledgment The authors would like to acknowledge the financial support from SQU-UAEU collaborative project CL/SQU-UAEU/2016/03 and to express their appreciation for the support of the Sultan Qaboos University for their works on the applications of e-nose and e-tongue in the food industry.

References

Apetrei, I. M., Rodriguez-Mendez, M. L., Apetrei, C., & de Saja, J. A. (2013). Fish freshness monitoring using an E-tongue based on polypyrrole modified screen-printed electrodes. *IEEE Sensors Journal, 13*(7), 2548–2554.

Beullens, K., Kirsanov, D., Irudayaraj, J., Rudnitskaya, A., Legin, A., Nicolai, B. M., & Lammertyn, J. (2006). The electronic tongue and ATR–FTIR for rapid detection of sugars and acids in tomatoes. *Sensors and Actuators B, 116*, 107–115.

Beullens, K., Meszaros, P., Vermeir, S., Kirsanov, D., Legin, A., Buysens, S., Cap, N., Nicolai, B. M., & Lammertyn, J. (2008). Analysis of tomato taste using two types of electronic tongues. *Sensors and Actuators B, 131*, 10–17.

Bleibaum, R. N., Stone, H., Tan, T., Labreche, S., Saint-Martin, E., & Isz, S. (2002). Comparison of sensory and consumer results with electronic nose and tongue sensors for apple juices. *Food Quality and Preference, 13*, 409–422.

Blott, S., Williams, J., & Haley, C. (1999). Discriminating among cattle breeds using genetic markers. *Heredity, 82*, 613–619.

Bueno, L., Araujo, W. R. D., Salles, M. O., Kussuda, M. Y., Thiago, R. L. C., & Paixão, T. R. L. C. (2014). Voltammetric electronic tongue for discrimination of milk adulterated with urea, formaldehyde and melamine. *Chem, 2*, 251–266.

Cozzolino, D., & Murray, I. (2004). Identification of animal meat muscles by visible and near infrared reflectance spectroscopy. *LWT- Food Science and Technology, 37*, 447–452.

Dias, L. A., Peres, A. M., Veloso, A. C. A., Reis, F. S., Vilas-Boas, M., & Machado, A. A. S. C. (2009). An electronic tongue taste evaluation: Identification of goat milk adulteration with bovine milk. *Sensors and Actuators B: Chemical, 136*(1), 209–217.

Du, H., Chen, Q., Liu, Q., Wang, Y., & Kong, B. (2021). Evaluation of flavor characteristics of bacon smoked with different woodchips by HS-SPME-GC-MS combined with an electronic tongue and electronic nose. *Meat Science, 182*, 108626.

Gil, L., Barat, J. M., Escriche, I., Garcia-Breijo, E., Martinez-Manez, R., & Juan Soto, J. (2008). An electronic tongue for fish freshness analysis using a thick-film array of electrodes. *Microchimica Acta, 163*, 121–129.

Han, F., Huang, X., & Teye, E. (2019a). Novel prediction of heavy metal residues in fish using a low-cost optical electronic tongue system based on colorimetric sensors array. *Journal of Food Process Engineering, 42*, e12983.

Han, F., Huang, X., Teye, E., Gu, F., & Haiyang Gu, H. (2014a). Nondestructive detection of fish freshness during its preservation by combining electronic nose and electronic tongue techniques in conjunction with chemometric analysis. *Analytical Methods, 6*, 529.

Han, F., Huang, X., Teye, E., Gu, H., Dai, H., & Yao, L. (2014b). A nondestructive method for fish freshness determination with electronic tongue combined with linear and non-linear multivariate algorithms. *Czech Journal of Food Sciences, 32*(6), 532–537.

Han, H., Wang, Z., Li, C., Ma, Z., Yang, Z., & Ma, X. (2019b). Purity detection of goat milk based on electronic tongue and improved artificial fish swarm optimized extreme learning machine. *IFAC PapersOnLine, 52*(30), 391–396.

Hong, X., & Wang, J. (2014a). Detection of adulteration in cherry tomato juices based on electronic nose and tongue: Comparison of different data fusion approaches. *Journal of Food Engineering, 126*, 89–97.

Hong, X., & Wang, J. (2014b). Use of electronic nose and tongue to track freshness of cherry tomatoes squeezed for juice consumption: Comparison of different sensor fusion approaches. *Food and Bioprocess Technology, 8*(1), 158–170.

Hruškar, M., Major, N., Krpan, M., Krbavčić, I. P., Šarić, G., Marković, K., & Vahčić, N. (2009). Evaluation of milk and dairy products by electronic tongue. *Mljekarstvo, 59*(3), 193–200.

Hu, G., Zheng, Y., Liu, Z., & Deng, Y. (2017). Effects of UV-C and single- and multiple-cycle high hydrostatic pressure treatments on flavor evolution of cow milk: Gas chromatography-mass spectrometry, electronic nose, and electronic tongue analyses. *International Journal of Food Properties, 20*(7), 1677–1688.

Kalit, M. T., Marković, K., & Kalit, S. (2014). Application of electronic nose and electronic tongue in the dairy industry. *Mljekarstvo, 64*(4), 228–244.

Kamruzzaman, M., Sun, D. W., Elmasry, G., & Allen, P. (2013). Fast detection and visualization of minced lamb meat adulteration using NIR hyperspectral imaging and multivariate image analysis. *Talanta, 103*, 130–136.

Kantor, D. B., Hitka, G., Fekete, A., & Balla, C. (2008). Electronic tongue for sensing taste changes with apricots during storage. *Sensors and Actuators B: Chemical, 131*, 43–47.

Mabuchi, R., Ishimaru, A., Tanaka, M., Kawaguchi, O., & Tanimoto, S. (2019). Metabolic profiling of fish meat by GC-MS analysis, and correlations with taste attributes obtained using an electronic tongue. *Metabolites, 9*, 1–11.

Paixão, T. R. L. C., & Bertotti, M. (2009). Fabrication of disposable voltammetric electronic tongues by using prussian blue films electrodeposited onto CD-R gold surfaces and recognition of milk adulteration. *Sensors and Actuators B, 137*, 266–273.

Pan, M., Tong, L., Chi, X., Ai, N., Cao, Y., & Sun, B. (2019). Comparison of sensory and electronic tongue analysis combined with HS-SPME-GC-MS in the evaluation of skim milk processed with different preheating treatments. *Molecules, 24*, 1650.

Paulos, K., Rodrigues, S., Oliveira, A. F., Leite, A., Pereira, E., & Teixeira, A. (2015). Sensory characterization and consumer preference mapping of fresh sausages manufactured with goat and sheep meat. *Journal of Food Science, 80*, S1568–S1573.

Pérez-González, C., Salvo-Comino, C., Martin-Pedrosa, F., Dias, L., Rodriguez-Perez, M. A., Garcia-Cabezon, C., & Rodriguez-Mendez, M. L. (2021). Analysis of milk using a portable potentiometric electronic tongue based on five polymeric membrane sensors. *Frontiers in Chemistry, 9*, 706460.

Rudnitskaya, A., Kirsanov, D., Legin, A., Beullens, K., Lammertyn, J., Nicolai, B. M., & Irudayaraj, J. (2006). Analysis of apples varieties—Comparison of electronic tongue with different analytical techniques. *Sensors and Actuators B, 116*, 23–28.

Rudnitskaya, A., Legin, A., Makarychev-Mikhailov, S., Goryacheva, O., & Vlasov, Y. (2001). Quality monitoring of fruit juices using an electronic tongue. *Analytical Sciences, 17*-(Supplement), 309–312.

Ruixue, D., Lijuan, G., Liyun, L., Xue, L., Haisu, S., & Junrui, W. (2019). Effects of storage temperature on pasteurized milk quality analyzed by electronic tongue combined with high throughput sequencing. *Food Science, 40*(22), 47–52. (in Chinese with English abstract).

Surányi, J., Zaukuu, J. L. Z., Friedrich, L., Kovacs, Z., Horváth, F., Németh, C., & Kókai, Z. (2021). Electronic tongue as a correlative technique for modeling cattle meat quality and classification of breeds. *Food, 10*, 2283.

Tan, J., & Xu, J. (2020). Applications of electronic nose (e-nose) and electronic tongue (e-tongue) in food quality-related properties determination: A review. *Artificial Intelligence in Agriculture, 4*, 104–115.

Tazi, I., Choiriyah, A., Siswanta, D., & Kuwat Triyana, K. (2017). Detection of taste change of bovine and goat milk in room ambient using electronic tongue. *Indonesian Journal of Chemistry, 17*(3), 422–430.

Tian, X., Wang, J., Ma, Z., Li, M., & Wei, Z. (2019). Combination of an e-nose and an e-tongue for adulteration detection of minced mutton mixed with pork. *Journal of Food Quality, 3*, 1–10.

Wadehra, A., & Patil, P. S. (2016). Application of electronic tongues in food processing. *Analytical Methods, 8*, 474–480.

Wei, Z., & Wang, J. (2011). Detection of antibiotic residues in bovine milk by a voltammetric electronic tongue system. *Analytica Chimica Acta, 694*, 46–56.

Wei, Z., Wang, J., & Zhang, X. (2013). Monitoring of quality and storage time of unsealed pasteurized milk by voltammetric electronic tongue. *Electrochimica Acta, 88*, 231–239.

Wu, H., Yue, T., & Yuan, Y. (2018). Authenticity tracing of apples according to variety and geographical origin based on electronic nose and electronic tongue. *Food Analytical Methods, 11*, 522–532.

Xu, S., Lia, J., Baldwin, E. A., Plotto, A., Rosskopf, E., Hong, J. C., & Bai, J. (2018). Electronic tongue discrimination of four tomato cultivars harvested at six maturities and exposed to blanching and refrigeration treatments. *Postharvest Biology and Technology, 136*, 42–49.

Zaukuu, J. L. Z., Gillay, Z., & Kovacs, Z. (2021). Standardized extraction techniques for meat analysis with the electronic tongue: A case study of poultry and red meat adulteration. *Sensors, 21*, 481.

Zhang, X., Zhang, Y., Meng, Q., Li, N., & Ren, L. (2015). Evaluation of beef by electronic tongue system TS-5000Z: Flavor assessment, recognition and chemical compositions according to its correlation with flavor. *PLoS One, 10*, e0137807.

The manufacturer's authorised representative in the EU is Springer Nature Customer Service Centre GmbH, Europaplatz 3, 69115 Heidelberg, Germany. If you have any concerns regarding our products, please contact ProductSafety@springernature.com

Printed and bound by CPI Group (UK) Ltd, Croydon, CR0 4YY
30/09/2026
02225239-0016